T0299348

Fundamentals of Environmental Law and Compliance

This textbook provides readers with the fundamentals and the intent of environmental regulations so that compliance can be greatly improved and streamlined. Through numerous examples and case studies, it explains concepts from how environmental laws are applied and work to why pollution prevention and sustainability are critical for the future of all life on Earth. It is organized to accommodate different needs of students with different backgrounds and career choices. It is also useful for site safety and environmental managers, researchers, technicians, and other young professionals with a desire to apply environmental regulations and sustainability measures to their facilities and stay up to date on recently changed regulations.

Features

- Introduces students to issues of global environmental and sustainability challenges and policy
- Explains the science behind issues such as climate change, how environmental policy is made at the national and international levels, and what role politics play in determining environmental resource use
- Focuses on fundamental principles that are applicable in all nations and legal contexts
- Addresses the planet as one biosphere and briefly discusses environmental laws and regulations of more than 50 countries
- Provides numerous case studies that demonstrate major concepts and themes, examples, questions, and exercises to strengthen understanding and promote critical thinking, discussion, and debate

This book will benefit students in advanced undergraduate and graduate programs in environmental sciences and environmental engineering. It will also be of use to new practitioners who are entering the field of environmental management and need an introduction to environmental regulations.

Fundamentals of Environmental Law and Compliance

Daniel T. Rogers

CRC Press
Taylor & Francis Group
Boca Raton London New York

CRC Press is an imprint of the
Taylor & Francis Group, an **informa** business

First edition published 2023
by CRC Press
6000 Broken Sound Parkway NW, Suite 300, Boca Raton, FL 33487-2742

and by CRC Press
4 Park Square, Milton Park, Abingdon, Oxon, OX14 4RN

CRC Press is an imprint of Taylor & Francis Group, LLC

© 2023 Taylor & Francis Group, LLC

ISBN: 978-1-032-00678-9 (hbk)
ISBN: 978-1-032-00804-2 (pbk)
ISBN: 978-1-003-17581-0 (ebk)

DOI: 10.1201/9781003175810

Dedication

To Danielle, even though you are gone from this place, your contribution lives on...

Contents

Preface

For the last 10,000 years, humans have modified the Earth through agriculture, building cities resource exploitation, and other related activities. Now with more than 7 billion people on the planet, we are feeling the effects of our own actions by running out of habitable land, polluting the entire planet, climate change, and ultimately being responsible for the sixth major extinction in Earth's 4.54 billion year history. Modifications to the environment have been conducted to suit our needs, make our lives safer and more comfortable, and support our ever-growing population. To avoid feeling guilty and taking responsibility for our own actions, we have fooled ourselves and tried to justify the destruction of the natural environment by creating clever labels such as taming nature, western expansion, progress, sustainable development, and falsely claiming that Earth was created just for us to exploit.

Although there is much we can be proud of for limiting pollutant effects through environmental regulation in the last 50 years, it clearly has not been enough and our species is now at a critical crossroads. What we choose to do in the next few years to address not only climate change, but other adverse impacts to water, land, and other living species, will have a profound impact on how we and other species survive in the future.

One thing is for sure, things are going to change and it's time to re-think how we have treated nature and adapt to a different world in the future. We have learned a hard lesson, instead of trying to change and tame nature, we must learn to understand and live in productive harmony with nature. This means that there will be a need for additional environmental scientists and engineers to address and help solve the anthropogenic impacts of pollution on the air, water, land, and living organisms. One significant aspect of our technological advances in the last few centuries, which have introduced a multitude of pollutants, is that they have greatly outpaced nature's ability to evolve and adapt to human's destructive behavior.

Many believe that we do not purposely treat our planet with malice and that much of the damage inflicted has been out of lack of knowledge or ignorance and that our labels for justifying our destructive behavior have an origin to avoid guilt and responsibility. If we do not learn from our past and continue our attempts to conquer and change nature, we will face the harsh reality that our actions will have a significant negative impact on Earth's atmosphere and climate, water, land, and threaten the survival of many species including our own. The choice is ours.

Sustainability has been defined as living in productive harmony with nature. It was first used in 1969 when the United States passed the National Environmental Protection Act (NEPA). The 1960s and 1970s are generally known as the beginning of the environmental movement. Earth Day had its origin in 1971 when pollution within the United States was perhaps at its height and could no longer be ignored. Since then, the United States has passed laws that address the air, water, land, and living species and include the Clean Air Act, Clean Water Act, Safe Drinking Water Act, Resource, Conservation, and Recovery Act (RCRA), Comprehensive, Environmental Response, Compensation, and Liability Act (CERCLA), Endangered Species Act, Marine Mammal Protection Act, and many others. All of the regulations of the United States and for that matter, the world, have sustainability-oriented connections.

This book addresses compliance with environmental regulations and sustainability. Achieving compliance with environmental regulations is not easy. Environmental regulations are complex and lengthy, can be misinterpreted, and rely on a large data set of scientific information from several interdisciplinary subject areas of study. Sustainability is much larger in scope than compliance with environmental regulations. Sustainability largely deals with how we live. The goal of this book is to provide the user with enough information to achieve and maintain compliance with environmental regulations and pursue sustainability measures.

Sadly, in conclusion, climate change is just one of many examples of how humans have negatively impacted Earth through pollution in its many forms on a global scale. Other impacts caused by pollution that may not be in the public eye but are just as significant include impacts on the land, oceans, surface water, groundwater, and living organisms. Each is related and affects each other. These impacts involve the physical and chemical composition and dynamics of pollutant behavior within the natural world. We must now accept that humans have adversely impacted the entire Earth and that our efforts to improve our environment since the enactment of environmental regulations in the United States and worldwide have been largely inadequate. To underscore this point, Earth Scientists are now convinced that we have now moved the needle of geologic time into a new period called the Anthropocene, which is defined as *the age of humans*. This is significant and means that scientists believe we are now in a time period that affects the entire Earth in a way that cannot be erased.

Author

Daniel T. Rogers is the Director of Environmental Affairs at Amsted Industries Incorporated in Chicago, Illinois. Amsted Industries is a diversified manufacturing company of industrial components serving railroad, vehicular, construction, and building markets. Amsted Industries has more than 70 manufacturing locations in 15 countries. Mr. Rogers participates in environmental due diligence for acquisitions and divestitures, investigation and remediation of contamination, creating and evaluating environmental compliance and sustainability programs, and providing environmental advice, oversight, training, and negotiation strategies at various levels within the organization.

Over the last few decades, Mr. Rogers has published nearly 100 research papers in professional and academic publications and peer-reviewed journals on subjects including environmental geology, hydrogeology, geologic vulnerability and mapping, contaminant fate and transport, urban geology, environmental site investigations, contaminant risk, brownfield redevelopment, remediation, pollution prevention, environmental compliance, and sustainability. He has authored *Urban Watersheds: Geology, Contamination, Environmental Regulations, and Sustainability* 2nd Edition (2020), *Environmental Compliance and Sustainability: Global Challenges and Perspectives* (2019), and *Environmental Geology of Metropolitan Detroit* (1996). Mr. Rogers has published surficial geologic and contaminant vulnerability maps of the Rouge River watershed in southeastern Michigan. In addition, he is a contributing author of *The Encyclopedia of Global Social Issues* (2013), *Urban Groundwater* (2007), *Geoenvironmental Mapping* (2002), and *Groundwater in the Urban Environment* (1997).

He has taught geology and environmental chemistry at Eastern Michigan University and the University of Michigan and has presented guest lectures at numerous colleges and universities both in the United States and internationally.

Acknowledgments

I wish to thank many regulatory agencies and scientific research organizations who assisted over the years providing research, opinions, and advice to many tough questions concerning protecting human health and the environment that have contributed to the development of this book. International regulatory agencies and countries I wish to thank include the World Health Organization, the United Nations, European Environmental Agency, Environment Canada, Mexico Environmental and Natural Resource Ministry, Norwegian Environment Agency, Turkey Ministry of Environment and Urbanization, India Ministry of Environment, Forestry, and Climate Change, Indonesia Ministry of Environment, Japan Ministry of the Environment, Korea Ministry of the Environment, Malaysia Department of Environment, Egyptian Environmental Affairs Agency, Kenya National Environment Management Authority, South Africa Department of Environmental Affairs, Tanzania Department of Environment, Brazil Ministry of the Environment, Chile Ministry of the Environment, Peru Ministry of the Environment, Argentine Secretariat for the Environment, Australia Department of Environment and Energy, and New Zealand Ministry for the Environment.

Other organizations and state agencies include International Union of Geological Sciences, Geological Society of America, International Association of Hydrogeologists (IAH), United States Environmental Protection Agency, Intergovernmental Panel on Climate Change (IPCC), United States Geological Survey (USGS), British Geological Survey, National Oceanic and Atmospheric Administration, Illinois State Geological Survey, California Environmental Protection Agency, California Department of Toxic Substances and Control, Delaware Department of Natural Resources and Environmental Control, Illinois Environmental Protection Agency, Indiana Department of Environmental Management, Iowa Department of Natural Resources, Kansas Department of Health and Environment, Maryland Department of Environment, Michigan Department of Environment, Great Lakes, and Energy, New Jersey Department of Environmental Protection, Ohio Environmental Protection Agency, Oregon Department of Environmental Quality, and the Wisconsin Department of Natural Resources.

Individuals who have provided guidance and advice from the research and academic field over the years include my colleagues at the University of Michigan, Dr. Martin Kaufman, Dr. Kent Murray, and Dr. Ken Howard at the University of Toronto, Dr. Jack Sharp at the University of Texas, Dr. Richard Berg at the Illinois State Geological Survey, Dr. Robert Bobrowsky at the University of British Columbia, Dr. Peter Kolesar at Utah States University, Dr. Bill Farrand at the University of Michigan, Dr. Robert Oaks at Utah State University, Dr. Krause formerly at Henry Ford Community College, Dr. Derek Wong at APEX Environmental, Dr. Fred Payne with Arcadis, Dr. Colin Booth at Northern Illinois University, Dr. Mike Barcelona at Western Michigan University, Dr. Rebecca Spearot formerly with Clayton Environmental, Dr. Mary Ann Thomas at USGS, Dr. Garth van der Kamp at Environment Canada, Dr. Vladimir Kovalevsky at the Russian Academy of Sciences, Dr. John Moore with IAH, Dr. Hugo Loaiciga with the University of California, and Dr. John Chilton and Dr. Craig Foster with the British Geological Survey.

When dealing with environmental regulations of the United States and the world one must also have strong guidance from many in the legal field. I wish to thank those special individuals who have been a source of inspiration, counsel, and debate concerning environmental regulations including Chris Athas, Esq., Ed Brosius Esq., James Enright Esq., Rick Glick Esq., Geneva Halliday, Esq., Brian Houghton Esq., Michael Krautner Esq., Stephen Lewis, Esq., Michael Maher, Esq., Chris McNevin Esq., Granta Y. Nakayama, P. C., Esq., Jeryl Olsen Esq., Ron Paterson, Esq., Tom Petermann Esq., Doug Schleicher, Esq., Stephen Smith Esq., and Tom Wilczak, Esq.

The author wishes to thank the following professional colleagues for their assistance, cooperation, and ideas over the many years that it took to prepare this book: Lauren Alkadis, Cathee Andrews, Tony Anthony, Ann Barry, Gary Blinkiewicz, Tom Buggey, Kristine Casper,

Chris Christensen Russ Chadwick, Bob Cigale, Tom Cok, Marc, DeLoecker, Sheryl Doxtader, Ian Drost, Rob Ellis, Rose Ellison, Rob Ferree, Jennifer Formoso, Orin Gelderloos, Steve Hoin, Heather Hopkins, George Karalus, Steve Kitler, Shannon Lian, Rick Linnville, Daniel J. Lombardi, Bill Looney, Kim Myers, Ernie Nimister, David O'Donnell, Tom O'Hara, Ray Ostrowski, Paul Owens, Bruce Patterson, Scott Pearson, Sara Pearson, Mark Penzkover, Robert Ribbing, Eric Ross, Jeanne Schlaufman, Dave Slayton, Dave Smith, Ed Stewart, Patricia Thornton, Mary Vanderlaan, Chris Venezia, Nick Welty, Cheryl Wilson, and Tom Wenzel.

Acronyms, Elements, Symbols, Molecules, and Units of Measure

ACRONYMS

ACGIH	American Conference of Governmental Industrial Hygienists
ADP	Atmospheric Decontamination Plan
APCP	Air Prevention and Control of Pollution Act
APP	Atmospheric Prevention Plan
AST	Aboveground storage tank
ASTM	American Society for Testing Materials
ATSDR	Agency for Toxic Substances and Disease Registry
BACT	Best Available Control Technology
BLM	Bureau of Land Management
BMP	Best management practice
BPA	Bisphenol A
BTEX	Benzene, toluene, ethyl benzene, and xylenes
CAA	Clean Air Act
CAAC	Clean Air Alliance of China
CAS	Chemical Abstract Service
CalEPA	California Environmental Protection Agency
CEPA	Canadian Environmental Protection Act
CESQG	Conditionally exempt small quantity generator
CFC	Chlorofluorocarbons
CMEE	China Ministry of Ecology and Environment
CONAGUA	Mexico National Water Commission
CERCLA	Comprehensive Environmental Response, Compensation and Liability Act
CFR	Code of Federal Regulation
CRF	Contaminant or pollution risk factor
CRF$_{air}$	Contaminant or pollution risk factor for air
CRF$_{gw}$	Contaminant or pollution risk factor for groundwater
CRF$_{soil}$	Contaminant or pollution risk factor for soil
CSO	Combined sewer overflow
CVOC	Chlorinated volatile organic hydrocarbon
CWA	Clean Water Act
DBP	Di-n-butyl phthalate
DCE	dichloroethylene, dichloroethane
DEHP	Bi or Bis(2-ethylhexyl) phthalate
DEP	Diethyl phthalate
DDT	Dichlorodiphenyltrichloroethane
DIVs	Dutch Intervention Values
DNAPL	Dense nonaqueous phase liquid
DOT	Department of Transportation
EEA	European Environment Agency
EEC	European Economic Community
EIA	Environmental Impact Assessment
EIS	Environmental Impact Statement

EHS	Extremely hazardous substance
EH&S	Environmental health and safety
EINETICS	European Information and Observation Network
EPCRA	Emergency Planning and Community Right-to-Know Act
EPA	Environmental Protection Agency
ESA	Environmental site assessment
ESPM	Elimination, Substitution, Prevention, and Minimization
ETS	Emission Trading System
EU	European Union
FDA	Food and Drug Administration
FEPCA	Federal Environmental Pesticide Control Act
FFDCA	Federal Feed, Drug, and Cosmetic Act
FIFRA	Federal Insecticide, Fungicide, and Rodenticide Act
GACT	Generally Available Control Technology
GHG	Greenhouse gas
GWP	Global warming potential
HAP	Hazardous air pollutant
HAZMAT	Hazardous materials
HMTA	Hazardous Materials Transportation Act
HCFC	Hydrochlorofluorocarbons
HFC	Hydrofluorocarbon
HMTA	Hazardous Materials and Transportation Act
HRS	Hazard Ranking System
IED	Industrial Emissions Directive
IPCC	Intergovernmental Panel on Climate Change
KDNREP	Kentucky Department of Natural Resources and Environmental Protection
LEPC	Local Emergency Planning Committee
LNAPL	Light nonaqueous phase liquid
LOAEL	Lowest-observed-adverse-effect level
LQG	Large quantity generator
MACT	Maximum Achievable Control Technology
MCL	Maximum contaminant level
MCLG	Maximum contaminant level goal
MDEP	Massachusetts Department of Environmental Protection
MDEQ	Michigan Department of Environmental Quality
MEK	Methyl ethyl ketone
MEP	China Ministry of Environmental Protection
MNR	Russian Ministry of Natural Resources
MS4S	Large Municipal Separate Storm Sewer Systems
MSDS	Material safety data sheet
MTBE	Methyl tertiary butyl ether
NAAQS	National Ambient Air Quality Standards
NAFTA	North American Free Trade Agreement
NAICS	North American Industry Classification System
NASA	National Aeronautics and Space Administration
NATO	North Atlantic Treaty Organization
NCZMP	National Coastal Zone Management Program
NERRS	National Estuarine Research Reserve System
NEA	Norwegian Environment Agency
NEPA	National Environmental Policy Act
NEPC	Australian National Environmental Protection Council

NESHAPS	National Emission Standards for Hazardous Air Pollutants
NGS	National Geographic Society
NHTSA	National Highway Traffic Safety Administration
NJDEP	New Jersey Department of Environmental Protection
NOAA	National Oceanic and Atmospheric Administration
NOM	Mexico Standards or Normas
NOV	Notice of violation
NPDES	National Pollutant Discharge Elimination System
NPDWS	National Primary Drinking Water Standards
NPL	National Priority List
NPS	Nonpoint source
NRC	National Response Center
NRC	Nuclear Regulatory Commission
ODS	Ozone-depleting substance
OECD	Organization for the Economic Co-operation and Development
OPP	Office of pesticide programs (USEPA)
OSHA	Occupational Safety and Health Administration
OSPA	Oil Spill Prevention Act
PAH	Polycyclic aromatic hydrocarbon
PFC	Perfluorocarbon
PBT	Persistent bio-accumulative toxic
PCB	Polychlorinated biphenyl
PCE	Tetrachloroethylene, tetrachloroethene, perchloroethylene, perchloroethene
PCP	Pentachlorophenol
PERC	Tetrachloroethylene, tetrachloroethene, perchloroethylene, perchloroethene
PEL	Permissible exposure limit
PFOA or PFAS	Per- and polyfluoroalkyl substances
PM	Particulate matter
PNA	Polynuclear aromatic hydrocarbon
POTW	Publicly owned treatment works
PPA	Pollution Prevention Act
PPE	Personal protective equipment
PREFEPA	Mexico Federal Attorney Generalship of Environmental Protection
PRP	Potential Responsible Party
PS	Point source
PSD	Prevention of Significant Deterioration
PTE	Potential to emit
PVC	Polyvinyl chloride
RCI	Reactivity, corrosivity, and ignitability
RCRA	Resource, Conservation, and Recovery Act
REC	Recognized environmental condition
RQ	Reportable quantity
SARA	Superfund Amendments and Reauthorization Act
SDWA	Safe Drinking Water Act
SDS	Safety datasheet
SEMARNAT	Mexico Secretariat of the Environment and Natural Resources
SEP	Supplemental environmental project
SERC	State Emergency Response Commission
SIC	Standard Industrial Code
SMG	Small Quantity generator
SPCC	Spill Prevention Control and Countermeasures Plan

SPLP	Synthetic precipitant leaching procedure
SRP	Spill response plan
SVOC	Semi-volatile organic compound
SWPPP	Stormwater pollution prevention plan
TCA	Trichloroethane
TCE	Trichloroethylene, trichloroethene
TCLP	Toxic characteristic leaching procedure
THM	Trihalomethane
TPQ	Threshold Planning Quantity
TSCA	Toxic Substance and Control Act
TSP	Total suspended particulates
UK	United Kingdom
UN	United Nations
UNFCCC	United Nations Framework Convention on Climate Change
UNESCO	United Nations World Heritage Site
USCB	United States Census Bureau
USCIA	United States Central Intelligence Agency
USDA	United States Department of Agriculture
USDC	United States Department of Commerce
USDOE	United States Department of Energy
USDHH	United States Department of Health and Human Services
USDOT	United States Department of Transportation
USEPA	United States Environmental Protection Agency
USFS	United States Forest Service
USGS	United States Geological Survey
USNRC	United States Nuclear Regulatory Commission
UST	Underground storage tank
VC	Vinyl chloride
VOC	Volatile organic compound
VSQG	Very small quantity generator
WDNR	Wisconsin Department of Natural Resources
WHO	World Health Organization

ELEMENTS

Ag	Silver
As	Arsenic
Ba	Barium
Br	Bromine
C	Carbon
Cd	Cadmium
Cl	Chlorine
Cr	Chromium
Cr III	Trivalent chromium
Cr VI	Hexavalent chromium
Cr^{+3}	Trivalent chromium
Cr^{+6}	Hexavalent chromium
Cu	Copper
F	Fluorine
H	Hydrogen
Hg	Mercury

I	Iodine
K	Potassium
N	Nitrogen
O	Oxygen
P	Phosphorus
Pb	Lead
S	Sulfur
Se	Selenium
Zn	Zinc

SYMBOLS

η	Effective porosity
ρb	bulk density
W_s	Water solubility
ATM	Atmosphere
°C	degrees Celsius
°F	degrees Fahrenheit
Foc	Organic carbon partition coefficient
H	Henry's Law constant
Kd	Distribution coefficient
Koc	Fraction of total organic carbon
M	Mobility
MW	Molecular weight
P	Persistence
R	Retardation factor
T	Toxicity
VP	Vapor pressure

MOLECULES

CO	Carbon monoxide
CO_2	Carbon dioxide
CH_4	Methane
H_2O	Water
NOx	Nitrogen oxides
N_2O	Nitrous oxide
O_3	Ozone
SO_2	Sulfur dioxide
SF_6	Sulfur hexafluoride

UNITS OF MEASURE

mg/kg	milligram per kilogram
ug/kg	microgram per kilogram
mg/L	milligram per liter
µg/L	microgram per liter
g/L	gram per liter
mg/m^3	milligram per cubic meter
$µg/m^3$	microgram per cubic meter
PM2.5	Particulate matter less than 2.5 microns

PM10	Particulate matter less than 10 microns
ppm	part per million
ppb	part per billion
TPY	tons per year
BTU	British Thermal Unit

1 Themes and Overview of Fundamentals of Environmental Law and Compliance

1.1 INTRODUCTION

The focus of this book is compliance with environmental laws and regulations and sustainability within the United States and throughout the world. The goal of environmental regulation is to protect human health and the environment. The goal of sustainability is to create an environment whereby humans live in productive harmony with nature. To achieve the goal of robust and effective environmental regulations, nations of the world have learned to concentrate their efforts that protect the air, protect the water, protect the land, and protect living organisms. One of the two goals of this book is to provide the user with enough technical information and practical experience so that compliance with environmental regulations and laws is achieved and maintained.

To achieve the goal of sustainability is more problematic in practice. Sustainability involves creating a balance between humans and nature that at this point does not exist. Creating a sustainable Earth will require enacting a set of proactive and coordinated measures that will require hard work and cooperation from everyone in every country. In order for the Earth to sustain a global human population of more than 7 billion people, which continues to increase daily, it is essential that we appreciate and comprehend the biosphere at a level that is greater than most, if not all, scientific levels in inquiry and understanding. The reason for such a high level of scientific inquiry is simple, this is where we live, and immediately beneath our feet is the contact between civilization and the natural world. Much of our human history has been spent in developing ways to conquer and change nature, which as we shall see in the course of reading this book, has not turned out well for nature, for us, and the organisms with which we share this planet. To underscore this point, Earth Scientists are now convinced that we have now moved the needle of geologic time into a new period called the Anthropocene. This is significant because the definition of a Geologic Age is a time period that effects the entire Earth and will be recorded in Earth's history and cannot be reversed.

This book is divided into three sections. The first section addresses the past. The second section addresses the present and the third section address the future. The first section examines the history of pollution and the birth of the environmental movement. The first section will also examine how investigation and remediation of legacy sites are conducted. This section will also provide the reader with a sense of our past mistakes and misperceptions of treating nature poorly and the huge cost of cleaning up the environment once it becomes impacted.

The second section of this book will examine the present by exploring the environmental regulations of the United States and the world. We will examine more than 50 different countries on 6 continents and will compare and contrast regulations from each country with respect to their effectiveness and how they compare to the United States.

The third and final section of this book is dedicated to the future framed as sustainability. This section is divided into chapters that address sustainability at the global level, national level,

DOI: 10.1201/9781003175810-1

local level, and individual level. We will learn the hard truth that contamination does not respect country boundaries. Changes required to existing environmental regulations so that they more adequately address inconsistencies and gaps within existing regulations that move our civilization toward the ultimate goal of instilling environmental stewardship and creating a sustainable planet.

This book concludes with guidelines, procedures, and experience-based practical advice on conducting an environmental audit.

1.2 MAJOR THEMES OF THIS BOOK

The following sections provide a brief introduction into the central themes that are covered in this book and include the following:

1. No Matter Where You Are, It's All the Same
2. The Past: Addressing and Repairing Nature
3. The Present: Environmental Compliance
4. The Future: Improving Environmental Regulations and Embracing Sustainability
5. The Importance of Conducting an Environmental Audit

1.2.1 THEME 1: NO MATTER WHERE YOU ARE, IT'S ALL THE SAME

Of any theme presented in this book, "No matter where you are, it's all the same," is probably the most significant. As stated in the introduction, pollution does not care about country borders. As we shall discuss, air pollution can and does pollute the entire planet. Therefore, it makes sense that when dealing with many types of air pollutants that there should be consistency in environmental regulations that deal with air pollution throughout the world. This becomes most apparent when we discuss environmental regulations of the world.

1.2.2 THEME 2: THE PAST: ADDRESSING AND REPAIRING NATURE

The first step in addressing and repairing nature from anthropogenic sources of pollution and land use requires an understanding of nature. Each day when we walk outside we are walking on a history book, yet most of us are unaware of this and are unaware of how this history book influences our lives. The arrangement, thickness, and composition of the soil and sediment layers just centimeters beneath our feet have a profound influence on our lives and all life on Earth. These soil and sediment layers don't just dictate where and how cities are built, where roads are built, and how buildings are constructed. Most significantly, these soil and sediment layers beneath our feet are the meeting place between human civilization and the natural world. This contact between human civilization and the natural world is the proving ground and point at which contaminants released into the environment begin their destructive journey that has now impacted all living things on Earth. We must understand our natural world in at least the same scientific detail that we need to understand pollution. The reason we must do this is simple, they interact with one other.

From a historical point of view, our treatment of the environment has not been kind. As recently as the early portion of the 20th century, environmental awareness and legislation were generally lacking not only in the United States but worldwide. There were a few international agreements that primarily focused on boundary waters, navigation, and fishing rights along shared waterways between countries. However, they ignored pollution and ecological issues. In fact, the most convenient and least costly method of waste disposal up until the middle of the 20th century was "up the stack or down the river" (Haynes 1954). It was not until the publishing of the Rachel Carson book, *Silent Spring* in 1962 described the environmental effects of polychlorinated biphenyls (PCBs) on birds that catapulted the environmental movement in the United States into the public eye (Carson 1962). The 1960s and 1970s are generally described as the age of environmentalism in the United States. This is when the United States realized that environmental degradation had significantly affected the air and water quality of many urban areas, especially the air in Los Angeles and water in Lake Erie, which is one of the Great Lakes. Starting on Earth Day in 1971, the Ad Council and Keep America Beautiful campaign aired public awareness commercials on television depicting a Native American shedding a tear when looking out over the polluted landscape of the United States. The commercial stated "people start pollution, people can stop pollution" (United States Advisory Council on Historic Preservation 2021).

Although laws focused on protecting human health or the environment can be traced back more than 2,000 years, it took the United States in the 1970s with the passage of the most significant environmental regulations and laws to set the example for the rest of the World (Hahn 1994).

Environmental regulations of the United States and the world are largely based on reactions to an incident that caused enormous harm to human health and the environment. A few of the many examples of incidents that resulted in the enactment of environmental regulations include Love Canal in New York, Times Beach in Missouri, Bhopal in India, the fire on the Cuyahoga River near Cleveland, the Exxon Valdez oil spill, and the recent incident on the Deepwater Horizon oil drilling platform in the Gulf of Mexico.

One of the many things we will learn and explore in the first section of this book is the huge cost of cleaning up the environment once it becomes contaminated. In fact, sometimes the environment has become so contaminated from a single incident that all the financial resources and technologies in the world will not restore the environment to pre-contaminant conditions. This theme will highlight the different remedial approaches for cleaning up our air, water, and soil. As we shall see, the approach for remediating the air and surface water is heavily weighted toward pollution prevention and permitting through environmental regulations. This approach has been effective at significantly reducing the amount of contaminants entering the air and surface waters of the United States. What may be surprising is that cleaning up the environment is not all equal. The most significant factors that control the cost of cleaning up the environment are related to the physical chemistry of the specific contaminants, the geologic environment in which the release occurs, and whether there is or could be harmful to human health and the environment.

1.2.3 THEME 3: THE PRESENT: ENVIRONMENTAL COMPLIANCE

Environmental compliance is sometimes perceived as just another set of rules and is easy to accomplish. Just follow the rules. This premise fails on several levels. First, the human population is ever-increasing and our thirst for land and modern conveniences increases as our standard of living is raised. Second, our technology is ever-changing with new products and chemicals. Third, the Earth is dynamic. There is change, ever getting more complex as more information and science is discovered, learned, and shared. The weather changes constantly and is perhaps the best example of how to imagine environmental management and risk. It also changes, as does everything. Everything obeys the second law of thermodynamics, namely entropy, in that nature tends to become more complex as time marches on. Factors that influence the effectiveness of environmental regulations vary from country to country and within each country as well. The United States is no different. Factors that influence the degree to which a country has effective environmental regulations include the following (Bates and Ciment 2013):

- Political will
- Environmental tragedies, harm to the environment, and lessons learned
- Cost
- Geography and climate
- Enforcement

- Incentives
- Lack of overriding social factors such as war, political structure, greed, poverty, hunger, lack of infrastructure, educational awareness, and corruption

Environmental laws and regulations in the United States and the world have continually become more numerous and complex to the point that even many scientists and other professionals need assistance in interpreting, understanding, and applying many environmental laws and regulations.

The growth of the number and complexity of environmental laws and regulations can be more easily understood by simply applying the principle of the scientific term called entropy. Entropy is a principle that states that as time progresses forward there is more disorder which increases complexity. This can be applied to our human civilization as well. Our civilization has become more complex with time as a result of our technological advances, population increase, and other factors. This in turn has put pressure on the natural world and our response has been the passage of environmental laws and regulations as an attempt to establish order in a disordered and out-of-equilibrium environment that humans caused. To make this point clear, a list of laws that are commonly attributed to protecting human health and the environment just in the United States include the following (USEPA 2021a):

- Antiquities Act of 1906
- Atomic Energy Act of 1946
- Atomic Energy Act of 1954
- Brownfield Revitalization Act of 2002
- Clean Air Act (CAA) of 1970
- Clean Water Act CWA) of 1972
- Coastal Zone Management Act of 1972
- Comprehensive Environmental Response, Compensation, and Liability Act (CERCLA) of 1980
- Emergency Planning and Community Right-to-Know Act of 1986
- Endangered Species Act of 1973
- Energy Policy Act of 1992
- Energy Policy Act of 2005
- Federal Power Act of 1935
- Federal Feed, Drug, and Cosmetic Act (FFDCA) of 1938
- Federal Insecticide, Fungicide, and Rodenticide Act (FIFRA) of 1947
- Fish and Wildlife Coordination Act of 1968
- Food Quality Protection Act of 1996
- Fisheries Conservation and Management Act of 1976
- Global Climate Protection Act of 1987
- Hazardous Materials Transportation Act (HMTA) of 1975
- Insecticide Act of 1910
- Lacey Act of 1900
- Low-Level Radioactive Waste Policy Act of 1980
- Marine Protection Act of 1972
- Marine Mammal Protection Act of 2015
- Medical Waste Tracking Act of 1980
- Migratory Bird Treaty Act of 1916
- Mineral Leasing Act of 1920
- National Environmental Policy Act of 1969
- National Forest Management Act of 1976
- National Historic Preservation Act of 1966
- National Parks Act of 1980
- Noise Control Act of 1974

- Nuclear Waste Policy Act of 1982
- Ocean Dumping Act of 1988
- Occupational Safety and Health Act (OSHA) of 1970
- Oil Spill Prevention Act of 1990
- Pollution Prevention Act (PPA) of 1990
- Refuse Act of 1899
- Resource, Conservation, and Recovery Act (RCRA) of 1976
- Rivers and Harbors Act of 1899
- Safe Drinking Water Act SDWA) of 1974
- Superfund Amendments and Reauthorization Act (SARA) of 1986
- Surface Mining Control and Reclamation Act of 1977
- Toxic Substance Control Act of 1976
- Wild and Scenic Rivers Act of 1968

Some of the most important conclusions reached in this book concerning environmental regulations of the world include the following (Rogers 2020):

- Each country has environmental regulations.
- No two countries are exactly the same.
- Each country has a unique set of attributes that influence the effectiveness of environmental protection that include climate, geography, geology, social and economic concerns, and politics.
- The platform of environmental regulations of each country is similar to that of the United States or the European Union.
- Although some countries are doing a better job than others, each country struggles with protecting human health and the environment for various reasons, two of which are universal and include a growing population and urban expansion.
- Every country has been significantly affected by pollution of the air, water, and land.

The principles of environmental compliance are significant lessons and themes to remember when evaluating and working with maintaining compliance with environmental laws and include the following (Rogers 2020):

1. It all begins with conducting an environmental audit.
2. Being in compliance with environmental regulations does have sustainability implications.
3. Work yourself out of environmental regulations.
4. There is little we can control.
5. Limit chemical use when possible.
6. When possible, eliminate chemicals that are very toxic or have a high chemical risk factor.
7. Fully Understand Environmental Permits.
8. Never Accept a Permit Term that the Facility Can't Achieve.
9. Always be accurate and truthful.
10. When in Doubt, It's Usually Always Better to Report a Spill or Other Incident Where Reporting May be Required.
11. The Importance of Waste Characterization and Points of Generation.
12. Any Operational Changes May Require Notice and Permit Changes or a New Permit.
13. Give yourself enough time for collecting additional compliance samples in case a data quality issue arises.
14. Keep well organized and communicate with management and employees regularly.
15. Housekeeping

16. Signage
17. Conduct spill drills.
18. Prepare a list of onsite chemicals with corresponding reportable quantities (RQs) and recommended cleanup procedures.
19. Conduct regular inspections.
20. Proper preparation prevents poor performance.
21. In many instances, proving a negative is required when evaluating a release
22. When a question arises, consult management, in-house environmental counsel, outside environmental counsel, or the regulatory authority, as appropriate.
23. Finally, remember that "no matter where you are, it's all the same."

We are learning that we live on one planet with no environmental boundaries and that contamination does not respect boundaries. Another point is that environmental regulations are built on the USEPA or EU platforms and those platforms rely on three basic principles:

1. Protect the air
2. Protect the water
3. Protect the land
4. Protect living organisms and cultural and historic sites

1.2.4   Theme 4: The Future: Improving Environmental Regulations and Embracing Sustainability

Improving environmental regulations and embracing sustainability is present in the third section of the book. The majority of subject areas where environmental regulations within the United States are either in need of improvement or where new regulations are needed to assist sustainability initiatives include the following subject areas (Rogers 2020):

- Modifying Environmental Enforcement Emphasis and Policy
- Science-based improvements to environmental regulations to account for regional risks and climate change
- Regulate the farming industry and farmland no different than other industries and regulate the farm industry through the USEPA
- Reducing and regulating pesticides and herbicides more strictly
- Reducing or eliminating some fertilizers
- Septic tank legislation
- Additional invasive species controls
- Urban Planning and Land use
- Residential and household waste
- Urban air improvements
- Further or new restrictions on harmful chemicals, especially those that are persistent and mobile in the environment
- Emerging contaminants, such as those described in Chapter 2
- Improving pollution prevent techniques and awareness
- Sustainability legislation
- Noise controls
- Plastic reduction or elimination

USEPA defines **sustainability** as creating and maintaining conditions under which humans can exist in productive harmony to support present and future generations (USEPA 2021b). Note that the definition states "productive harmony." One may ask, in productive harmony with what? The answer is with the environment. Therefore, in order to achieve some level of sustainability, we must understand the environment and also understand the aspects of how facility operations impact the environment. Sustainability is then the outcome of analyzing the aspects of facility operations with the natural environment and developing and engineering methods to either minimize or eliminate harmful potential impacts.

Sustainability is based on a simple principle that states that (USEPA 2021b):

Everything we need for our survival and well-being depends, either directly or indirectly, on our natural environment

The United Nations defines sustainability as (United Nations 2021a):

development that meets the needs of the present without compromising the ability of future generations to meet their own needs

Sustainability had its origin in 1969 in the incorporation of the United States National Environmental Policy Act (NEPA):

to foster and promote the general welfare, to create and maintain conditions under which humans and nature can exist in productive harmony and fulfill the social, economic, and other requirements of our present and future

The term "sustainable development" was first used by the United Nations in 1987 (United Nations 2021b).

Points of contention exist between countries concerning sustainability and include differences in power and responsibility, environmental, and economic concerns that must be solved before sustainability can proceed on a path to become a reality on a global level (United Nations 2021b). Additional points of contention are religious conviction and population, which are difficult and sensitive to many and are often not openly discussed. They include the following:

1. Power
2. Responsibility
3. Economic
4. Religious

The United Nations (2021b) has framed sustainability goals termed "Global Goals," which include the following:

• No Poverty	• Zero Hunger	• Good Health
• Well Being	• Quality Education	• Gender Equality
• Clean Water	• Effective Sanitation	• Affordable Energy
• Clean Energy	• Economic Growth	• Satisfactory Employment
• Industry	• Innovation	• Infrastructure
• Reduced Inequalities	• Sustainable Cities	• Responsible Consumption
• Climate Action	• Life Below Water	• Life on Land
• Peace and Justice	• Partnerships	• Life in the Air

At least five actions at the global level should be undertaken to improve sustainability at every level and include the following:

- Cohesive environmental policy and enforcement
- Controlling the human population
- Modifying the human diet
- Financial fairness and Equity Distribution
- Modifying business models

1.2.5 THEME 5: THE IMPORTANCE OF CONDUCTING AN ENVIRONMENTAL AUDIT

The final chapter in the book presents guidelines, procedures, and experience-based practical advice on conducting an environmental audit. Conducting an environmental audit forms the cornerstone of any environmental program whether the focus is on compliance or sustainability. An analogy of conducting an environmental audit is similar to going to the doctor for a complete physical.

1.3 SUMMARY

The goal of this book is to provide the user with enough information so that compliance with environmental regulations is achieved and is maintained. In addition, it is also a goal of this book to introduce the reader to the meaning and action items required to set human civilization on a path to sustainability.

Most should conclude after reading this book that there are many ways that we can individually and collectively improve our relationship with Earth and move toward reaching a sustainable balance with nature. Many believe that we do not purposely treat our planet with malice and that much of the damage inflicted has been out of lack of knowledge or ignorance. If we do not learn from our past and continue our attempts to conquer and change nature, we may face the reality that the climate may continue to change because of us, water will continue to be polluted because of us, and we will no longer have wildlands because of us. If we collectively do not take appropriate action, we threaten our species and many others with which we share our home. The choice is ours!

REFERENCES

Bates, C. G. and Ciment, J. 2013. Encyclopedia of Global Social Issues. M.E. Sharpe Publishers: New York. 1450 p.

Carson, R. 1962. Silent Spring. Houghton Mifflin: Boston, MA.

Hahn, R. W. 1994. United States Environmental Policy: Past, Present and Future. Natural Resource Journal. John F. Kennedy School of Government, Harvard University. Cambridge, Massachusetts. Vol. 34. no 1. pp. 306–348.

Haynes, W. 1954. American Chemical Industry – A History. Vols. I–IV. Van Nostrand Publishers: New York, NY.

Rogers, D. T. 2020. Urban Watersheds: Geology, Contamination, Environmental Regulations, and Sustainability. CRC Press: Boca Raton, FL. 608 p.

United Nations. 2021a. Sustainability Goals. https://www.un.org/sustainabledevelopment/ (accessed July 27, 2021).

United Nations. 2021b. Human Development Report. United Nations Development Programme. http://www.hdr.undr.org/sites/2017 (accessed July 28, 2021).

United States Advisory Council on Historic Preservation. 2021. National History Preservation Act. https://www.achp.gov (accessed July 25, 2021).

United States Environmental Protection Agency. 2021a. Summary of Environmental Regulations of the United States. https://epa.gov/summary-environmental-laws (accessed July 25, 2021).

United States Environmental Protection Agency (USEPA). 2021b. What is Sustainability? https://www.epa.gov/sustainability (accessed July 27, 2021).

2 The History of the Environmental Movement in the United States

2.1 INTRODUCTION

Environmental regulations form the foundation of protecting human health and the environment. Many also believe that environmental regulations form the foundation of sustainability. In the United States, the National Environmental Policy Act (NEPA) of 1969 is considered by many as the seed of environmental protection and sustainability (USEPA 2021a). Therefore, this means that laws protecting the environment in the United States and much of the world are only 50 years old. This chapter addresses the early beginnings of the environmental movement and regulations in the United States. Keep in mind a major theme while working your way through this book, which is to protect the air, protect the water, protect the land, and protect living organisms and cultural and historic sites.

As we shall see throughout this book, protecting human health and the environment is difficult if the regulations are not comprehensive and all-encompassing. This is almost an impossible task since environmental regulations are in large part based on single historical incidents that, in most cases, have caused large-scale harm to human health and the environment. Therefore, they are reactive rather than proactive, which tends to narrow the focus of regulations and does not fully address other circumstances that cause similar harm. This is where Sustainability takes over. In order for Sustainability to be effective, it must be proactive. As amazing as it may seem, there are no specific environmental regulations in the United States that target sustainability.

Why is this the case? The answer is not clear and likely involves several underlying factors. Some of the underlying factors include the following:

- Sustainability is largely a proactive response, and environmental regulations in the United States are almost solely reactive.
- Lack of political will. Without a disaster to point to, why should elected officials pass a law that promotes sustainability.
- Environmental degradation that sustainability measures can address is always gradual. For instance, climate change is a good example in that it originates from numerous anthropogenic activities and has taken many decades of study and a few hundred years of environmental degradation to become clear and convincing.
- Enacting sustainability measures require some level of inconvenience, sacrifice, and change. This can cause negative feedback from the public and might be perceived as political suicide by elected officials.
- The full measure of environmental impact from anthropogenic activities is difficult to predict because they are complex.

Factors that influence the effectiveness of environmental regulations and sustainability vary from country to country and within each country as well. The United States is no different. Factors that influence the degree to which a country has effective environmental regulations and sustainability measures include the following (Bates and Ciment 2013 and United Nations 2021a):

DOI: 10.1201/9781003175810-2

- Political will
- Environmental tragedies, harm to the environment, and lessons learned
- Cost
- Geography and climate
- Enforcement
- Incentives
- Lack of overriding social factors such as war, political structure, greed, poverty, hunger, lack of infrastructure, educational awareness, corruption, and other social issues

This chapter is organized by history and subject. The first section addresses geography and climate, which greatly influence how the environment responds to stresses such as pollution and also how it responds to sustainability efforts. The second section is a brief history of the environmental movement in the United States. We then review the National Environmental Response Act, which may actually be the most important legislation that encompasses how we regulate ourselves and established the objective and purpose of sustainability in the United States and, as we shall discover, has been copied by most countries of the World.

2.2 DONORA PENNSYLVANIA 1948 AIR POLLUTION INCIDENT

In many instances in the United States, there is a single incident that shocks us into action. There have been numerous environmental incidents that have caused enormous damage to the environment and loss of life. Some include the following:

- Love Canal soil and groundwater contamination in 1978
- Valley of drums soil, groundwater, and surface water contamination in 1978
- Woburn Massachusetts
- Times Beach dioxin contamination in 1983
- Bhopal, India, toxic gas release in 1984
- Garbage barge of 1988
- Exxon Valdes oil spill in 1989
- Deepwater Horizon explosion and oil spill in 2010

A simple question always then seems to follow, "How could this have happened?" The next question is, "How do we prevent this from happening in the future?" This is where new environmental regulations are born. Each of the listed incidents above paved the way for new environmental regulations that included the following:

- Resource Conservation and Recovery Act (RCRA)
- Comprehensive Environmental Response, Compensation, and Liability Act (CERCLA)
- Superfund Amendments and Reauthorization Act (SARA) and Emergency Planning and Community Right-to-Know Act
- Ocean Dumping Act
- Oil Spill Prevention Act

What about air pollution? Was there an incident that influenced the development of the Clean Air Act? The answer is yes but it took time. The incident occurred in the fall of 1948 in Donora, Pennsylvania. A yellow fog appeared on October 26 and lasted until October 31 when a rainstorm dissipated the yellow fog (Smithsonian 2021). The fog grew so bad that 20 residents died of asphyxiation and over 7,000 residents were hospitalized as a result of the severe air pollution (USEPA 2021b). Figure 2.1 is a photograph taken around noon on October 29, 1948, of the smog that got so bad that it looked like the photo was taken at night.

FIGURE 2.1   Donora Pennsylvania air pollution incident.

USEPA. 2021b. Donora Smog of 1948. EPA Alumni Association: News and Views (accessed July 9, 2021).

The incident drew national attention when Walter Winchell broadcast news of the disaster on his national radio show. During the incident, the Donora Zinc Works shut down its smelters to eliminate as much emissions as possible. Rain began to fall at this point as well which dissipated the smog and the Zinc works returned to its normal operations the next day (USEPA 2021b).

The incident was investigated by several governmental agencies at the time and the conclusion was that the main air pollutant that caused the health concerns was sulfur dioxide and the concentrations were estimated to be at a concentration that is more than 125 times what is recommended by the World Health Organization (2021). However, this incident did not have the impact that later incidents had on developing new environmental regulations.

The Donora incident provided the scientific community with data that related to

- Toxicity of certain pollutants
- Human population sensitivity variability
- Exposure data
- Expanded the knowledge and methods of toxicology

The lessons learned from this incident, in part, showed that air pollution can inflict enormous harm and even death to humans. It also helped to identify the types of pollution most dangerous that would later be part of the Clean Air Act Criteria Pollutants, identified a need for environmental regulations, and demonstrated that geography and climate can be very important in influencing the behavior of contaminants and human health (USEPA 2021b).

2.3   GEOGRAPHY AND CLIMATE

Geography, geology, hydrology, and climate influence the types and behavior of pollution throughout the world. Therefore, it's only fitting that we include an overview of the geography and climate of the United States. This will provide some context as to the unique attributes the Untied States has when the connection between pollution and specific geographic and climatic influences is realized. The United States encompasses an area of 9,826,675 square kilometers with 93% being land and 7% being rivers, streams, and lakes. The climate of the United States is highly variable ranging from subtropical to tundra (see Figure 2.2).

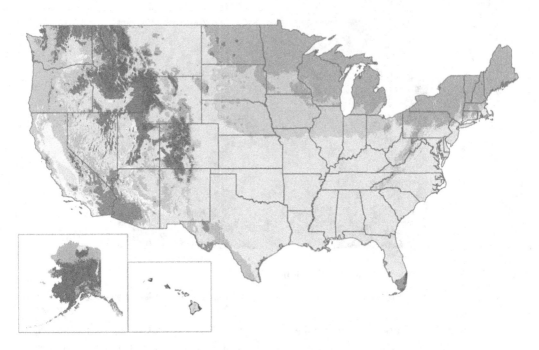

Köppen climate type

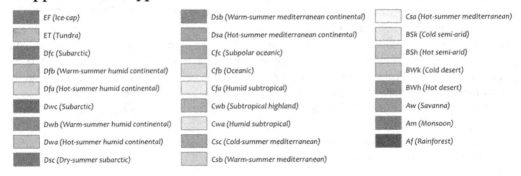

FIGURE 2.2   Climate map of the United States.

National Oceanic and Atmospheric Administration (NOAA). 2021. Koppen-Geiger Climate Changes. https://sos.noaa.gov/datasets/koppen-geiger-climate-changes-1901-2100/ (accessed March 31, 2021).

Land use in the United States consists of the following (United States Department of Agriculture 2011):

- 2,748,717 square kilometers or 30% forest
- 2,473,845 square kilometers or 27% pasture and rangeland
- 1,6,492,230 square kilometers or 18% cropland
- 1,282,734 square kilometers or 14% parks and wildlife areas
- 824,615 square kilometers or 8.5% swamps and tundra
- 274,871 square kilometers or 2.5% urban

A significant portion of the 30% of land in the United States made up of forest is under the jurisdiction of the National Forest Service or the Department of the Interior. This forested land is routinely harvested for lumber and other pulp and paper products (United States Forest Service 2021). The United States population is estimated at 329 million (United States Census Bureau 2021 and United Nations 2021b). The economy of the United States is the third-largest in the world, just behind China and the European Union.

The estimated gross domestic product of the United States is $19.3 trillion (United States Department of Commerce 2021). Per capita income in the United States is approximately $60,000 and ranks 19th in the world. The United States' economic drivers are natural resources (e.g., mining, petroleum, timber), agriculture, manufacturing, well-developed infrastructure, and high productivity (United States Department of Commerce 2021).

2.4 OVERVIEW OF ENVIRONMENTAL REGULATIONS IN THE UNITED STATES

To begin our journey through environmental regulations, we must first consider the meaning of the term **environment**. According to the United Nations (2021a), the environment is a broad term that encompasses all living and nonliving things on Earth or some regions thereof. This definition identifies the natural world existing within the atmosphere, hydrosphere, lithosphere, and biosphere, and also includes the built or developed world (defined as the **anthrosphere**). Second, we must define the difference between a law and a regulation. A **law** is a written statute passed by an appropriate governing body.

A **regulation**, which is often used interchangeably with a law, is a set of standards and rules adopted by administrative agencies that govern how laws will be enforced. In our case that primarily means the USEPA (USEPA 2021a). Regulations often have the same force as such laws, since, without them, regulatory agencies wouldn't be able to enforce laws. Therefore, the USEPA is called a regulatory agency because the United States Congress authorized the USEPA to write regulations that explain the technical, operational, and legal details necessary to implement laws passed by the United States Congress (USEPA 2021c, 2021d).

Although laws focused on protecting human health or the environment can be traced back more than 2,000 years, it was the United States in the early 1970s that set the example when it enacted the most significant and comprehensive environmental regulations and laws in the World (Hahn 1994). Based on the example set by the United States, over 50 countries have established laws that protect human health and the environment on the same platform as USEPA. Table 2.1 is a short list of the countries that have well-established laws that protect human health and the environment (United Nations 2021a).

Environmental laws and regulations in the United States have methodically become more numerous and complex to the point that even many scientists and other professionals need assistance in interpreting, understanding, and applying many of these environmental laws and regulations. The current cost of protecting human health and the environment in the United States is generally considered to be 2% of the Gross National Product (USEPA 2021c).

The growth of the number and complexity of environmental laws and regulations can be more easily understood by simply applying the principle of the scientific term called entropy. Entropy is a principle that states that as time progresses forward there is more disorder which increases complexity. This can be applied to our human civilization as well. Our civilization has become more complex with time as a result of our technological advances, population increase, and other factors. This in turn has put pressure on the natural world and our response has been the passage of environmental laws and regulations as an attempt to establish order in a disordered and out-of-equilibrium environment that humans caused.

The basic overriding principle of environmental law in the United States is to protect human health and the environment. This principle is broken down into categories that track some of the laws and how humans impact the environment and other living species with which we share our planet (Lazarus 2004). These principles include the following:

TABLE 2.1

Selected List of Countries with Established Environmental Regulations Based on the United States

AFRICA	EUROPE	MALTA
Egypt	Austria	Netherlands
Kenya	Belgium	Norway
South Africa	Bulgaria	Poland
ASIA	Cyprus	Portugal
China	Czech Republic	Romania
India	Denmark	Russia
Japan	Estonia	Slovakia
Kyrgyzstan	Finland	Slovenia
Pakistan	France	Spain
NORTH AMERICA	Germany	Sweden
Canada	Greece	Switzerland
Mexico	Iceland	Turkey
United States	Ireland	United Kingdom
SOUTH AMERICA	Italy	**OCEANIA**
Bolivia	Latvia	Australia
Brazil	Lithuania	New Zealand
Chile	Luxembourg	Indonesia

Source: United Nations. 2021a. The United Nations Environment Programme. https://www.unenvironment.org/environment-you (accessed March 31, 2021).

- Environmental Impact Assessment
- Water Quality
- Contaminant Investigation
- Risk Assessment
- Chemical Transport
- Mineral Resources
- Wildlife
- Soil
- Hunting
- Equity
- Public Transparency
- Prevention
- Air Quality
- Waste Management
- Contaminant Cleanup
- Chemical Safety
- Water Resources
- Forest Resources
- Plants
- Fish
- Sustainable Development
- Transboundary Responsibility
- Public Participation
- Polluters Pay

2.5 THE BIRTH OF THE ENVIRONMENTAL MOVEMENT AND REGULATIONS IN THE UNITED STATES

The term **environmental regulation** refers to a set of laws, sometimes referred to as an amalgam of principles, treaties, laws, regulations, directives, ordinances, policies, at the local, State, or Federal level that are enacted to protect the environment in some shape or form (USEPA 2021c). Environmental regulations or laws can be traced to the Roman Empire when the Roman Senate passed legislation to protect the city's supply of water which addressed concerns over fresh water and sewage disposal (Journal of Roman Studies 2021).

In the 14th century, England prohibited the burning of coal in London and the disposal of waste into waterways, specifically the Thames River (Smithsonian 2021). In December 1682, in what would become the United States, William Penn ordered that one acre of forest be preserved for every five acres cleared for settlement (Pennsylvania Historical Association 2021). In addition, Benjamin Franklin widely wrote about the adverse effects of lead and dumping of waste (Environmental Education Associates 2021). However, it was not until 1899 that the United States government passed what is commonly referred to as the first environmental law in the United States, the Rivers and Harbors Act which was intended to clear the harbors of garbage so that ships would not be impeded in transporting goods (USEPA 2021c, 2021d).

It was not until the industrial revolution starting around 1760 that environmental degradation of the air, water, and land became a significant issue. The industrial revolution marked a time of rapid transition from making products by hand to making products with machines. The stationary steam engine invented by James Watt led to widespread use of steam power that used coal. New chemical manufacturing methods, iron production processes, the development of machine tools, and the rise of the factory system all came about in a relatively short time, which sparked the rise of the industrial age and resulted in the discharge of large amounts of waste emitted into the air, into waterways, and discarded onto the land. Figure 2.3 shows an example of this through air emissions in the late 1880s at an industrial facility located in Pennsylvania (United States Research Archives, 2021).

FIGURE 2.3 Air emissions from industrial plants in Pennsylvania in the 1880s.

United States Research Archives. 2021. https://www.UnitedStates.research.archives.gov (accessed March 31, 2021).

As recently as the first half of the 20th century, environmental awareness and legislation were generally lacking not only in the United States but worldwide. There were a few international agreements that primarily focused on boundary waters, navigation, and fishing rights along shared waterways between countries. However, they ignored pollution and ecological issues. In fact, the most convenient and least costly method of waste disposal at the time was "up the stack or down the river" (Haynes 1954). One of the first international agreements was between 12 European countries for the protection of valuable birds called the Convention for the Protection of Valuable Birds Useful to Agriculture in 1902.

It was not until the publishing of Rachel Carson's book *Silent Spring* in 1962 which described the environmental effects of dichlorodiphenyltrichloroethane, which is more commonly known as DDT, on birds that catapulted the environmental movement in the United States into the public eye (Carson 1962).

The 1960s and 1970s are generally described as the age of environmentalism in the United States. **Environmentalism** is defined as a value system that seeks to redefine humankind's relationship with nature (Tarlock 2001). This is when the United States realized that environmental degradation had significantly affected the air and water quality of many urban areas, especially the air in Los Angeles and the water in Lake Erie, which is one of the Great Lakes. Starting on Earth Day in 1971, the Ad Council and Keep America Beautiful campaign aired public awareness commercials on television depicting a Native American shedding a tear when looking out over the polluted landscape of the United States. The commercial stated, "people start pollution, people can stop pollution" (see Figure 2.4).

Two significant sites of environmental contamination were discovered about this time and were Times Beach in Missouri and Love Canal near Buffalo, New York. These two sites would become the first Superfund Sites in the United States. Currently, there are over 1,500 Superfund Sites (USEPA 2021e).

2.6 FORMATION OF NATIONAL POLICY AND THE ENVIRONMENTAL PROTECTION AGENCY

On December 2, 1970, then-President Richard Nixon signed an executive order that established the United States Environmental Protection Agency (USEPA) and a cavalcade of environmental legislation ensued. The creation of the USEPA originated from the passage of the United States National Environmental Policy Act (NEPA) of 1969. NEPA required the federal government to give environmental issues priority when planning any type of large-scale project and also provided for the establishment of councils and agencies to monitor and protect the environment. NEPA was developed to establish consistency in the application of environmental laws and standards throughout the United States (Eccleston 2008 and USEPA 2021a).

FIGURE 2.4   Pollution: it's a crying shame.

United States Advisory Council. 2021. https://www.adcouncil.org/Our-Campaigns/The-Classics/pollution-keep-america-beautiful-iron-eyes-cody (accessed March 31, 2021).

Prior to its passage and the creation of the USEPA, some states had pollution statutes, but many did not. For example, Ohio in particular had an active and highly regarded environmental program but surrounding states did not. The result created significant confusion because what was considered permissible at a location in one State that resulted in the release of pollutants to the environment was not permissible in another State and in which they often shared a common border.

NEPA developed standards for conducting Environmental Impact Statements (EIS) (Eccleston 2008). An EIS describes investigation and remediation methods necessary in order to mitigate or minimize any and all potentially detrimental effects of any new development. Another method of evaluation is environmental auditing, modeled similar to that of a financial audit and which examines processes and outcomes of environmental impacts.

Environmental laws, along with all other federal laws, are organized by subject into what is termed United States Code. The United States Code (U.S.C.) contains all the current enacted statutory language. The official United States Code is maintained by the Office of the Law Revision Counsel in the United States House of Representatives (USEPA 2021a).

USEPA is divided into 10 regions (see Figure 2.5). Table 2.2 shows each region and the States within each region.

2.7 ENVIRONMENTAL LAWS OF THE UNITED STATES

Laws that are commonly attributed to protecting human health and the environment include the following (USEPA 2021c):

- Antiquities Act of 1906
- Atomic Energy Act of 1946
- Atomic Energy Act of 1954
- Brownfield Revitalization Act of 2002
- Clean Air Act (CAA) of 1970
- Clean Water Act (CWA) of 1972
- Coastal Zone Management Act of 1972
- Comprehensive Environmental Response, Compensation, and Liability Act (CERCLA) of 1980
- Emergency Planning and Community Right-to-Know Act of 1986
- Endangered Species Act of 1973

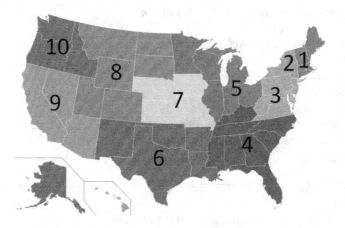

FIGURE 2.5   USEPA regions.

United States Environmental Protection Agency. 2021c. Summary of Environmental Regulations of the United States. https://epa.gov/summary-environmental-laws (accessed March 31, 2021).

TABLE 2.2
USEPA Regions

Region	States
Region 1	Maine, New Hampshire, Vermont, Rhode Island, Massachusetts, Connecticut
Region 2	New York, New Jersey
Region 3	Pennsylvania, West Virginia, Virginia, Maryland, Delaware, District of Columbia
Region 4	Tennessee, Kentucky, Georgia, North Carolina, South Carolina, Florida, Alabama, Mississippi
Region 5	Illinois, Wisconsin, Michigan, Ohio, Indiana, Minnesota
Region 6	Texas, Oklahoma, New Mexico, Arkansas, Louisiana
Region 7	Kansas, Nebraska, Iowa, Missouri
Region 8	North Dakota, South Dakota, Wyoming, Colorado, Utah, Montana
Region 9	California, Arizona, Nevada, Hawaii
Region 10	Washington, Oregon, Idaho, Alaska

Source: United States Environmental Protection Agency. 2021c. Summary of Environmental Regulations of the United States. https://epa.gov/summary-environmental-laws (accessed March 31, 2021).

- Energy Policy Act of 1992
- Energy Policy Act of 2005
- Federal Power Act of 1935
- Federal Feed, Drug, and Cosmetic Act (FFDCA) of 1938
- Federal Insecticide, Fungicide, and Rodenticide Act (FIFRA) of 1947
- Fish and Wildlife Coordination Act of 1968
- Food Quality Protection Act of 1996
- Fisheries Conservation and Management Act of 1976
- Global Climate Protection Act of 1987
- Hazardous Materials Transportation Act (HMTA) of 1975
- Insecticide Act of 1910
- Lacey Act of 1900
- Low-Level Radioactive Waste Policy Act of 1980
- Marine Protection Act of 1972
- Marine Mammal Protection Act of 2015
- Medical Waste Tracking Act of 1980
- Migratory Bird Treaty Act 1916
- Mineral Leasing Act 1920
- National Environmental Protection Act of 1969
- National Forest Management Act of 1976
- National Historic Preservation Act of 1966
- National Parks Act of 1980
- Noise Control Act 1974
- Nuclear Waste Policy Act of 1982
- Ocean Dumping Act of 1988
- Occupational Safety and Health Act (OSHA) of 1970
- Oil Spill Prevention Act of 1990
- Pollution Prevention Act (PPA) of 1990
- Refuse Act of 1899
- Resource, Conservation, and Recovery Act (RCRA) of 1976

- Rivers and Harbors Act of 1899
- Safe Drinking Water Act (SDWA) of 1974
- Superfund Amendments and Reauthorization Act (SARA) of 1986
- Surface Mining Control and Reclamation Act of 1977
- Toxic Substance Control Act (TSCA) of 1976
- Wild and Scenic Rivers Act of 1968

2.8   NATIONAL ENVIRONMENTAL POLICY ACT

The National Environmental Policy Act (NEPA) of 1969 is considered the "Environmental Magna Carta," in that it established a pathway for the development of the national environmental policy and the emergence of the United States Environmental Protection Agency. NEPA grew out of the increased awareness and concern for the environment from human events that caused harm to the environment included (USEPA 2021a) the following:

- Rachel Carson's chronicled the impact of DDT on the environment, especially birds, in her book published in 1962.
- An oil spill off the coast of Santa Barbara in 1969.
- Freeway revolts in the 1960s consisted of a series of protests in response to the destruction of many communities and ecosystems during the construction of the Interstate Highway System.

The purpose of NEPA was to establish three goals which were (USEPA 2021a) as follows:

1. The Nations' environmental policies and goals
2. Provisions for the federal government to enforce such policies
3. The creation of the Council of Environmental Quality in the Executive Branch of the President

The declaration of the national environmental policy contained in Section 2, Title I of the NEPA states (USEPA 2021a):

The Congress, recognizing the profound impact of man's activity on the interrelations of all components of the natural environment, particularly the profound influences on population growth, high-density urbanization, industrial expansion, resource exploitation, and new expanding technological advances and recognizing further the critical importance of restoring and maintaining environmental quality to the overall welfare and development of man, declares that is the continuing policy of the Federal Government in cooperation with State and local governments, and other concerned public and private organizations, to use all practicable means and measures, including financial and technical assistance, in a manner calculated to foster and promote the general welfare, to create and maintain conditions under which man and nature can exist in productive harmony, and fulfill the social, economic, and other requirements of present and future generations of Americans.

Of special note in the paragraph above is the phrase "..., *particularly the profound influences on population growth...*". This will be discussed in greater detail in subsequent chapters when we discuss the influence of continued population growth on sustainability. Specifically, the question to be discussed is how can humans maintain sustainability with the environment with an increasing population and when will the stresses caused by human population growth on the quality of air, water, land, and other species become too great. Just since NEPA was written our population in the United States has risen from 202 million (in 1969) to over 323 million (in 2019) or an increase of more than 37% (United States Census Bureau 2021). During this same period of time, the world population has increased from 3.6 billion to more than 7.2 billion or 100% (United Nations 2021b).

In essence, the purpose of NEPA was to ensure that environmental factors are weighted equally compared to other factors in the decision-making process undertaken by federal agencies and to establish a national environmental policy. NEPA also promotes the Council of Environmental Quality (CEQ) to advise the President in the preparation of an annual report on the progress of federal agencies in implementing NEPA. A significant environmental benefit as a result of NEPA is that it required federal agencies to prepare an **Environmental Impact Statement** (EIS) to accompany reports and recommendations for Congressional funding. An EIS is a systematic scientific evaluation of the relevant environmental effects of a federal project or action, such as building a dam or an interstate highway (USEPA 2021d).

2.9   ENVIRONMENTAL JUSTICE

Environmental Justice is the fair treatment and meaningful involvement of all people regardless of race, color, national origin, or income during environmental activities that include the development, implementation, and enforcement of environmental laws, regulations, and policies (USEPA 2021f). Environmental justice had its beginning as far back as 1968 when sanitation workers in Memphis went on strike because of unfair treatment and environmental justice concerns. The strike advocated for fair pay and better working conditions for Memphis garbage workers. It was the first time that African Americans mobilized a national, broad-based group to oppose environmental injustices (USEPA 2021f). The position of the workers was summed up by Professor Robert Bullard who stated, "whether by conscious design or institutional neglect, communities of color in urban ghettos, in rural poverty pockets, or on economically impoverished Native-American reservations face some of the worst environmental devastation in the nation."

A second incident worthy of note occurred in 1983 in Warren, North Carolina, where a polychlorinated biphenyl (PCB) landfill was proposed to be constructed. Over 500 environmentalists and civil rights activists were arrested during a peaceful demonstration. The protests and demonstrations were ultimately unsuccessful but this incident is widely considered the catalyst for the environmental justice movement (USEPA 2021d). Since 1983, there are perhaps hundreds of instances and examples of environmental justice throughout the United States. So many in fact, that the USEPA created the Office of Environmental Justice (USEPA 2021f).

2.10   ENVIRONMENTAL ENFORCEMENT

Enforcing environmental laws is a central part of USEPA's strategic plan to protect human health and the environment. To ensure that environmental regulations are taken seriously by the regulated community, the USEPA has established an Enforcement Policy and has published several guidance documents that provide detail so that enforcement measures are applied equally and consistently (USEPA 2021h, 2021i). In addition, USEPA has a top priority to protect whole communities disproportionately affected by pollution through what is termed the environmental justice network, which was discussed in the previous section (USEPA 2021h).

In large part, the reason why the United States has been so successful at improving the environment and protecting human health has been because USEPA has an effective enforcement program. Purposeful acts that harm the environment have been greatly reduced in large part because of education and the threat of enforcement. No longer is there a mentality that "It's not a violation until I get caught." In fact, it may be a good time to rethink enforcement policies that have not changed in 30 years.

There are generally two levels of enforcement penalties in the United States which are civil and criminal (USEPA 2021i). The criminal enforcement program at USEPA was established in 1982 and was granted full law enforcement authority by Congress in 1988 and employs special agents and investigators, forensic scientists and technicians, lawyers, and support staff. Civil enforcement arises most often through an environmental violation. Civil enforcement does not take into consideration what the responsible party knew about the law or regulation that was violated. Environmental criminal liability is triggered through some level of intent (USEPA 2021h). A simple explanation of the difference between a civil or criminal violation is that a criminal act is characterized and a "knowing violation" where a person or company is aware of the facts that create the violation. A conscious and informed action brought about the violation. In contrast, a civil violation may be caused by an accident or mistake (USEPA 2021h, 2021i, 2021j).

Types of enforcement results from a civil violation include settlements, civil penalties, injunctive relief, and supplemental environmental projects (SEPs) (USEPA 2021h). To evaluate the level of seriousness of an environmental violation, one must answer three basic questions:

1. Who first discovered the environmental violation? For the most part, environmental regulations are based on self-reporting. It is generally a worse situation if the regulatory authority discovers a violation especially one that should have been self-reported.
2. Did the violation cause harm? It is generally a worse situation if the violation caused harm to human health or the environment. A permit violation or exceedance is generally considered a situation that has the potential to cause harm.
3. Was the violation a purposeful act? It will always be a worse situation if the violation was a purposeful act.

To evaluate the level of a potential fine, the following are generally considered (USEPA 2021i):

- How long did the violation last? Under many circumstances, fines are determined on per day violation occurred and can ranged from $7,500 per day to as much as $37,500 per day.
- What was the economic benefit of the violation considered from a point of view of delayed cost and avoided cost? In addition, compelling public concerns are considered.
- Did the violation cause harm to human health and the environment? To evaluate the level of risk, the toxicity of the pollutant, sensitivity of the environment, pollutant mass released, and length of time of the violation are each considered.
- Did the violation continue to occur after discovery?
- Size of the violator
- Litigation risk
- Ability to pay
- Offsetting penalties such as those paid to State and Local governmental agencies or citizen groups.
- Did more than one violation occur?

Settlement of violations often includes the following (USEPA 2021j):

- Payment of a fine. As stated above, fines are accessed on how many days the violation occurred and the level of seriousness of the violation itself, such as, did the violation cause harm to human health and the environment.
- Permit modification. Usually, when a permit modification is justified, a stricter emission standard is applied.
- Supplemental environmental project (SEP). If a facility must upgrade equipment or invest capital expense in order to return to compliance, a portion of that expense may be used to offset the fine amount. However, the amount of offset is generally capped at 50%.

Certainly, circumstances surrounding any violation are considered. However, any violation must be taken as a serious matter and may determine whether any facility continues to operate.

2.11   SUMMARY OF THE HISTORY OF ENVIRONMENTAL REGULATIONS AND SUSTAINABILITY OF THE UNITED STATES

After reading this chapter, you should start to have an understanding of just how much environmental regulation we have in the United States and why. It should also be apparent that staying in compliance is a complex undertaking requiring much skill and expertise.

There is much we can be proud of when we examine the improvements in our air quality, water quality, and clean-up of sites of environmental contamination, which now number in the tens of thousands of sites across the United States. Looking back over the sections of different environmental regulations, you may have noticed that much of the legislation was oftentimes a reaction to an incident or environmental tragedy. Laws and regulations originated from many circumstances because of unintentional actions that ended up making quite a mess and causing significant harm to human health, the environment, or both. Some of the incidences that we will touch upon include the Exxon Valdez spill in 1990, Deepwater Horizon in 2010, Love Canal, Bhopal India in 1984, Valley of Drums, and Toms River to the killing and extinction of animal species that also nearly included the American Bison, California Condor, Grizzly Bear, and numerous others. However, this is a reactionary approach, not a proactive approach. When we discuss Sustainability later in this book, we will discuss the need to move beyond reactive measures to proactive sustainability measures to improve our environment in the future. If there are delays in enacting Sustainability measures, it may be too late.

The amount and complexity of environmental laws and regulations in the United States are due to the fact that we are an industrialized society and consume enormous amounts of energy and goods and our hunger for those goods and energy continues to rise as we continue to improve our standard of living. The United States produces large volumes of waste of all types and as our population continues to increase, the amount of waste produced will also continue to increase.

These central facts combined with the dynamic and innate complexities of the natural world add an almost infinite number of negative outcomes in a world with over 7 billion people all wanting to improve their standard of living. The fact is, if we do not have effective and comprehensive environmental regulation, we simply threaten our existence and numerous other species as we are slowly devoured by our own garbage.

Our advances in science have created many technologies that have improved our standard of living and in some circumstances have improved our relationship with nature. But our technological advancements have also created many unintended negative side effects that question the benefit of some technologies. An example involves the creation of synthetic compounds that nature has difficulty breaking down. Some of these compounds include polychlorinated biphenyls, chlorinated hydrocarbons, and many pesticides, including DDT which began the environmental movement in the United States in earnest over 50 years ago. This is to say nothing about plastics which we all should know about since we commonly use and discard huge amounts of plastic in our everyday life.

Although efforts are underway to reduce our energy needs and our voracious hunger for consuming nature, one thing seems certain, our thirst for consumer goods and energy needs will continue to grow, and as our population increases our energy needs will increase even faster. This means that environmental regulation will have to grow with us in volume and complexity as we consume more and more of Earth's resources and replace it with our waste.

What may be a surprise to many is that agriculture is now the biggest polluter in the United States and the world (United Nations 2021a). The United Nations estimates that most land pollution is caused by agricultural activities, such as grazing, use of pesticides, fertilizers, irrigation, plowing, and confined animal facilities (United Nations 2021a). Agricultural activities release significant amounts of greenhouse gases from plowing, decaying vegetation, and from livestock (USEPA 2021g). Fugitive emissions and erosion are also significant and release significant amounts of chemicals into the air and the water (see Figure 2.6). In addition, fertilizers used in agriculture are a significant source of pollution to the oceans of the world (USEPA 2021g).

FIGURE 2.6 Example of fugitive air emissions (Photograph by Daniel T. Rogers).

Over the next several chapters, we will evaluate pollution and contaminants in the United States and will touch upon contaminant impacts throughout the world, including Antarctica and the Oceans and examine how other countries' environmental laws compare to the United States. Some findings might be rather surprising in that it will become apparent that a few countries are more advanced at protecting human health and the environment than the United States. Surprising still might be the impression that some countries either lack environmental laws or the will to enforce the environmental laws they have due to other overriding social or economic challenges. As we shall see, it is not enough just to have environmental regulations. Many other factors are required to have an effective environmental regulatory framework that actually accomplishes the ultimate goal of protecting human health and the environment, which paves the path toward sustainability.

Questions and Exercises for Discussion

1. Who is Rachel Carson and why is she important to environmental science?
2. What is the overriding principle of environmental regulation?
3. How does this principle of environmental regulation apply to sustainability?
4. Compare and contrast the differences between a law and a regulation?
5. Do you see a need for additional environmental regulation? If so, please explain.

REFERENCES

Bates, C. G. and Ciment, J. 2013. Encyclopedia of Global Social Issues. M.E. Sharpe Publishers. New York. New York. 1450 p.

Carson, R. 1962. Silent Spring. Boston, MA. Houghton Mifflin.

Eccleston, Charles S. 2008. NEPA and Environmental Planning: Tools, Techniques, and Approaches for Practitioners. CRC Press. New York, New York. 432p.

Environmental Education Associates. 2021. The Famous Benjamin Franklin Letter on Lead Poisoning. http://environmentaleducation.com/wp-content/uploads/userfiles/Ben%20Franklin%20Letter%20on%20EE%281%29.pdf (accessed March 31, 2021).

Hahn, R. W. 1994. United States Environmental Policy: Past, Present and Future. Natural Resource Journal. John F. Kennedy School of Government, Harvard University: Cambridge, Massachusetts. Vol. 34. no 1. pp. 306–348.

Haynes, W. 1954. American Chemical Industry – A History. Vols. I – IV. Van Nostrand Publishers: New York, NY.

Journal of Roman Studies. 2021. Fresh Water in Roman Law: Rights and Policy. Vol. 107. https://www.cambridge.org/core/journals/journal-of-roman-studies/article/fresh-water-in-roman-law-rights-and-policy/548B1C559B3D6ACEDF50C4576DD14603/core-reader#fns01 (accessed March 31, 2021).

Lazarus, R. 2004. The Making of Environmental Law. Cambridge Press. Cambridge, United Kingdom. 367 p.

National Oceanic and Atmospheric Administration (NOAA). 2021. Koppen-Geiger Climate Changes. https://sos.noaa.gov/datasets/koppen-geiger-climate-changes-1901-2100/ (accessed March 31, 2021).

Pennsylvania Historical Association. 2021. The Vision of William Penn. http://explorepahistory.com/story.php?storyId=1-9-3&chapter=1 (accessed March 31, 2021).

Smithsonian. 2021. Air Pollution Goes Back Further than You Think. https://www.smithsonianmag.com/science-nature/air-pollution-goes-back-way-further-you-think-180957716/ (accessed March 31, 2021).

Tarlock, D. A. 2001. History of Environmental Law. Encyclopedia of Life Support Systems. University of Chicago Press. Chicago, IL. 1450 p.

United Nations. 2021a. The United Nations Environment Programme. https://www.unenvironment.org/environment-you (accessed March 31, 2021).

United Nations. 2021b. World Population through Time. https://www.unitednations.org/population (accessed March 31, 2021).

United Nations. 2021b. https://www.un.org/sustainabledevelopment/ (accessed July 27, 2021).

United States Advisory Council. 2021. Pollution, It's a Crying Shame. https://www.adcouncil.org/Our-Campaigns/The-Classics/pollution-keep-america-beautiful-iron-eyes-cody (accessed March 31, 2021).

United States Census Bureau. 2021. United States Population through Time and World Populations Clock. https://www.UScensus.gov/USpopluation (accessed March 31, 2021).

United States Department of Agriculture. 2011. Land Use in the United States. https://www.ers.usda.gov/data-products/major-land-use (accessed March 31, 2021).

United States Department of Commerce. 2021. Principle Federal Economic Indicators. Bureau of Economic Analysis. https://www.bea.gov (accessed March 31, 2021).

United States Environmental Protection Agency. 2021a. National Environmental Policy Act. www.epa.gov/NEPA (accessed March 31, 2021).

United States Environmental Protection Agency. 2021b. Donora Smog of 1948. EPA Alumni Association: News and Views (accessed July 9, 2021).

United States Environmental Protection Agency (USEPA). 2021c. Laws and Regulations. https://www.epa.gov/laws-regulations/regulations (accessed March 31, 2021).

United States Environmental Protection Agency. 2021d. Summary of Environmental Regulations of the United States. https://epa.gov/summary-environmental-laws (accessed March 31, 2021).

United States Environmental Protection Agency (USEPA). 2021e. CERCLA Sites in the United States. https://www.epa.gov/CERCLA-sites (accessed March 31, 2021).

United States Environmental Protection Agency. 2021f. Environmental Justice. Environmental Justice|US EPA (accessed July 22, 2021).

United States Environmental Protection Agency. 2021g. Agriculture 101. https://www.epa.gov/sites/production/files/2015-07/documents/ag_101_agriculture_us_epa_0.pdf (accessed February 15, 2021).

United States Environmental Protection Agency (USEPA). 2021h. Enforcement Policy, Guidance and Publications. https://www.epa.gov/enforcement/enforement-policy-guidance-publications (accessed February 15, 2021).

United States Environmental Protection Agency (USEPA). 2021i. Enforcement Policy, Guidance and Publications. https://www.epa.gov/enforcement/criminal-enforement-overview (accessed February 15, 2021).

United States Environmental Protection Agency (USEPA). 2021j. Basic Information in Enforcement. https://www.epa.gov/enforcement/basic-information-enforement (accessed February 20, 2021).

United States Forest Service. 2021. Forest Products Cut and Sold from the National Forests and Grasslands. https://www.fs.fed.us/forestmanagement/products/cut-sold/index.shmtl (accessed March 31, 2021).

United States Research Archives. 2021. https://www.UnitedStates.research.archives.gov (accessed March 31, 2021).

World Health Organization (WHO). 2021. Nine Out of Ten People Breathe Unhealthy Air. Geneva. Switzerland. https://www.who.int/news-room/detail/02-02-2018 (accessed February 6, 2021).

3 Nature's Response to Contamination

3.1 INTRODUCTION

In this chapter, we will turn our attention to how nature reacts to contamination. As you can imagine, nature rarely has a positive reaction when contaminated. However, nature's reaction widely varies. The reason why there is a large variation is dependent on two factors that include the physical chemistry of the contaminant and the geological environment in which the release occurs. Other potentially significant factors are related to the amount released and over what period of time.

Each day when we walk outside, we are walking on a history book, and most of us are unaware of this and of its significance. The arrangement, thickness, and composition of the soil and sediment layers just centimeters deep have a profound influence on our lives and all life on Earth. These soil and sediment layers dictate where cities are built, where roads are built, and how buildings are constructed. Perhaps most significantly to our civilization at this juncture in our collective history, these soil and sediment layers are the meeting place between human civilization and the natural world. This contact between human civilization and the natural world is the proving ground and point at which contaminants released into the environment begin their destructive journey and eventually can detrimentally impact all living things on Earth. This is why we must understand our natural world in at least the same scientific detail that we need to understand pollution, and that is because they interact with one another.

Throughout the world, the largest cities share a geologic environment dominated by unconsolidated sedimentary deposits and are located near water (see Table 3.1). Most of those sedimentary deposits are saturated with water very near the surface, and function as sources of drinking water and/or as hydraulic connections to surface water, ecosystems, and ultimately to the oceans. Moreover, water is considered the universal solvent, so any pollution released into the environment from anthropogenic or natural sources has the potential to migrate and cause harm. Scientific factors that control the severity of harm to the environment from pollutant releases are as follows: (1) the geologic and hydrogeologic environment, (2) the physical chemistry of the contaminants and amounts released, and (3) the mechanism in which the release occurs (Kaufman et al. 2005; Murray and Rogers 1999a; Rogers 1996, 2018). There are several techniques available for investigating and managing the complexity of water pollution (Rogers 2016a, 2016b). One important and critical concept to keep in mind when evaluating groundwater is that the hydrologic cycle is modified and altered by human development (Wong et al. 2012). To achieve any level of success in mitigating environmental contamination, it becomes a prerequisite to understand the geology, hydrology, and the fate and migration of contaminants within its specific geology (Rogers 2014, 2020).

The focus of this chapter is to identify regions vulnerable to contamination and those areas where widespread contamination is less likely. To accomplish this task, an additional level of interpretation consisting of a comprehensive vulnerability analysis is added to the near-surface geologic maps. This chapter also explains why certain types of geology may be especially susceptible to contamination which can be utilized when developing a sustainability program for any region or location (Rogers 2019).

3.2 SUBSURFACE VULNERABILITY AND VULNERABILITY MAP DEVELOPMENT

The concept of vulnerability of the subsurface to contamination originated in France during the 1960s and was introduced into the scientific literature by Albinet and Marget (1970). Since then,

DOI: 10.1201/9781003175810-3

TABLE 3.1

Water Bodies and Near-Surface Geology Near the Major Cities of the World

City	Location	Estimated Metropolitan Population (millions)	Geology	Water Body
Tokyo	Japan	37.8	Unconsolidated	Pacific Ocean
Shanghai	China	34.8	Unconsolidated	Pacific Ocean
Jakarta	Indonesia	31.7	Unconsolidated	Pacific Ocean
Delhi	India	26.4	Unconsolidated	Yamuna River
Seoul	Korea	25.5	Unconsolidated	Han River
Beijing	China	24.9	Unconsolidated	Yongding River
New York City	USA	23.8	Unconsolidated	Atlantic Ocean, Hudson River
Mexico City	Mexico	21.6	Unconsolidated	Lerma River, Santiago River
Sao Paulo	Brazil	21.2	Unconsolidated	Atlantic Ocean
Cairo	Egypt	20.5	Unconsolidated	Nile River
Los Angeles	USA	18.7	Unconsolidated	Pacific Ocean
Moscow	Russia	16.9	Unconsolidated	Moskve River
Istanbul	Turkey	15.2	Unconsolidated	Turkist Straits
London	England	14.2	Unconsolidated	Thames River
Buenos Aires	Argentina	13.1	Unconsolidated	Atlantic Ocean
Paris	France	12.6	Unconsolidated	Seine River
Rio de Janeiro	Brazil	12.3	Unconsolidated	Atlantic Ocean
Chicago	USA	9.8	Unconsolidated	Lake Michigan
Johannesburg	South Africa	9.6	Unconsolidated	Jukskei River
Riyadh	Saudi Arabia	7.7	Unconsolidated	Simbacom River
Santiago	Chile	6.7	Unconsolidated	Pacific Ocean
Berlin	Germany	6.1	Unconsolidated	Spree River
Toronto	Canada	5.9	Unconsolidated	Lake Ontario
Sydney	Australia	5.0	Unconsolidated	Pacific Ocean

Source: United Nations. 2021. World Population Prospects. United Nations Department of Economic and Social Affairs. Population Division. https://esa.un.org/unpd/wpp/data (accessed March 31, 2021).

the concept of subsurface vulnerability has evolved to include both a distinction between and a combination of vulnerability and risk assessment. Groundwater vulnerability is currently interpreted as a function of the natural properties of the overlying soil or sediments of the unsaturated zone, aquifer properties (e.g., effective porosity and recharge area), and aquifer material (Foster and Hirata 1988; Robins et al. 1994; Rogers et al. 2007).

Geologic vulnerability mapping can be divided into two groups: subjective rating methods and statistical and process-based methods (Focazio et al. 2001). The subjective rating methods are characterized by numerical scales representing low to high vulnerabilities. Typically, the results are applied to large areas and used for policy and management objectives. By contrast, the statistical process-based methods produce finite values, such as areas exceeding specific water quality values. With these methods, the results are usually not applied to large areas due to data gaps and variable geology. In addition, the results are generally obtained under more detailed site-specific assessments and used for purely scientific purposes (Focazio 2001). In practice, the subjective rating methods are preferred for conducting vulnerability assessments on a watershed scale (Murray and Rogers 1999a). The concept of geologic vulnerability relies on the assessment and representation of various hydrogeologic parameters such as vadose zone characteristics (e.g., thickness and infiltration capacity), depth to water, and amount of recharge (Eaton and Zaporozec 1997; Zaporozec and Eaton 1996). The utility of this concept, however, becomes more important when the geologic data are supplemented with environmental, economical, and political insight gained through past environmental cleanup efforts (Foster et al. 1994; Loague et al. 1998).

Urbanization and artificial infrastructure (e.g., sewers and detention ponds) can have a profound influence on the regional hydrogeology (Kibel 1998; Vuono and Hallenbeck 1995; Zaporozec and Eaton 1996). Basic processes affecting surface water and groundwater are modified, including surface water drainage patterns and velocities, evaporation rates, infiltration, and aquifer recharge (Burn et al. 2007; Fresca 2007; Howard et al. 2007; Mohrlok et al. 2007). Geologic vulnerability mapping provides a starting point for quantifying anticipated environmental risk at a particular site and also highlights locations where additional information is required. In lieu of specific site information, this risk assessment can serve, if necessary, as a proxy for anticipated future cleanup costs. Additionally, this method can be used by other interested parties during the recycling of industrial sites to estimate the liability of sites (Murray and Rogers 1999b; Stiber et al. 1995). With respect to water resource allocation, since surface and groundwater interact, mapping groundwater also provides valuable information concerning their respective distributions (Pierce et al. 2007; Rogers and Murray 1997).

3.3 METHODS

The evaluation of geologic vulnerability in an urban watershed using a subjective rating method requires a combination of geologic and hydrogeologic data, identification of the potential receptor sites, and political and economic information. The first and most crucial step is mapping the near-surface geology. Once the geologic map is created the process of developing a geologic vulnerability map can be initiated. Murray and Rogers (1999a, 1999b) and Kaufman et al. (2003, 2005) have developed a method for geologic vulnerability mapping using a modified DRASTIC model (Aller et al. 1987). This method contains a subjective numerical rating system and uses different weighting coefficients for various geologic and hydrogeologic parameters of concern and incorporates potential receptors and political data into the model (Kaufman et al. 2005; Murray and Rogers 1999a; Rogers 1992, 1996, 2002, 2020).

3.4 DEMONSTRATING THE SIGNIFICANCE OF VULNERABILITY MAPPING

As noted by Foster et al. (1994) and Loague et al. (1998), the significance of geologic vulnerability is not appreciated until it can be put into environmental, economic, or political perspective by actually cleaning up sites of environmental contamination. Therefore, to evaluate whether certain geologic units are vulnerable to contamination, a comparison between specific sites located in low geologic vulnerability areas to sites located in high geologic vulnerability areas must be conducted. If valid, the geologic vulnerability mapping should confirm that the sites situated above high vulnerability locations pose a greater risk of exposure than sites located above low vulnerability locations. For this analysis, a site of low vulnerability located in a geological environment predominantly composed of clay sediments (Site 1) is compared to a high vulnerability site located in a geological environment predominantly composed of sand (Site 2). However, Site 1 is (1) significantly larger than Site 2 (approximately twice the size), (2) had a much longer heavy industrial operational history (operated approximately 40 years longer), and (3) had significantly more contamination and types of contaminants (nearly 10 times the mass and three times as many contaminants) released into the environment (see Table 3.2).

TABLE 3.2
Geologic Vulnerability Matrix and Scoring System

Parameter Identification Number	Parameter Description	Rating Strength
1	**Depth to Groundwater**	
	Less than 10 feet below the ground surface	10
	10–30 feet	5
	Greater than 30 feet	1
2	**Composition, areal extent, and thickness of soil units in the unsaturated zone**	
	Thick and extensive sequence of sand and gravel	10
	Interbedded sands and clay deposits	5
	Thick and extensive sequence of clay	1
3	**Composition, areal extent, and thickness of saturated zone**	
	Thick and extensive sequence of sand and gravel	10
	Interbedded sands and clay deposits	5
	Thick and extensive sequence of clay	1
4	**Occurrence and relative abundance of groundwater**	
	25% or less likelihood before encountering an aquiclude	10
	25%–74% likelihood	5
	Greater than 75% likelihood	1
5	**Area of groundwater recharge**	
	Significant area of recharge	10
	Moderate area of recharge	5
	Not a significant area of recharge	1
6	**Areas of groundwater discharge**	
	Significant area of recharge	10
	Moderate area of recharge	5
	Not a significant area of recharge	1
7	**Travel time and distance to point of potential exposure.**	
	Less than 10 years	10
	10–25 years	5
	Greater than 25 years	1
8	**Source of potable water**	
	Current source of potable water	10
	Potential source of potable water	5
	Not a potential source of potable water	1

Source: Rogers, D. T. 2020. Urban Watershed: Geology, Contamination, Environmental Regulations, and Sustainable Development. CRC Press. Boca Raton, FL. 606 p.

Without considering the geology of each site, it would be logical to assume the environmental risks were higher at Site 1, and the associated cleanup costs would be higher and reflect its contamination history. We now determine if the vulnerability map predicts these outcomes.

3.4.1 Site 1 – Low Vulnerability Site

Site 1 is a former heavy manufacturing facility located on approximately 16 acres of land that operated for approximately 70 years. A Phase I environmental site assessment was required by the lending institution and conducted due to a real estate transaction involving the property. This initial assessment identified six Recognized Environmental Conditions (RECs). During the next investigational period (Phase II), several subsurface investigations were conducted at the facility, and four main sources of contaminant release were identified that required remediation and included (1) surface spills, (2) an above-ground storage tank, and, (3) spills and leaks of hazardous liquids located in waste storage areas. Other sources or releases were identified during the course of evaluating the site but were not severe enough to warrant further action. The six RECs identified during the Phase I investigation included the following:

- Former chemical storage areas. Evidence of surface staining indicating some spillage of liquids was observed on bare ground near the two former storage areas. No staining was observed near the current storage area.
- Current storage area. The current storage area was located inside the main manufacturing building (northern building). The concrete flooring was heavily cracked, providing a potential pathway for spills and leaks to contaminate the ground beneath the building.
- A former above-ground storage tank that stored gasoline. A limited amount of surface staining was observed at the general locations the tank had been located.
- Surface soil staining and stressed vegetation near a back door of the facility and close to an inside location where maintenance activities were conducted and DNAPL solvents were used.
- Stained soil and stressed vegetation were observed at a location of bare ground where deliveries to the facilities were conducted, and where materials including liquids were offloaded from trucks.

The initial Phase II investigation involved drilling 15 soil borings in areas with the highest likelihood of detecting contamination. This initial subsurface investigation had the following objectives: (1) begin to characterize the subsurface geology of the site, (2) identify the contaminants that were suspected to have been released, and (3) evaluate the magnitude of contaminants present by collecting and analyzing "worst-case" samples from each area of suspected contamination. The results confirmed the presence of contamination at five of the six RECs identified during the Phase I environmental site assessment. Those compounds detected were at significant concentrations and required further evaluation. The current storage area was eliminated as an area of concern because contamination could not be confirmed through the drilling of three soil borings into the most visibly vulnerable areas of the concrete and the analysis of six soil samples.

Three additional investigation phases were conducted to define the nature and extent of the contamination. During the subsequent phases, a total of 40 additional soil borings were drilled, and six monitoring wells were installed to evaluate whether there was enough groundwater present for analysis and to establish the direction of groundwater flow. In many of the soil borings, more than two soil samples were analyzed to gather data on the vertical extent of contamination. The maximum depth of the soil borings was 15 feet beneath the surface in impacted areas, yet the vertical extent of contamination did not exceed a depth of 5 feet. The monitoring wells indicated damp to moist soils existed at some locations within very thin layers of silt. These silty layers were just a few millimeters thick and were also observed in soil samples collected from some of the soil borings drilled during investigative activities. After the monitoring wells failed to detect any groundwater seepage, they were pulled from the ground and the boreholes sealed with a bentonite clay grout. Characterization of the geology at the site was accomplished by drilling a soil boring to a depth of 40 feet in a non-impacted area. In general, the site was underlain by a clay deposit from a Pleistocene age glacial lake that occupied the region more than 12,000 years before the present. Historical geological literature of the region indicates a ground moraine or lodgment till deposit extended to depths of approximately 180 feet beneath the ground at the site.

Other pertinent technical and geological information concerning the site included the following:

- Storm sewers in the immediate vicinity did not intersect any of the contamination.
- Surface water drainage was controlled by storm sewers.
- No buried utilities intersected contaminated areas.
- Potable water was supplied by the municipality and the source was more than 10 miles away.
- The contamination did not extend beyond the property boundary.

The types of contaminants detected at the facility included the following:

- Volatile organic compounds (VOCs) include
  - Dense nonaqueous phase liquids (DNAPLs) commonly referred to as chlorinated solvents used to degrease and clean metal surfaces.
  - Light nonaqueous phase liquids (LNAPLs) used as solvents, paint thinners, and cleaning products, and are common constituents in fuels such as gasoline.
- Polynuclear aromatic hydrocarbons (PNAs or PAHs) commonly used as lubricants and motor oils and cutting fluids.
- Polychlorinated biphenyls (PCBs) used in electrical equipment.
- Heavy metals (arsenic, chromium, and lead) commonly used in paints and pigments, batteries, and metal plating.

A list of specific chemical compounds, highest concentrations detected, and estimated contaminant mass remediated are listed in Table 3.3.

Soil excavation and disposal of the contaminated soils at a licensed landfill was the remedial method of choice for the contamination at this site. The overriding considerations for selecting this method were the contaminated areas were less than 5-feet deep and were not located beneath any buildings; the activities could be conducted quickly without disturbing ongoing facility operations, and it represented the lowest cost alternative. Approximately 7,000 cubic yards –equivalent to 10,000 tons of soil – was excavated and transported to a local landfill for disposal. The total cost for investigation and remediation was approximately $400,000. This translates into a remediation cost per kilogram of contaminant of nearly $2,000. After the remediation was verified by the regulatory authority through the collection and analysis of the soil samples taken from each area remediated, closure was granted and a "No Further Action Required" letter was issued for the site. The closure was deemed unrestricted, meaning the site had been remediated to comply with residential land-use requirements. Eighteen months had elapsed since the Phase I environmental site assessment.

3.4.2 SITE 2 – HIGH VULNERABILITY SITE

Site 2 is a former heavy manufacturing facility approximately 8 acres in size that operated for 30 years. The Phase I environmental site assessment identified four RECs:

- Two former chemical storage areas. Evidence of surface staining indicated some spillage of liquids at one storage area located on bare ground near an area storing waste paints. An additional waste storage area was located on stained asphalt pavement that was heavily cracked and broken.
- A former underground storage tank. This underground storage tank once stored gasoline and was identified as a REC because during its removal: (1) it was not physically inspected by a qualified professional; and (2) confirmatory soil samples were not collected and analyzed from the excavation pit to verify the tank did not leak.
- Surface staining. Surface staining and stressed vegetation were observed near the former location of a back door near the location where solvents were used inside the manufacturing building.

TABLE 3.3

Contaminant Types, Concentration, Mass Remediated for Site 1

Contaminant	Maximum Concentration (µg/kg)	Estimated Contaminant Mass Remediated (kilograms)
Volatile Organic Compounds (VOCs)		80
DNAPL Compounds		
Tetrachloroethene (PCE)	60,100	
Trichloroethene (TCE)	45,000	
Cis-1,2-dichoroethene	20,000	
Trans-1,2-dichloroethene	3,000	
Methylene chloride	800	
LNAPL Compounds		
Ethyl benzene	10,000	
Xylenes	10,000	
Acetone	280	
Carbon disulfide	200	
Polynuclear Aromatic Hydrocarbons		115
Naphthalene	339,000	
Acenaphthalene	18,000	
Fluorene	22,000	
Phenanthrene	280,000	
Fluoranthene	156,000	
Pyrene	13,000	
Benzo(a)anthracene	4,800	
Benzo(a)pyrene	3,200	
Dibenzo[a.h]anthracene	1,800	
Indeno[1,2,3-cd]pyrene	1,400	
Chrysene	11,000	
Polychlorinated Biphenyls (PCBs)	16,000	2
Heavy Metals		23
Arsenic	23,000	
Chromium	530,000	
Lead	930,000	

Source: Rogers, D. T. 2020. Urban Watershed: Geology, Contamination, Environmental Regulations, and Sustainable Development. CRC Press. Boca Raton, FL. 606 p.

µg/kg = micrograms per kilogram

The multiple Phase II subsurface investigations conducted at the facility identified four main sources of contamination resulting in the release of contaminants and requiring remediation. These sources of contamination included four areas with a track record of prior spills. Other sources or releases were identified during site evaluation but were not severe enough to require further action.

The initial subsurface investigation conducted involved drilling 12 soil borings in areas with the highest likelihood of detecting contamination, and was performed with similar objectives to those at Site 1: characterize the geology; identify the contaminants; and analyze the worst-case samples. The results confirmed the presence of contamination at three of the four RECs identified in the

Phase I environmental site assessment. Contaminant concentrations in near-surface soil were also detected at sufficient levels to require further investigation. The one location not pursued for further investigation was near the former underground storage tank. Four soil borings were made at the location of this tank and the analysis of four soil samples taken from the soil beneath the tank did not confirm the presence of contamination above detectable concentrations.

Groundwater was encountered at a depth of 10–12 feet beneath the surface of the ground during the initial investigation. Temporary monitoring wells were installed at select locations to evaluate the possible presence of groundwater impacts and estimate the direction of groundwater flow. The analytical results suggested the presence of groundwater impacts likely originating from onsite sources. This finding was confirmed because there were levels of several contaminants exceeding applicable cleanup criteria, and the same contaminants were detected in near-surface unsaturated soil at the locations where the RECs were identified, but were not detected in soil or groundwater at upgradient locations.

Six additional investigation phases were conducted to define the nature and extent of contamination. During these subsequent phases, a total of 132 additional soil borings were made, with many of the soil borings having multiple samples analyzed to help characterize the vertical extent of contamination. A total of 80 monitoring wells were installed to define the nature and extent of impacts to groundwater.

The general subsurface geology of the site immediately beneath the surface consisted of a sand deposit originating from a Pleistocene age glacial lake that occupied the region more than 12,000 years before the present. Specific geology beneath the site consisted of sand from the surface to a depth of 26–32 feet. Beneath this glacial lacustrine beach sand deposit was a ground moraine or lodgment till deposit extending to a depth of least 75 feet. Historical geological literature of the region indicated that the ground moraine or lodgment till deposit extended to depths of approximately 200 feet beneath the ground in the area.

During the investigation, multiple groundwater monitoring wells were installed at the same location but were screened at different depths within the saturated zone to evaluate the vertical distribution of contaminants within the aquifer. Several samples of the ground moraine deposit beneath the aquifer were also analyzed for the presence of contamination and for certain hydrologic parameters (such as hydraulic conductivity and grain size analysis) to evaluate whether the ground moraine deposit was an effective aquiclude preventing contaminant migration to deeper aquifers. In addition, three deep soil borings were drilled to a depth of 75 feet in non-impacted areas of the site to verify the horizontal distribution and thickness of the ground moraine deposit. Other technical and geologically-related information relevant to the site analysis included the following:

- Storm sewers in the immediate vicinity did not intersect any of the contamination.
- Surface water drainage was controlled by storm sewers.
- No buried utilities intersected contaminated areas.
- Potable water was supplied by the municipality and the source was more than 15 miles away. However, some local residences used groundwater within the same aquifer for irrigation purposes.
- Contaminated groundwater extended beyond the property boundary by approximately 1,200 feet.

The types of contaminants detected at the facility included the following:

- Volatile organic compounds (VOCs) include
  - Dense nonaqueous phase liquids (DNAPLs) commonly referred to as chlorinated solvents used to degrease and clean metal surfaces.
  - Light nonaqueous phase liquids (LNAPLs) used as solvents, paint thinners, and cleaning products, and are common constituents in fuels such as gasoline.
- Polychlorinated biphenyls (PCBs) used in electrical equipment.

TABLE 3.4

Contaminant Types, Concentration, Mass Remediated for Site 2 High Vulnerability Site

Contaminant	Maximum Concentration in Soil (µg/kg)	Maximum Concentration in Groundwater (µg/L)	Estimated Contaminant Mass Remediated (kg)
Volatile Organic Compounds (VOCs)			45
DNAPL Compounds	22,000	3,250	
Tetrachloroethene (PCE)	8,000	2,200	
Trichloroethene (TCE)	5,000	1,800	
Cis-1,2-dichoroethene	480	280	
Trans-1,2-dichloroethene	640	220	
1,1-Dichloroethene	2,800	2,100	
1,1,1-Trichloroethane	1,100		
Vinyl chloride			
LNAPL Compounds			
Ethyl benzene	32,000	240	
Xylenes	28,000	130	
Polychlorinated Biphenyls (PCBs)	7,000,000	Not Detected	12

Source: Rogers, D. T. 2020. Urban Watershed: Geology, Contamination, Environmental Regulations, and Sustainable Development. CRC Press. Boca Raton, FL. 606 p.

µg/kg = micrograms per kilogram

µg/L = microgram per liter

The specific chemical compounds with their highest concentrations detected and estimated contaminant mass remediated are listed in Table 3.4.

The highly permeable soil and shallow groundwater depth (i.e., the vulnerable geology) allowed VOCs to rapidly infiltrate and migrate to groundwater. In addition, the high groundwater seepage velocities resulted in a VOC plume extending one-third of a mile to a downgradient spring, and ultimately discharged into a recreational lake and the Rouge River.

PCBs were not detected in groundwater. Therefore, excavation and disposal of the PCB-contaminated soils at a licensed landfill was the remedial method of choice because the impacted soils were less than 3 feet in depth, were not located beneath any buildings, could be conducted quickly, did not disturb ongoing facility operations, and was the lowest cost alternative.

The excavation was also the remedial action of choice for soils highly impacted with VOCs. This method was chosen because there was not a substantial volume of impacted soil with VOCs, as the sandy soils at the site had a low capacity for retaining VOCs. As a result, the VOCs tended to migrate downward through the soil column and contaminate groundwater without adsorbing to soil grains.

The remedial method chosen for groundwater was air sparging and soil vapor extraction since the contaminants in groundwater were VOC compounds and did not extend past the mid-portion of the saturated zone. Air sparging involved the injection of air beneath the impacted groundwater and then letting the air rise naturally through the saturated zone. As the air migrated upward through the saturated zone it volatilized the contaminants, and the vapors containing the VOCs were removed using a soil vapor extraction system in the vadose zone. The vapors were removed from the air by passing them through a granular activated carbon tank.

The VOCs contaminating groundwater at this site were DNAPL compounds having a specific gravity slightly greater than water. Therefore, when present at sufficient concentrations, DNAPL compounds may sink through the water column and contaminate lower portions of an aquifer. This sinking action did not occur at this site because its geology and hydrogeology – in effect a stratigraphic control – prevented the VOCs from migrating to the bottom of the aquifer. As listed and described in Table 3.4, the composition of the aquifer gradually became fine-grained with depth and the hydraulic conductivity decreased proportionately. This reduction of hydraulic conductivity, combined with low contaminant mass in groundwater resulted in restricting contaminants to the upper portion of the aquifer. With the VOCs restricted to the upper portion of the aquifer, air sparging became the most practical remediation technique.

The time duration between the Phase I environmental site assessment and the receipt of the closure letter was approximately 14.5 years. The dollar costs for remediating this site stacked up this way: the PCB-contaminated soil, including investigation, was $1.1 million; $0.1 million for the VOC-contaminated soil; VOC-contaminated groundwater, including investigation, was $6.6 million. The total cost for investigation and remediation of this site was $7.8 million dollars, which translates into a cost of $60,000 per pound. Remediation was verified by the regulatory authority through the collection and analysis of the soil samples from each area remediated. In addition, four iterations of groundwater samples over a period of one year were made until cleanup levels were achieved. Closure was then granted and a "No Further Action Required" letter was issued for the site. A risk assessment was also conducted for the site and was designed to determine whether the residual contamination at the site would pose an ongoing and unacceptable risk to human health and the environment. Locations included in this evaluation were the onsite areas with persistent soil contamination, and the onsite and offsite areas with contaminated surface water and groundwater. The results of the risk assessment indicated that a deed restriction was appropriate for the site property and banned the use of groundwater for any purpose. A cap consisting of asphalt pavement was also required for portions of the property where some soil contamination remained in place, and if any soil were to become disturbed or exposed, another evaluation must be conducted to evaluate the need for further remedial actions. Land use for the property was restricted to industrial use.

3.4.3   Site Comparison Analysis

A profound difference between these two examples is seen in the cost per pound to remediate the contaminants. Despite the smaller acreage, shorter time of industrial operations, and low contaminant mass released to the environment, the cost to remediate a pound of contaminant at the site of high vulnerability (Site 2) is 75 times greater than to remediate a pound of contaminant at the low vulnerability site (Site 1). This cost differential is solely due to the geology present at each site. The high vulnerability site is located in a geologic area (a Sand) that is environmentally vulnerable to contamination because (1) the highly permeable soil allows contaminants to infiltrate readily and migrate to groundwater and (2) contaminant plumes are transported at relatively high seepage velocities in the sand aquifer to potentially sensitive receptors that include potable water wells and surface water. The costs for the high vulnerability site would have been much higher had the contaminants migrated along the bottom portion of the aquifer; this is sometimes the case with the type of contaminants present (DNAPLs) since they are denser than water. Luckily, the decreased hydraulic conductivity within the lower portion of the aquifer prevented this from happening at the higher vulnerability site. In addition, the costs would have been significantly higher had there been a completed human pathway represented by the ingestion of contaminated groundwater. Table 3.5 summarizes the major differences between these two sites of environmental contamination. Please note that the vulnerability map ranking accurately predicted the relative costs of remediation.

3.5   SUMMARY AND CONCLUSION

Through the construction of a vulnerability map, we have combined the knowledge gained from earlier chapters to explain the relationship between the natural environment and human influence in

TABLE 3.5
Site Comparison Table

Parameter	Site 1 – Low Vulnerability	Site 2 – High Vulnerability
Predominant Geology	Clay	Sand
Presence of Shallow Groundwater	No	Yes
Size of Site	16 acres	8 acres
Length of Operation	70 years	30 years
Land Use	Heavy Industry	Heavy Industry
Types of Contaminants	VOCs PNAs, PCBs, and heavy metals	VOCs and PCBs
Number of Contaminants Remediated	27	9
Contaminant Mass Remediated	500 pounds	130 pounds
Cleanup criteria	Same as Site 2 for overlapping compounds	Same as Site 1 for overlapping compounds
Cleanup Cost*	$400,000	$7,800,000
Remedial Methods (Soil)	Excavation	Excavation
Remedial Methods (groundwater)	Remediation Not Required	Air Sparging and Soil Vapor Extraction
Other Remedial Control Measures	None, unrestricted closure	Restricted closure included: • Deed restriction • Institutional controls • Industrial land use only
Timeframe	18 months	14.5 years
Cost per kilogram of Contaminant	$2,000	$150,000
Vulnerability Ranking Using Table 6.2.	13	65

Source: Rogers, D. T. 2020. Urban Watershed: Geology, Contamination, Environmental Regulations, and Sustainable Development. CRC Press: Boca Raton, FL. 606 p.
*Cleanup costs include costs for investigation and remediation.

urban areas. The arrangement, thickness, and composition of the sediment layers beneath our feet have a profound influence on where cities are located, how buildings are constructed, where roads are built – and perhaps most important to the development and redevelopment of our urban centers – how contaminants behave and how they affect the environment and people. The two case studies presented in this chapter have highlighted this relationship – but they are just a small subset of the thousands of examples of this important concept and its multi-faceted connections.

As demonstrated by comparing the two sites in this chapter and as we shall see in the next section of this book – once the environment has been contaminated at levels that pose a human or ecological risk – it is often very expensive to remediate, especially when groundwater is affected. Furthermore, it may be impossible to fully remediate some sites even with the most advanced technology. Therefore, minimizing waste and preventing pollution have proven to be the most effective methods for reducing costs, and ultimately, preserving our environment. The two examples highlighted in this chapter are not uncommon. Tens of thousands of industrial and even commercial and residential sites in the United States have contaminated soil and groundwater to levels requiring one or more expensive remedial actions.

The realization that certain locations or areas within urban regions are especially vulnerable to contamination offers even greater promise for resolving future environmental issues. Geologic vulnerability analysis of urban regions produces essential information for evaluating the environmental and financial risks associated with development and redevelopment. By minimizing the impact of pollution once a release has occurred, certain geological features may play, if we so choose, a central role in the development and redevelopment of any urban area.

Questions and Exercises for Discussion

1. **Develop a list of five different areas near your current residence that exhibit low to high vulnerability and provide facts that support your conclusion?**
2. **Do you feel that environmental vulnerability is evaluated as an item of consideration in land-use planning where you live?**
3. **Make a list of locations within 0.5 kilometers of your current residence that could potentially contaminate the environment? (Hint: Don't overlook your garage or under your kitchen sink).**

REFERENCES

Aller, L., Bennett, T., Lehr, J. H., Petty, R. J., and Hackett, G. 1987. DRASTIC: A Standardized System for Evaluating Ground Water Pollution Potential Using Hydrogeologic Settings. *United States Environmental Protection Agency USEPA-600/2-87-35.* USEPA. Ada, OK.

Albinet, M. and Marget, J. 1970. Cartographic de la Vulnerabilite a la Pollution des Nappes d"neau Souterrians. *Bulletin BRGM Sect 3 (Fr.).* Vol. 2. pp. 13–22.

Burn, S., Eiswirth, M., Correll, R., Cronin, A., DeSilva, D., Diaper, C., Dillon, P., Mohrlok, U., Morris, B., Rueedi, J., Wolf, L., Vizintin, G., and Vott, U. 2007. Urban Infrastructure and Its Impact on Groundwater Contamination. In: Howard, K. W. F., editor. *Urban Groundwater – Meeting the Challenge.* Taylor & Francis. London, England.

Eaton, T. T. and Zaporozec, A. 1997. Evaluation of Groundwater Vulnerability in an Urbanizing Area. In: Chilton, J. et al., editors. *Groundwater in the Urban Environment.* Balkema Publishers. Rotterdam, The Netherlands.

Focazio, M. J., Reilly, T. E., Rupert, M. G., and Helsel, D. R. 2001. Assessing Ground-Water Vulnerability to Contamination: Providing Scientifically Defensible Information for Decision Makers. *United States Geological Survey Circular 1224.* Denver, CO.

Foster, S. S. D. and Hirata, R. 1988. *Groundwater Risk Assessment: A Methodology Using Available Data.* CEPIS. Lima, Peru.

Foster, S. S. D., Morris, B. L., and Lawrence, A. R. 1994. Effects of Urbanization on Groundwater Recharge. In: Wilkinson, W. B., editor. Groundwater Problems in Urban Areas: ICE Conference Proceedings. London, England.

Fresca, B. 2007. Urban – Enhanced Groundwater Recharge: Review and Case Study of Austin, Texas, USA. In: Howard, K. W. F., editor. *Urban Groundwater – Meeting the Challenge.* Taylor & Francis. London, England.

Howard, K. W. F., Di Biase, S., Thompson, J., Maier, H., and Van Egmond, J. 2007. Stormwater Infiltration Technologies for Augmenting Groundwater Recharge in Urban Areas. In: Howard, K. W. F., editor. *Urban Groundwater – Meeting the Challenge.* Taylor & Francis. London, England.

Kaufman, M. M., Rogers, D. T., and Murray, K. S. 2003. Surface and Subsurface Geologic Risk Factors to Ground Water Affecting Brownfield Redevelopment Potential. *Journal of Environmental Quality.* Vol. 32. pp. 490–499.

Kaufman, M. M., Rogers, D. T., and Murray, K. S. 2005. An Empirical Model for Estimating Remediation Costs at Contaminated Sites. *Journal of Water, Air and Soil Pollution.* Vol. 167. pp. 365–386.

Kibel, P. S. 1998. The Urban Nexus: Open Space, Brownfields, and Justice. *Boston College Environmental Affairs Law Review.* Vol. 25. no. 3. pp. 589–618.

Loague, K. Corwin, D. L., and Ellsworth, T. R. 1998. The Challenge of Predicting Nonpoint Source Pollution. *Environmental Science and Technology.* Vol. 26. pp. 127–154.

Mohrlok, U., Cata, C., and Bucker-Gittel, M. 2007. Impact on Urban Groundwater by Wastewater Infiltration into Soils. In: Howard, K. W. F., editor. *Urban Groundwater – Meeting the Challenge.* Taylor & Francis. London, England.

Murray, K. S. and Rogers, D. T. 1999a. Groundwater Vulnerability, Brownfield Redevelopment and Land Use Planning. *Journal of Environmental Planning and Management.* Vol. 42. no. 6. pp. 801–810.

Murray, K. S. and Rogers, D. T. 1999b. Evaluation of Groundwater Vulnerability in an Urban Watershed. Proceedings of the 2nd International Congress on Water Resources and Environmental Research. Brisbane, Australia. pp. 877–883.

Pierce, S. A., Sharp, J. M., and Garcia-Fresca, B. 2007. Evaluating Groundwater Allocation Alternatives in an Urban Setting Using a Geographic Information System Data Model and Economic Valuation Technique. In: Howard, K. W. F., editor. *Urban Groundwater – Meeting the Challenge*. Taylor & Francis. London, England.

Robins, N., Adams, B., and Foster, S. S. D. 1994. Groundwater Vulnerability Mapping: The British Perspective. *Hydrogeologie*. Vol. 3. pp. 35–42.

Rogers, D. T. 1992. The Importance of Site Observation and Followup Environmental Site Assessments – A Case Study. Proceedings of the National Ground Water Association Phase I ESA Conference. Orlando, FL. pp. 218–227.

Rogers, D. T. 1996. *Environmental Geology of Metropolitan Detroit*. Clayton Environmental Consultants. Novi, MI.

Rogers, D. T. 2002. The Development and Significance of a Geologic Sensitivity Map of the Rouge River Watershed in Southeastern Michigan, USA. In: Bobrowsky, P. T., editor. *Geoenvironmental Mapping: Methods, Theory, and Practice*. A. A. Balkema Publishers. The Netherlands. pp. 295–319.

Rogers, D. T. 2014. Scientists Call for a Renewed Emphasis on Urban Geologic Mapping. *American Geophysical Union. Earth and Space News. Eos*. Vol. 95. no. 47. pp. 431–432.

Rogers, D. T. 2016a. Next Generation of Urban Hydrogeologic Investigations. *Journal of the Italian Geological Society*. Rome, Italy. Vol. 39. no 1. pp. 349–353.

Rogers, D. T. 2016b. Scientific Advancements that Improve the Conceptual Site Model in Urban Hydrogeological Site Investigations. 35th International Geological Congress. Paper 3019. Cape Town South Africa.

Rogers, D. T. 2018. Derivation of a Comprehensive Environmental Risk Model for Urban Groundwater Protection. *International Association of Hydrogeologists Congress*. Vol. 1. Daejeon, Korea.

Rogers, D. T. 2019. *Environmental Compliance and Sustainability: Global Challenges and Perspectives*. CRC Press. Boca Raton, FL. 583 p.

Rogers, D. T. 2020. *Urban Watersheds: Geology, Contamination, Environmental Regulations, and Sustainability*. CRC Press. Boca Raton, FL. 606 p.

Rogers, D. T. and Murray, K. S. 1997. Occurrence of Groundwater in Metropolitan Detroit, Michigan, USA. In: Chilton, J. et al., editors. *Groundwater in the Urban Environment*. Vol. 1. Balkema Publishers. The Netherlands. pp. 155–160.

Rogers, D. T., Murray, K. S., and Kaufman, M. M. 2007. Assessment of Groundwater Contaminant Vulnerability in an Urban Watershed in Southeast Michigan, USA. In: Howard, K. W. F., editor. *Urban Groundwater – Meeting the Challenge*. Taylor & Francis. London, England.

Stiber, N. A., Small, M. J., and Fischbeck, P. S. 1995. The Relationship Between Historic Industrial Site Use and Environmental Contamination. *Journal of the Air and Waste Management Association*. Vol. 48. no. 9. pp. 809–818.

United Nations. 2021. World Population Prospects. United Nations Department of Economic and Social Affairs. *Population Division*. https://esa.un.org/unpd/wpp/data (accessed March 31, 2021).

Vuono, M. and Hallenbeck, R. P. 1995. Redeveloping Contaminated Properties. *Journal of Risk Management*. Vol. 42. pp. 58–69.

Wong. C. I., Sharp, J. M., Hauwert, N., Landrum, J., and White, K.M. 2012. Impact of Urban Development on Physical and Chemical Hydrogeology. *Journal of Elements*. Vol. 8. pp. 429–434.

Zaporozec, A. and Eaton, T. T. 1996. Ground-water Resource Inventory in Urbanized Areas. In: Hydrology and Hydrogeology of Urban and Urbanizing Areas. American Institute of Hydrology (AIH) Annual Meeting Proceedings, St. Paul, MN.

4 Conducting Science-Based Environmental Investigations

4.1 INTRODUCTION

This chapter presents the sequence of science-based procedures used to evaluate a specific land parcel or larger area for environmental contamination and assess its environmental risk. We conclude with specific methods available to environmental professionals for collecting samples from soil, groundwater, sediments, surface water, and air. First, we define and present some basic concepts necessary for understanding and framing the concept of a subsurface environmental investigation.

The **environment** is a broad term encompassing all living and nonliving things on the Earth or some region thereof (United Nations 1987). This definition identifies the natural world existing within the atmosphere, hydrosphere, lithosphere, and biosphere, and also includes the built or developed world (anthrosphere). **Environmental risk** is defined as the probability of an event resulting in an adverse impact on the environment or humans (Fletcher and Paleologos 2000). Environmental risk increases when a completed human pathway occurs or a sensitive ecological system is impacted (USEPA 2002).

An environmental assessment contributes to our understanding of human impacts on the environment and is considered the first step in evaluating and ensuring the long-term viability of Earth as a habitable planet. The practice of conducting assessments to determine adverse environmental impacts was never a significant priority until the last 50 years. Since the last quarter of the 20th century, however, significant amounts of human and financial resources have been mobilized to evaluate the actual or potential environmental damage incurred from human activities of the past, present, or potential adverse impact from future human actions.

Environmental investigations are generally conducted when there is a likelihood certain environmental contaminants exist at a specific property or site having the potential to cause material harm to human health or the environment historically, presently, or in the foreseeable future. Some subjectivity is inherent in the decision to proceed with an environmental investigation at any given property. This uncertainty occurs because most properties contain, store, or at some time, have used substances posing potential harm to human health and/or the environment, including most households. Subsurface environmental investigations exist as a subset of a larger set of environmental investigation types. Throughout the chapter when we refer to an environmental investigation, the reference is to a subsurface environmental investigation, unless otherwise noted.

4.2 SUBSURFACE ENVIRONMENTAL INVESTIGATIONS, STUDIES, PLANS, OR REPORTS

Environmental investigations are conducted for many different purposes and objectives but generally follow a step-wise, logic- and progressive-based approach where the results of one investigation are used to evaluate the need and scope of subsequent investigations if necessary. The most common types of environmental investigations include the following:

- Phase I Environmental Site Assessment
- Phase II Investigation

DOI: 10.1201/9781003175810-4

- Geophysical Investigation
- Remedial Investigation
- Feasibility Study
- Ecological and Human Health Risk Assessment
- Natural Resource Damage Assessment
- Environmental Impact Statement
- Remedial Action Plan
- Completion Evaluation Study or Closure Study

Each type of investigation is discussed in greater detail in the following sections.

4.2.1   PHASE I ENVIRONMENTAL SITE ASSESSMENT

The Phase I Environmental Site Assessment (ESA) is typically the first environmental investigation conducted at a specific property. The Phase I ESA is conducted with the objective of qualitatively evaluating the environmental condition and potential environmental risk of a property or site. A **property** is defined here as a parcel of land with a specific and unique legal description. A **site** is defined here as a parcel of land including more than one property or easement and typically refers to an area of contamination potentially affecting more than one property.

As the first environmental investigation, the Phase I ESA is often regarded as the most important activity because all subsequent decisions concerning the property are, in part, based on the results of the Phase I ESA (Rogers 1992). Therefore, great care, scrutiny, scientific inquiry, and objectivity should be exercised while conducting the Phase I ESA. Standards for conducting Phase I ESAs were published by the American Society for Testing Materials (ASTM) in 1993 and were revised in 1997, 2000, and 2005 (ASTM 2005a). On November 1, 2006, the United States Environmental Protection Agency (USEPA) established federal standards for conducting Phase I ESAs termed "Standards and Practices for All Appropriate Inquiries" (AAI) under the Comprehensive, Environmental Response, Compensation and Liability Act (CERCLA) of 1980, commonly known as Superfund (USEPA 1980).

Under CERCLA, United States courts have ruled that a buyer, lessor, or lender may be held responsible for remediation of hazardous substance residues, even if a prior owner caused the contamination. However, the performance of a Phase I ESA may create a safe harbor or protection from liability, known as the "Innocent Landowner Defense" for new purchasers or lenders. Therefore, Phase I ESAs have become the standard type of environmental investigation employed when initially investigating a property. According to the USEPA (2005) requirements, an environmental professional, such as a geologist, environmental scientist, or engineer must conduct the Phase I ESA. General requirements for conducting a Phase I ESA include the following:

- Extensive review of current and historical written records, operations, and reports
- Extensive site inspection
- Interviews with knowledgeable and key onsite personnel
- Assessment of potential environmental risks from offsite properties
- Data gap or data failure analysis

A Phase I ESA is typically a noninvasive assessment, performed without sampling or analysis. In some instances, limited sampling may be conducted on a case-by-case basis if in the professional judgment of the person conducting the assessment – or due to other requests or mitigating factors – sampling is justified. In most cases, collecting and analyzing samples is usually deferred to the Phase II investigation, but if it does occur, the sampling conducted during a Phase I ESA may include the following:

- Sediment
- Surface water
- Waste material
- Lead-based paint
- Mold

- Drinking water
- Groundwater
- Soil
- Suspected asbestos-containing material
- Radon

The environmental professional conducting the Phase I ESA must perform extensive research and review all available written records. These researching activities apply not only to the property in question but also to the surrounding properties within a radius of up to one mile of the investigated property or site. The research/review process typically encompasses the following items and actions:

- Title search and environmental liens
- Historical chemical use and ordering
- Engineering reports and diagrams
- Previous inspection and audit reports
- Safety data sheets (SDS)
- Fire insurance maps and documents
- USGS investigation reports
- Soil Service maps

- Historical operation documents
- Historical photographs
- Previous environmental reports
- Agency inspection documents
- Hazardous substance inventory reports
- USGS topographic maps
- USGS geological maps
- Contaminant pathway analysis

The site inspection consists of a walk-through of the property or site. Items to evaluate and document during the site inspection include (modified from Rogers 1992):

- Employee interviews
- Stressed vegetation
- Chemical storage areas
- Signage
- Special labeling
- General topography
- Animal scat
- Floor drains
- Catch basins
- Weather conditions
- Onsite drainage
- Offsite inspections

- Area of no vegetation
- Aboveground storage tanks
- Back doors
- Storage sheds
- Fill material
- Wetlands
- Insects
- Roof drains
- Broken concrete
- Recent precipitation
- Offsite drainage
- Soil type(s)

- Stain soil or pavement
- Belowground storage tanks
- Back of property
- Refuse storage area
- Land depressions
- Mold
- Pits Trenches
- Storm drains
- Inaccessible areas
- Nearest water body
- Evidence of dumping
- Migration pathways

Once the environmental professional has completed the data collection and site inspection portion of the Phase I ESA, an evaluation of whether there is evidence of an existing release, a past release, or a material threat of a release of any hazardous substance or petroleum is made. If a product has been released and made its way into structures on the property or into the ground, groundwater, or surface water of the property, then this situation is termed a Recognized Environmental Condition

FIGURE 4.1   Recognized environmental condition (Photograph by Daniel T. Rogers).

(REC). If an REC is discovered, further investigation will likely be recommended to evaluate its potential significance. Many sites have more than one REC, and many of these RECs may require further evaluation. An example of an REC with significant amounts of oil-stained soil, stressed, and dead vegetation is presented in Figure 4.1.

RECs are intended to exclude *de minimis* conditions generally not presenting a threat to human health or the environment and typically would not be the subject of an enforcement action if brought to the attention of appropriate governmental agencies. However, there may be impacts encountered whose perceived severity falls between a *de minimus* condition and an REC. In these situations, the term **environmental concern** is applied. Items listed as environmental concerns become RECs if left unattended or lead to a release. Figure 4.2 is a photograph of a paved parking space with a residual amount of what appears to be a petroleum product discharged from an automobile. In the opinion of the environmental professional who conducted a Phase I ESA at this property, this condition was not evaluated to be an REC but was characterized as *de minimis*.

Historical aerial photographs are effective sources of information, and often help with environmental investigations. In urban areas especially, aerial photographs from several sources are readily available. These sources include the following:

FIGURE 4.2   Example of a de minimis release (Photograph by Daniel T. Rogers).

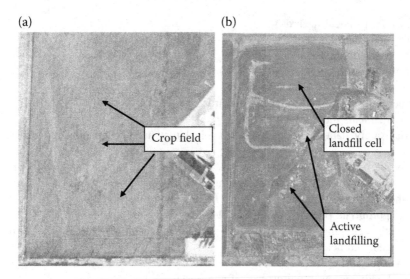

FIGURE 4.3   (a) Aerial photograph showing farmland and (b) earlier aerial photograph showing landfilling activities.

- Private local companies specializing in aerial photography
- Private national companies specializing in aerial photography
- Local and state historical societies
- Local and state agencies
- Utility companies
- Local companies
- Federal agencies, such as USEPA, USGS, Soil Conservation Service, National Forest Service, Bureau of Land Management, National Park Service.

Figure 4.3 demonstrates how historical aerial photographs can help to identify RECs. On the left, Figure 4.3a shows a particular property as farmland with no identifiable REC. Analysis of an earlier aerial photograph taken a few years earlier of the same property (Figure 4.3b to the right) indicates that the property was used as a landfill.

4.2.2   HEALTH AND SAFETY

If the result of the Phase I ESA indicates that further investigation is needed, a health and safety plan must be prepared before the Phase II investigation can begin (Occupational Safety and Health Administration [OSHA] 1989, as revised). This plan can only be prepared by a qualified professional because onsite Phase II investigations are conducted at locations containing physical and chemical hazards. The potential for exposure to these hazards requires specialized health and safety training for personnel going into the field – an absolute necessity for anybody doing this work – and especially critical in urban settings with their additional array of hazards.

The items below represent the minimum level of health and safety planning required before fieldwork begins:

- List of activities, or scope of work to be conducted.
- List of emergency contact names, titles, and contact information.
- Identification of potential risks excluding chemical exposure includes the following:

- Potential physical hazards
- Confined space entry
- Overhead hazards
- Traffic hazards
- Weather
- Other natural and anthropogenic hazards
- List of chemicals that may be present.
- Material safety data sheet for every chemical that may be encountered during onsite activities.
- Review medical procedures for each chemical that may be encountered during onsite activities to ensure that proper safety equipment is onsite and readily available.
- Map showing nearest medical treatment facility with directions.
- List of appropriate personal protective clothing.
- List of other clothing requirements such as hard hat, steel-toed shoes, gloves, reflective safety vests, and hearing protection.
- Review safety procedures for nonroutine activities.
- Review buddy system requirements, if appropriate.
- Review emergency hand signals, if required.
- Check safety equipment to ensure that it has been properly decontaminated and in working order.
- Contingency plans in case an incident occurs.
- Update the health and safety plan as appropriate as new information is obtained.

Specific health and safety equipment may be required when conducting Phase II investigations. The list below also contains quality control and quality assurance considerations:

- Hard hat
- Steel-toed boots
- Specialized gloves
- Hearing protection
- Safety glasses
- Reflective vests, if appropriate
- Face protection, when appropriate
- Boot covers, when appropriate
- Vapor monitors and detectors for compounds or conditions include the following:
  - Explosive conditions
  - Oxygen levels
  - Hydrogen sulfide
  - Carbon monoxide
  - Volatile organic compounds
  - Chlorine gas
  - Temperature
  - Other potential hazards, as appropriate

Depending on the specific physical and chemical hazards encountered, increased levels of health and safety procedures and personal protective equipment may be required to ensure the health and safety of personnel conducting the investigation. The highest level of protection would include a specialized suit and breathing apparatus supplying purified air along with several layers of specialized gloves and boots. In many cases, personal protective equipment is promptly removed and disposed of properly after field activities to prevent the inadvertent spread of any contamination beyond the impacted area.

4.2.3 DEMOLITION

In many instances, evaluating options for demolition of onsite structures should be made before starting an anticipated robust subsurface investigation. The advantages of conducting demolition activities include the following:

- Lowering the potential for trespass and injury
- Increases potential for discovery of sources of potential contamination such as historical floor drains, sumps, pits, trenches, tanks, etc.
- Access to the subsurface by not having any obstructions that could impede evaluating the nature and extent of contamination or interim remedial activities, if necessary
- An improved Conceptual Site Model
- Removing and proper disposal of potential residual wastes not previously discovered
- Lowers maintenance costs

In many instances, the demolition of onsite structures provides the investigation team with direct access to the developmental and operational history of the site. This provides the investigation team with valuable information as to the storage, use, and waste disposal practices of hazardous substances at the site, which improves the reliability of the Conceptual Site Modal, which we will discuss in greater detail later in the chapter.

4.2.4 PHASE II ENVIRONMENTAL SITE ASSESSMENT

When a Phase I ESA has been completed and there are one or more RECs identified, a Phase II investigation will likely be required to confirm the existence of the REC(s). The scope of Phase II investigations varies widely and largely depends on the RECs identified in Phase I ESA. Under normal circumstances, a Phase II investigation will involve the collection and analysis of samples collected from the area identified as an REC in Phase I ESA. These samples are collected and analyzed to evaluate the specific contaminants present, and if they exceed Federal- or State-mandated concentrations for the protection of human health and the environment.

Phase II investigations are designed to use the scientific data gathered from each step to evaluate the need for or to plot the direction of the subsequent step(s). The United States Environmental Protection Agency (USEPA) has published guidance for conducting investigations under the CERCLA (USEPA 1980, 1988) and under the Resource Conservation and Recovery Act (RCRA) (USEPA 1976). In addition, many states have published guidelines for conducting investigations; e.g., New Jersey (New Jersey Department of Environmental Protection [NJDEP] 2005), Ohio (Ohio Environmental Protection Agency [OEPA] 1995, 2006), and California Environmental Protection Agency [CalEPA] (1995).

The investigative process follows three steps consisting of find, define, and refine. Each step in the sequence is described in greater detail below (Rogers 2016a, 2016b, Rogers and Welty 2017):

- **Find** – This type of investigation has the dual purpose of finding and identifying specific contaminants, and is accomplished by collecting and analyzing biased or "worst-case" samples. Examples of "worst-case" samples are those: nearest the source of the release, directly beneath an outfall, appearing the most heavily stained, exhibiting the strongest odor, or those registering the highest readings on field screening equipment.
- **Define** – After contamination is found, the purpose of the define-type of investigation is to evaluate the nature and extent of the contamination. Define-type investigations may consist of several phases requiring a greatly expanded scope compared to the find-type of investigation. The resulting scope depends upon the following factors:
  - Federal- or State-specific requirements for each contaminant of concern

- The media being evaluated (e.g., soil, groundwater, sediment, surface water)
- The pathways being considered (e.g., ingestion, inhalation, dermal contact)
- Background concentrations of heavy metals such as lead or arsenic

The extent is derived laterally and vertically by these two equivalencies: (1) to the concentration equal to or less than the lowest applicable cleanup criteria for non-naturally occurring compounds, and (2) to background concentrations for those compounds naturally occurring in the environment, such as heavy metals.

- **Refine** – The purpose of the "refine" type of investigation is to gather additional information about the contamination and determine whether an unacceptable risk to human health and the environment exists and requires remedial action. Refine-type investigations include collecting and analyzing samples from locations of highest risk for specialized analytical parameters to evaluate certain remedial options, or to gather more specific data on the nature and extent of the contamination.

Depending on several co-dependent factors, many phases of investigation may be necessary to fully understand the nature and extent of contamination detected at any one site. The three main factors that influence the depth of scientific inquiry required include three factors that include the following:

1. The nature of the release(s). This includes information that may be included in the Phase I ESA but may require further evaluation based on Phase II results. In basic terms, scientific information contained in the Phase I should generally match the information in Phase II. If not, the **Conceptual Site Model** (CSM) must be revised. A CSM is a model that tells the science story concerning all aspects of the release of hazardous substance(s) into the environment. Items that the investigation team needs to consider are questions that are listed below. If the investigation team suspects that the CSM that was developed does not match the scientific evidence collected, the CSM must be revised and further scientific data will likely be required until the CSM and the scientific data are in general agreement. The questions to ask include the following (Rogers and Welty 2017 and Rogers 2016a):
   a. Was the source(s) of the contamination properly identified?
   b. Were other potential points of contaminant release(s) ruled out?
   c. Was the estimate of contaminant mass within reason?
   d. Was the time-frame of contaminant release(s) reasonable?
   e. Did the investigation team properly rule out other potential contaminants?
   f. Did the investigation team evaluate all potential contaminant pathways and media?
   g. Was the extent of contamination defined including the vertical extent?
2. The nature of the contaminant(s) released. Evaluating the nature of the contaminants involves understanding the physical chemistry of each contaminant and how each behaves in differing geologic environments.
3. The nature of the natural environment. This involves where and how to evaluate the nature and extent of contamination in the specific geologic environment in which the release occurred.

4.2.5 GEOPHYSICAL INVESTIGATION

There are occasions when direct sampling under a traditional Phase II investigation approach does not adequately characterize a specific site. Examples may include areas not accessible to sampling equipment, extremely large sites, or locations where buried objects are of great concern. In these cases, surface geophysical techniques may be used to fill in data gaps. Since geophysical investigations provide supplemental data, they rarely are the only type of environmental investigation conducted at a site. There is a wide range of geophysical techniques to accommodate different objectives, investigative requirements, and site limitations, and these techniques offer certain advantages, including:

- Increasing the accuracy and area of coverage of subsurface investigations
- Increasing data density
- Collecting data on subsurface geology and hydrogeology at locations inaccessible to other investigative techniques, or areas considered dangerous to investigate using other methods
- Decreasing the cost and time necessary to characterize certain sites, and are generally noninvasive

Geophysical investigations include the following techniques and benefits:

- Electrical resistivity: Electrical resistivity measures the apparent resistivity averaged over a volume of material (ASTM 1999a).
- Electromagnetics: Electromagnetic methods measure the conductivity of subsurface materials and are frequently used to detect buried metal objects (ASTM 2000a).
- Gravity survey: Gravity or microgravity surveys measure changes in subsurface density (ASTM 1999b).
- Ground-penetrating radar: Ground-penetrating radar (GPR) uses high-frequency electro-magnetic waves to evaluate subsurface strata (ASTM 2005b).
- Borehole geophysics: Borehole geophysics uses instruments to measure and record different properties outside of a well or borehole as the instrument is lowered down the borehole (ASTM 2005c, Keys 1990).
- Seismic refraction and reflection. Seismic refraction and reflection measurements involve the measurement of seismic waves traveling through the subsurface (ASTM 2000b).

4.2.6 Ecological and Human Health Risk Assessment

After the investigation phase has been completed and the nature and extent of contamination have been defined, an environmental **risk assessment** is conducted. A common objective for conducting a risk assessment is to evaluate whether there is an unacceptable risk posed by the presence of contamination at a given site for the future intended use of the property (USEPA 2002). Intended land use usually fits into the three broad categories of industrial, commercial, and residential. In general, if the intended land use is industrial or commercial, higher concentrations of contaminants are allowed if certain site conditions are satisfied to minimize the exposure potential to the contamination. An environmental risk assessment follows scientific procedures for evaluating the risk to humans and the environment posed by the contamination present at a specific site. Three basic components are used to evaluate risk: (1) toxicity or potency of contaminants present; (2) exposure pathways; and (3) receptors (California Environmental Protection Agency 1996, Oregon Department of Environmental Quality [ODEQ] 1998, USEPA 2002).

Toxicity: Toxic substances or toxicants are health threats because of the resultant effects on biotic receptors when exposure to these toxicants occur. Taking into account not only the contaminant itself but exposure level, duration of exposure is required (Yong 2002).

Exposure pathways fall into three broad categories: inhalation, dermal adsorption, and ingestion (USEPA 1989).

Receptors imply biotic receptors, but can also pertain to the physical land environment, including surface water and groundwater. Consequently, the nature and extent of the threat of a particular pollutant will not only depend on the nature and distribution of the pollutant but also on the target that is threatened.

Risk assessment is accomplished using a four-step process:

1. Hazard identification – answers this question: Does exposure to a chemical or agent cause an increase in the incidence of an adverse health effect (e.g., cancer or birth defects)?
2. Dose-Response Assessment – quantitative characterization of the relationship between the dose of a chemical or agent and the incidence of an adverse health effect.
3. Exposure assessment – evaluation of the intensity, frequency, duration, and routes of exposure to the chemical or agent.
4. Risk Characterization – estimation of the potential incidence of a health effect, calculated by obtaining information from the dose-response assessment along with information from the exposure assessment.

The outcome of a risk assessment usually produces a hazard quotient (HQ), which is the result of evaluating each chemical of concern for potential carcinogenic risk and chronic health risk. USEPA (2002) currently has an acceptable carcinogenic risk of 1 in a 100,000 (1×10^{-5}) cancer incidence, with values under 1 being an acceptable non-carcinogenic chronic health risk. However, many risk assessments evaluate cumulative risk. For instance, there may be 10 different chemicals present, and one chemical alone may not result in an unacceptable risk, but exposure to more than one may result in an unacceptable exposure. In most circumstances, this is an acceptable methodology especially if contaminants overlap. USEPA and most States have published guidelines for conducting ecological and human risk assessments, and these address whether a cumulative risk evaluation is warranted. A review of federal, local, and State requirements is recommended before conducting an ecological or human health risk assessment.

4.2.7 FEASIBILITY STUDY

A feasibility study is conducted after the nature and extent of contamination have been defined. The timing of this study corresponds with the risk assessment if contamination is present at sufficient levels to justify the possibility of remedial action. Selecting the most appropriate remedial technology or technologies – if more than one is necessary – is a key objective of the feasibility study. Technology selection can be evaluated within this framework: (1) the nature and extent of contamination at a specific site; (2) the future intended land use or uses; (3) the results of the risk assessment; and (4) other sites conditions, if present (New York State Department of Environmental Conservation 2021, USEPA 1988).

4.2.8 REMEDIAL INVESTIGATION

Remedial investigations represent the actions taken at sites of environmental concern based on a combination of the Phase I ESA, Phase II investigation, Risk Assessment, and Feasibility Assessment outcomes. Remedial investigations generate large amounts of geological and hydrogeological information about the area and often the region being investigated, but often require two or more years to complete and may cost several million dollars. These time and cost factors dictate that these investigations are conducted only when it is generally accepted the site poses an obvious and significant risk to human health and the environment. Combining the investigation and evaluation process can shorten the time required to investigate a site and save money, provided the necessary amount of environmental risk is removed to protect human health (USEPA 1998).

4.2.9 NATURAL RESOURCE DAMAGE ASSESSMENT

A Natural Resource Damage Assessment (NRDA) is a study that evaluates the damage or injuries to the environment (USEPA 2021). The term injury refers to an actual adverse impact to, or loss of a natural resource. Damage refers to the monetary cost of restoration or replacement of the natural resource. The USEPA (2021) defines a natural resource as:

- Land
- Air
- Water
- Fish
- Biota
- Wildlife
- Groundwater
- Drinking water supplies
- Other potentially identifiable natural resources

NRDAs are usually conducted after the extent of contamination has been defined and before any final remedy is implemented. Typically, an NRDA is conducted at large sites where there is clear evidence a human-induced environmental impact significantly degraded the ecological and natural resources of a specific region.

4.2.10 Environmental Impact Statement

An Environmental Impact Statement (EIS) is conducted to identify and evaluate the positive and negative biophysical, social, and other environmental effects that a proposed development action may have on the environment (Glasson et al. 2005). If the EIS process concludes the proposed action has a high potential to cause significant environmental degradation, other options may be required to reduce the environmental impact. The process thus does not prohibit the proposed action from causing any harm to the environment, it merely requires disclosure and understanding of the potential impacts prior to taking any action (Glasson et al. 2005).

4.2.11 Remedial Action Plan

A remedial action plan (RAP) describes the actions implemented to lower the risk posed by the presence of contamination at a specific site to an acceptable level. RAPs include the following (New Hampshire Department of Environmental Services 2008, USEPA 1988):

- A summary of each investigation completed
- A list of the contaminants of concern (COC)
- Cleanup objectives and goals for each COC
- The technology selected to reduce concentrations of each COC
- Methodology and justification for technology selection
- A timeframe to complete each activity
- A plan outlining methods to confirm the remedial action has achieved the desired result and has met the cleanup standard
- Contingency plan in case the selected remedial action or actions does not achieve the desired result

An RAP includes detailed engineering and planning drawings to precisely depict the methods and systems proposed for lowering the contaminant concentrations at a particular site to an acceptable level.

4.2.12 Completion Evaluation Study or Closure Study

A Completion Evaluation Study is conducted after the site is remediated. The purpose of the Completion Evaluation Study is to evaluate the effectiveness of the remediation and to confirm the desired result of permanently reducing contaminant concentrations to acceptable levels has been achieved. Samples of previously affected media, including soil, groundwater, sediment, surface

TABLE 4.1

Types and Purposes of Environmental Investigations

Investigation Type	Purpose
Phase I ESA	Qualitative review of a specific site to evaluate if there is any scientific basis the site poses any environmental risk
Phase II	Quantitative study that finds, defines, and refines the nature and extent of contamination
Geophysical Investigation	Used to supplement and add geologic and hydrogeologic information to Phase II investigations; helps to identify and locate buried objects, utilities, or other buried structures.
Risk Assessment	Evaluate whether the presence of contamination poses an unacceptable risk to human health and the environment
Feasibility Study	Evaluate potential remedial alternatives with the objective of selecting the most appropriate technology(s) to lower risk to an acceptable level
Remedial Investigation	Defines the nature and extent of contamination at large sites likely posing a significant risk to human health and the environment
NRDA	Assess damages to natural resources
EIS	Evaluate potential impacts of future development on natural resources and the environment
RAP	Plan outlining what actions will be conducted to lower risk to an acceptable level
Completion Study	Quantitative investigation conducted to confirm the remediation at a site is sufficient and complete

water, and air are collected to help confirm the remediation was successful. To avoid bias, requirements for the Completion Evaluation Study are usually outlined in the RAP. In addition, many individual states have general guidelines for determining when remediation has been completed. A Completion Evaluation Study may require several years to finish, especially if groundwater is a media of concern and has undergone an active remedial action.

4.2.13 SUMMARY OF ENVIRONMENTAL INVESTIGATIONS

Table 4.1 lists each type of environmental investigation and its major purpose. The Phase II Investigation, Remedial Investigation, and Geophysical Investigation typically generate the most geological and hydrogeological information – and in some instances – generate a large volume of detailed information. In addition, the Phase I ESA often has valuable information useful for geological purposes, especially if a Phase II or other investigation is subsequently conducted.

4.3 COMMON ENVIRONMENTAL SAMPLING METHODS

Sampled items during an environmental investigation typically include soil, groundwater, sediment, surface water, or air. The primary objectives when collecting any sample during an environmental investigation are (1) collecting a representative sample and (2) ensuring the integrity of samples collected by taking measures not to contaminate or cross-contaminate the samples, or altering any of the original properties of the sample (USEPA 1988). With these objectives in mind, acceptable sampling techniques have been developed by USEPA. In addition, most States have published protocols for the collection of samples for environmental purposes. Therefore, applicable local, State, and Federal guidelines should be followed when conducting an environmental investigation at a specific site.

4.3.1   Soil Sampling and Description

Soil sampling is usually initiated during the Phase II investigation and involves collecting surface and subsurface soil samples from soil borings drilled within and perhaps adjacent to RECs identified in Phase I ESA. The number and depth of the borings are dependent upon the source and nature of contamination suspected to have been released. Soil samples are collected in a wide variety of established methods, the most common include surface samples using a stainless steel trowel, hand auger, from test pits, and soil samples collected from drilling rigs.

4.3.1.1   Surface Sampling

Many investigations begin simply by collecting surface soil samples in each area of concern. When surface sampling for the presence of contamination, select a location to represent a "worst-case" sample and fingerprint the contaminant's release area depicted by the area of heavy staining or other sign such as stressed vegetation or actual free product that represented the released substances themselves. A surface soil sample should be collected using a stainless steel trowel, then placed into a stainless steel bowl. The sample should then be transferred into appropriate sample containers before transport to the laboratory for analysis.

4.3.1.2   Excavating Test Pits

When possible, excavating test pits using a backhoe is a preferred method of collecting and characterizing shallow subsurface geology. Test pits expose much more of the subsurface than other methods and give the geologist greater opportunity to observe and gather geologic information. An added advantage with test pits is the ability to select optimal locations for collecting soil samples for analysis. Recording observations in the field in real-time is a critical and important step in the data collection process. While test pits may be the preferred method of characterizing subsurface geology and providing optimal sample collection points, excavating test pits in an urban setting is uncommon. In many cases, test pits cannot be excavated because they disturb much larger areas than drilling a soil boring using a Geoprobe™ or other similar method. In addition, many sampling locations are located near utilities, buildings, or beneath paved areas making test pits very difficult, dangerous, or impossible to excavate.

4.3.1.3   Drilling Using a Hand Auger

Using a hand auger to characterize subsurface geology has severe limitations. The twisting required to advance the auger deeper into the subsurface causes the loss of most, if not all depositional structures and features in the recovered samples. For this reason, a hand auger is used to characterize soil or sediment type and is not used to interpret most complex geological features or structures. A hand auger may also be used to drill the first few feet to explore and avoid buried utilities before a drill rig completes the rest of the boring.

4.3.1.4   Mechanical Drilling Methods

For drilling in unconsolidated materials, preferred mechanical drilling methods include a Geoprobe™, hollow-stem auger, or roto-sonic methods. Characterizing subsurface geology from soil borings obtained by mechanical methods is much more difficult than test pits because these methods create a long and very narrow sample (four to ten feet long, and only two or four inches in diameter). These dimensions of sample size increase the difficulty of the analysis. In addition (1) the outer edges of soil samples collected from a mechanical method may be smeared from the sample collection method; and (2) the sample probe must be re-inserted into the borehole after each sample is collected, and this action may lead to cross-contamination or result in a gap in the depositional history of the area. Additional care and scrutiny must be undertaken when characterizing subsurface geologic features from samples collected using these methods.

Using a Geoprobe™ for soil sampling is the preferred technology in states like Michigan. A Geoprobe™ is typically faster than other methods, can penetrate hard surfaces such as pavement and concrete, and can be used inside buildings and other locations with limited access and space. Some Geoprobe™ drill rigs are capable of directional drilling, providing advantages in some unique situations. Soil samples collected using a Geoprobe™ are typically two inches in diameter and either two or four feet long. The sample is collected inside a plastic sleeve inserted into the bottom of a metal rod hydraulically pushed and vibrated into the ground. The Roto-sonic method is similar to the geoprobe method of advancing a boring into unconsolidated subsurface deposits. One advantage of the Roto-sonic method when compared to the Geoprobe is its ability to reach greater depths with a larger diameter borehole. This capability enables a larger volume of samples to be collected and also results in improved monitoring well installation success rates.

4.3.1.5   High-Resolution Sampling

There are certain instances where more intense investigation techniques should be conducted that involve the collection of perhaps an order of magnitude of more environmental samples and using additional sampling techniques to more fully understand the subsurface and contaminant distribution and extent. This occurs when certain types of contaminants are released into a geologic environment that favors the spread of contaminants quickly and has the potential to transport the contaminants to sensitive locations where humans or other life forms are put at a higher exposure risk. Under these circumstances, high-resolution sampling may be justified to better understand contaminant migration pathways. High-resolution sampling involves the collection and analyses of significantly more data points and using several different investigative techniques and methods to fully characterize the nature and extent of contamination (Rogers 2016a, 2016b). High-resolution sampling should be considered when the one or more of the following factors has been satisfied (Rogers 2016a, 2016b, Welty et al. 2016) include evaluating the following factors:

- Types of contaminant present: In general when the site being investigated has known releases of chlorinated volatile organic compounds (CVOCs), also commonly known as chlorinated solvents, or hexavalent chromium, especially from multiple sources.
- Subsurface geology: If subsurface geology is such that contaminant migration is not impeded through either natural conditions or from anthropogenic pathways.
- The site is located near sensitive or complex ecological exposure pathways.
- The site is located near human routes of exposure.

High-resolution sampling essentially provides a higher magnification of existing subsurface conditions that enhance the interpretation of the fate and migration of contaminants being evaluated. The advantages of conducting high-resolution sampling are many, some include the following:

- More accurate characterization
- Greatly improved and more accurate Conceptual Site Model
- Ensures that remediation and risk reduction measures are properly located and sized
- Lowers risk to the human health and the environment by a greater understanding

The potential disadvantage is that the cost of the investigation is initially higher. However, this is offset by the other factors listed above. Using high-resolution techniques also must include enhanced data organization and analysis techniques which may include electron data management and analysis and three-dimensional representations of the data, and computer modeling.

4.3.1.6   Sample Containers

Soil samples using any of the methods described above are transferred into appropriate sample containers following their collection. Most containers consist of glass having a Teflon inner lid designed to minimize the loss of vapors and are filled to capacity. Specification of sample containers and the methods, protocols, and equipment required for sample transfer varies with the specific contaminants being evaluated. In addition, sampling procedures and methodology differ from State to State for certain analytes. To avoid confusion, a work plan identifying the appropriate sampling procedures should be prepared prior to initiation of sampling activities

4.3.1.7   Documentation

Proper description of each soil sample by a qualified scientist is crucial. A field log book houses this information. Information recorded in a field log book when each soil boring is being advanced should include the following:

- Drilling method
- Size of auger or drill bit
- Drilling or advance rates
- Drilling or advance difficulties
- Equipment failures
- Drilling difficulties
- Weather conditions

After extraction, each soil sample is inspected and information related to the site's geology and soil properties is recorded in the field log book, and shown below:

- Soil or sediment type (i.e., sand, clay, etc.) using the Unified Soil Classification System
- Grain size
- Grain shape
- Grain sorting
- Grain composition
- Consistency
- Plasticity
- Stratigraphy
- Observable depositional features and bedding
- Presence of fill or anthropogenic materials or substances
- Color using the Munsell Color Chart (developed by Albert Munsell 1905)
- Color changes
- Degree of saturation
- Field screening results
- Odors
- Location of sample collection for laboratory analysis, if any

Consistency refers to the relative ease a sediment can be deformed.

Plasticity is the property of soil or rock allowing it to be deformed beyond the point of recovery without cracking or exhibiting appreciable change in volume – its "plastic limit."

Additional physical parameters may be used to characterize subsurface solid materials, but these cannot be reliably measured in the field, and most often require determination at a laboratory. These additional parameters include the following:

- Bulk density
- Cation exchange capacity
- Organic carbon content
- pH
- Specific gravity
- Mineral content

4.3.2 Groundwater Sampling Methods

If groundwater is encountered during the drilling of soil borings or excavating test pits, then the environmental investigation must also evaluate the potential for groundwater contamination. Sampling groundwater usually requires the installation of a monitoring well, but this is not always the case. Occasionally, a water sample is collected directly from the borehole or excavation pit, or temporary monitoring well is installed and removed after a groundwater sample is collected. A groundwater sample collected directly from an open borehole is called a grab a sample. Groundwater samples collected from temporary monitoring well may also be called grab samples. In general, groundwater samples collected from open boreholes or temporary monitoring wells installed without great care are used as a rough screening of groundwater to evaluate the need for more permanent monitoring well. Collecting and analyzing groundwater samples from an open borehole or temporary monitoring well is not always recommended because the analytical results may not be representative of the actual groundwater in the aquifer.

4.3.2.1 Installing a Monitoring Well

Collecting a representative groundwater sample requires the installation of a monitoring well. The objectives of installing a monitoring well include the following:

- Collecting a representative sample of groundwater for analytical testing to evaluate groundwater quality
- Evaluating the direction of groundwater flow
- Estimating the hydraulic conductivity
- Evaluating specific yield
- Monitoring seasonal fluctuations in flow direction
- Future monitoring of groundwater quality

In many cases, monitoring wells should not be used for evaluating the nature and extent of groundwater contamination. Monitoring wells should be used to monitor groundwater, not to define the nature and extent of the contamination. Evaluating the nature and extent of contamination in complex contaminants and geological strata should be evaluated using high-resolution techniques in most instances. After defining the nature and extent of contamination, monitoring wells should then be considered to monitor groundwater quality at strategic locations based on the results of the defined stage of investigation. To further demonstrate the limitations of installing monitoring wells too early in an investigation or using monitoring wells to define the extent of contamination is depicted in Figure 4.4.

Monitoring wells are typically constructed of polyvinyl chloride (PVC), but may also be constructed of stainless steel (ss), Teflon, or steel. The **casing** in a monitoring well is the hollow tube placed inside the borehole and provides access to the subsurface. The **screen** at the bottom of the well allows water to seep into the well from the aquifer. The slits have a very small diameter so the water enters the well but not the filter pack around the outside of the well screen. A filter pack is placed around the well screen and serves to prevent any fine-grain sediment in the aquifer formation from entering the well. The **filter pack** contains granular sand or gravel of uniform size and is placed between the aquifer formation and the well screen.

Many states require that the filter pack and the slot size of the well-screen be "engineered" based on the textural characteristics of the aquifer material. In addition, some states require a permit to install a monitoring well. Therefore, local requirements should be reviewed and well understood before installing a monitoring well.

4.3.2.2 Groundwater Sampling

Collecting groundwater samples consists of a process involving the following nine steps (modified from Harter 2003):

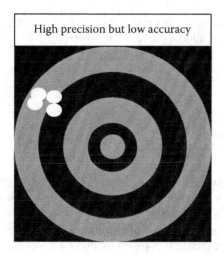
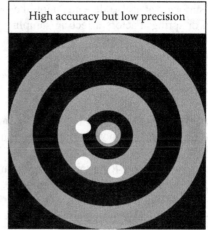

FIGURE 4.4 Comparing data precision and accuracy between monitoring wells and high-resolution aquifer profiling.

1. Preparation
2. Inspecting and accessing the monitoring well
3. Measuring the water level
4. Purging the well
5. Letting the well recover
6. Measuring the water level
7. Collecting the water sample
8. Decontamination
9. Securing the well after sampling

4.3.3 SEDIMENT SAMPLING

Sediment sampling is conducted with several types of specialized equipment. One instrument commonly used is called a Ponar Dredge. A Ponar dredge is equipped with a set of jaws and is deployed in the open position while being lowered through the water column until it becomes embedded into the bottom sediment. Once the instrument is pulled upward with minimal force to avoid pulling it off the bottom, the jaws are engaged and come together as the instrument is raised. This sequence of events traps sediment in the instrument bucket. These types of samplers allow the collection of small or large samples depending on the size of the bucket (USEPA 2007).

Another popular sediment sampler is called a tube or grab sampler; it consists of a hollow tube typically constructed of stainless steel, PVC, or Teflon (Capri et al. 2005). These types of samplers are intended for collecting shallow sediments samples within water columns of less than 5 to 10 feet. Proper function of these devices depends upon the presence of a strong water current or other associated factors that may cause the sampler to drift and not collect an appropriate sample size, or sample the wrong location.

4.3.4 SURFACE WATER SAMPLING

Sampling surface water uses many of the same procedures as groundwater sampling. The first step in sampling surface water is evaluating the morphology and hydrology of the surface water body so appropriate sampling location(s) can be identified. Collecting an appropriate and representative sample of surface water also depends upon the type of surface water body; e.g., a lagoon, stream, lake, river, or spring. Factors affecting sampling location include the following:

- Chemical compounds of concern
- Depth of water body
- Flow rate
- Size
- Whether there is a specific location of concern
- Topography and composition of the bottom of the water body

Each surface water sample location should be marked and surveyed in case a future sampling event is required. Placements of stakes, flags, or buoys are common ways to mark surface water sampling locations. The elevation of a surface water body is also measured during every sampling event to assess the current hydrologic conditions; e.g., drought or flood. Placing a benchmark on a stationary object, such as the bottom or side of a bridge, edge of a paved roadway, or building corner usually works.

Surface water samples can be collected using a glass tube, pumps, bailers, or obtained directly using a sample container. Specialized surface water instruments include the Kemmerer Bottle, Bacon Bomb Sampler, or Dip Sampler. These types of instruments are used from a boat or a bridge and are lowered into the water column and can collect a water sample from any depth. The exception here is the Dip Sampler, which only collects samples from the surface and is essentially a sample container dipped into the water (USEPA 1994). Dip samplers are useful where access is limited.

4.3.5   Air Sampling

Conducting subsurface investigations sometimes requires sampling air. Some contaminants migrate through subsurface soil and groundwater and may come into contact with subsurface structures such as basements. When this event occurs, certain contaminants have the potential to accumulate in these subsurface void spaces and contaminant the air. Additionally, contaminants are released directly into the air and may become deposited on the ground or water surfaces.

Under certain conditions, air sampling is justified when conducting a subsurface environmental investigation when (1) evaluating whether the source of soil or water contamination originated from air deposition, (2) protecting the health and safety of people during an investigation, and (3) as part of conducting a risk assessment.

Sampling for contaminants in air fall into two broad categories: contaminants in the vapor or gas phase, and contaminants as particulates or sorbed to particulates (USEPA 1991). The objectives of air sampling include the assessment of exposure for health reasons, and for normal or ambient air quality. Exposure monitoring usually is performed with indoor air, and ambient air monitoring is typically done with outdoor air.

4.3.5.1   Indoor Air Sampling

Before conducting any air sampling, a site inspection should be completed –especially if the sampling will be conducted indoors. The inspection seeks to identify two conditions: anything affecting or interfering with the proposed testing, and obvious sources of air contaminants. The site inspection should include the following (New York Department of Health 2005):

- Type of structure
- Physical conditions
- Odors such as solvents, mold, sweet, sour
- Airflow and ventilation engineering design of the building, structure, or residence

- An analysis of potential contaminants sources including
  - Chemicals used
  - Chemical storage areas
  - Chemical use areas
  - Chemical disposal methods
  - Chemical disposal areas
- Weather conditions

Conducting a chemical inventory is recommended prior to any indoor sampling. This inventory will greatly assist in providing valuable information concerning the identification of the contaminants of concern and whether air sampling is necessary. Material Safety Data Sheets for each chemical at the site should be reviewed. At least 24 hours prior to conducting any indoor air sampling, there are typically established standards for ensuring a representative sample is collected. (New York Department of Health 2005). Air samples should be collected from an adequate number of locations to understand the likely sources of potential chemical exposure, and to evaluate their potential exposure to occupants at various locations. In private residences, air samples should be collected from all floors including the basement and outdoors.

4.3.5.2   Outdoor Air Sampling

Outdoor air sampling for particulates is conducted using high-volume air sampling equipment (USEPA 1998). The ambient air sampling device shown pumps air from all directions across a 7-inch by 9-inch exposed filter inside the shelter at flow rates ranging from 39 to 60 cubic feet of air per minute. The roof design of the shelter is a standard design acting as a plenum above the filter to permit the free flow of air into the plenum space. The sizes of particles collected using the sampler range from 0.3 microns to as large as 25 to 50 microns in aerodynamic diameter, and are dependent upon wind direction and speed (USEPA 2008).

Filters used for collecting outdoor air sampling for particulates are composed of either glass or cellulose fibers. Glass fiber samples have been extensively used and cellulose fiber filters are commonly used when sampling for heavy metals. One drawback to cellulose fibers is their potential to increase the potential for adsorption of water and enhance artifact formation of nitrates and sulfates on the filter. The filters are kept in a clean and sterile environment prior to sampling at a constant temperature and relative humidity of approximately 3% and are precisely weighed (USEPA 1998). After the sample is collected, several parameters must be recorded, including

- Starting time and date
- Ending time and date
- Airflow rate
- Temperature ranges during sampling
- Relative humidity difference during sampling
- Summary of conditions that may affect results (construction activities in the area, etc.)
- Barometric pressure at start and end

4.4   SUMMARY AND CONCLUSION

Environmental subsurface investigations are very detailed scientific investigations. These investigations collect enormous amounts of geologic and hydrogeologic information, and significant amounts of other information. All of this information allows the professionals conducting the investigation to determine the existence of contamination at a site or property, and if the level of any contamination present requires remedial action because it presents an unacceptable risk to human health or the environment.

Questions and Exercises for Discussion

Of the following examples, which are Recognized Environmental Conditions (RECs) and which are not, and why?

1a. A 1,000-liter aboveground storage tank containing nitric acid is located inside a manufacturing building within secondary containment. The floor inside the facility is concrete with no observed cracks. The tank is observed to be dripping nitric acid slowly onto the concrete floor. The floor appears to be eroding slightly where the dripping nitric acid hits the floor.

1b. A 55-gallon drum is observed in a wooded area with no lid and is half full of an unknown substance.

1c. A small pond is located near some industrial manufacturing plants and you observe what appears to be a discharge pipe into the pond and directly beneath the pipe, you observe a small area of stained soils and stress vegetation. There does not appear to be any sheen on the water surface near the discharge pipe. There is no available information on where the pipe originates.

1d. You observe an area where garbage, including tires, plastics, furniture, and other related items were burned in a wooded area.

1e. Some used computer monitors and hard drives are observed in a closet inside a manufacturing facility.

REFERENCES

American Society for Testing Materials (ASTM). 1999a. *Standard Guide for Using the Direct Current Resistivity Method for Subsurface Investigation*. D6431-99. ASTM. West Philadelphia, PA.

American Society for Testing Materials (ASTM). 1999b. *Standard Guide for Using the Gravity Method for Subsurface Investigation*. D6430-99. ASTM. West Philadelphia, PA.

American Society for Testing Materials (ASTM). 2000a. *Standard Guide for Selection of Methods for Assessing Ground Water or Aquifer Sensitivity and Vulnerability*. ASTM. West Philadelphia, PA.

American Society for Testing Materials (ASTM). 2000b. *Standard Guide for Using the Seismic Refraction Method for Subsurface Investigation*. D5777-00. ASTM. West Philadelphia, PA.

American Society for Testing Materials (ASTM). 2005a. *Standard Practice for Environmental Site Assessments: Phase I Environmental Site Assessment Process*. E1527-05. ASTM. West Philadelphia, PA.

American Society for Testing Materials (ASTM). 2005b. *Standard Guide for Using the Surface Ground Penetrating Radar Method for Subsurface Investigation*. D6432-99. ASTM. West Philadelphia, PA.

American Society for Testing Materials (ASTM). 2005c. *Standard Guide for Planning and Conducting Borehole Geophysical Logging*. D5753-05. ASTM. West Philadelphia, PA.

California Environmental Protection Agency (CalEPA). 1995. *Guidelines for Hydrogeologic Characterization of Hazardous Substance Release Sites*. Volume 1: Field Investigation Manual. California Environmental Protection Agency: Sacramento, CA.

California Environmental Protection Agency. 1996. *Guidance for Ecological Risk Assessment at Hazardous Waste Sites and Permitted Facilities*. Department of Toxic Substances and Control. Sacramento, CA.

Capri, J., Schumacher, B. A., Wanning, S., Smith, E., Zimmerman, J., and Vanover, J. D. 2005. *Collecting Undisturbed Surface Sediments*. United States Environmental Protection Agency. EPA/600/R-05/076. Washington, D.C.

Fletcher, C. D. and Paleologos, E. K. 2000. *Environmental Risk and Liability Management*. American Institute of Professional Geologists. Westminster, CO.

Glasson, J., Therivel, R., and Chadwick, A. 2005. *An Introduction to Environmental Impact Assessments*. Routledge Publishing. London, England.

Harter, T. 2003. *Groundwater Sampling and Monitoring. University of California ANR Publication 8085*. University of California, Division of Agriculture and Natural Resources. Oakland, CA.

Keys, W. S. 1990. Borehole Geophysics Applied to Ground-Water Investigations. *United States Geological Survey Techniques of Water-Resources Investigation*. Book 2. United States Geological Survey: Washington, DC.

Munsell, A. H. 1905. *A Color Notation*. Geo. H. Ellis and Company Publishers. Boston, MA.

New Jersey Department of Environmental Protection (NJDEP). 2005. *Field Sampling Procedures Manual*. Trenton, NJ. 574 p.

New Hampshire Department of Environmental Services. 2008. *Contaminated Site Management Remedial Action Plan Checklist*. Concord, NH.

New York Department of Health. 2005. *Indoor Air Sampling and Analysis Guidance. New York Department of Health. Division of Environmental Health Assessment*. Center for Environmental Health. Albany, NY.

New York State Department of Environmental Conservation. 2021. *Remedial Investigation and Feasibility Study Fact Sheet*. http://www.dec.ny.gov/chemical/8658.html (accessed March 31, 2021).

Occupational Safety and Health Administration (OSHA). 1989. Hazardous Waste Operations and Emergency Response (HAZWOPER). *Code of Federal Regulations 40 (CFR) 1910.120*. Washington, D.C.

Ohio Environmental Protection Agency (OEPA). 1995. *Technical Guidance Manual for Hydrogeologic Investigations and Ground Water Monitoring*. Division of Drinking and Ground Waters. Columbus, OH. 255 p.

Ohio Environmental Protection Agency (OEPA). 2006. *Technical Guidance Ground Water Investigations*. Columbus, OH. 67 p.

Oregon Department of Environmental Quality. 1998. *Guidance for Ecological Risk Assessment: Levels I, II, III and IV*. Oregon Department of Environmental Quality. Portland, OR.

Rogers, D. T. 1992. The Importance of Site Observation and Follow-up Environmental Site Assessment – A Case Study. *Groundwater Management Book 12*. National Groundwater Association. Dublin, Ohio. pp. 563–572.

Rogers, D. T. 2016a. Next Generation of Urban Hydrogeologic Investigations. *Journal of Italian Geological Society*. Rome, Italy. Vol 39. no. 1. p. 349.

Rogers, D. T. 2016b. Scientific Advancements that Improve the Conceptual Site Model in Urban Hydrogeologic Investigations. *35th International Geological Congress*. Paper No. 3019. International Union of Geologic Sciences:Cape Town, South Africa. p. 325–337.

Rogers, D. T. and Welty, N. 2017. From Characterization to Closure: A Large TCE Plume. *Michigan Department of Environmental Quality and American Institute of Professional Geologists*. Lansing, MI.

United Nations. 1987. Report on the World Commission on Environment and Development. http://www.un.org/documents/ga/res/42/ares42–187.tm (accessed March 31, 2021).

United States Environmental Protection Agency (USEPA). 1976. *Resource Conservation and Recovery Act (RCRA). U.S.C. 6901*. United States Government Printing Office. Washington, D.C.

United States Environmental Protection Agency (USEPA). 1980. *Comprehensive, Environmental Response, Compensation and Liability Act U.S.C.* 9601–9675. United States Government Printing Office. Washington, D.C.

United States Environmental Protection Agency (USEPA). 1988. Guidance for Conducting Remedial Investigations and Feasibility Studies Under CERCLA. *Office of Emergency and Remedial Response*. EPA 540/G-89-004. United States Printing Office. Washington, D.C. 173 p.

United States Environmental Protection Agency (USEPA). 1989. Risk Assessment Guidance for Superfund. Volume 1: Human Health Evaluation Manual (Part A). *United States Environmental Protection Agency, Office of Emergency and Remedial Response*. OSWER Directive 9285.7 Ola. Washington, D.C.

United States Environmental Protection Agency (USEPA). 1991. Evaluating Exposures to Toxic Air Pollutants. *United States Environmental Protection Agency EPA Document 450/3-90-23*. Washington, D.C.

United States Environmental Protection Agency (USEPA). 1994. *Standard Operating Procedure for Surface Water Sampling*. S.O.P. 2013. Washington, D.C.

United States Environmental Protection Agency (USEPA). 1998. Guidance for Using Continuous Air Monitors. *United States Environmental Protection Agency Office of Air Quality Planning and Standards*. Washington, D.C.

United States Environmental Protection Agency (USEPA). 2002. *Risk Assessment Guidance for Superfund; EPA-540/R-02/002*. United States Government Printing Office. Washington, D.C.

United States Environmental Protection Agency (USEPA). 2005. *Standards and Practices for All Appropriate Inquiries. 40 Code of Federal Regulation (CFR), Part 312*. United States Government Printing Office. Washington, D.C.

United States Environmental Protection Agency (USEPA). 2007. Sediment Sampling. *USEPA Region 4, Science and Ecosystem Support Division*. Athens, GA.

United States Environmental Protection Agency (USEPA). 2008. List of Designated Reference and

Equivalent Methods for Air Sampling. *United States Environmental Protection Agency National Exposure Research Laboratory.* Washington, D. C.

United States Environmental Protection Agency (USEPA). 2021. Natural Resource Damage Assessment. http://www.epa.gov/superfund/programs/nrd/nrda2.htm#pagetop (accessed March 31, 2021).

Welty, N., Rogers, D. T., and Quinnan, J. 2016. High Resolution Profiling of Aquifer Permeability and Contaminant Mass. Battelle Chlorinated Remediation Conference. Monterey, CA.

Yong, R.N. 2002. *Geoenvironmental Engineering, Contaminated Soils, Pollution Fate, and Mitigation.* CRC Press. Boca Raton, FL.

5 Methods and Cost of Restoring Historical Anthropogenic Damage to Nature

5.1 INTRODUCTION

Restoring the environment once it becomes contaminated is expensive and time-consuming. A sad and disturbing fact is that sometimes when the environment becomes contaminated, any amount of financial resource or technology will not restore the environment to precontamination levels (Rogers 2018). Examples include the Exxon Valdez incident in Alaska, Deep Water Horizon incident in the Gulf of Mexico, numerous sites where groundwater becomes contaminated with chlorinated volatile organic compounds or hexavalent chromium, and polychlorinated biphenyls (PCBs) in sediments of large rivers such as the Hudson River in New York and the Willamette River in Oregon. Therefore, in many of these instances, the contamination must be managed through other means so that the spread of contamination is halted or minimized, and that humans and other life forms are not exposed to unacceptable levels of contaminants. An additional item, which perhaps is even more important, is that prevention is key to creating a sustainable balance with nature especially since there are over 7 billion humans that inhabit Earth.

Under many scenarios, contaminants released into the environment requires some sort of clean-up. The clean-up, termed **remediation**, meaning remedy, varies widely and depends on many different factors. However, as stated above and as we shall discover, typical remediation does not result in returning the affected area(s) or media to precontamination levels. Remediation of contamination is typically conducted to specific targets for each chemical of concern and is based on ecological and human risk. Remediation goals can be site-specific or to generic standards established by the appropriate regulatory agency. USEPA estimates that more than 100 billion dollars are spent each year investigating and remediating contaminated sites just in the United States (USEPA 2008a). Estimates on the number of sites remediated or under remediation total over 1 million sites. In addition, there are an estimated 500,000 to 1 million abandoned industrial facilities termed brownfield sites, that are waiting to be investigated. This is because a cost estimate to investigate and remediate brownfield sites has not been conducted. Most of these abandoned sites are located in urban areas of the United States. These sites possess some degree of contamination and will require some degree of assessing risk or remediation before they can be redeveloped. Thus, it is likely that the environmental costs to investigate and remediate sites of environmental contamination in the United States will exceed the latest available USEPA estimate of $250 billion.

The following sections describe some of the more common remedial technologies for soil, groundwater, surface water, sediment, and air. As with any successful remedial strategy, controlling the source of contamination is critical for achieving the objective. If controlling or eliminating the source is not achieved, any remedial strategy will fail because re-contamination will occur. The chapter concludes with an analysis of the cost of remediating common contaminants in the environment.

DOI: 10.1201/9781003175810-5

5.2 OVERVIEW

The objective of any remediation project is to prevent, remove, treat, change, destroy, or transform the potentially harmful contaminants from the media or medias of concern so that the risk posed by the presence of the contamination has been effectively eliminated or reduced to an acceptable level (USEPA 2005a). As we discussed in Chapter 4, there are several different studies that are conducted at a site of environmental contamination that outline and assist the navigation process through the required procedures and scientific studies that are routinely conducted so that the selected remedy or combination of remedies is successful at achieving the desired goal or objective. However, even with these many checks and balances, many remedial projects are unsuccessful and must be either revised or abandoned in favor of a different technological approach.

The most common reason for unsuccessful remediation to occur is because the site has not been properly characterized. Not identifying a source of contamination, underestimating contaminant mass, not identifying a complete list of contaminants of concern, incomplete sampling plan, and improper sampling techniques and methods all can lead to remediation being unsuccessful and may lead to increased cost, time, and effort to achieve the remedial objective. Other remediation failures include technology misapplication and design, environmental changes such as a change in groundwater flow direction, sudden changes in local climatic conditions such as a drought or flood, changes in geochemistry, and additional releases. Some of the factors that control or influence the degree to which remediation must be conducted depends upon the following factors:

- Contaminant or contaminants released
- Concentration
- Nature and extent
- Media impacted (e.g., air, surface water, groundwater, soil, sediment, etc.)
- Geology
- Hydrogeology
- Hydrology
- Location
- Potential receptors
- Land use
- Future land use
- Climate
- Time necessary to achieve objective(s)
- Results of risk assessment and feasibility study
- Cost

Driven by the need for more effective and less costly cleanup costs, new remedial technologies are constantly being developed (USEPA 2007). Over the last few decades, several remedial technologies have been developed to remediate contaminated soil and groundwater. In fact, USEPA (2007) has estimated that there are approximately 2,000 remedial technologies for environmental contamination. New technologies were developed due to several factors that included (1) ineffectiveness of early remedial methods, (2) excessive cost of early methods, and (3) regulatory goals that required remediation to standard-based levels that in many instances were to pristine precontaminant or background conditions. In the mid-1990s a transition occurred from the standard-based remediation cleanup goals toward risk-based cleanup goals (ASTM 1995). Remediating a site to risk-based criteria involves calculating site-specific cleanup levels that are based on the risks posed by the presence of specific contaminants and involve evaluating (1) the toxicity and nature of each contaminant, (2) the site's geological and hydrogeological setting, (3) fate and transport mechanisms, (4) future land use, and (5) analysis of potential receptors. In most cases, applying risk-based cleanup goals raised the concentration required to achieve site closure for specific contaminants at a particular site, which in turn, lowered the cost and time required to remediate a site (USEPA 2002). To begin our journey of an overview of the most common available remedial technologies, we begin first with a description of methods to remediate contaminated soil.

5.3 REMEDIATION CRITERIA

Most states have passed their own versions of CERCLA so that sites that do not qualify to become Superfund sites but have had a release of hazardous substances above cleanup levels get addressed. Cleanup levels established at the state level vary from state to state depending on various factors. However, much is based on USEPA exposure scenarios and toxicological data. States have established cleanup levels for different types of land use, typically residential and nonresidential (industrial or commercial), and by media or exposure pathway includes the following:

- Groundwater or drinking water
- Soil
- Groundwater–surface water interface
- Inhalation to indoor air
- Inhalation to outdoor or ambient air

Table 5.1 lists the generic residential cleanup levels for select chemical compounds for the state of Michigan (MDEQ 2016). The exposure pathways listed in the table are residential direct contact, commercial-industrial direct contact, and construction worker direct contact. The complete list can be viewed at https://www.michigan.gov/documents/deq/deq-rrd-chemCleanupCriteriaTSD_527410_7.pdf (MDEQ 2016). The values listed under the Groundwater-surface water interface in Table 5.1 are values within a zone where groundwater becomes surface water before mixing with surface water or exposure to the atmosphere. The values listed for air are presented in micrograms per kilogram in the soil immediately beneath a building for indoor air and immediately beneath the groundwater surface if an obstruction is not present (MDEQ 2016). The values listed for indoor and ambient air should be considered screening values since they are not actual air samples collected from a typical breathing zone. In Michigan, groundwater is defined as any water encountered beneath the surface of the ground regardless of quantity or if it occurs in a naturally occurring formation.

As a comparison, Table 5.2 presents soil cleanup criteria for New Jersey (New Jersey Department of Environmental Protection [NJDEP] 2008, 2015, and 2017), Illinois (Illinois Environmental Protection Agency [IEPA] 2017, California Department of Toxic Substances [DTSC], Control Chemical Look-Up Table [California Department of Toxic Substances and Control (DTSC)] 2013, California Environmental Protection Agency [CalEPA] 2005), and USEPA Regional Soil Screening Levels (RSLs) (USEPA 2021) Criteria presented in Table 12.2 generally correspond to a carcinogenic risk of 1 in 100,000 and a Hazard Quotient of 1 for residential properties. The values presented for New Jersey are not for the protection of groundwater. Soil remediation values for New Jersey protective of groundwater are calculated by NJDEP on a site-by-site basis. The lessons from examining Tables 5.1 and 5.2 are the following:

- Cleanup or remediation criteria vary substantially from by media which include soil, groundwater, surface water, indoor air, and outdoor air
- Cleanup or remediation criteria vary substantially based on land use which typically includes residential, commercial, industrial, construction worker, or recreational
- Cleanup or remediation criteria vary between each state and with USEPA

TABLE 5.1

Michigan Generic Residential Cleanup Levels for Select Compounds

Contaminant	CAS No.	Soil (µg/kg)	Groundwater (µg/L)	Surface Water Interface (µg/L)	Indoor Air (µg/kg)	Ambient Air (µg/kg)
Volatile Organic Compounds						
Benzene	71432	100	5	4,000	1,600	13,000
Bromobenzene	108861	550	18	NA	3.10 E+5	4.50 E+5
Bromochloromethane	74755	1,600	80	ID	1,200	9,100
Bromodichloromethane	75274	1,600	80	ID	1,200	9,100
Bromoform	75252	1,600	80	ID	1.50 E+5	9.0 E+5
Bromomethane	74839	200	10	35	860	11,000
n-Butylbenzene	104518	2.60 E+5	80	44,000	5.4 E+7	2.90 E+7
sec-Butylbenzene	135988	1,600	80	ID	1.0 E+5	1.0 E+5
tert-Butylbenzene	98066	78,000	80	ID	3.1 E+8	9.70 E+9
Carbon tetrachloride	56235	100	5	900	190	3,500
Chlorobenzene	108907	2,000	100	25	1.3 E+5	7.70 E+5
Chloroethane	75003	8,600	430	1,100	2.9 E +6	3.0 E+7
Chloroform	67663	1,600	80	350	7,200	45,000
Chloromethane	74873	5,200	260	ID	2,300	40,000
2-Chlorotoluene	95498	3,300	150	ID	2.70 E+5	1.20 E+6
4-Chlorotoluene	106434	900	150	360	4.3 E+5	9.60 E+6
1,2-Dibromo-3-chloropropane	96128	100	0.2	ID	260	260
1,2-Dibromoethane	106934	20	80	ID	1.1 E+5	2.60 E+6
Dibromomethane	74953	1,600	80	ID	7,000	7,000
1,2-Dichlorobenzene	95501	14,000	600	13	1.1 E+7	3.90 E+7
1,3-Dichlorobenzene	541731	170	6.6	28	26,000	79,000
1,4-Dichlorobenzene	106467	1,700	75	17	19,000	77,000
Dichlorodifluoromethane	75718	95,000	80	ID	5.3 E+7	5.50 E+8
1,1-Dichloroethane	75353	18,000	880	740	2.30 E+5	2.10 E+6
1,2-Dichloroethane	10762	100	5	360	2,100	6,200
1,1-Dichloroethene	75354	140	7	130	62	1,100

cis-1,2-Dichloroethene	156592	1,400	70	620	22,000	1.80 E+5
trans-1,2-Dichloroethene	156592	2,000	100	1,500	23,000	2.80 E+5
1,2-Dichloroporopane	78875	100	5	230	4,000	25,000
1,4-Dioxane	123911	1,700	85	2,800	NLV	NLV
Ethylbenzene	100414	1,500	74	18	87,000	7.20 E+5
Hexachlorobutadiene	87673	26,000	50	ID	1.30 E+5	1.30 E+5
Isopropylbenzene	98828	91,000	800	28	4.0 E+5	1.70 E+6
Methylene chloride	75092	100	5	1,500	45,000	2.10 E+5
Methyl-*tert*-butyl ether (MTBE)	163404	800	40	7,100	9.9 E+6	2.50 E+7
1,1,1,2-Tetrachloroethane	630206	1,500	8.5	78	36,000	54,000
1,1,2,2-Tetrachloroethane	79345	170	8.5	35	4,300	10,000
Tetrachloroethene (PCE)	127184	100	5	60	11,000	1.70 E+5
Toluene	108883	16,000	790	270	3.3 E+5	2.80 E+6
1,2,4-Trichlorobenzene	120821	4,200	70	99	9.6 E+6	2.80 E+7
1,2,3,-Trichlorobenzene	87616	4,200	70	99	9.6 E+6	2.80 E+7
1,1,1-Trichloroethane (TCA)	75556	4,000	220	89	2.50 E+5	3.80 E+6
1,1,2-Trichloroethane	79005	100	5.0	330	4,600	17,000

Source: Michigan Department of Environmental Quality. 2016. Cleanup Criteria and Screening Levels Development and Application. https://www.michigan.gov/documents/deq/deq-rrd-chem-CleanupCriteriaTSD_527410_7.pdf (accessed March 31, 2021).
μg/kg, microgram per kilogram; μg/L, microgram per liter; NA, Not available; ID, Insufficient data; NLV, Not likely to volatilize.

TABLE 5.2

Soil Cleanup Levels for New Jersey, Illinois, New York, California, and USEPA

Contaminant	CAS ID No.	New Jersey[1] (mg/kg)	Illinois[2] (mg/kg)	California[3] (mg/kg)	USEPA[4] (mg/kg)
Acetone	67641	1,000	25	20	2.9
Benzene	71432	3	0.03	5	0.0023
Bromobenzene	108861	NL	NL	NL	0.042
Bromochloromethane	74755	11	NL	NL	0.021
Bromodichloromethane	75274	NL	0.6	NL	0.00036
Bromoform	75252	86	0.8	NL	0.00087
Bromomethane	74839	79	NL	NL	0.0019
Methyl ethyl ketone	789833	1,000	17	NL	1.2
tert-Butylbenzene	98066	NL	NL	NL	1.6
Carbon tetrachloride	56235	2	0.07	NL	0.00018
Chlorobenzene	108907	37	1	NL	0.053
Chloroethane	75003	NL	NL	NL	0.081
Chloroform	67663	19	0.6	NL	0.000061
Chloromethane	74873	520	NL	NL	0.049
2-Chlorotoluene	95498	NL	NL	NL	0.23
4-Chlorotoluene	106434	NL	NL	NL	0.24
1,2-Dibromo-3-chloropropane	96128	NL	NL	NL	0.081
1,2-Dibromoethane	106934	NL	NL	NL	0.0000021
Dibromomethane	74953	NL	NL	NL	0.0021
1,2-Dichlorobenzene	95501	5,100	17	NL	0.03
1,3-Dichlorobenzene	541731	5,100	NL	NL	0.00082
1,4-Dichlorobenzene	106467	570	2	NL	0.00046
Dichlorodifluoromethane	75718	NL	NL	NL	0.03
1,1-Dichloroethane	75353	8	0.06	NL	0.00078
1,2-Dichloroethane	10762	6	0.02	NL	0.000048
1,1-Dichloroethene	75354	8	0.06	5	0.00078
cis-1,2-Dichloroethene	156592	4	0.4	5	0.011
trans-1,2-Dichloroethene	156592	4	0.7	NL	0.11
1,2-Dichloroporopane	78875	10	0.03	NL	0.00015
2,2-Dichloropropane	594207	NL	NL	NL	0.013
1,3-Dichloropropane	142289	NL	2NL	NL	0.13
1,4-Dioxane	123911	NL	NL	10	0.000094
Ethylbenzene	100414	1,000	13	5	0.0017
Hexachlorobutadiene	87673	1	NL	5	0.00027
Isopropylbenzene	98828	NL	NL	NL	0.084
Methylene chloride	75092	49	0.02	10	0.0029
Methyl-*tert*-butyl ether (MTBE)	1634044	NL	0.32	NL	0.0032
1,1,1,2-Tetrachloroethane	630206	170	NL	NL	0.00022
1,1,2,2-Tetrachloroethane	79345	34	NL	NL	0.00003
Tetrachloroethene (PCE)	127184	4	0.06	5	0.0051
Toluene	108883	1,000	12	5	0.76
1,2,4-Trichlorobenzene	120821	68	5	NL	0.0034

TABLE 5.2 (Continued)
Soil Cleanup Levels for New Jersey, Illinois, New York, California, and USEPA

Contaminant	CAS ID No.	New Jersey[1] (mg/kg)	Illinois[2] (mg/kg)	California[3] (mg/kg)	USEPA[4] (mg/kg)
1,2,3,-Trichlorobenzene	87616	NL	NL	NL	0.021
1,1,1-Trichloroethane (TCA)	75556	210	2	NL	2.8
Trichloroethene (TCE)	75694	23	0.06	5	0.00018
Trichlorofluoromethane	96184	NL	NL	NL	3.3

Notes

[1] New Jersey Department of Environmental Protection (NJDEP). 2008. Introduction to Site-Specific Impact to Ground Water Soil Remediation Standards Guidance Document. Trenton, New Jersey. 10 pages. https://www.nj.gov/dep/srp/guidance/rs/igw (accessed May 5, 2017); New Jersey Department of Environmental Protection (NJDEP). 2015. Site Remediation Program: Remediation Standards. http://www.nj.gov/dep/srp/guidance (accessed May 4, 2017); and New Jersey Department of Environmental Protection (NJDEP). 2017. Soil Cleanup Criteria. https://www.nj.gov.srp/guidance/scc (accessed May 5, 2017).

[2] Illinois Environmental Protection Agency (IEPA). 2017. Tiered Approach to Corrective Action Objectives (TACO). Springfield, Illinois. 284 p.

[3] California Department of Toxic Substances and Control (DTSC). 2013. Chemical Look-Up Table. Technical Memorandum. Sacramento, California. 5 p.; and California Environmental Protection Agency (CalEPA). 2005. Use of Human Health Screening Levels (CHHSLs) in Evaluation of Contaminated Properties. Sacramento, California. 65 p.

[4] United States Environmental Protection Agency (USEPA). 2021. Regional Screening Levels. https://www.epa.gov/RSLs (accessed October 29, 2021).

mg/kg, milligram per kilogram; NL, Not listed.

The World Health Organization and United Nations have not established guidelines for soil or groundwater. However, most countries that have not developed standards for the remediation of land that includes, soil, sediment, or groundwater commonly refer to a set of standards established by the Dutch in 1987 and were revised in 1994, 2001, and 2015 and are commonly referred to as the Dutch Intervention Values (DIVs) (Lijzen et al. 2001, Netherlands Ministry of Infrastructure and the Environment 2015).

The DIVs are based on potential risks to human health and ecosystems and are technically evaluated. The values presented usually have two integers for each chemical listed. One is a target level that is considered optimal and likely will result in no or limited health or ecological effect and the other is a level of concentration above which some intervention is necessary to lower risk posed by site-specific exposure pathways. The DIVs have been used worldwide as a benchmark for the cleanup of contaminated sites in many areas of the world where cleanup values have not been established. In addition, even if there are established values, the Dutch Intervention Values are used for comparison purposes. Therefore, in many instances, the Dutch Intervention Values represent the benchmark for cleanup standards worldwide.

Table 5.3 presents a partial list of Dutch Intervention Values. A complete list can be viewed at http://www.sanaterre.com/guidelines/dutch.html (Lijzen et al. 2001 and Netherlands Ministry of Infrastructure and the Environment 2015).

5.4 COMMON SOIL REMEDIATION TECHNOLOGIES

Common soil remediation technologies include (USEPA 2002, 2007, Rogers et al. 2008, 2009):

TABLE 5.3

Dutch Intervention Values for Select Compounds for Soil and Groundwater

Compound	Soil		Groundwater	
	Target Value (mg/kg)	Intervention Value (mg/kg)	Target Value (µg/L)	Intervention Value (µg/L)
Arsenic (As)	29.0	55.0	10	60
Barium (Ba)	160	625	50	625
Cadmium (Cd)	0.8	12	0.4	6
Chromium (Cr)	100.0	380	1	30
Copper (Cu)	36.0	190	15	75
Nickel (Ni)	35.0	210	15	75
Lead (Pb)	85.0	530	15	75
Mercury (Hg)	0.3	10.0	0.05	0.3
Silver (Ag)	None	15	None	40
Selenium (Se)	0.7	100	None	40
Zinc (Zn)	140	720	65	800
Chloride	None	None	100,000	None
Cyanide	None	20	5	1,500
Cyanide (complex)	None	50	10	1,500
Thiocyanate	None	20	None	1,500
Benzene	None	1.1	0.2	30
Ethyl benzene	None	110	4	150
Toluene	None	320	7	1,000
Xylenes (sum)	None	17	0.2	70
Styrene (vinylbenzene)	None	86	6	300
Phenol	None	14	0.2	2.000
Cresols (sum)	None	13	0.2	200
Naphthalene	None	None	0.01	70
Phenanthrene	None	None	0.003	5
Anthracene	None	None	0.0007	5
Vinyl chloride	None	0.1	0.01	5
Dichloromethane	None	3.9	0.01	1,000
Trichloroethene (TCE)	None	2.5	24	500
Tetrachloroethene (PCE)	None	8.8	0.01	40
Pentachlorophenol	None	12	0.04	3
Polychlorinated biphenyl (PCB) (sum)	None	1	0.01	0.01
Monochloroanilines (sum)	None	50	None	30
Dioxin (sum)	None	0.00018	None	None
Chloronaphthalene (sum)	None	23	None	6
Chlordane	None	4	0.00002	0.2
DDT (sum)	None	1.7	None	None
DDE (sum)	None	2.3	None	None
DDD (sum)	None	34	None	None
DDT/DDE/DDD sum	None	None	0.000004	None
Aldrin	None	0.32	0.000009	None
Dieldrin	None	None	0.0001	None
Lindane	None	1.2	0.009	None

TABLE 5.3 (Continued)
Dutch Intervention Values for Select Compounds for Soil and Groundwater

Compound	Soil		Groundwater	
	Target Value (mg/kg)	Intervention Value (mg/kg)	Target Value (µg/L)	Intervention Value (µg/L)
Atrazine	None	7.1	0.00029	150
Carbofuran	None	0.017	0.0009	100
Asbestos	None	100	None	None

Source: Netherlands Ministry of Infrastructure and the Environment. 2015. Dutch Pollutant Standards. https://www.government.nl/ministry-of-infrastructure-and-the-environment or at http://www.sanaterre.com/guidelines/dutch.html (accessed October 29, 2021).
mg/kg, milligram per kilogram; µg/L, microgram per liter.

- Excavation: Excavation and disposal of impacted material within a landfill is the most common soil remediation technique (USEPA 2002).
- Bioremediation: Bioremediation involves the introduction of microorganism to degrade targeted contaminants. Many LNAPL compounds can be degraded using microorganisms (USEPA 2001a).
- Monitored natural attenuation (MNA): MNA relies on natural processes to attenuate contaminants which does commonly occur in soil (USEPA 2001b).
- Mechanical soil aeration (MSA): MSA involves turning or tilling impacted soil so that contaminants can more easily evaporate and are exposed directly to sunlight so that a process called photolysis can also add in degrading contaminants (USEPA 2002, 2003).
- Capping: Capping is considered an engineering control because it does not break down contaminants. Capping is simply the placement of an engineered barrier over the contamination to avoid direct exposure. Capping also typically requires long-term maintenance to ensure the contamination does not migrate and the cap remains stable (USEPA 2002, 2003).
- Land use restrictions: Land use restrictions are common and involve the placement of restrictions on an impacted site that may involve maintaining site use (e.g., industrial use only), prohibiting groundwater use, and restricting access to certain areas. Soil vapor extraction (SVE). SVE can only be used if the contaminants can readily evaporate, such as VOCs. In addition, the geology must typically be composed of material that will permit the movement of air through the zone where the contaminants exist, such as sand or gravel. The vapors are typically captured whereby the contaminants in the air are removed before the air is discharged into the atmosphere (USEPA 2002, Federal Remediation Technologies Roundtable 2021).
- Phytoremediation: Phytoremediation uses plants to remediate contaminated soil. Certain plants can degrade, destroy, or transform from many types of contaminants such as copper, some pesticides, perchlorates, and certain PAH compounds (USEPA 1999a).
- Soil washing: Soil washing essentially "scrubs" soil to remove and separate the contaminant from soil particles using detergents or any number of chemicals depending on the type of contaminant, concentration, and soil type. Many contaminants sorb to soil particles and can be removed or separated using a detergent or chemicals that lower the sorptive potential of the contaminants. A variety of contaminants including some heavy metals, fuels, and pesticides can be remediated using this technique (USEPA 2001c).

- In situ thermal desorption: In situ thermal treatment is a method that injects a form of heat into subsurface soils in an effort to mobilize the contaminants so that they can be more easily recovered. Common heat sources include steam, hot air, hot water, electrical resistance, and others (USEPA 2001d).
- Ex situ thermal desorption or incineration: Ex situ thermal treatment is similar to in situ thermal treatment with the only difference being that the contaminated soil is excavated and then thermally treated at the surface. After treatment, the soil is either returned to the ground or is transported and disposed in a landfill (USEPA 2002).
- Electrokinetics: Electrokinetics is a remedial technology that separates and extracts heavy metals and other contaminants from saturated and unsaturated soil, sludges, and sediments by inducing the migration of contaminants in subsurface soil through an imposed electrical field via electroosmosis. Migration of contaminants occurs when the soil is electrically charged with a low voltage current. The configuration involves the application of an electrical potential between electrode pairs that have been implanted in the ground on each side of a contaminated soil mass (Cauwenberghe 1997).
- Solidification/Stabilization: Solidification/Stabilization is conducted in situ and ex situ and involves the addition or injection of a material, such as concrete or bentonite clay, to permanently immobilize or entomb the contaminants. This technique does not remove contaminant mass but rather eliminates or minimizes the potential of the contaminants from migrating and potentially causing harm to human health or the environment (USEPA 2002).
- Fracturing: Fracturing is a method to crack dense contaminated soils or rock so that other remedial technologies can work more efficiently. The cracks, which are called *fractures*, create pathways through which contaminants can be more quickly and efficiently removed from the ground. The fractures are created by injecting either high-pressure water or air into the contaminated zone which then creates the fractures (USEPA 2001e).
- Vitrification: Vitrification is a form of solidification/stabilization that uses electrical power to transform contaminated subsurface soil into a glass-like substance (USEPA 2001f).
- Chemical oxidation: Chemical oxidation involves the addition of an oxidizing agent to contaminated soil that results in either the complete destruction of the contaminants or can be used as a source of oxygen to more rapidly induce bioremediation as long as the correct dosage is used (USEPA 2007).
- Chemical dehalogenation: Chemical dehalogenation is a remedial method that removes halogens from contaminants. Common halogenated contaminants include several DNAPL VOCs, PCBs, and dioxins. There are two common types of chemical dehalogenation and include (1) glycolate dehalogenation and (2) base-catalyzed dehalogenation (USEPA 2001g).

Each soil remediation technology has advantages and disadvantages depending upon the numerous different conditions that exist at any given site. What most often occurs at most sites is a combination of technologies are used to achieve the objective. At sites where there are widespread impacts with different types of contaminants, rarely if ever, is one technology used (USEPA 2007). A total of 53% of the technologies used ex situ technologies and 47% used in situ technologies. Not listed in the figure is soil excavation. Soil excavation was purposely not included because it was the technique most commonly employed. Soil excavation remains the most common technology used for the remediation of soil. This is in part because soil excavation is the least costly method when small relative volumes (less than 5,000 tons) are present that are located in an accessible location. In addition, soil excavation may be the only economical technology available, especially when different types of contaminants are present (USEPA 2007, Rogers et al. 2009).

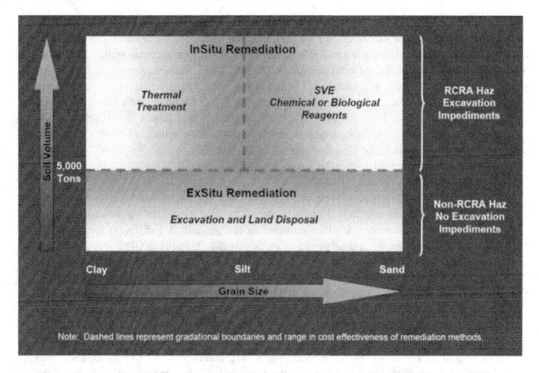

FIGURE 5.1   Selection of VOC soil remediation methods based on geology and volume.

For the most part, soil remediation technologies heavily rely on two factors that control which technology will work most efficiently and include (1) the geological environment in which the contaminants reside, and (2) the physical chemistry of the contaminants themselves. An additional important factor is whether a mixture of different types of contaminants and geology exists. As long as the geology and contaminant type and distribution are straightforward, selecting an appropriate soil remedial technology is much more simple a task. Figure 5.1 shows the relationship between geology and soil remediation technology for VOCs (Rogers et al. 2009).

5.5   COMMON GROUNDWATER REMEDIATION TECHNOLOGIES

The first accepted method to remediate impacted groundwater was to pump the contaminated water from beneath the ground to the surface where it was treated and then to either re-inject the treated water or discharge the water to the sanitary sewer or surface water body. Early treatment technologies were adopted from known techniques developed in treating wastewater from industrial sources and from wastewater treatment at publicly owned treatment plants (POTWs). These techniques commonly included, solids removal, flocculation, bioreactors, activated carbon filtration, and reverse osmosis among others. In the middle to late 1980s, it became apparent that groundwater pump and treat technologies, as they became known, were very expensive when applied to remediate contaminated groundwater.

After many years, it was discovered that pumping and treating groundwater was not effective. Essentially, groundwater pump and treat was effective at ensuring a contaminant plume did not spread if it were designed and operated correctly but was not at all cost-effective at restoring an aquifer's groundwater to precontaminated levels. Since this discovery, several new technologies have been developed that are termed *in situ technologies*. Instead of bringing the contaminated water to the surface to be treated, the treatment is delivered to where the contamination resides – in the saturated zone. These new technologies included air sparging, installation of permeable re-active barriers, injection of biological agents, and injection of chemical reagents. The goal of the in situ technologies was to destroy, chemically reduce, or transform the contaminants into harmless compounds within the plume of contaminants (USEPA 2007). However, pumping and treating groundwater can potentially be effective at removing significant quantities of contaminant mass under certain favorable conditions. Young and immature groundwater plumes that have not had a sufficiently long period of time to allow significant diffusion and hence significant sorption of contaminants into lower permeable zones is a potential situation where a focused groundwater pump and treat system could be effective (Payne et al. 2008).

Common groundwater remediation technologies include the following (USEPA 2002, 2007):

- Pump and Treat: Pump and Treat is a groundwater remediation technology that uses pumps to transport contaminated groundwater to the surface where the contamination is removed. The clean water is either returned to the ground at a different location, dis-charged to the sewer, or is discharged to surface water (USEPA 2001h).
- In situ soil flushing: In situ soil flushing is a remedial method that pumps or percolates water or chemicals into the ground with the objective of flushing or driving the con-taminants present in the saturated soil to a location where they can be removed, typically to a pumping well. In situ soil flushing is a modification to a traditional pump and treat system and is employed to increase efficiency (USEPA 2001i).
- Air sparging: Air sparging is considered an in situ groundwater remediation technology that uses injected air to volatize contaminants in groundwater. As the injected air rises through the saturated zone and reaches the unsaturated zone, the vapors containing the contaminants are typically removed from the ground using a soil vapor extraction system (USEPA 2001j).
- Permeable reactive barriers: A permeable reactive barrier or PRB is a wall or fence-like structure that is constructed beneath the surface of the ground within the saturated zone downgradient of groundwater contamination. The wall is composed of chemicals that degrade or destroy the targeted contaminants. PRBs are considered passive remedial systems and generally take a long period of time to achieve remedial objectives because it relies on the natural flow of groundwater to pass through the reactive barrier (USEPA 2001k).
- Injection of biological agents: Injection of biological agents is essentially the same technology as remediating contaminated soil with microorganisms. The exception or modification is that the same technology is applied to groundwater but the delivery systems are different. Typically, the injection of biological agents is not microorganisms themselves, rather the injection is commonly chemical that promotes microbial growth such as oxygen or other food sources such as molasses or other carbohydrate sources (USEPA 2007).
- Injection of chemical agents: Injection of chemical agents is similar to injecting biological agents. The only difference is that injecting chemical agents is not intended to promote microbial growth. Injecting chemical agents attempts to create or promote a chemical reaction to transform or destroyed contaminants (USEPA 2007).
- Monitored natural attenuation: Monitored natural attenuation of contaminants in groundwater is similar to that of natural attenuation in soil. However, natural attenuation in groundwater relies upon dilution to assist in lowering contaminant concentrations and can be effective with certain types of contaminates as long as no continuing source of groundwater contamination exists. In addition, natural attenuation also relies on natural physical degradation and natural biodegradation to reduce contaminant concentrations (USEPA 1998, 2001b).

- Institutional controls: Institutional controls are used to remove potential receptor pathways, such as restricting or prohibiting the use of groundwater in impacted areas. These are usually used at locations where large areas of groundwater impacts have occurred or when remediating groundwater is not practical or cost-prohibitive. Using institutional controls usually requires State and local municipality approval.
- Multi-phase extraction: Some sites may have such high concentrations of contaminants that they may be present in the subsurface as pure products. In these cases, it's typical that more than one contaminant phase is present that includes vapor phase, dissolved phase, pure product, or free phase (USEPA 2005a).

Selecting the appropriate groundwater remediation technology is dependent upon the same set of criteria as selecting the most appropriate soil remediation technology, the two most important factors include (1) the geology and hydrogeology in which the contaminants reside, and (2) the physical chemistry of contaminants targeted for remediation. As with soil remediation, if a mixture of different types of contaminants are present in groundwater and are required to be remediated, selecting more than one technology may be required to achieve the remedial objectives (USEPA 2005b). Certain technologies are more efficient than others at remediating particular contaminants. For instance, air sparging may be an appropriate remedial technology for remediating LNAPL VOCs in groundwater if the soil is permeable enough (e.g., sands and gravels) (USEPA 1995). On the other hand, air sparging is not an appropriate technology for remediating chrome VI because the chrome and most other metals do not readily evaporate (Rogers et al 2009). Figure 5.2 presents some of the groundwater remediation decision parameters.

An additional factor not always considered when examining the remediation of groundwater is the time it requires to complete the remediation. Remediating groundwater may require more than 20 to 30 years or more to complete. Factors that influence the time required, which may greatly impact cost, to remediate groundwater include (USEPA 1999b, 2000, Rogers et al. 2006):

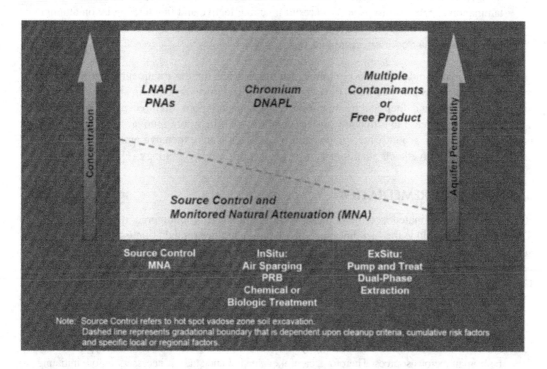

FIGURE 5.2 Remedial matrix of geology, contaminant type, and remedial technology.

- Geology and hydrogeology: Various types of geology influence the technology selected and also dictate the rate of contaminant recovery, such as low permeability aquifers may require longer periods of time to remediate.
- Contaminant concentration and distribution: Groundwater plumes that are extensive in size and have had sufficient time to mature typically have contaminants that have diffused into lower permeable zones within an affected aquifer. In these situations, remediation time may be extended by decades.
- Clean up goals: In basic terms, the lower the cleanup goals the more time is typically necessary to achieve that goal.
- Technology selected: PRBs generally take longer because they are more passive remedial systems and heavily rely on the natural groundwater flow to deliver contaminants to the remediation area.
- Remedial system design and maintenance: System design may not always be for speed of cleanup but must also consider other factors such as structural impediments such as buildings and roads. In addition, spacing of extraction or injection wells may not be optimized due to engineering factors and other factors that cannot be controlled.
- Whether there are multiple types of contaminants presents. Contaminant plumes with more than one phase of contaminant, for instance, dissolved phase and free phase, may have to be remediated in steps or phases which may require additional time. In addition, if a contaminant plume has more than one contaminant type, such as DNAPL VOCs and metals, additional time may be necessary because remediation may require additional steps to complete.
- Location: Urban areas often dictate or limit available preferred technologies that can shorten the time required for groundwater remediation. For instance, building and roads and other structures may result in lengthening the time required to investigate and conduct source control measures. The result is that remediation of groundwater is initiated later and subsequently takes longer to complete.
- Investigative errors and delays: Not fully characterizing the nature and extent of contamination can be disastrous and can result in an ineffective and failed remediation attempt.
- Source control: If all the sources have not been identified and abated, remediation of groundwater may be ineffective and fail.

Groundwater pump and treat is a common method selected for the treatment of groundwater at Superfund sites and is selected more than 70% of the time. However, other in situ methods such as air sparging, PRBs, and chemical treatment are being selected more often recently (USEPA 2007). This is due to an increase in efficiency and acceptability of other more innovative technologies and is also because many groundwater pumps and treat systems are expensive and have not achieved cleanup goals and have been abandoned in favor of a different technology (USEPA 2007).

5.6 SEDIMENT REMEDIATION

Remediating contaminated sediment presents difficulties not encountered in remediating soil. These difficulties include the fact that much of what may require remediation is located beneath the surface of water bodies and is usually in especially sensitive ecosystems. Due to these two complicating factors, remediating contaminated sediment requires careful study and planning. Common remedial methods for remediating contaminated sediment include the following (USEPA 2005c, 2019a, Sediments Focus Group 2007):

- Source control: Source control may be more difficult when evaluating contaminated sediments. This is because the source or sources may have originated a significant distance or typically may be from numerous sources. Therefore, careful evaluation and study is necessary before initiating remedial activities to ensure that re-contamination does not occur (USEPA 2005c).

- Excavation: Excavation of contaminated sediments is usually conducted within floodplains, stream and river banks, and beach areas when water does not cover the contaminated materials. These areas, if contaminated and require remediation, can be easily accessed. Excavation is usually preferred over other technologies if the volume of contamination is manageable because the impacted sediments can be removed quickly before they become covered with water making excavation potentially impossible and much more costly (USEPA 2005c).
- Dredging: Dredging involves excavating contaminated sediments that are submerged beneath the water. Several techniques and different types of dredging equipment have been developed to address many kinds of situations encountered during dredging. Dredging sediment that is submerged often causes contaminated sediment particles to become suspended in the water column. This may cause contamination to spread in the downgradient direction of water flow. To avoid this situation, extreme measures are sometimes conducted such as USEPA (2005c) and Sediments Focus Group (2007).
- Bioremediation: Bioremediation of contaminated sediments involves the same processes involved in bioremediating soil and groundwater. However, many contaminants that are often present in sediments, such as PCBs and mercury, are not readily remediated using microbes (USEPA 2005c).
- Capping: Many situations occur where no active remedial method is available to effectively remove all contaminated sediment without potentially causing more harm, such as suspending sediments in the water column. In these situations, capping may be preferred. Capping contaminated and submerged sediments must be conducted slowly so that disturbance is kept to a minimum (USEPA 2005c).
- Natural attenuation: Natural attenuation may be an effective remedial option in situations where the source(s) have been eliminated and either the contaminated sediments are not heavily impacted, or other methods will not substantially reduce the risk posed by the presence of the contaminated sediments (USEPA 2005c).

There are limited technologies for the remediation of sediment. Remediation of sediment once contaminated is also very expensive compared to soil remediation. One of the difficulties in remediation of sediment is access and not making things worse, especially if dredging is the selected technology. This is because dredging has the tendency to suspend contaminated particles within the water column and thereby re-exposing the environment to contamination and increasing the potential to contaminate locations that were not previously impacted.

5.7 SURFACE WATER REMEDIATION

The preferred approach to remediating surface water has been through the enactment of pollution prevention regulations. Since the National Pollutant Elimination System or NPDES was enacted in 1972, significant improvements have been made in improving the surface waters of the United States (USEPA 2008a, 2019b). The reason this is a preferred method is because there are no effective methods to remediate surface water and often when a technology is attempted, the probability that the situation will become worse is high. The most common methods to address surface water include the following:

- Source control: Simply stated, the NPDES process requires municipalities and industries to meet specified standards in order to discharge wastewater or stormwater to the surface waters of the United States (USEPA 2019b).
- Engineered wetlands: Engineered wetlands or Engineered Natural Systems (ENS) for water treatment is an additional and especially effective method for remediation of surface water given favorable conditions involves a combination of natural biological, chemical, and physical processes to remediate common surface water contaminants.

Constructing an effective ENS involves matching contaminant properties to the natural conditions of the site, and available degradation methods. A typical ENS utilizes several different natural methods to destroy contaminants including photolysis, bioremediation, phytoremediation, chemical degradation, settling, and many others (Eifert 2010).

- Direct removal: Direct remediation methods for surface water are available, such as when there has been an accidental spill. Reported spills to surface water are usually petroleum related. Remediating impacted surface water is most effective when there is a quick response so that the release is confined to as small an area as possible. Common types of releases to surface water include petroleum products such as fuels, refined oil, and unrefined or crude oil. Common techniques for responding and addressing a surface water fuel or oil spill include doing nothing, containment, Applying chemicals to disperse contaminants, and introducing biological agents to break down contaminants (NOAA 2021).

Groundwater often discharges to surface water and accounts for much of the base flow in surface streams. Therefore, surface water can also become contaminated by impacted groundwater that discharges to surface water. This may be and often is, poorly understood and underestimated as a pollutant source to surface water (Rogers 1997, Rogers and Murray 1997).

This potential scenario (USEPA 2000) and is commonly referred to as nonpoint source pollution (USEPA 2000). Remediation of nonpoint source pollution is challenging and can only be effective if:

- A monitoring network is established
- Continuous monitoring is conducted using mass balance techniques to establish what fraction of the total contaminant load originates from nonpoint sources of pollution.
- The geology and hydrogeology of a region or watershed is well understood. Knowing where base flow is received is critical.
- Potential sources of pollution are identified and may not be restricted to just soil and groundwater but may also include air.
- Potential contaminants identified. Identifying potential contaminants will assist with where and how to monitor potential contaminant pathways (e.g., shallow groundwater, deep groundwater, air, or sediment).
- Urban factors that include stormwater discharge and modifications to the natural geological setting
- Historical factors such as rebuilding urban centers and areas of historical landfilling.
- Creating pollution prevention programs and initiatives to minimize or eliminate future impacts
- Creating public awareness

Many of the techniques described above are the subject of the next section of this book which provides numerous practical ideas and guidelines on improving the natural environment and lowering the risk of exposure to contaminants in the environment. Remediation of surface water once impacted is not only expensive but sometimes is not possible. This is because contamination once in surface water has the potential to spread quickly and may be inaccessible. Therefore, the key to success is prevention.

5.8 AIR REMEDIATION

The approach to remediating air is similar to the approach used to address surface water, that being pollution prevention. However, accidental releases do occur. They are most often addressed through emergency response actions which may involve evacuation in the down wind direction, air monitoring, and source control (e.g., stopping the release). Releases of air contaminants are regulated under the Clean Air Act which is administered by the USEPA and are enforced by State and local agencies. In addition, many States may have enacted their own stricter regulations. Air pollution is regulated through a permit process that requires numerous operations to obtain a permit or inspection, such as motor vehicle emission inspections and testing, that show proof that emission standards are not exceeded. Along these same lines, numerous industrial operations that emit atmospheric pollutants often require a permit and often require that an air pollution control device be installed to capture the pollutant or pollutants before they are emitted to the atmosphere (USEPA 2008b).

According to USEPA (2008b), there are two types of air pollutant sources: (1) mobile sources, such as automobiles, trucks, buses, farm machinery, airplanes, and (2) stationary sources, such as industrial manufacturing facilities, chemical production facilities, pharmaceutical companies, and refineries. According to USEPA (2008b), 90% of air contamination is attributable to motor vehicle exhaust. However, as described in Chapter 8, there have been significant improvements in air quality have been achieved in the last few decades through many programs and initiatives to reduce the amount and type of contaminants from motor vehicle exhaust include the following:

- Increasing fuel efficiency
- Decreasing the amount of emissions
- Installing air pollution control devices on vehicles (such as catalytic converters)
- Increasing public awareness (e.g., driving less and not refueling during critical periods)
- Reformulating fuels (e.g., eliminating the use of lead as an additive)
- Developing more efficient engines
- Requiring inspections and routine maintenance, if necessary (e.g., vehicle emissions inspections)

Stationary sources are divided into two categories that include (1) gaseous contaminants and (2) particulate matter. Removal of contaminants from the gaseous phase is accomplished through various technologies that include the following (USEPA 2008b):

- Contact condenser
- Surface condenser
- Thermal incinerator
- Catalytic incinerator

Removal of particulates is usually accomplished with the installation of a dust collector (also referred to as a baghouse), or a similar device called a wet scrubber. The process of collecting dust (particulates) using a baghouse involves capturing the particulate matter at or very near the emission source by placing the source under a vacuum. The particulates are captured using a filter, such as a fabric before the air stream is exhausted into the atmosphere. Capture efficiencies can be very high and may routinely exceed 95%.

5.9 COST OF REMEDIATION

Remediation is costly and not just financial. Remediation takes time, patience, and scientific understanding. Some of the factors that control whether remediation is conducted and how it's conducted and subsequently the cost includes the following (Rogers et al. 2006, 2009):

- Geology: Different types of geology can influence the extent of contamination and methods of remediation
- Hydrogeology: The presence of groundwater often plays a significant role in assessing risk, migration pathways, and technology(s) selection.
- Contaminant type(s): The type(s) of contaminants influence migration pathways, mobility, risk, and cleanup targets.

- Nature and extent: Nature and extent of contamination impact receptor analysis and potential routes of exposure and cleanup targets.
- Contaminant mass and concentration: The mass and concentration of contaminants and whether there is the presence of free phase influences remedial technology(s) selection.
- Media(s) to be remediated (e.g., soil, groundwater, surface water, or sediment). Some media are more difficult and costly to remediate than others.
- Land use: Residential land typically requires the most stringent cleanup targets.
- Cleanup targets: Contaminant type, location, potential receptors, and exposure pathways have the most significant influence over establishing cleanup targets.
- Technology(s) selected: Some technologies may be more costly than others (i.e., pumping and treating groundwater vs. natural attenuation).
- Time: Time may influence cost, especially if the release involves actual human exposure or migration to a sensitive ecological receptor that requires immediate remedial action.

To evaluate (1) whether certain contaminants cost more to remediate than others and (2) what influences the geology has on remediation, let us examine remediation that has been conducted at sites of environmental contamination. A total of 370 sites where investigations and remediation had been conducted were compiled and evaluated (Rogers 2018; Rogers et al. 2006, 2009). A few characteristics of each site included the following:

- Of the 370 total sites, 268 were located in the United States, and 102 were located in other countries.
- Sites were located in 36 different States and included AL, AK, AR, CA, CO, DE, FL, GA, IL, IN, IA, KS, KY, MA, MD, MI, MN, MO, MS, NC, NH, NJ, NV, NY, OH, OK, OR, RI, SC, TN. TX, UT, VA, WA, WI, WV.
- International sites were located in 14 countries and included Australia, Belgium, Brazil, Canada, China, England, France, Germany, Italy, Malaysia, Mexico, Russia, South Africa, and Ukraine.
- Property size ranged from 1.8 acres to more than 300 acres.
- All were located in urban areas.
- Many of the facilities operated for more than 100 years.
- Operations at most facilities included heavy manufacturing.

The data from each site included (Rogers 2018, Rogers et al. 2006, 2009):

• Contaminant discover facts	• Topography
• Estimated date of release	• Hydrological setting
• Release circumstances	• Fate and transport analysis
• Release operation location	• Analysis of potential receptors
• Source of release	• Remedial technology selected
• Contaminants(s) of concern	• Contaminant mass remediated
• Media(s) impacted	• Contaminant mass left in place
• Media(s) remediated	• Timeframe from discovery to remediation
• Surface and subsurface geology	• Projected future cost to closure
• Cost of investigation	• Projected future time to closure
• Cost of Remediation	• Regulatory involvement

Contaminants remediated included the following:

• Arsenic	• Chlordane
• Chrome VI	• LNAPL VOCs
• Lead	• DNAPL VOCs
• Mercury	• PAHs
• Perchlorate	• PCBs

Several contaminants were detected but were not at sufficiently high enough concentrations to be the focus or target of remedial action. Some of these contaminants also naturally occur. However, concentrations detected exceed background concentrations and were therefore considered anthropogenic. These contaminants included the following:

• Antimony	• Manganese	• Phenols
• Barium	• Nickel	• Phthalates
• Beryllium	• Selenium	• Other SVOCs
• Cadmium	• Silver	• Asbestos
• Copper	• Zinc	

Of the 370 sites evaluated:

- 259 or 70% had soil contamination only.
- 111 or 30% had soil and groundwater contamination.
- 24 or 6.5% had significant free product.
- All the sites were located on unconsolidated sediments that consisted of glacial, fluvial, lacustrine, alluvial, or marine sediments that were composed of gravel, sand, silt, clay, or mixtures of these materials.

Table 5.4 is a summary of the frequency of detection of contaminants, the frequency of when remediation was required, and the percent of the total cost. Table 5.5 presents a breakdown of sites by media that required remediation by contaminant types. Table 5.6 shows the frequency of detection in the soil of the different contaminant groups evaluated and the frequency of when remediation by contaminant type was required. Table 5.7 shows the frequency of detection in groundwater of the different contaminant groups evaluated and the frequency of when remediation by contaminant types was required. Table 5.8 shows a breakdown of the percent of total cost by media remediated. Table 5.9 shows a breakdown of the cost to remediate a kilogram of contaminant by soil type.

Examination of Tables 5.4 through 5.9 reveals that

- PAHs were the group of compounds most often detected but were remediated just one-third of the time and accounted for only 1% of the total cost. This indicates that while PAHs were often detected, remediation was not always required, and when it was required, the cost was less.
- LNAPL VOCs were the group of compounds detected second most often. Remediation of LNAPL VOCs was required approximately as often as PAHs but at a much higher cost. However, over 90% of the cost to remediate LNAPL VOCs was associated with free product.

TABLE 5.4

Summary of Frequency of Detection, Remediation, and Cost

Contaminant	Frequency of Detection (%)	Frequency of Remediation (%)	Percent of Total Cost (%)
PAHs	85	22.6	1.5
LNAPL VOCs	61	25.9	42.7
DNAPL VOCs	31.4	72.1	39.4
Lead	22	20.0	1.2
PCBs	8	67.0	3.0
Chrome VI	9	80.5	9.7
Mercury	3	82.5	1.8
Perchlorate	0.6	100	0.1
Arsenic	5.2	67	0.4
Chlordane	1.2	100	0.1
Total			100

TABLE 5.5

Site Breakdown by Contaminant Type and Cost

Contaminant	Total Number of Sites	Number of Groundwater Sites	Number of Free Product Sites	Cost (Millions $)
PAHs	103	9	7	11.2
LNAPL VOCs	90	37	17	260.9
DNAPL VOCs	104	65	1	290.2
Lead	26	0	0	7.5
PCBs	19	0	0	28.9
Chrome VI	13	9	0	59.5
Mercury	8	0	0	11.3
Perchlorate	3	2	0	1.1
Arsenic	2	1	0	4.8
Chlordane	2	0	0	0.2
Total	370	113	25	675.6

- Excavation of soil was the preferred remedial option, especially when the volume of soil was less than 5,000 tons and there was no excavation impediment.
- For VOCs, SVE was the preferred remedial option for coarse-grained soils with a large impacted soil volume contaminated or with excavation impediments.
- A combination of alternatives including, risk assessment, institutional controls, capping, and thermal treatment were preferred alternatives for large volumes of fine-grained soil or contaminated soil with excavation impediments.
- A combination of alternatives is most often selected when groundwater was impacted, especially with Cr+6 or DNAPL.
- Groundwater pump and treat is not currently a preferred remedial method for groundwater unless there is a free product or an incompatible overlapping multi-compound contaminant plume (e.g., heavy metals and DNAPL).
- In situ chemical treatment is currently the preferred remedial method for Cr+6 and DNAPL contaminated groundwater (e.g., ZVI, calcium polysulfide, ferrous sulfate).

TABLE 5.6

Breakdown of Soil Remediation

Soil Remediation Technology	Frequency	Frequency of Remediation (%)
Excavation	323	94.5
Risk Assessment	123	35.9
Institutional Controls	108	30.2
SVE	55	16.2
Capping	67	19.6
Thermal Treatment	5	0.9
In Situ Chemical Treatment	5	0.9

TABLE 5.7

Breakdown of Groundwater Remediation

Soil Remediation Technology	Frequency	Frequency of Remediation (%)
Risk Assessment	287	84.1
Institutional Controls	248	72.4
In Situ Chemical Treatment	161	47.1
Pump and Treat	137	40.2
MNA or NA	121	35.6
Air Sparging	11	1.8
Thermal Treatment	3	0.5

TABLE 5.8

Total Cost and Mass Remediated by Media

Remediated Media or Phase	% of Total Cost	% of Total Mass Recovered
Soil	23.4	17.6
Groundwater	42.5	2.1
Free Product	34.1	80.3
Total	100	100

- Cost-effectiveness in remediating groundwater has improved but still remains high.
- Remediating groundwater increased the cost by an average sixfold compared to soil.
- Remediating contaminated groundwater with Cr^{+6} or DNAPL increased the cost more than 10-fold compared to soil.
- The average time to investigate and remediate a site with only soil was 2 years.
- The average time to investigate and remediate a site with groundwater contamination was 13 years.
- Groundwater which accounted for 42.5% of the total cost but recovered only 2.1% of total contaminant mass is by far the least efficient compared to free product and soil.

TABLE 5.9
Cost of Remediating a Kilogram of Contaminant

Contaminant	Geology	Remedial Cost ($/kg)
Chrome VI	Clay	1,000
	Silty Clay	1,500
	Sandy and Silty Clay	1,500
	Sand	81,333
DNAPL	Clay	698
	Silty Clay	1,276
	Sandy and Silty Clay	3,019
	Sand	68,363
LNAPL	Clay	311
	Silty Clay	489
	Sandy and Silty Clay	765
	Sand	567
PCBs	Clay	4,500
	Silty Clay	4,800
	Sandy and Silty Clay	5,600
	Sand	5,900
PAHs	Clay	942
	Silty Clay	817
	Sandy and Silty Clay	309
	Sand	27
Lead	Clay	62
	Silty Clay	218
	Sandy and Silty Clay	535
	Sand	462

The cost of remediating chrome VI and DNAPL VOCs in geologically vulnerable areas results in a cost of remediating a kilogram of contaminant that is greater than $81,000 and $68,000, respectively. The cost of remediating chrome VI and DNAPL VOCs in the least geologically vulnerable areas is more than 50 times less.

Table 5.7 also highlights the differential and cost inefficiency of remediating groundwater. The total cost of remediating groundwater at the sites evaluated totaled $260 million or 42.5% of the total cost yet only recovered or remediated 2.1% of the total contaminant mass. This cost disparity highlights the difficulty in cost-effective remediation of groundwater once it has become impacted to such an extent that active remediation is required.

The cost to remediate a kilogram of PCB is the third-highest for the contaminants evaluated. This elevated cost was due to two factors: (1) PCB remediation is heavily regulated and requires intensive investigation and review and (2) PCBs are often detected in sediments. When PCBs are detected in sediments of a river or stream that supports fish and other aquatic life, the expenses for remediation are usually very high due to the expensive nature of cleaning up river and stream sediments with minimal environmental impact.

The remaining types of contaminants evaluated including LNAPL VOCs, lead, and PAHs are not influenced by geology to the extent that geology influences the remedial cost for chrome VI and DNAPL VOCs. As discussed above and in Chapter 8, these contaminants do not exhibit sufficiently high mobility or persistence which results in a lower CRF_{GW}. Therefore, these contaminants are more cost-effective to remediate on a per kilogram basis.

5.10   SUMMARY AND CONCLUSION

This chapter has highlighted the different remedial approaches for cleaning up our air, water, and soil. The approach for remediating the air and surface water is heavily weighted toward pollution prevention and permitting. This approach has been effective at significantly reducing the amount of contaminants entering the air and surface waters of the United States.

This chapter has also highlighted the fact that chrome VI and DNAPL VOCs are expensive to remediate. A synergistic effect is realized when chrome VI or DNAPL VOCs are released in a geologically vulnerable area. This phenomenon combines elements from his book that have established the significant role that geology and nature play in environmental risk with the other significant factor that influences contaminant risk, that being the physical chemistry of contaminants themselves. Analysis of remedial cost at sites of environmental contamination also demonstrates the importance and predictive power of using CRFs to assist in future green development and redevelopment of urban areas across the United States.

The last section of this book addresses a set of practical watershed approaches for sustainable development using the elements of geology and contamination that we have examined up to this point.

A logical first step in sustainable development involves pollution prevention techniques that can be implemented on a watershed-wide and site-specific basis that greatly depend on geology and contaminant-specific physical chemistry that assist in determining what chemicals should be used, where they should be used, and how they should be used. But first, we must introduce and discuss watersheds as ecosystems, which is the subject of the next chapter.

Questions and Exercises for Discussion

1. A concentration of 90 micrograms per liter is discovered in a municipal water supply well. The compound detected is trichloroethene. The maximum contaminant level (MCL) established by USEPA for trichloroethene is 5 micrograms per liter. List several future actions that should be conducted?

2. A concentration of 4 micrograms per liter is discovered in a municipal water supply well. The compound detected is trichloroethene. The maximum contaminant level (MCL) established by USEPA for trichloroethene is 5 micrograms per liter. List several future actions that should be conducted?

3. You have been requested to review an engineering report that outlines proposed remedial actions that are to take place at a former municipal landfill. What are items in the report that you should look for during your evaluation?

REFERENCES

American Society for Testing Materials (ASTM). 1995. *Standard Guide for Risk-Based Corrective Action.* ASTM Publication E1739-95. West Philadelphia, PA.

California Environmental Protection Agency (CalEPA). 2005. *Use of Human Health Screening Levels (CHHSLs) in Evaluation of Contaminated Properties.* Sacramento, California. 65 p.

California Department of Toxic Substances and Control (DTSC). 2013. Chemical Look-Up Table. Technical Memorandum. Sacramento, California. 5 p.

Cauwenberghe, L. V. 1997. *Electrokinetics. Ground-Water Remediation Technologies Analysis Center.* Pittsburgh. PA.

Eifert, W. H. 2010. *Simple, Reliable and Cost-Effective Solutions for Water Management and Treatment: A Technology Overview with Case Studies.* Roux Associates. New York.

Federal Remediation Technologies Roundtable. 2021. *Soil Vapor Extraction.* http://www.frtr.gov/matrix2/section1/list-of-fig.html (accessed March 31, 2021).

Illinois Environmental Protection Agency (IEPA). 2017. *Tiered Approach to Corrective Action Objectives (TACO).* Springfield, Illinois. 284 p.

Lijzen, J. P., Baars, A. J., Otte, P. F., Rikken, M. G., Swartjes, F. A., Verbruggen, E. M., and van Wezel, A. P. 2001. Technical Evaluation of the Intervention Values for Soil/Sediment and Groundwater. *Institute of Public Health and the Environment.* The Netherlands. 147 p.

Michigan Department of Environmental Quality (MDEQ). 2016. Cleanup Criteria and Screening Levels Development and Application. https://www.michigan.gov/documents/deq/deq-rrd-chem-CleanupCriteriaTSD_527410_7.pdf (accessed March 31, 2021).

National Oceanic and Atmospheric Administration (NOAA). 2021. *Spill Response and Restoration. Office of Response and Restoration.* NOAA's National Ocean Service. http://response.retoration.noaa.gov (accessed March 31, 2021).

Netherlands Ministry of Infrastructure and the Environment. 2015. Dutch Pollutant Standards. https://www.government.nl/ministry-of-infrastructure-and-the-environment. or at http://www.sanaterre.com/guidelines/dutch.html (accessed October 29, 2021).

New Jersey Department of Environmental Protection (NJDEP). 2008. *Introduction to Site-Specific Impact to Ground Water Soil Remediation Standards Guidance Document.* Trenton, New Jersey. 10 p. https://www.nj.gov/dep/srp/guidance/rs/igw (accessed March 31, 2021).

New Jersey Department of Environmental Protection (NJDEP). 2015. Site Remediation Program: Remediation Standards. http://www.nj.gov/dep/srp/guidance (accessed March 31, 2021).

New Jersey Department of Environmental Protection (NJDEP). 2017. Soil Cleanup Criteria. https://www.nj.gov.srp/guidance/scc (accessed March 31, 2021).

Payne, F. C., Quinnan, J. A. and Potter, S. T. 2008. *Remediation Hydraulics.* CRC Press. Boca Raton, FL.

Rogers, D. T. 1997. The Influence of Groundwater on Surface Water in Michigan's Rouge River Watershed. Proceedings of the American Water Resources Association, Conjunctive Use of Water Resources; Aquifer Storage and Recovery Conference, Long Beach, CA. pp. 173–180.

Rogers, D. T. 2018. Derivation of a Comprehensive Environmental Risk Model for Urban Groundwater Protection. *International Association of Hydrogeologists Congress.* Vol. 1. p. 879. Daejeon, Korea.

Rogers, D. T. and Murray, K. S. 1997. Occurrence of Groundwater in Metropolitan Detroit. In: J. Chilton et al. *Groundwater in the Urban Environment.* Volume 1. Problems, Processes, and Management. Balkema Publishers. The Netherlands. pp. 155–161.

Rogers, D. T. Kaufman, M. M., and Murray, K. S. 2006. Improving Environmental Risk Management Through Historical Impact Assessments. *Journal of Air and Waste Management.* Vol. 56. pp. 816–823.

Rogers, D. T., Kaufman, M. M., and Murray, K. S. 2008. Empirical Analysis of Contaminant Risk at Brownfield Sites. United States Environmental Protection Agency National Brownfield's Conference Proceedings. Detroit, MI.

Rogers, D. T., Kaufman, M. M., and Murray, K. S. 2009. An Analysis of Remedial Technology Effectiveness at Brownfield Sites. Association of American Geographers Annual Meeting. Abstracts with Programs. Las Vegas, NV.

Sediments Focus Group. 2007. Guide to the Assessment and Remediation of State-Managed Sediment Sites. *Association of State and Territorial Solid Waste Management Officials.* USEPA. Washington, D.C.

United States Environmental Protection Agency (USEPA). 1995. *Remediation Technologies for LNAPL Contaminated Groundwater.* USEPA. Office of Underground Storage Tanks. Washington, D.C.

United States Environmental Protection Agency (USEPA). 1998. *Technical Protocol for Evaluating Natural Attenuation of Chlorinated Solvents.* USEPA. EPA/600/R-98/128. Office of Research and Development: Washington, D.C.

United States Environmental Protection Agency (USEPA). 1999a. *Phytoremediation Resource Guide.* USEPA. EPA 542-B-99-003. Office of Solid Waste and Emergency Response. Washington, D.C.

United States Environmental Protection Agency (USEPA). 1999b. *Groundwater Cleanup: Overview of Operating Experience at 28 Sites.* USEPA. EPA 542-R-99-006. Office of Solid Waste and Emergency Response. Washington, D.C.

United States Environmental Protection Agency (USEPA). 2000. Proceedings of the Ground-Water Surface-Water Interactions Workshop. USEPA. EPA/542/R-00–007. Office of Solid Waste and Emergency Response. Washington, D.C.

United States Environmental Protection Agency (USEPA). 2001a. *A Citizen's Guide to Bioremediation.* USEPA. EPA 542-F-01-001. Office of Solid Waste and Emergency Response. Washington, D.C.

United States Environmental Protection Agency (USEPA). 2001b. *A Citizen's Guide to Monitored Natural Attenuation.* USEPA. EPA 542-F-01-004. Office of Solid Waste and Emergency Response. Washington, D.C.

United States Environmental Protection Agency (USEPA). 2001c. *A Citizen's Guide to Soil Washing.* USEPA. EPA 542-F-01-008. Office of Solid Waste and Emergency Response. Washington, D.C.

United States Environmental Protection Agency (USEPA). 2001d. *A Citizen's Guide to In Situ Thermal Treatment*. USEPA. EPA 542-F-01-0012. Office of Solid Waste and Emergency Response. Washington, D.C.

United States Environmental Protection Agency (USEPA). 2001e. *A Citizen's Guide to Fracturing*. USEPA. EPA 542-F-01-0015. Office of Solid Waste and Emergency Response. Washington, D.C.

United States Environmental Protection Agency (USEPA). 2001f. *A Citizen's Guide to Vitrification*. USEPA. EPA 542-F-01-0015. Office of Solid Waste and Emergency Response. Washington, D.C.

United States Environmental Protection Agency (USEPA). 2001g. *A Citizen's Guide to Chemical Dehalogenation*. USEPA. EPA 542-F-01-0010. Office of Solid Waste and Emergency Response. Washington, D.C.

United States Environmental Protection Agency (USEPA). 2001h. *A Citizen's Guide to Pump and Treat*. USEPA. EPA 542-F-01-0025. Office of Solid Waste and Emergency Response. Washington, D.C.

United States Environmental Protection Agency (USEPA). 2001i. *A Citizen's Guide to Soil Flushing*. USEPA. EPA 542-F-01-0011. Office of Solid Waste and Emergency Response. Washington, D.C.

United States Environmental Protection Agency (USEPA). 2001j. *A Citizen's Guide to Soil Vapor Extraction and Air Sparging USEPA*. EPA 542-F-01-006. Office of Solid Waste and Emergency Response. Washington, D.C.

United States Environmental Protection Agency (USEPA). 2001k. *A Citizen's Guide to Permeable Reactive Barriers*. USEPA. EPA 542-F-01-005. Office of Solid Waste and Emergency Response. Washington, D.C.

United States Environmental Protection Agency (USEPA). 2002. *A Citizen's Guide to Cleanup Methods*. USEPA. EPA 542-F-01-007. Washington, D.C.

United States Environmental Protection Agency (USEPA). 2003. *Evaportranspiration Landfill Cover Systems Fact Sheet*. USEPA. EPA 542-F-03-015. Office of Solid Waste and Emergency Response. Washington, D.C.

United States Environmental Protection Agency (USEPA). 2005a. *Road Map to Understanding Innovative Technology Options for Brownfield Investigation and Cleanup, Fourth Edition*. USEPA. EPA 542-B-05-001. Office of Solid Waste and Emergency Response. Washington, D.C.

United States Environmental Protection Agency (USEPA). 2005b. *Cost and Performance Report for LNAPL Recovery*. Office of Solid Waste and Emergency Response. USEPA. EPA 542-R-05-016. Washington, D.C.

United States Environmental Protection Agency (USEPA). 2005c. *Contaminated Sediment Remediation Guidance for Hazardous Waste Sites*. USEPA. EPA-540-R-05-012. Office of Solid Waste and Emergency Response. Washington, D.C.

United States Environmental Protection Agency (USEPA). 2007. *Treatment Technologies for Site Cleanup: Annual Status Report*. 12th Edition. Office of Solid Waste and Emergency Response. USEPA. EPA-542-R-07-012. Washington, D.C.

United States Environmental Protection Agency (USEPA). 2008a. *EPA's Report on the Environment*. EPA/600/R-07/045F. Washington, D.C.

United States Environmental Protection Agency (USEPA). 2008b. *Latest Findings on National Air Quality: Status and Trends Through 2006*. USEPA. EPA454/R-07-007. Research Triangle Park. NC.

United States Environmental Protection Agency (USEPA). 2019a. *Hudson River PCBs Superfund Site*. http://hunsondredgingdata.com/Monitoring/Water?currentweek=08–16-2002 (accessed March31, 2021).

United States Environmental Protection Agency (USEPA). 2019b. *National Pollutant Discharge Elimination System (NPDES)*. http://www.epa.gov/npdes/index.cfm (accessed March 31, 2021).

United States Environmental Protection Agency (USEPA). 2021. *Regional Screening Levels*. https://www.epa.gov/RSLs (accessed October 29, 2021).

6 The Challenges and Opportunities of Addressing Climate Change

6.1 INTRODUCTION

For the last few decades, the scientific community in the United States and the world did not generally have difficulty in accepting the science of climate change. However, many governments and politicians did have difficulty. This difficulty did not necessarily focus on disputing the science as much as it focused on dealing with potential economic consequences, fairness between countries, coordinating efforts at the global level, and threats to political systems and perhaps individual careers. This is evidenced by the fact that the United States did not join the Kyoto Protocol more than 20 years ago. The United States did sign on to the Paris Agreement in 2015 under the President Obama administration only to be taken back under the Trump administration and then returned under the Biden administration in January 2021. The result has been lost opportunities to lower greenhouse gas (GHG) emissions and will likely be significant because the most recent report on climate change from the Intergovernmental Panel on Climate Change (IPCC) (IPCC 2021a), states:

> "It is unequivocal that human influence has warmed the atmosphere, ocean and land. Widespread and rapid changes in the atmosphere, ocean, cryosphere and biosphere have occurred."

The above statement has significant implications because it more than suggests that GHG emissions from humans have negatively impacted life on Earth. From a historical perspective, the United States is the world's largest emitter of GHGs (USEPA 2021a). Pressure is mounting for the world to take significant action to reduce GHG emissions so that the consequences can be limited. If not, the result could be cataclysmic.

In this chapter, we shall first examine the history of GHG emissions in the United States and the world, and then we will discuss regulations and treaties that address GHGs. We shall conclude with a summary of GHG regulations and what potentially may occur in the near term with respect to future alternatives and regulations that will likely impact all of us.

6.2 THE CHALLENGES AND HISTORY OF CLIMATE SCIENCE

A Newsweek article in April 1975 titled "The Cooling World" described scientific evidence for an impending ice age. Global temperatures had cooled by a half of degree Fahrenheit since 1940 and there was support from some scientific studies that suggested that the Earth was cooling. However, this was not the case. This example demonstrates the difficulties in doing good science. While the data used in the article may have been correct, it did not tell the whole story, and the conclusion was wrong. The basic reason was that not all potential climate influences were examined and the data set covered time a period that was too short to conclude a long-term climate shift (International Society of Environmental Ethics 2021).

Perhaps, the most significant challenge of climate science over the past few decades relates to the complexity of climate science, collecting reliable data worldwide, collecting the right data, and collecting the data over a long enough period of time such that the influences of cause and effect factors become evident and can be studied directly. Arguably, reliable worldwide data on

DOI: 10.1201/9781003175810-6

numerous individual variables have only existed for a few decades. In addition, variations in GHG emissions are not the only anthropogenic factors. Other factors such as land use and aerosols also play a role in the net change in Earth's energy balance from anthropogenic sources (USEPA 2021b). Factors that have influenced climate in the past from non-anthropogenic sources are numerous, a few include the following (USGS 2021):

- Position of the continents
- Patterns of ocean currents
- Changes in reflectivity of solar energy
- Solar energy variations
- Snow pack
- Rainfall
- Rainfall patterns
- Cloud and weather patterns
- Volcanic activity
- Methane sinks and releases
- Patterns of air current
- Presence or absence of mountains
- Types and amounts of vegetation
- Wildfires
- Meteor impacts
- Amount of water vapor in the atmosphere
- Amount of particulate matter in the atmosphere
- Other factors

Paleoclimate studies have evaluated atmospheric climate variations in temperature and carbon dioxide atmospheric content that with some reliably reach back 800,000 years. The results show that there have been significant natural variations (see Figure 6.1). Figure 6.2 shows the amount of solar energy compared to observed temperatures.

Monitoring and measuring anthropogenic influences on climate have recently been possible based on an examination of the entire data set and is presented in Figure 6.3 (USEPA 2021b).

As additional data is collected, evaluated, and carefully validated, climate models will continue to improve. However, it seems clear and convincing that anthropogenic influences go well beyond affecting just our climate but also impact air quality, water quality, the oceans, and land.

6.3 HISTORY OF GHGS AND CLIMATE CHANGE REGULATIONS IN THE UNITED STATES

As discussed in Chapter 2, GHGs include carbon dioxide (CO_2), methane (CH_4), nitrous oxide (N_2O), and fluorinated gases (F-gases). GHGs cause the Earth to warm up and since the start of the industrial revolution in the early 18th century, levels of GHGs in the Earth's atmosphere have increased (USEPA 2021b). Figure 6.4 shows this increase in GHGs from the National Oceanic and Atmospheric Administration (NOAA) global air sampling network since 1975 (NOAA 2021a).

In the United States and as discussed briefly in Chapter 2, the significant sources of GHG emissions include the following:

- Transportation
- Electrical generation plants
- Industry
- Commercial and residential sources
- Agriculture

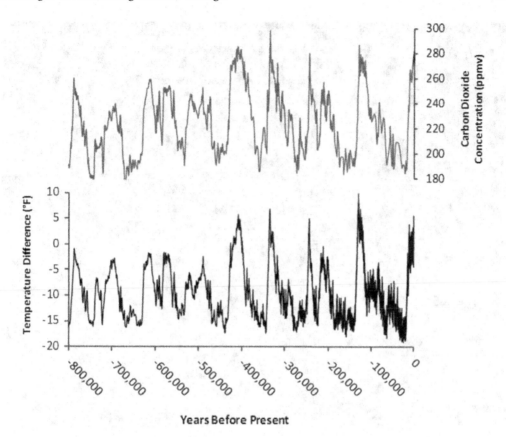

FIGURE 6.1 Paleoclimate carbon dioxide and temperature variations.

USGS. 2021. Climate Research and Development Program. Climate Research and Development Program (usgs.gov) (accessed May 10, 2021).

Figure 6.5 shows the contribution of each measure in 2018 (USEPA 2021a).

The NOAA Earth System Research Laboratory is a cooperative air sampling network where GHGs are measured weekly in ambient air. Figure 6.6 shows the monitoring and sampling stations across the globe (NOAA 2021a).

Globally, GHG emissions began to emit in earnest from Anthropogenic sources at the beginning of the industrial revolution in the early 18th century when coal (a fossil fuel) became the fuel of choice. Since around 1900, gasoline (a fossil fuel) was added and quickly became the preferred fuel, especially with the invention of the internal combustion engine and the automobile. The automobile then became affordable and with the construction of the interstate transportation system in the 1950s, GHGs from mobile sources greatly increased. This is certainly depicted in Figure 6.7, which is a graph that shows global carbon emissions from fossil fuel use since 1900 (USEPA 2021b). Globally since 1970, carbon dioxide emissions have increased by approximately 90%, with emissions from fossil fuel combustion and industrial processes contributing approximately 78% of the total GHG emissions increase from 1970 to 2011. Agriculture, deforestation, and other land-use changes have been the second-largest contributors (USEPA 2021b).

The United States is no longer the largest emitter of GHGs. That dubious distinction belongs to China. The United States as of 2014, emits the second most GHG with the European Union in third place. Emissions and sinks related to changes in land use are not included in these estimates. However, changes in land use can be important. Current estimates indicate that the net global GHG

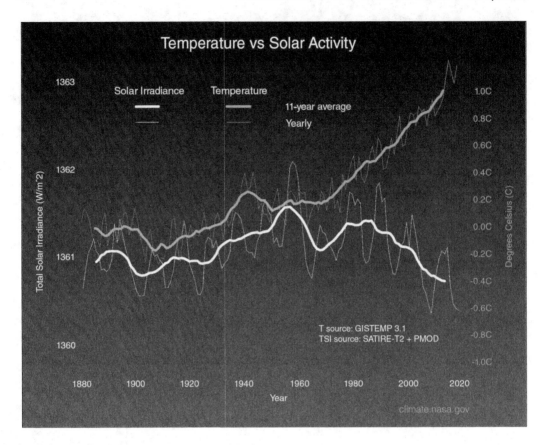

FIGURE 6.2   Global temperature vs. solar activity.

NASA. 2020. Global Temperature and Solar Activity. click to see the linkhttps://www.climate.nasa.gov (accessed December 10, 2020).

emissions from agriculture, forestry, and other land use exceed 8 billion metric tons of carbon dioxide equivalent or about 24% of total GHG emissions. Specifically, in the United States and in Europe, changes in land use associated with human activities have the net effect of absorbing carbon dioxide, which partially offsets the emission from deforestation and rapid urbanization in other parts of the world (USEPA 2021b).

The path leading up to regulating GHGs in the United States began within the growing concern within the scientific community since the 1970s but did not gain enough traction until 1999 when the International Center for Technology Assessment (ICTA) and several other organizations filed a petition for rulemaking to the USEPA requesting that the agency regulate GHG emission from new automobiles under the CAA. USEPA then argued that GHG emissions did not meet the definition of "air pollutant" under the CAA and that GHG emissions contributed to the pollution that is reasonably anticipated. It was not until a decision by the Supreme Court on April 2, 2007, that found that the USEPA had the authority to regulate GHG emissions from new motor vehicles under the CAA (National Association of Clean Air Agencies 2013, USEPA 2021a).

In April 2009, USEPA proposed a finding that GHG emissions endangered public health and welfare and that GHG emissions from new motor vehicles and new motor vehicle engines were contributing to air pollution that endangers public health the welfare. In December 2009, USEPA published the final finding that has come to be known as "The Endangerment Finding" (USEPA 2021c).

The Endangerment Finding not only had consequences for mobile sources of GHGs but also for stationary sources. In 2011, USEPA declared that GHGs were subject to regulation under the CAA

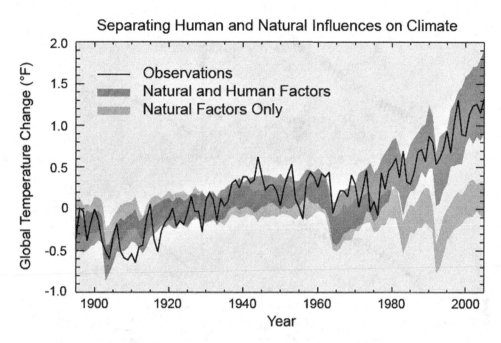

FIGURE 6.3 Separating human and natural influences on climate.

USEPA. 2021b. Global Greenhouse Gas Emissions Data. Global Greenhouse Gas Emissions Data | Greenhouse Gas (GHG) Emissions | US EPA (accessed February 3, 2021).

for stationary sources for PSD and Title V permitting (USEPA 2021a). The PSD permitting process applies to any source on a specified list of 28 source categories, which will be discussed in the next chapter.

6.3.1 GLOBAL CLIMATE PROTECTION ACT

The Global Climate Protection Act was passed in January 1987. The Act directed the President of the United States to establish a task force to conduct research to assist in the development and implementation of a national strategy on global climate. The task force was to report to the President within one year and the President was to inform specified members of Congress on the content of the information. The Act also directed the President to appoint an ambassador-at-large to coordinate Federal efforts in multilateral activities that relate to climate change and global warming. In addition, the Act also urged the President to give climate protection a high priority in dealing with the United States and Russia relations (United States Congress 2021).

Perhaps the most significant contribution of the Global Climate Change Protection Act was to bring to the eyes of the nation from the appointed task force was that GHG emissions may be affecting and substantially increasing Earth's average temperature and, though the consequences may not be fully known, the impact may be irreversible. From this finding from the task force, the President and Congress notified the USEPA to develop a coordinated national policy on global climate change and also ordered the Secretary of State to coordinate diplomatic efforts to combat global warming (United States Congress 2021).

In a report to Congress in 1991, USEPA stated that in order to address global climate change the following elements must be addressed (USEPA 1991):

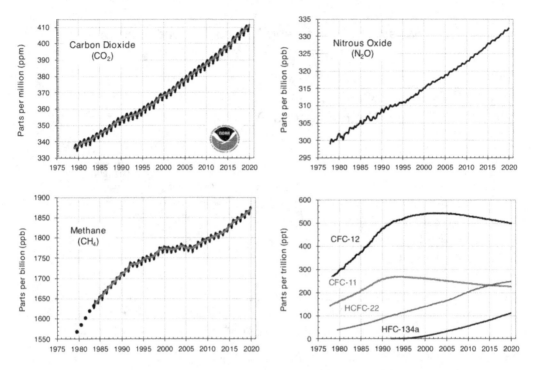

FIGURE 6.4   Global average of GHGs in earth's lower atmosphere.

NOAA. 2021a. The NOAA Annual Greenhouse Gas Index (AGGI). NOAA/ESRL Global Monitoring Laboratory - The NOAA Annual Greenhouse Gas Index (AGGI) (accessed February 4, 2021).

- Scientific research on climate change to reduce uncertainties with respect to operation of the climate system and to improve understanding of the impacts of human activities on future climate conditions and responses
- Assessment of environmental, social, and economic impacts of climate change
- Evaluation of policy options and practices to limit, mitigate, or adapt to climate change, including assessment of effectiveness and social and economic impacts
- Development and implementation of feasible and cost-effective policies and practices

USEPA stated in its report to Congress in 1991 that the nature of human existence on the Earth has changed dramatically during the last century. The world population has increased threefold. However, this was in 1991, now in 2017 (26 years later) the human population is estimated at over 7.5 billion people as opposed to just over 5 billion in 1991, which is an increase of 2.5 billion humans (United States Census Bureau 2017). USEPA (1991) further states that industrial production has increased by a factor of 50 and the world economy has increased by a factor of 20. In all, USEPA concluded that human activity is changing the features of Earth and the chemistry of the atmosphere.

Since the Act was passed in 1987, USEPA has supported and produced hundreds of reports and academic journal articles that have been valuable in building our understanding of causes, impacts, and potential solutions to climate change. The CAA gave USEPA the authority to regulate emissions from power plants and other large sources of GHGs, and now requires facilities that are large emitters of GHGs to report emissions to USEPA.

USEPA has developed and implemented voluntary programs to cut GHG emissions through the following (USEPA 2021d):

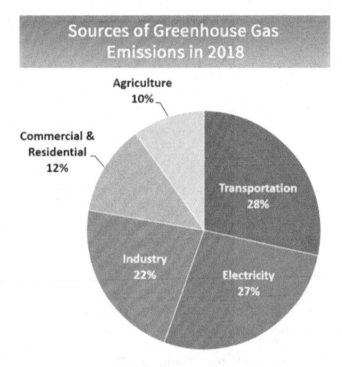

FIGURE 6.5 Source of GHG emissions in the United States as compiled in 2018.

USEPA. 2021a. Inventory of U.S. Greenhouse Gas Emissions and Sinks. Inventory of U.S. Greenhouse Gas Emissions and Sinks | Greenhouse Gas (GHG) Emissions | US EPA (accessed February 2021).

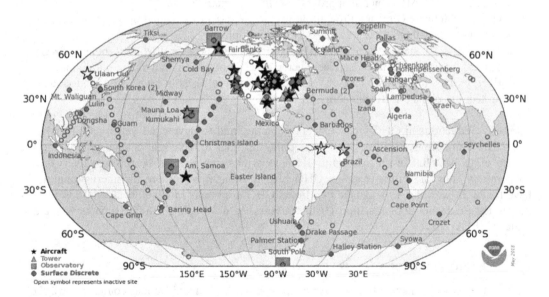

FIGURE 6.6 Global monitoring and sampling locations for GHGs.

NOAA. 2021a. The NOAA Annual Greenhouse Gas Index (AGGI). NOAA/ESRL Global Monitoring Laboratory – The NOAA Annual Greenhouse Gas Index (AGGI) (accessed February 4, 2021).

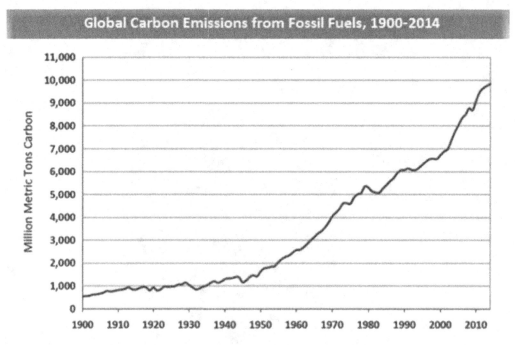

FIGURE 6.7 Global carbon emissions from fossil fuels, 1900–2014.

**USEPA. 2021b. Global Greenhouse Gas Emissions Data. Global Greenhouse Gas Emissions Data |
Greenhouse Gas (GHG) Emissions | US EPA (accessed February 3, 2021).**

- Natural Gas STAR, which is a program to limit methane emissions
- Coalbed Methane Outreach Program to encourage mine owners and operators to capture
  methane rather than allow its escape in the atmosphere
- Environmental stewardship partnership programs to address the most potent GHGs
  emitted from the aluminum, semiconductor, refrigerant, power, and magnesium industries
- Energy Star Program, which is a voluntary program to assist businesses and individuals
  protect the climate through energy efficiency

Currently, reductions in GHG emissions are voluntary and Congress has yet to pass legislation
to cut GHG emissions. USEPA (1991) stated that in order to have real change in reducing the
effects of global climate change from GHG emissions that it must be through international
cooperation. Thus far, efforts to reduce GHG emissions under an enforceable international treaty
have failed, in part, due to language that limits GHG emissions in developed countries such as
the United States but does not limit GHG emissions in developing nations, such as China and
India (USEPA 2021c).

6.3.2 Kyoto Protocol

The Kyoto Protocol is named for the Japanese city in which a United Nations conference was held
and where the protocol was adopted in 1997. The full name of the treaty is referred to as the United
Nations Framework Convention on Climate Change (UNFCCC) (United Nations 2021a).

The treaty is aimed at reducing emissions of GHGs that contribute to global warming and
climate change and has been in force since 2005. The treaty called for the reduction of six GHGs to
a level of 5.2% below 1990 levels in 41 countries and the European Union.

The treaty commits industrialized countries and economies in transition to limit and reduce GHG emissions (United Nations 2021a).

According to the Intergovernmental Panel on Climate Change, which was established by the United Nations Environment Programme and the World Meteorological Organization in 1988, the long-term effects of global warming include the following (United Nations 2021a):

- A rise in sea level worldwide results in the inundation of low-lying coastal areas and the possible disappearance of some island states
- The melting of glaciers, sea ice, and Arctic permafrost
- An increase in the number and severity of climate-related events, such as floods and droughts, and changes in their distribution
- An increased risk of extinction for 20 or more percent of all plant and animal species

The protocol provided several means for countries to reach their targets. One such approach was to create or enhance natural processes, called "carbon sinks," that remove GHGs from the atmosphere. The planting of trees, which removes carbon dioxide from the atmosphere, is just but one example.

Another approach was the international program called Clean Development Mechanism (CDM), which encouraged developed countries to invest in technology and infrastructure in less-developed countries, where there were often significant opportunities to reduce GHG emissions. Under the CDM, the investing country could claim the effective reduction in emissions as a credit for meeting its own obligations under the protocol. An example would be an investment in a cleaner-burning natural gas power plant or in zero-emission technology, such as wind or solar, to replace a coal-fired power plant (United Nations 2021a).

A third approach was the concept of emission trading, which allowed participating countries to purchase and sell emissions rights and thereby placed an economic value on GHG emissions. Many European countries initiated an emission-trading market as a mechanism to work toward meeting their commitments under the protocol. Countries that failed to meet their emissions targets would be required to make up the difference between their targeted and actual emissions, plus a penalty amount of 30%. In addition, each country not in compliance would be prevented from participating in emissions trading united they were evaluated to be in compliance with the protocol (United Nations 2021a).

The Kyoto Protocol represents a significant diplomatic accomplishment but was doomed to fail before it had a chance since China and the United States, the two countries with the most GHG emissions, were not bound by the Protocol. This was because China was a developing nation and the United States did not ratify the protocol. In addition, some within the scientific community believed that reductions from other countries would not significantly reduce overall GHG emissions to a level that would be meaningful. Lastly, some developing countries argued that improving adaptation to the effects of climate change variability and change was just as important as reducing GHG emissions (United Nations 2021a).

Two important points to note are through an examination of Figures 6.8 and 6.9, where as depicted in both figures, GHG emissions from the United States in 1990 compared to GHG emissions in 2018 are nearly identical. This is significant and indicates that the rise in GHG emissions worldwide has not been from the United States over the last 28 years. The second point is that the United States appears to be within approximately 10% of the GHG emissions allowed under the Kyoto Protocol.

6.3.3 Paris Climate Agreement

The Paris Climate Agreement or COP21 is named for the city in which a United Nations Convention on Climate Change took place. Delegates at the conference adopted a concept to

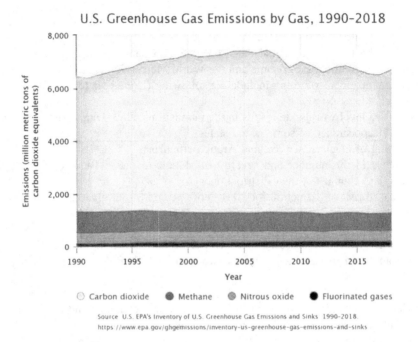

FIGURE 6.8   United States GHG emissions by compound.

USEPA. 2021a. Inventory of U.S. Greenhouse Gas Emissions and Sinks. Inventory of U.S. Greenhouse Gas Emissions and Sinks | Greenhouse Gas (GHG) Emissions | US EPA (accessed February 20, 2021).

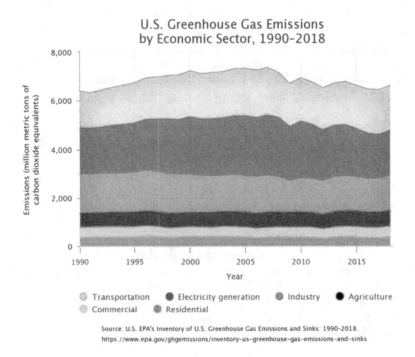

FIGURE 6.9   Global carbon emissions from fossil fuels from 1900 to 2018.

USEPA. 2021a. Inventory of U.S. Greenhouse Gas Emissions and Sinks. Inventory of U.S. Greenhouse Gas Emissions and Sinks | Greenhouse Gas (GHG) Emissions | US EPA (accessed February 20, 2021).

improve on the Kyoto Protocol to reduce the emission of gases that contribute to climate change. It entered on November 4, 2016, and has been signed by 195 countries and ratified by 190 as of January 2021. The objective of the Paris Climate Agreement was to limit GHG emissions to levels that would prevent the global temperature from increasing more than 2°C (3.6°F) above the temperature benchmark set before the beginning of the Industrial Revolution in the 18th century (United Nations 2021b).

Before the conference in Paris, the United Nations asked participating countries to submit detailed plans on how each country intended to reduce GHG emissions. A total of 185 countries submitted plans. The United States stated that its intention was to reduce GHG emissions from between 26 and 28% below 2005 levels and that this was to achieve by the year 2025. The United States stated that accomplishing this goal was to be conducted through a national Clean Power Plan that would set limits on existing and planned power plant emissions. China set its target for peaking in GHG emissions on or before 2030 and would make its best efforts to peak as early as possible. China's plan was to lower GHG emissions per unit of gross domestic product (GDP) by 60 to 65% from their 2005 level of GHG emissions. India's plan was to reduce GHG emissions by 33 to 5% by 2030 as compared to 2005 GHG emission levels (United Nations 2021b). This was viewed as a robust plan and target especially given that India is considered a developing nation with 24% of the global human population and that the plan would require an estimated investment of 2.5 trillion US dollars (United Nations 2021b).

The ultimate goal was to achieve a balance in GHG emissions by 2050 between atmospheric inputs and GHGs from Anthropogenic sources and removal of GHGs into sinks. The Paris Climate Agreement also recognized that the developing countries need to improve their economies and reduce poverty, which made reducing GHG emissions immediately more difficult. The test of the Paris Climate Agreement had no mechanism of enforcement but emphasized cooperation, transparency, flexibility, and progress updates (United Nations 2021b).

Three main issues arose not necessarily from the time of the Paris Climate Agreement itself in 2015 but in the time since the 2015 conference and include the following (United Nations 2021b):

- Many countries, especially those that are considered island states and are developing nations, which are the most vulnerable nations to sea-level rise from an economic and location point of view, wanted to restrict warming to 1.5°C (2.7°F).
- Developed nations did not want to be the only nations paying the cost of reducing GHG emissions.
- Many in the scientific community believe that in order to successfully achieve a global temperature rise of only 2°C (3.6°F) above the temperature benchmark set before the beginning of the Industrial Revolution in the 18th century, the carbon dioxide concentration globally must be below 400 parts per million (ppm). However, the carbon dioxide concentration of 400 ppm had been exceeded shortly after the Paris Climate Agreement was put into place and continues to rise as of 2020 (NOAA 2021b).

6.3.4 THE CLEAN AIR ACT

As briefly discussed above, the USEPA concluded that under section 202(a) of the CAA that GHGs threaten both public health and public welfare and that GHG emissions from motor vehicles have contributed to that threat (USEPA 2011). This action by USEPA had two distinct findings which are (USEPA 2021c):

1. The **Endangerment Finding** in which the Administrator concluded that the mix of atmospheric concentrations of six key, well-mixed GHGs threatens both the public health and the public welfare of current and future generations. These six GHGs are considered

by USEPA to be the compounds that constitute the air pollutants that threaten public health and welfare. The compounds include the following:

a. Carbon dioxide (CO_2)
b. Methane (CH_4)
c. Nitrous oxide (N_2O)
d. Hydrofluorocarbons (HFCs)
e. Perfluorocarbons (PFCs)
f. Sulfur hexafluoride (SF_6)

2. The **Cause or Contribute Finding** in which the Administrator concluded that the combined GHG emissions from new motor vehicles and motor vehicle engines contribute to the atmospheric concentrations of the six listed GHGs and hence to the threat of climate change.

The USEPA issued these endangerment findings in response to a 2007 United States Supreme Court case of Massachusetts v. EPA, when the court ruled that GHGs are air pollutants according to the Clean Air Act (USEPA 2021c). In 2010 and in response to the Endangerment Finding, USEPA required industry to track facility emissions and report GHG emissions if the facilities if the sources of GHG emissions were greater than 25,000 metric tons of carbon dioxide equivalent to the newly formed USEPA Greenhouse Gas Reporting Program under 40 CFR Part 98 (United States Environmental Protection Agency 2013).

GHGs originate from both stationary sources and mobile sources. An example of a stationary source of GHG emissions would be a coal-fired power plant. An example of a mobile source would be an automobile. GHG emissions from mobile sources are now greater than any other source passing electrical generation as of December 2018 (USEPA 2021d). USEPA and the National Highway Traffic Safety Administration (NHTSA) have established economy standards for light-duty vehicles, commercial trucks and buses, aircraft, and federal fleets that reduce the generation of GHG emissions over time (USEPA 2021e).

6.3.5 ENERGY POLICY ACT

The Energy Policy Act of 1992 was enacted to sets goals and mandates to increase clean energy use, reduce GHG emissions, improve overall energy efficiency, and lessen the nation's dependence on imported energy. The Act set standards for many different sectors of the economy including establishing standards for (National Energy Institute 2005):

- Buildings: To improve energy efficiency within existing buildings such as improved insulation, to improve building design for new structures with energy efficiency as a priority
- Utilities: Encouraging efficiency efforts within energy distribution systems.
- Equipment standards: Improving standards for electric motors, lamps, bulbs, commercial heating and air conditioning units, water heaters, etc.
- Renewable energy: Established a program for providing federal support for renewable energy technologies such as wind (see Figure 6.7).
- Alternative fuels: Provided for the use of alternative fuels for automobiles, trucks, and buses such as electric, natural gas, and ethanol additives.

The Act was amended in 2005 to provide tax incentives for alternative energy development such as wind and solar energy (see Figures 6.10 and 6.11). The Act also made it easier for new technology to expand which was oil fracking by exempting waste fluids generated from the Clean Air Act, Clean Water Act, the Safe Drinking Water Act, and CERCLA (National Energy Institute 2005).

An unforeseen disadvantage to wind turbines is that they kill birds, especially raptors, migratory birds, and bats. Some estimates range from 250,000 to 500,000 birds are killed annually in the

FIGURE 6.10   Wind farm in Washington state (Photograph by Daniel T. Rogers).

FIGURE 6.11   Solar array in the California desert east of Los Angeles (Photograph by Daniel T. Rogers).

United States and this will likely increase as more wind turbines are installed and old ones are replaced with wind turbines that are much larger and kill even more birds because of increased reach and spin speed of the turbine (Eveleth 2013, Curry 2017).

6.3.6   Summary of GHG Regulations in the United States

At this point, GHG regulation in the United States has not yet become a reality for many industrial operating facilities but will likely become a reality soon. The exception has been a changing national policy to move away from coal-burning power plants and move toward more renewable sources of energy, which include wind and solar. In addition, electric automobiles are beginning to replace internal combustion engines, which have been part of our transportation needs for more than a century.

Currently, USEPA only requires reporting of GHG emissions for those facilities that emit more than 25,000 metric tons of GHGs.

The majority of the actions that have taken place to this point have not necessarily been mandated under law but have been amending business strategies and individual decisions to limit GHG emissions. Most Individual opportunities to lower impacts from GHGs include the following (USEPA 2021e):

- Conserve energy – turn off appliances and lights when you leave a room
- Recycle paper, plastic, glass bottles, cardboard, and aluminum cans
- Keep woodstoves and fireplaces well maintained. Consider replacing old wood-burning stoves with an EPA-certified model.
- Plant deciduous trees in locations around your home to provide shade in the summer and light into your home during the winter
- Purchase green electricity, if possible, from your utility company
- Wash clothes with warm and cold water instead of hot water
- Lower the thermostat in your home by even just a couple of degrees
- Lower the thermostat on your hot water heater
- Connect your outside lights to a timer
- Use solar lighting when and where possible
- Use low-VOC or water-based paints, stains, finishes, and paint strippers
- Test your home for radon
- Do not smoke in your home
- Purchase "Energy Star" products
- Choose vehicles to purchase that are low-polluting
- Choose consumer products that have less packaging and that are recyclable
- Shop with a reusable canvas bag instead of using paper or plastic bags
- Purchase rechargeable batteries
- Keep vehicle tires properly inflated
- Fill gas tanks in the evening during the summer months
- Avoid spilling gasolines or topping off the tank while refueling
- Attempt to limit or avoid drive-thru lines
- Use public transportation
- Walk
- Ride a bike
- Get regular engine tune-ups and regular maintenance checks
- Use an energy-conserving motor oil
- Request flexible work hours
- Request to work from home, if possible
- Report vehicles with excess exhaust
- Join a carpool
- Check daily air quality forecasts
- Remove indoor asthma triggers
- Avoid outdoor asthma triggers
- Minimize sun exposure

6.4 GLOBAL INVOLVEMENT

The United Nations has played perhaps the most significant role in educating and promoting actions for reducing GHGs throughout the world. The Intergovernmental Panel on Climate Change (IPCC) was established by the United Nations Environment Programme (UNEP) and the World Meteorological

Organization in 1988 (IPCC 2021b). The IPCC is the United Nations body for assessing the science related to climate change. The IPCC is widely considered the world leader in setting the standards for assessing, mitigating, and reducing GHG emissions (IPCC 2021b and NOAA 2021b).

6.4.1 GHG Protocol

In 1998, a multi-stakeholder group that included, IPCC, businesses, nongovernmental organizations (NGOs), and governments formed with the goal of developing an internationally accepted GHG accounting and reporting standards. The first standard was published in 2001 and was subsequently updated in 2015 (World Resource Institute 2015, USEPA 2005, 2020a, 2020b).

The standard, now commonly known as the GHG Protocol, developed a methodology that measures GHG from a company or other organization that involved three steps termed Scopes. Scope 1 is direct emissions of GHG. Scope 2 is upstream GHG producing activities such as electrical generation. Scope 3 involves downstream activities such as transportation, product use, waste disposal, travel, and others. Figure 6.12 is a diagram that demonstrates and shows each scope of the GHG Protocol (World Resource Institute 2015).

6.4.2 Opportunities for Change

Many businesses in the United States and the world now measure and track GHG emissions using the GHG Protocol. USEPA has provided guidance in conducting self-assessments and setting GHG reduction targets for businesses (USEPA 2020b). GHG reduction targets range between 40% and 50% by the year 2030 and carbon neutral by 2050 (USEPA 2020b).

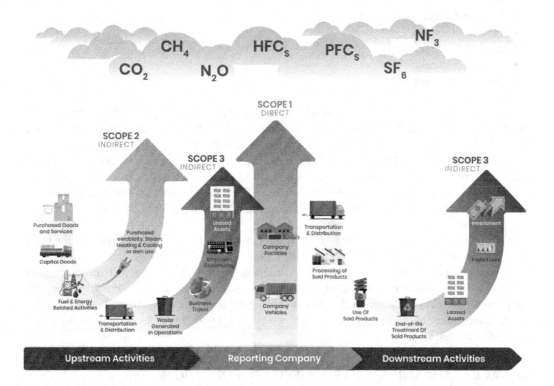

FIGURE 6.12   GHG protocol scope 1, 2, and 3.

World Resource Institute. 2015. Greenhouse Gas Protocol. World Resource Institute. Washington, D.C. 114 p. (accessed July 30, 2021).

Achieving these goals first requires tracking and measuring GHG emissions for each facility using the GHG Protocol. The second step requires identifying those areas where GHG emissions can be reduced. The third step, if necessary, requires providing offsets that remove or reduce carbon emissions, such as planting trees (USEPA 2020b).

Reducing GHG emissions have been a priority for many countries including the United States. USEPA has been instrumental at reducing GHG emissions from automobiles and from electrical generating plants. However, much more needs to be conducted and will likely involve (1) further increases in renewable energy sources such as wind, solar, and others, (2) eliminating carbon-producing transportation and replacement with electric cars, trucks, trains, and planes, (3) modifying the human diet, (4) reducing GHG emissions from agriculture practices.

6.5   HIGH ALTITUDE OZONE

Ozone in the stratosphere, a layer of the atmosphere 10–30 miles above the Earth, serves as a shield, protecting life on the surface of the Earth from ultraviolet radiation. The ozone hole is a human-caused hole in the ozone layer above the south pole, especially during the Southern Hemisphere's spring (National Aeronautics and Space Administration 2021a).

The 1990 amendments to the CAA required USEPA to establish a program for phasing out production and use of ozone-destroying compounds termed ozone-depleting substances (ODS), which were used as aerosol propellants in consumer products such as hairspray and deodorants, and as coolants in refrigerators and air conditioners (USEPA 2021f).

Specific ODS compounds include the following (USEPA 2021f):

- Chlorofluorocarbons (CFCs)
- Hydrochlorofluorocarbons (HFCs)
- Halons
- Methyl bromide
- Carbon tetrachloride
- Methyl chloroform

Scientists have been monitoring ozone levels in the upper atmosphere since the 1970s and in 1987, 190 countries, including the United States and other industrialized nations, signed the Montreal Protocol which called for an elimination of chemicals that destroy the ozone layer.

In 1996, production of ODS in the United States ceased for many of the compounds capable of the most harm such as CFCs, halons, and methyl chloroform. Scientists estimate that it will take another 60 years for the upper atmosphere to recover. The largest observed ozone hole attributed to ODS was estimated to be 28.3 million square kilometers in September 1998 and has been shrinking ever since (NASA 2021b).

In addition to phasing out ODS compounds, the CAA includes other steps to protect the ozone layer. The CAA encourages the development of ozone-friendly substitutes, and because of this, many products have been reformulated. For instance, aerosol propellants and refrigerators no longer use ODS compounds. However, there are a few examples where alternatives for ODS compounds have not been developed. These include some applications in the medical field and an effective substitute for methyl bromide which is a pesticide that is used by farmers in the United States (USEPA 2021f).

6.6   ACID RAIN AND SMOG FORMING AIR POLLUTION

The CAA Amendments of 1990 established what is called an allowance market system is known today as the Acid Rain Program. The Acid Rain Program is a market-based initiative created by the USEPA with the purpose of reducing overall atmospheric levels of the principal acid rain-causing chemicals, which are sulfur dioxide (SOx) and nitrogen oxides (NOx). The program is an implementation of emission trading that primarily targets coal-burning power plants, allowing them to buy and sell emission credits called "allowances" according to individual plant needs and costs (USEPA 2008).

Phase I of the program required that significant reductions be achieved by January 1, 1995, largely requiring 110 electric power generating plants to reduce sulfur dioxide emission rates by 2.5 pounds per million British thermal unit (BTU). Each of the electric power generating plants was identified in the statute by name, address, and quantity of emissions allowances in tons of allowable sulfur dioxide emissions per year. New generating plants built after 1978 were required to limit sulfur dioxide to an emission rate of approximately 0.6 pounds per million BTUs (USEPA 2008). As an incentive for reducing emissions, for each ton of sulfur reduced below the applicable emissions limit, owners of the generating unit received an emissions credit that they could use at another unit or sell. This legitimized a market for sulfur dioxide emission credits and was administered by the Chicago Board of Trade. Power generating units that installed flue-gas desulfurization equipment (e.g., scrubbers) or other qualified equipment, which reduced sulfur emission by 90% qualified for an extension of two years, provided that they already owned allowances to cover their total actual emissions for each year of the extension period (USEPA 2008, 2021g).

Phase II of the program required that all-fossil-fired unit over 75 million megawatts in size were required to limit emissions of sulfur dioxide to 1.2 pounds per million BTUs generated by January 1, 2000. Thereafter, they were required to obtain an emissions credit for each ton of sulfur emitted and were subject to a mandatory fine of $2,000 for each ton emitted in excess of the credits held. USEPA distributes credits equivalent to nearly 9 million tons each year (USEPA 2021g).

To reduce nitrogen oxide (NOx) emissions, many electric power generating plants installed low NOx burner retrofits and reduced NOx emissions by approximately 50%. This technology was readily available, so installation of NOx control was considerably less expensive and easy to install and comply with the required CAA amendments. Since the program was initiated, SOx emissions have been reduced by 40% and acid rain levels have dropped by 65% when compared to 1976 levels. USEPA has estimated that the cost of continued compliance ranges from 1 to 2 billion US dollars per year (USEPA 2021g).

6.7   SUMMARY OF CLIMATE CHANGE AND ITS INFLUENCE ON SUSTAINABILITY REGULATIONS

In the United States, one can say that regulations to reduce the effects of acid rain and high-altitude ozone have been effective. However, GHG emission regulations in the United States are just now being considered and have not yet become a reality for many industrial operating facilities outside of power generating plants and automobile manufacturers. However, this will likely change soon. The exception has been a changing national policy to move away from coal-burning power plants and toward more renewable sources of energy, which include wind and solar, as evidenced by the Energy Policy Act. In addition, electric automobiles are beginning to replace internal combustion engines, which have been part of our transportation needs for more than a century.

When examining the Anthropogenic sectors that emit GHGs, namely transportation and electrical generation, more than 50% of GHG emissions in the United States have been addressed. The remaining Anthropogenic sectors, including residential, commercial, industry, and agriculture have yet to be addressed significantly, if at all. There is now a sense of urgency in reducing GHG emissions globally, especially with the publication of the IPCC (2021) report on climate change and the NOAA State of the Climate report (2021b). Both reports conclude that if reduction does not occur immediately the effects of climate change will be significant.

A likely near-term goal will be a 50% reduction in GHG emissions for businesses in the United States by the year 2030. Some actions that businesses can implement to reduce GHG emissions can

- Reduce business travel
- Change to electric vehicles
- Purchase electricity from renewable sources
- Encourage more employees to work from home
- Conduct more virtual meetings rather than in-person meetings
- Encourage implementing offsets such as planting trees and other actions to remove carbon dioxide from the atmosphere

Furthermore, USEPA has required automobile manufacturers to improve engine emissions which has made a major impact on reducing air pollution and has reduced the amount of GHG emissions. These accomplishments are evidenced by the fact that GHG emissions from the United States have remained relatively constant when comparing GHG emissions from 1990 compared to 2018 (see Figure 6.9). This is significant and can't be understated. However, much improvement can and must be made and soon if humans are to limit climate change to levels agreed upon under the Paris Climate Agreement. USEPA must play a larger role in reducing GHG in the United States along with the IPCC worldwide in the future.

Questions and Exercises for Discussion

1. **What three issues are the most important to you after reading this chapter? Why are they important to you?**
2. **What actions have you undertaken or will undertake in the future to reduce your carbon footprint?**
3. **What organization should be delegated the responsibility to lower GHG emissions worldwide and why?**
4. **What do you think the world will look like 100 years from now?**

REFERENCES

Curry, A. 2017. Will Newer Wind Turbines Mean Fewer Bird Deaths. *The National Geographic*. https://www.nationalgeographic.com/news/energy/2014/04/140427altamont-pass (accessed February 4, 2021).

Eveleth, R. 2013. How Many Birds do Wind Turbines Really Kill? Smithsonian Magazine. Washington, D.C. https://www.smithsoniation.com/smart-new/how-many-birds-do-wind-turbines-kill (accessed July 31, 2021).

Intergovernmental Panel on Climate Change (IPCC). 2021a. Climate Change 2021: The Physical Science Basis. *IPCC AR6 WGI*. 3949 p.

Intergovernmental Panel on Climate Change (IPCC). 2021b. *The History of IPCC. History — IPCC* (accessed July 4, 2021).

International Society for Environmental Ethics (ISEE). 2021. The Cooling World – Newsweek April 28, 1975. The Cooling World – Newsweek, April 28, 1975 | ISEE – International Society for Environmental Ethics (enviroethics.org) (accessed May 10, 2021).

National Aeronautics and Space Administration (NASA). 2020. *Global Temperature and Solar Activity*. https://www.climate.nasa.gov (accessed December 10, 2020).

National Aeronautics and Space Administration (NASA). 2021a. *Largest-Ever Ozone Hole Over Antarctica*. https://visibleearth.nasa.gov/view.php?id=54991 (accessed February 4, 2021).

National Aeronautics and Space Administration (NASA). 2021b. Is the Ozone Layer Causing Climate Change? Is the Ozone Hole Causing Climate Change? – Climate Change: Vital Signs of the Planet (nasa.gov) (accessed February 4, 2021).

National Association of Clean Air Agencies. 2013. Background and History of EPA Regulations of Greenhouse Gas (GHG) Emissions Under the Clean Air Act. Washington, D.C. 19. Microsoft Word – Background and History of EPA Regulation of Greenhouse Gas – Aug 2013 – FINAL (4cleanair.org) (accessed February 3, 2021).

National Energy Institute. 2005. *Summary of the Energy Policy Act of 1992 and 2005*. https://www.nei.org/summary-of-energy-policy-act (accessed February 4, 2021).

National Oceanic and Atmospheric Administration (NOAA). 2021a. The NOAA Annual Greenhouse Gas Index (AGGI). NOAA/ESRL Global Monitoring Laboratory – The NOAA Annual Greenhouse Gas Index (AGGI) (accessed February 4, 2021).

National Oceanic and Atmospheric Administration (NOAA). 2021b. *State of the Climate in 2020. Special Supplement to the Bulleting of the American Meteorology Society.* Vol. 102. no. 8. August 2021. American Meteorology Society, Washington, D.C. 475 p.

United Nations. 2021a. What is the Kyoto Protocol? What is the Kyoto Protocol? I UNFCCC (accessed February 2, 2021).

United Nations. 2021b. The Paris Agreement. The Paris Agreement I United Nations (accessed February 4, 2021).

United States Congress. 2021. Global Climate Protection Act of 1987. S. 420 – 100th Congress (1987–1988): Global Climate Protection Act of 1987 I Congress.gov I Library of Congress (accessed February 4, 2021).

United States Census Bureau. 2017. United States Population through Time and World Populations Clock. https://www.UScencus.gov/USpopulation (accessed July 26, 2021).

United States Environmental Protection Agency (USEPA). 1991. *U.S. Efforts to Address Global Climate Change.* USEPA. Washington, D.C. 87 p.

United States Environmental Protection Agency (USEPA). 2005. Climate Leaders Greenhouse Gas Inventory Protocol Design Principles. *Climate Leaders Greenhouse Gas Inventory Protocol (epa.gov)* (accessed July 30, 2021).

United States Environmental Protection Agency. 2008. *Acid Rain.* https://www.epa.gov/acidrain (accessed February 4, 2021).

United States Environmental Protection Agency. 2011. *PSD and Title V Permitting Guidance for Greenhouse Gases.* Office of Air and Radiation. Washington, D.C. 94 p.

United States Environmental Protection Agency. 2013. *USEPA Greenhouse Gas Reporting Program.* https://epa.gov/green-house-gas-reporting (accessed March 3, 2017).

United States Environmental Protection Agency (USEPA). 2020a. Causes of Climate Change. Causes of Climate Change I Climate Change Science I US EPA. (accessed May 10, 2021).

United States Environmental Protection Agency (USEPA). 2020b. GHG Inventorying and Target Setting. GHG Inventorying and Target Setting Self-ASsessment: V1.0 (epa.gov) (accessed July 30, 2021).

United States Environmental Protection Agency (USEPA). 2021a. Inventory of U.S. Greenhouse Gas Emissions and Sinks. Inventory of U.S. Greenhouse Gas Emissions and Sinks I Greenhouse Gas (GHG) Emissions I US EPA (accessed February 20, 2021).

United States Environmental Protection Agency (USEPA). 2021b. Global Greenhouse Gas Emissions Data. Global Greenhouse Gas Emissions Data I Greenhouse Gas (GHG) Emissions I US EPA (accessed February 3, 2021).

United States Environmental Protection Agency (USEPA). 2021c. Endangerment and Cause or Contribute Findings for Greenhouse Gases. Section 202(a) of the Clean Air Act. https://www.epa.gov.climatechange/endangerment (accessed February 4, 2021).

United States Environmental Protection Agency (USEPA). 2021d. *Regulations for Greenhouse Gas (GHG) Emissions.* https://www.epa.gov/regulations-emissions-vehicles-and-engines/regulations-greenhouse-gas-ghg-emissions (accessed February 20, 2021).

United States Environmental Protection Agency (USEPA). 2021e. *Air Quality – 2018 National Summary.* https://www.epa.gov/air-trends/air-quality-national-summary (accessed February 20, 2021).

United States Environmental Protection Agency (USEPA). 2021f. *Laws and Regulations.* https://www.epa.gov/laws-regulations/regulations (accessed February 4, 2021).

United States Environmental Protection Agency (USEPA). 2021g. Acid Rain Emission Reductions. *USEPA Archive.* https://www.epa.gov/acid-rain-reductions (accessed February 4, 2021).

United States Geological Survey (USEPA). 2021. Climate Research and Development Program. Climate Research and Development Program (usgs.gov). (accessed May 10, 2021).

World Resource Institute. 2015. *The Greenhouse Gas Protocol.* World Resource Institute. Washington, D.C. 114 p.

7 Protecting the Air through Environmental Regulations

7.1 INTRODUCTION TO AIR POLLUTION REGULATIONS

Many believe that environmental regulations in the United States are the most comprehensive and protective on Earth. In many respects this is true but there are significant exceptions and gaps and we will examine many in this chapter.

The purpose of environmental regulation is to protect human health and the environment. However, protecting human health and the environment is difficult if the regulations are not comprehensive and all-encompassing. To assist in explaining this point, it is best to demonstrate the need for comprehensive environmental regulations through example scenarios. For example, if a person were placed in a situation where there are ten detrimental environmental circumstances that may cause harm to that person and the person is only protected through regulations against nine of the items, that person is still harmed. Some may interpret that the regulations were ineffective in this example because the person was still harmed. This is a common frustration of applying environmental regulations; there always seems to be something no one thought of that leads to an unacceptable exposure. This is one of the reasons why environmental regulations are complex, they have to be in order to be truly protective.

Another challenge is that environmental regulations of the United States only apply to the United States. What if the person in the example above was protected from nine of the ten potential hazards but the one hazard that the person was not protected originated in a different country. This is more difficult to address because our laws only apply to our country. This scenario is central to one of our themes in this book which is *pollution does not respect boundaries*. This is another reason why environmental regulations are so complex and sometimes, ineffective.

The third challenge from our example above would be that the person was harmed not because the regulations were inadequate, but because something new was introduced into the situation that was not present when the regulations took effect which modified the situation or environment enough to cause harm. This is why environmental regulations always seem to need to be updated and improved. New information and facts or new circumstances arise that must be accounted for and addressed. This is yet another example of why environmental regulations are so complex and become more complex with time as our civilization moves forward.

Lastly, environmental regulations are in large part based on single incidents that cause large-scale harm to human health and the environment. Therefore, they are reactive rather than proactive, which tends to narrow the focus of regulations and does not fully address other circumstances that may cause similar harm.

In this chapter, we shall discuss the state of environmental regulations pertaining to air in the United States. We shall also discover that there are areas where additional regulations are necessary or, are totally lacking and are in need of more immediate attention. Factors that influence the effectiveness of environmental regulations vary from country to country and within each country as well. The United States is no different. Factors that influence the degree to which a country has effective environmental regulations include the following (Bates and Ciment 2013, United Nations 2021):

- Political will
- Environmental tragedies, harm to the environment, and lessons learned

DOI: 10.1201/9781003175810-7

- Cost
- Geography and climate
- Enforcement
- Incentives
- Lack of overriding social factors such as war, political structure, greed, poverty, hunger, lack of infrastructure, educational awareness, and corruption

Along with our example situations, each of the above-listed factors either influences the effectiveness of environmental regulations in a positive way or undermines environmental regulations. While reading this chapter, keep in mind the four basic environmental fundamentals, which are

- Protect the air
- Protect the water
- Protect the land
- Protect living organisms and cultural and historical sites

7.2 SIGNIFICANCE OF THE CLEAN AIR ACT

At the time of its inception, The Clean Air Act (CAA) was by far the most complex and lengthy piece of environmental legislation of its kind in the world. The CAA has grown enormously since 1970 into a complex behemoth that comprises tens of thousands of pages of statutory requirements, regulations, and rules (Cornell University Law 2021). The CAA and its amendments, including a significant amendment in 1990, cover numerous topics that have generated intense political controversy. The CAA had its beginnings in the Air Quality Act of 1967 which initially focused on the regulation of ambient air quality to protect public health and welfare. The purpose of the initial law was to (USEPA 2007a, 2007b):

- Protect and enhance the quality of the Nation's resources so as to promote the public health and welfare and the productivity capacity of its population
- Initiate and accelerate a national research and development program to achieve the prevention and control of air pollution
- Provide technical and financial assistance to state and local governments in connection with the development and execution of their air pollution and prevention and control programs
- Encourage and assist the development and operation of regional air pollution control programs

7.3 CLEAN AIR ACT

The centerpiece of this legislation was the development of air quality "criteria," which at the time was headed up by the Department of Health, Education, and Welfare since USEPA had not yet been formed and would not be for a few more years. The goal of the "criteria" was to accurately reflect the latest in scientific knowledge on the health and welfare effects of individual pollutants, such as sulfur dioxide (SOx), nitrogen oxide (NOx), and particulate matter (PM). These criteria would eventually evolve into what will be later described as "criteria pollutants," which are discussed in greater detail later in this chapter (USEPA 2021a, 2021b).

These early experiences with attempting to regulate air pollution revealed many scientific uncertainties and technical difficulties that challenged early regulatory agencies. The criteria development process proved long and cumbersome and States had difficulty translating the information made available into source-specific standards. Among other things, techniques for relating air quality concentrations to source-specific emissions (e.g., atmospheric dispersion models) were not well-developed or were not developed at all. In addition, it became apparent that since air quality is influenced by regional as well as local factors, more sophisticated regulatory and control techniques and technologies were required. Hence, the CAA of 1970 was developed and passed and the United States established the Environmental Protection Agency to administer the regulations with the purpose of protecting human health and the environment (USEPA 2021b).

The CAA attempts to resolve how areas in the United States that have been unable to attain the goal of clean air should be brought into compliance. The Act also describes the penalties that should be applied to communities that do not achieve these goals. It addresses how much more we as a nation should spend on controlling emissions from automobiles, and in what ways the fuels that we burn in our automobiles should be improved. It attempts the almost impossible task to allocate among different regions of the United States the multi-billion dollar burden of reducing emissions that emit or contribute to hazardous air pollutants, acid rain, smog, ozone, ozone layer depletion, and lastly climate change, all the while with an increasing population. Simply put, the CAA has provided standards that govern the day-to-day operation of virtually every industry, car, truck, train, plane, refrigerator, furnace, air conditioner, light bulb, and power plant in the United States (USEPA 2021b).

The primary responsibility for implementation and assuring air quality in their region are each individual state. Each state must submit a plan termed **State Implementation Plan** (SIP) that describes just how each state will achieve requirements mandated in the CAA. The SIP includes detailed information regarding the permitting process for individual facilities and outlines specific source emission requirements, reduction in emissions, record keeping, inspections, and monitoring and testing. Under the CAA, States may further delegate permitting authority to local regulatory authorities, and tribes as long as each qualifies (USEPA 2021c).

The CAA essentially required that every industrial facility that had air emissions to obtain a permit to operate. It required that industrial facilities be designed and operated to prevent or minimize releases of hazardous air pollutants. Those facilities that own or operate larger sources of emissions must not modify those sources or add new sources without first obtaining pre-construction approvals. In addition, industries that offered electricity for sale must purchase or otherwise obtain pollution allowances in order to operate. In addition, the CAA required new or modified emission sources to meet New Source Emission Standards (NSPS) (USEPA 2021c).

The CAA is a comprehensive law that regulates air emissions from stationary and mobile sources. Among other things, the law authorizes the EPA to establish National Ambient Air Quality Standards (NAAQS) to protect public health and public welfare and to regulate emissions of what are considered Criteria Pollutants and Hazardous Air Pollutants (HAPs) (USEPA 2021d). Other provisions in the CAA that we will examine include ozone-depleting substances (ODS), smog forming chemicals, acid rain, and climate change. A more detailed discussion of climate change regulations will be addressed in a stand-alone section later in this book.

7.3.1 Criteria Pollutants and NAAQS Standards

The CAA identifies two types of NAAQS standards and includes **primary standards** which are provided for public health protection, including the protection of sensitive populations such as asthmatics, children, and the elderly. **Secondary standards** provide for public welfare protection, including protection against decreased visibility and damage to animals, crops, vegetation, and buildings (USEPA 2021d). An additional subdivision is **primary and secondary pollutants**. A primary pollutant is an air pollutant emitted directly from a source. A secondary pollutant is a pollutant that is not directly emitted from a source but forms when other pollutants (primary pollutants) react with the atmosphere, such as acid rain, ozone, or what is commonly referred to as smog (USEPA 2021d). Figure 7.1 shows a diagram of primary and secondary air pollution formation.

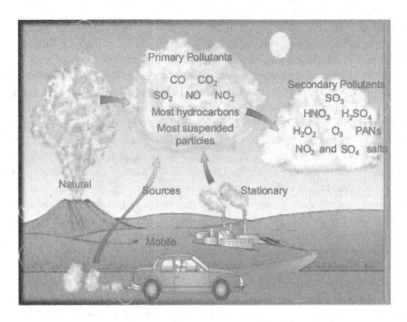

FIGURE 7.1 Source and formation of primary and secondary air pollution.

USEPA Non-attainment areas of the United States. 2021e. www.eps.gov/nonattainmentmap (accessed February 15, 2021).

Criteria pollutants or Criteria Air Pollutants (CAPs) were the first set of pollutants recognized by USEPA as requiring standards established on a national level. USEPA uses the term Criteria Pollutants because it regulates them by developing human health-based or environmentally based criteria (science-based guidelines) for establishing permissible levels. The set of limits based on human health comprises primary standards. Another set of limits intended to prevent environmental and property damage is called secondary standards. A geographic area with air quality lower than the established standards is termed an "attainment area." Conversely, a geographic area with air quality exceeding one or more of the stands is termed "nonattainment area" (USEPA 2021e).

The six criteria pollutants include (USEPA 2021d)

- Ozone (ground level)
- Particulate matter (PM)
- Lead
- Carbon monoxide
- Sulfur oxides (SOx)
- Nitrogen oxides (NOx)

USEPA has established standards for these six principal pollutants under NAAQS. Periodically, the standards are reviewed and are revised when appropriate or necessary. The current standards are presented in Table 7.1. Units of measure for the standards presented in Table 7.1 are parts per million (ppm) by volume, parts per billion (ppb) by volume, and micrograms per cubic meter of air ($\mu g/m^3$) (USEPA 2021d). Particle pollution, also known as particulate matter (PM) includes very fine dust soot, smoke, and droplets that are formed from chemical reactions and are produced when fuels such as coal, wood, or oil are burned (Brownwell and Zeugin 1991).

With the CAA amendments of 1990, PM was divided into two parts, PM10 and PM2.5. PM10 refers to a particle size greater than 10 micrometers in diameter, and PM2.5 refers to a particle size less than 2.5 micrometers in diameter. The logic behind this distinction is that these smaller

TABLE 7.1

NAAQS Table of Select Criteria Air Pollutants

Pollutant		Primary/ Secondary	Time Period	Level	Form
Carbon Monoxide (CO)		Primary	8 hours	9 ppm	Not to be exceeded more than once per year
			1 hour	35 ppm	
Nitrogen Dioxide (NO$_2$)		Primary	Rolling 3-month average	0.15 μg/m^3	98th percentile of 1-hour daily maximum concentrations averaged over 3 years
			1 hour	100 ppb	
		Primary and Secondary	1 year	53 ppb	Annual mean
Ozone		Primary and Secondary	8 hours	0.070 ppm	Annual fourth-highest daily maximum 8-hour concentration averaged over 3 years
Particle Pollution (PM)	PM 2.5	Primary	1 year	12.0 μg/m^3	Annual mean averaged over 3 years
		Secondary	1 year	15.0 μg/m^3	Annual mean averaged over 3 years
		Primary and Secondary	24 hours	35 μg/m^3	98th percentile averaged over 3 years
	PM 10	Primary and Secondary	24 hours	150 μg/m^3	Not to be exceeded more than once per year on average over 3 years

Source: United States Environmental Protection Agency (USEPA). 2021d. NAAQS Table. https://www.epa.gov/criteria-air-pollutants/naaqs-table (accessed February 15, 2021).

particles are more likely to harm our health (USEPA 2021d). One of the goals of the act was to set and achieve NAAQS in every state by 1975 in order to address the public health and welfare risks posed by certain widespread pollutants. The setting of these pollutant standards was coupled with directing the States to develop State implementation plans (SIPs), applicable to appropriate industrial sources in such State, in order to achieve these standards (USEPA 2021e).

Particulate matter includes dust particles that can also be generated through mechanical movements such as vehicles or other equipment. An example is presented in Figure 7.2.

7.3.2 HAZARDOUS AIR POLLUTANTS

The Act was amended in 1977 and in 1990 primarily to set new goals (dates) for achieving the attainment of NAAQS since many areas of the country had failed to meet the deadlines. Figure 7.3 shows counties within the United States that are in nonattainment for one or more hazardous air pollutants as of February 2017 (USEPA 2021e). In addition to the Hazardous Air Pollutants (HAPs) listed above, the 1990 amendments to the CAA added a list of 188 compounds on the USEPA termed **Hazardous Air Pollutants (HAPs)** which are addressed in Section 112 of the amended Act. The list can be viewed in Table 7.2 and at https://www.epa.gov/haps (USEPA 2021f). The list of HAPs is important for air permits as we shall see when we discuss Title V permits. Haps are those pollutants that are known or suspected to cause cancer or other serious human health effects, such as reproductive effects, birth defects, or adverse environmental effects (USEPA 2021f).

FIGURE 7.2 Example of particulate matter generation from mechanical equipment use (Photograph by Daniel T. Rogers).

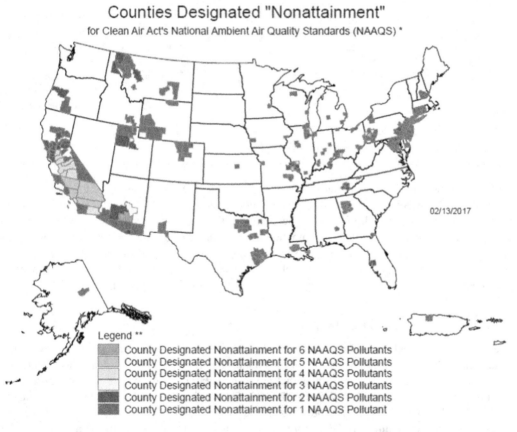

FIGURE 7.3 US counties that are in nonattainment with one or more hazardous air pollutants as of February 2017.

United States Environmental Protections Agency (USEPA). 2021e. Non-attainment areas of the United States. www.eps.gov/nonattainmentmap (accessed February 15, 2021).

TABLE 7.2

Compounds Regulated under the Clean Air Act as Hazardous Air Pollutants (HAPs)

CAS No.	Compound	CAS No.	Compound	CAS No.	Compound
75070	Acteteldehyde	79447	Dimethyl carbomoyl chloride	598920	N-Nitrosomorpholine
60355	Acetamide	68122	Dimethyl formamide	56382	Parathion
75058	Acetonitrile	57147	1,1-Dimethyl hydrazine	82688	Pentachloronitrobenzene
98862	Acetophenone	131113	Dimethyl phthalate	87865	Pentchlorophenol
53963	2-Acetylaminofluorene	77781	Dimethyl sulfate	108952	Phenol
107028	Acrolein	534521	4,6-Dinitro-o-cresol	106503	p-Phenylenediamine
79061	Acrylamide	51285	2,4-Dinitrophenol	75445	Phosgene
79107	Acrylic acid	121142	2,4-Dinitrotoluene	7803512	Phosphine
107131	Acrylonitrile	123911	1,4-Dioxane	7723140	Phosphorus
107051	Allyl chloride	122667	1,2-Diphenylhydrazine	85449	Phthalic anhydride
92671	4-Aminobiphenyl	106898	Epichlorohydrin	1336363	Polychlorinated biphenyls
62533	Aniline	106887	1,2-Epoxybutane	1120714	1,3-Propane sultone
90040	o-Anisidine	140885	Ethyl acrylate	57578	Beta-Propiolactone
1332214	Asbestos	100414	Ethyl benzene	123386	Propionaldehyde
71432	Benzene	51796	Ethyl Carbamate	114261	Propoxur
92875	Benzidine	75003	Ethyl chloride	78875	Propylene dichloride
98077	Benzotrichloride	106934	Ethylene dibromide	75569	Propylene oxide
100447	Benzyl chloride	107062	Ethylene dichloride	75558	1,2-Propylenimine
92524	Biphenyl	107211	Ethylene glycol	91226	Quinoline
117817	Bis(2-ethylhexyl) phthalate	151564	Ethylene imine	106514	Quinone
542881	Bis(chloromethyl) ether	75218	Ethylene oxide	100425	Styrene
75252	Bromoform	96457	Ethylene thiourea	96093	Styrene oxide
106990	1,3-Butadiene	75343	Ethylidene dichloride	1746016	2,3,7,8-Tetrochlorodibenzo-p-dioxin
156627	Calcium cyanamide	50000	Formaldehyde	79345	1,1,2,2-Tetrachloroethane
105602	Caprolactam	76448	Heptachlor	127184	Teterachloroethylene
133062	Captan	118741	Hexachlorobenzene	7550450	Titanium tetrachloride
63252	Carbaryl	87683	Hexachlorobutadiene	108883	Toluene
75150	Carbon disulfide	77474	Hexachlorocyclopentediene	95807	2,4-Toluene diamine
56235	Carbon tetrachloride	67721	Hexachloroethane	584849	2,4-Toluene diisocyanate
463581	Carbonyl sulfide	822060	Hexamethylene-1,6-diisocyanate	95534	o-Toluidine
120809	Catechol	680319	Hexamethylphosphoramide	8001352	Toxaphene
133904	Chloramben	110543	Hexane	120821	1,2,4-Trichlorobenzene
57749	Chlordane	302012	Hydrazine	79005	1,1,2-Trichlrorethane
7782505	Chlorine	7647010	Hydrochloric acid	79016	Trichloroethylene
79118	Chloroacetic acid	7664393	Hydrogen fluoride	95954	2,4,5-Trichlorophenol
532274	2-Chloroacetophenone	7783064	Hydrogen sulfide	88062	2,4,6-Trichlorophenol
108907	Chlorobenzene	123319	Hydroquinone	121448	Triethylamine
510156	Chlorobenzilate	78591	Isophorone	1582098	Trifluralin
67663	Chloroform	58899	Lindane	540841	2,2,4-Trimethylpentane
107302	Chloromethyl methyl ether	108316	Maleic anhydride	108054	Vinyl acetate

(Continued)

TABLE 7.2 (Continued)
Compounds Regulated under the Clean Air Act as Hazardous Air Pollutants (HAPs)

CAS No.	Compound	CAS No.	Compound	CAS No.	Compound
126998	Chloroprene	67561	Methanol	593602	Vinyl bromide
1319773	Cresols/Cresylic acid	72435	Methoxychlor	65014	Vinyl chloride
95487	o-Cresol	74839	Bromomethane	75354	1,1-Dichloroethylene
108394	m-Cresol	74873	Chloromethane	1330207	Xylenes
106445	p-Cresol	71556	1,1,1-Trichloroethane	95476	o-Xylenes
98828	Cumene	78933	Methyl ethyl ketone	108383	m-Xylenes
94757	2,4-D, salts, and esters	60344	Methyl hydrazine	106423	p-Xylenes
3547044	DDE	74884	Methyl iodide	0	Antimony compounds
3347044	Diazomethane	108101	Methyl isobutyl ketone	0	Arsenic compounds
132649	Dibenzofurans	624839	Methyl isocyanate	0	Beryllium compounds
96128	1,2-Dibromo-3-chloropropane	80626	Methyl methacrylate	0	Cadmium compounds
84742	Dibutylphthalate	1634044	Methyl-tert-butyl ether	0	Chromium compounds
106467	1,4-Dichlorobenzene(p)	101144	4,4-Methylene bis(2-chloroaniline)	0	Cobalt compounds
91941	3,3-Dichlorobenzidene	75092	Methylene chloride	0	Coke oven emissions
111444	Dichloroethyl ether	101688	Methylene diphenyl diisocyanate	0	Cyanide compounds
542756	1,3-Dichloropropene	101779	4,4'-Methylenedianiline	0	Glycol ethers
72737	Dichlorvos	91203	Naphthalene	0	Lead compounds
111422	Diethanolamine	98953	Nitrobenzene	0	Manganese compounds
121697	N,N-Dimethylaniline	92933	4-Nitrobiphenyl	0	Mercury compounds
64675	Diethyl sulfate	100027	4-Nitrophenol	0	Fine mineral fibers
119904	3,3-Dimethoxybenzidine	79469	2-Nitropropane	0	Nickel compounds
60117	Dimethyl aminoazobenzene	684935	N-Nitroso-N-methylurea	0	Polycyclic organic matter
119937	3,3'-Dimethyl benzidine	62759	N-Nitrosodimethylamine	0	Radionuclides (Radon)
				0	Selenium Compounds

Source: United States Environmental Protection Agency (USEPA). 2021f. Hazardous Air Pollutants (HAPs). https://www.epa.gov/haps (accessed February 15, 2021).
Note that the listing above which contains the word "compounds" and for glycol ethers are defined as including any unique chemical substance that contains the named chemical (i.e., antimony, arsenic, etc.) as part of that chemical's infrastructure.

Prior to 1990, the CAA established a risk-based program under which only a few standards were developed. The 1990 amendments to the CAA revised Section 112 to first require the issuance of technology-based standards for major sources and certain area sources. The term **major source** is defined as a stationary source or group of stationary sources that emit or have the potential to emit 10 tons per year or more of a hazardous air pollutant or 25 tons per year or more of a combination of hazardous air pollutants. An **area source** is defined as any stationary source that is not a major source (USEPA 2021g). The 1990 amendments to the CAA also expanded the hazardous air pollution program and shifted its focus from a risk-based regulatory provision to one that is primarily based on control technology regulation (i.e., installing a baghouse or other air pollutant capture system). For major sources, Section 112 of the CAA requires that

EPA establish emission standards that require the maximum degree of reduction in emission of hazardous air pollutants. These emission standards are commonly referred to as **Maximum Achievable Control Technology** or **MACT** standards (USEPA 2021g).

Like other permitted facilities, stationary sources regulated under the CAA may also be subject to **New Source Performance Standards** (NSPS). NSPS are uniform standards throughout the United States for new or modified emission sources. The NSPS includes both equipment specifications and operating and monitoring requirements. Depending on where an individual facility is located, stricter standards may apply if the location is deemed to be within a nonattainment area for criteria pollutants. If stricter emission standards are deemed applicable, they are written into the operating permit (USEPA 2021c).

7.3.3 NATIONAL EMISSION STANDARDS FOR HAZARDOUS AIR POLLUTANTS

Eight years after the technology-based MACT standards are issued for a source category, EPA is required to review those standards to determine whether any residual risk exists for that source category, and if necessary, revise the standards to address the risk. The 1990 amendments to the CAA define MACT standards for both existing and new sources. MACT must be equal to at least the level of control achieved by the best performing twelve percent of regulated sources within the same category. Where a category has less than 30 sources, MACT is defined as at least the level of control activity achieved by the best five sources. New source standards may be more stringent than existing source standards, and existing sources are subject to statutory schedules for retrofitting with upgraded technology, as necessary (USEPA 2021h).

Nonmajor sources subject to National Emission Standards for Hazardous Air Pollutants (NESHAPS) are required to install and operate air pollution control equipment subject to MACT (USEPA 2021h). These types of operations include the following (USEPA 2021h):

- Hazardous waste combustors
- Portland cement manufacturers
- Mercury cell chlor-alkali plants
- Secondary lead smelters
- Carbon black production
- Chemical manufacturing: chromium compounds
- Primary copper smelting
- Secondary copper smelting
- Nonferrous metals area sources: zinc, cadmium, and beryllium
- Glass manufacturing
- Electric arc furnace (EAF) steelmaking facilities
- Gold Mine Ore Processing and Production

In addition to MACT standards, there is also an additional set of standards called Generally Available Control Technology or GACT standards which are less stringent standards and are discretionary for area sources by the appropriate regulatory authority (USEPA 2021c).

7.3.4 STATIONARY SOURCES

A **stationary source or point source** is defined as a fixed pollution source such as a coal-fired electrical power generation plant or a refinery. Just in the State of California alone, there are over 13,000 permitted stationary sources (California Environmental Protection Agency 2021). Stationary sources are not all the same and are not treated equally by the CAA.

In general, the more pollution a stationary source emits into the air, the more likely it will be regulated. However, there are a couple of other variables that may also apply. Geography is a

TABLE 7.3
Title V PTE Emission Threshold Levels

Pollutant	Level (tons per year [tpy])
Volatile Organic Compounds (VOCs)	10
Nitrogen Oxides (NOx)	10
Sulfur Oxides (SOx)	100
Carbon Monoxide (CO)	50
Single Hazardous Air Pollutant	10
Combination of Hazardous Air Pollutants	25
Particulate Matter less than 10 microns (PM10)	70

Source: United States Environmental Protection Agency (USEPA). 2021i. Title V Operating Permits. https://www.epa.gov.title-v-operating-permits (accessed February 15, 2021).

variable that can influence the degree a facility is regulated. If a facility is located in an area that is considered heavily polluted, commonly referred to as located in a nonattainment area, the more likely the facility will be regulated. Lastly, another variable is what chemical compounds are being released into the air. This is important since not all pollutants are the same. Some pollutants are more toxic to the environment and humans than others. Therefore, it seems logical that the more a facility emits a very toxic compound, the more heavily regulated that facility becomes.

7.3.4.1 Major Source

A major stationary source is a facility that emits, or has the potential to emit (PTE), any criteria pollutant, or hazardous air pollutant (HAP) at levels equal to or greater than specific emission thresholds defined by USEPA. **PTE** is defined for functional purposes as the maximum capacity of a facility to emit any air pollutant under its physical and operational design. Table 7.3 presents the Title V PTE emission threshold levels established by USEPA (USEPA 2021i). If a facility has the PTE above the thresholds, the facility will require what is termed a Title V Air Permit (USEPA 2021h). Title V permits are re-discussed in the next section.

7.3.4.2 Minor Source and Synthetic Minor Source

Facilities that cannot emit as many criteria pollutants or HAPs as a major source are called a **Minor Source** (USEPA 1998a). A **Synthetic Minor Source** means a source that otherwise has the potential to emit regulated New Source Review pollutants in amounts that are at or greater than the thresholds for major sources, but has taken a restriction so that its Potential to Emit is less than such amounts for major sources. In addition, the restrictions must be enforceable (USEPA 2021h, 2021j).

A **Synthetic Minor Hazardous Air Pollutant Source** means a source that otherwise has the potential to emit HAPs in amounts that are at or exceed those for major sources of HAPs but has taken a restriction so that its Potential to Emit is less than amounts for major sources. In addition, the restrictions must be enforceable (USEPA 2021j).

7.3.4.3 Area Source

As required by the 1990 amendments of the CAA, the USEPA is required to regulate emissions of HAPs and criteria pollutants from a published list of industrial source categories. The USEPA has developed a list of source categories that must meet the control technology requirements for these toxic air pollutants that are known or suspected to cause cancer or other serious health problems commonly known as National Emission Standards Hazardous Air Pollutants (NESHAPs) (USEPA 2021g).

Under the 1990 amendments of the CAA, if a facility was not defined as a major source of HAPs or criteria pollutants meaning that a facility did not emit more than 10 tons of a single HAP or 25 tons of one or more HAPs and were below the thresholds for criteria pollutants but the facility still emitted one or more criteria pollutant or HAP, the facility may be subject to Area Source Rules and would be required to obtain an air permit, but not a Title V permit (USEPA 2021g).

USEPA is required under the CAA to identify 30 hazardous air pollutants (HAPs) posing the greatest health threat in urban areas, which USEPA has identified in its Urban Air Toxic Strategy. The CAA also requires USEPA to identify sufficient area source categories to ensure that area sources representing 90% of area source missions of the thirty urban HAPs are subject to regulations (USEPA 2021g)

The list of Area Source Categories includes the following 70 types of operations or sites (USEPA 2021g, 2021h):

- Acrylic and Modacrylic Fibers Production
- Agricultural Chemicals and Pesticides Manufacturing
- Aluminum Foundries
- Asphalt Processing and Asphalt Roofing Manufacturing
- Autobody Refinishing
- Brick and Structural Clay
- Carbon Black Production
- Chemical Manufacturing: Chromium Compounds
- Chemical Preparations
- Chromic Acid Anodizing
- Clay Ceramics
- Commercial Sterilization Facilities
- Copper Foundries
- Cyclic Crude and Intermediate Production
- Decorative Chromium Electroplating
- Dry Cleaning Facilities
- Fabricated Metal Products, Electrical, and Electronic Products-Finishing Operations
- Fabricated Metal Products not elsewhere classified
- Fabricated Metal Products (Boiler Shops)
- Fabricated Metal Products, Structural Metal Manufacturing
- Fabricated Metal Products, Heating Equipment, Except Electric
- Fabricated Metal Products, Industrial Machinery, and Equipment-finishing operation
- Fabricated Metal Products, Iron and Steel Forging
- Fabricated Metal Products, Primary Metal Products Manufacturing
- Fabricated Metal Products, Valves, and Pipe Fitting
- Ferroalloys Production: Ferromanganese and Silicomanganese
- Flexible Polyurethane Foam Production
- Flexible Polyurethane Foam Fabrication
- Gasoline Distribution Stage I
- Halogenated Solvent Cleaners
- Hard Chromium Electroplating
- Hazardous Waste Incineration
- Hospital Sterilizers
- Industrial Boilers
- Industrial Inorganic Chemical Manufacturing

- Industrial Organic Chemical Manufacturing
- Inorganic Pigments Manufacturing
- Institutional/Commercial Boilers
- Iron Foundries
- Lead Acid Battery Manufacturing
- Medical Waste Incinerators
- Mercury Cell Chlor-Alkali Plants
- Miscellaneous Coatings
- Miscellaneous Organic Chemical Manufacturing
- Municipal Landfills
- Municipal Waste Combustors
- Nonferrous foundries
- Oil and Natural Gas Production
- Other Solid Waste Incineration
- Paint Stripping
- Paints and Allied Products Manufacturing
- Pharmaceutical Production
- Plastic Materials and Resins Manufacturing
- Plating and Polishing
- Polyvinyl Chloride and Copolymers Production
- Portland Cement Manufacturing
- Prepared Feeds Manufacturing
- Pressed and Blown Glass and Glassware Manufacturing
- Primary Copper Smelting
- Primary Nonferrous Metals-Zinc, Cadmium, and Beryllium
- Publicly Owned Treatment Works
- Secondary Copper Smelting
- Secondary Lead Smelting
- Second Nonferrous Metals
- Sewage Sludge Incineration
- Stainless and Non-Stainless Steel Manufacturing: Electric Arc Furnaces (EAF)
- Stationary Internal Combustion Engines
- Steel Foundries
- Synthetic Rubber Manufacturing
- Wood Preserving

7.3.4.4   Air Permitting Process for Stationary Sources

The following sections describe the most common types of air permits and basic information concerning each. However, each State will vary and may refer to permits differently. Therefore, it is recommended that the person responsible for air permitting at a specific facility contact the local environmental authority for further direction. In addition, based on the level of complexity and specific geographical requirements, it may be advisable to retain a consulting firm or outside counsel. A short list of the terminology universe for the types of air permits includes the following (USEPA 2021c):

- Title V Permit
- Major Source Permit
- Minor Source Permit
- Synthetic Minor Permit
- De Minimis Permit
- Relocation Permit

- Construction Permit
- Intent to Construct Permit
- Preconstruction Permit
- New Source Review Permit
- Temporary Permit
- Permit-by-Rule
- Lifetime Operating Permit
- PSD Permit

Many States provide assistance in the form of classes and online worksheets to assist and guide the business community in evaluating and determining the appropriate for almost all types of operations and locations. The following sections provide more detailed information pertaining to the most common nomenclature and most common air permits.

7.3.4.4.1   Title V Permit

The 1990 amendments to the CAA introduced the concept of a Title V permit. A Title V permit is an operating permit issued to each large stationary source of air emissions. A Large Source is a facility that has actual or potential emissions at or above the major source threshold for any air pollutant and the default value for a major threshold is 100 tons (USEPA 2021i). In addition, the major thresholds for HAPs are 10 tons per year for a single HAP and 25 tons per year for any combination of HAPs per year for any air pollutant. Previous to the 1990 amendments, each emission source at a particular operating facility had its own permit. As one can imagine, this led to much confusion from a regulatory perspective and a facility perspective. For example, a large facility may contain 50 or more emission sources each requiring a separate and usually, a distinctive operating permit.

The concept of a Title V permit is to combine each emission source at a particular facility under one comprehensive permit. The intent from a regulatory perspective was that this concept would streamline the permit process and would simplify compliance inspections and would improve compliance of the regulated community (USEPA 2021i).

Most Title V permits are issued by the state or local regulatory authority. A few are issued directly from USEPA. It should be noted that other permits, such as Preconstruction or Construction, Prevention of Significant Deterioration (PSD), and New Source Review (NSR) still remain required and may also require that the Title V permit for any given facility may need to be revised (USEPA 2021i).

Title V permits are comprehensive in nature and include all requirements for controlling air emissions at a stationary source that is required to have a Title V permit. This amended requirement streamlined the process requiring just one permit. Previous to the amendments in 1990, an air permit was required for each source of air pollution, which at some stationary sources, meant that perhaps 50 or more permits were necessary, each being different and with unique compliance monitoring and reporting obligations. The Title V permit process brought all sources of air pollutants from a particular facility under one permit, which simplified compliance and regulatory enforcement (USEPA 2021i).

Title V permits generally include the following terms (USEPA 2021i):

- Permit duration (usually 5 years)
- List of all emissions sources covered under the permit
- Emission source diagram showing the location of each source
- Emission limits for each source
- List air pollution control equipment for each source
- Acceptable operating ranges for each source
- Monitoring requirements for each source

- Record-keeping requirements
- Reporting requirements
- Provision stating that noncompliance constitutes a violation
- Authorization for modifying, revoking, reopening, reissuing, or terminating a permit
- Identification of terms that are federally enforceable
- Compliance certification
- Authorization for emergencies to constitute an affirmative defense to enforcement action

Facilities regulated under a Title V permit must submit periodic, typically quarterly but not less often than annually, compliance and monitoring reports to the regulatory authority to demonstrate compliance or noncompliance. These reports inform the regulatory authority whether the facility is in compliance or if not, it informs the agency of noncompliance. If the reports document an instance of noncompliance other actions must be undertaken by the facility to return to compliance and to document what actions were conducted. An official responsible for the facility must certify under perjury of law whether or not a facility is in compliance. Documentation of noncompliance, especially multiple violations, or misinformation of noncompliance can lead to the regulatory authority to take legal action, typically in the form of a **Notice of Violation** (NOV) to the offending facility. It should be noted that the USEPA is not the only authority that can initiate an enforcement action. Public citizens can also pursue penalties against facilities that do not comply with the CAA (USEPA 2021c).

As you can well imagine at this point, the application process for a Title V permit is extensive and detailed. To assist in becoming familiar with the magnitude of the application process, an example table of contents is presented as the following (Iowa Department of Natural Resources 2021):

1.0 Introduction
 Part 1. Emission Information
        Form 1.0 Facility Identification
        Form 1.1 Plant Location and Layout Diagram
        Form 1.2 Schematic – Process flow Diagram
        Form 1.3 Insignificant Activities – Potential Emissions
        Form 1.4 Potential Toxic Emissions – Significant Activities
        Form 1.5 Potential Emission – Significant Activities
        Form CA-01 Calculations
        Form 2.0 Emission Point Information
        Form 3.0 Emission Unit Description – Potential Emissions
        Form 4.0 Emission Unit – Actual Operations and Emissions
        Form CE-01 Pollution Control Equipment Data Sheet
        Form ME-01 Continuous Monitoring Systems
        Form 5.0 Title V Annual Emission Summary/Fee

 Part 2. Requirements and Compliance
        Part 2. General Facility Requirements
            Part 61 NESHAP Information
            Boiler and Process Heater Information
            Engine Information

 Part 3. Application Certification
 Appendices:
        Hazardous Air Pollutants
        Accidental Release Prevention

Part 61 NESHAP Reference List
Stratospheric Ozone Depleting Chemicals
Acid Rain and CAIR
Prevention of Significant Deterioration (PSD) Worksheet
Proposed Limits and Alternative Operating Scenarios
NSPS Reference List
Part 63 NESHAP Reference List
Compliance Assurance Monitoring

Note that in the above outline under Part 1 will require an emission inventory to be conducted and that each emission source be carefully evaluated. In addition, under Part I Form 3.0 that each emission source is evaluated for actual and potential to emit. The potential to Emit an emission source is generally calculated assuming that the equipment is running at maximum capacity while operating at the maximum hours of operation under its physical and operational design. Usually, the maximum hours of operation are 8,760 hours per year unless enforceable limitations on hours of operation have been incorporated within a construction permit or an enforcement order for the equipment. Bottlenecks in a production line do not constitute an enforceable limitation on production unless those bottlenecks are included as an operating condition in a federally enforceable permit. Only enforceable limitations on raw materials, fuels, capacity, or hours of operation can be used to limit potential emissions (IDNR 2021).

Since 1972, USEPA has published a compilation of air emission factors and is commonly referred to as AP-42. AP-42 contains emissions factors and process information for more than 200 air pollution source categories. A **Source Category** is a specific industry sector or group of similar emitting sources. The emission factors have been developed and compiled from source test data, material balance studies, and engineering estimates (USEPA 2019). Alternatively, emission factor can be obtained by measuring the emission directly. The next section presents a brief summary of emissions testing for stationary sources.

Other related items that may be listed in a Title V permit may include the following (USEPA 2021i):

- Provision for alternative operating scenarios
- Provisions for emission caps
- Provisions for a permit shield

In addition to major sources, other types of sources are required to obtain a Title V permit as well. These include solid waste incineration units of the following (USEPA 2021i):

- Municipal waste combustors
- Hospital/medical/infectious waste incinerators
- Commercial and industrial solid waste incinerators
- Other solid waste incinerators
- Sewage sludge incinerators

Once a facility has been built and has obtained a Title V permit and a modification to operations is considered, the facility must conduct an evaluation to assess if the physical or operational change considered will result in a significant net increase in emissions of a regulated pollutant. That consideration must include all aspects of operations potentially affected by such modification. If the modification results in increased emissions for pollutants listed in Table 7.4, the facility must apply for a permit called a Prevention of Significant Deterioration (PSD) Permit. Table 7.4 presents the PSD thresholds. If the modification results in increased emissions but are below the thresholds

TABLE 7.4
PSD Permit Thresholds

Pollutant	Threshold (tons per year [tpy])
Carbon Monoxide (CO)	100
Nitrogen Oxides (NOx)	40
Sulfur Dioxide (SOx)	40
Total Particulate Matter (PM)	25
Particulate Matter (PM10)	10
Ozone (VOCs)	40
Lead	0.6
Asbestos	0.007
Beryllium	0.0004
Mercury	0.1
Vinyl Chloride	1
Fluorides	3
Sulfuric Acid Mist	7
Hydrogen Sulfide	10
Total Reduced Sulfur (including H_2S)	10
Reduced Sulfur Compounds (including H_2S)	10

Source: United States Environmental Protection Agency (USEPA). 2021i. Title V Operating Permits. https://www.epa.gov.title-v-operating-permits (accessed February 15, 2021).

listed in Table 7.4, the facility must still apply for what is termed a Permit to Construct or a New Source Review (USEPA 2021i).

These thresholds may be locally lower if the facility is located in an area of nonattainment. Nonattainment areas also vary depending on the degree to which they are in nonattainment. USEPA classifies nonattainment as moderate, significant, severe, and extreme. Nonattainment pollutants include (USEPA 2021e)

- Carbon Monoxide (CO)
- Nitrogen Oxides (NOx)
- Sulfur Dioxide (Sox)
- Lead
- Particulate Matter (PM)
- Ozone

Therefore, contacting the local air quality management district should be the first step in evaluating whether a Title V permit is required because emissions limits that trigger the need for a Title V permit vary from location to location (USEPA 2021c).

7.3.4.4.2   Minor Source Permit

To qualify for a minor source permit under the CAA, a facility must emit, or has the potential to emit, regulated New Source Review pollutants in amounts less than the major source thresholds under either the Prevention of Significant Deterioration program or the Major NSR program for Nonattainment Areas (USEPA 2021h).

7.3.4.4.3 New Source Review and PSD

New Source Review (NSR) is a broad term for a collection of federal laws and regulations, as well as state-run programs, that require new sources of air pollution to be as clean as possible. NSR applies to new facilities and existing facilities that plan to significantly expand or upgrade their operations. An NSR permit is required before construction begins on a new facility or pollution-causing modification (USEPA 2021k). NSR Permits are sometimes termed Construction Permits since they must be obtained prior to beginning construction. In general, an existing facility will need to evaluate the NSR Permitting process if the planned modification results in increased emissions of a pollutant (USEPA 2021k).

USEPA's New Source Review (NSR) permitting program protects air quality when an industrial facility, industrial boiler, or power plant is newly built or undergoes modification. As stated above, NSR permitting also assures that a new or modified facility is as clean as possible and that advances in pollution control technologies are installed concurrently with expansion. Hand in hand with NSR is an additional review process that evaluates whether the newly built or modified facility has the potential to deteriorate air quality in the region during its operation. Therefore, USEPA requires a review to be conducted termed **Prevention of Significant Deterioration (PSD).** PSD applies to major sources or major modifications at existing sources for pollutants where the area the source is located is in attainment or unclassified with the NAAQS standards (USEPA 2021l). The source is a major source of the emissions of any pollutant that exceeds applicable thresholds regardless of the area designation (i.e., attainment, nonattainment, or noncriteria pollutants). If an individual unit is classified as one of 28 regulated source categories and its emission exceeds 100 tons per year, then the unit is a major source. The 28 regulated source categories include the following (USEPA 2021l):

1. Coal cleaning plants (with thermal dryers)
2. Kraft pulp mills
3. Portland cement plants
4. Primary zinc smelters
5. Iron and steel mills
6. Primary aluminum ore reduction plants
7. Primary copper smelters
8. Municipal incinerators capable of charging more than 250 tons of refuse per day
9. Hydrofluoric acid plants
10. Sulfuric acid plants
11. Nitric acid plants
12. Petroleum refineries
13. Lime plants
14. Phosphate rock processing plants
15. Coke oven batteries
16. Sulfur recovery plants
17. Carbon black plants (furnace process)
18. Primary lead smelters
19. Fossil-fuel boilers (or a combination thereof) totaling more than 250 tons
20. Fuel conversion plants
21. Secondary metal production plants
22. Chemical process plants
23. Petroleum storage and transfer units with a total storage capacity exceeding 300,000 barrels
24. Taconite ore processing plants
25. Glass fiber processing plants
26. Charcoal production plants

27. Fossil fuel-fired steam electric plants of more than 250 million British thermal unit (BTU) per hour of heat input
28. Sintering plants

PSD requires the following (USEPA 2021l):

1. Installation of the BACT
2. An air quality analysis
3. An additional impacts analysis
4. Public involvement

The main purpose of the air quality analysis is to demonstrate that new emissions emitted from a proposed major stationary source or major modification, in conjunction with other applicable emissions increases and decreases from existing sources, will not cause or contribute to a violation of any applicable NAAQS or PSD increment. Generally, the analysis involves the following (USEPA 2021l):

1. An assessment of existing air quality, which may include ambient monitoring data and air quality dispersion modeling
2. Predictions, using dispersion modeling, of ambient pollutant concentrations that will result from the proposed activity and future potential growth associated with the project.

Areas of special national or regional natural, scenic, recreational, or historic value for which the PSD consideration and regulations provide special protection is termed a Class I area. In this case, any increase in emissions from a major source of 1 microgram per cubic meter as a 24-hour average or more for sources within 10 kilometers of the Class I area will likely require further evaluation (USEPA 2021l).

A PSD increment is the amount of pollution is allowed to increase. PSD increments prevent the air quality in clean areas from deteriorating to the level established by the NAAQS. The NAAQS is the maximum allowable concentration "ceiling". A PSD increment, on the other hand, is the maximum allowable increase in pollutant concentration that is allowed to occur above the baseline pollutant concentration. The baseline pollutant concentration is defined for each pollutant, and in general, is the ambient pollutant concentration existing at the time the first PSD permit application affecting the area is submitted. Significant deterioration occurs when the mass of new pollution exceeds the applicable PSD increment. It should be noted that the air quality cannot deteriorate beyond the pollutant concentration allowed by the applicable NAAQS, even if not all the PSD increment is consumed (USEPA 2021l).

PSD does not prevent sources from increasing emissions. Instead, PSD is designed to (USEPA 2021l):

1. Protect human health and the environment
2. Preserve, protect, and enhance the air quality of national parks, national wilderness areas, national monuments, national seashores, and other areas of special national or regional natural, recreational, scenic, or of historic value.
3. Ensure that economic growth will occur in a manner consistent with the preservation of existing clean air resource
4. Assure that any decision to permit increased air pollution in any area is made only after careful evaluation and after adequate procedural opportunity for informed public participation in the process.

The additional impacts analysis assesses the impacts of air, ground, and water pollution on soils, vegetation, and visibility caused by an increase in emission of any regulated pollutant from the source or modification under review, and from associated growth. Associated growth is industrial, commercial, and residential increases that will occur in the area due to the source.

7.3.4.4.4   Synthetic Minor Permit

The designation of the synthetic minor source is allowed for both regulated NSR pollutants and HAPs and although a facility may choose to obtain emission limitation at their own discretion, once a facility has accepted an enforceable emission limitation, the facility must then comply with that limitation according to Synthetic Minor Source Rule 201.2 (USEPA 2021m). USEPA states in Rule 201.2 that this is necessary to ensure that the facility is legally prohibited from operating as a major source. In addition, if the facility submits an application for a synthetic minor source or a synthetic minor HAP source, the facility must comply with the same participation requirements and the same procedures for final permit issuance and administrative and judicial review as required by the CAA (USEPA 2021m).

In general, synthetic minor permit applications will undergo a rigorous review process that will include the following (USEPA 2021m):

- Completeness determination
- Designation of Federally Enforceable Conditions
- Public Notification and Review
- USEPA review
- Final Action and Determination

Once a determination has been approved for a synthetic source permit, a draft permit will be issued that will include reporting requirements, monitoring requirements, noncompliance provisions, and other requirements, if necessary. If a facility that is issued a synthetic minor permit does not remain in compliance with any permit condition identified as federally enforceable or with any requirement of Rule 201.2, or that submits false information to obtain a synthetic minor source permit, would be considered in violation with the CAA. A noncompliant determination may result in enforcement actions that can include civil or criminal penalties, permit termination, permit renewal denial, or other enforcement actions (USEPA 2021m).

7.3.4.4.5   Permit Exemptions and Permits-by-Rule

There are exemptions to air permitting requirements. Some sources of air pollution are exempt because the air emissions from certain sources are low or below what is termed "de-minimis" amounts. Other sources are specifically listed as exempt sources. Finally, some sources are eligible for coverage under a "permit by rule" (PBR) which means that the facility would be required to submit a notification form instead of a more complex permit application. Contacting the local air pollution regulatory authority is recommended since each air pollution control district in each state can be different, especially if it has recently been designated as nonattainment (Ohio Environmental Protection Agency 2021). Given the rather extensive operating conditions and circumstances of operating with or without a permit, conducting an inventory of all air emission sources and then contacting the local air permitting authority is recommended.

Some examples of each type of exemption are listed below. However, contacting the local air pollution regulatory authority is recommended since each air pollution control district in each state can be different, especially if it has recently been designated as nonattainment. For instance, California requires that any facility that emits air pollutants must obtain an operating permit from the local Air Pollution Control District (APCD) with few exceptions.

In general, the de minimis exemption excludes sources of air pollution from requiring a permit if the source emits less than 10 pounds per day of particulate matter, nitrogen oxides, organic compounds, carbon monoxide, lead, or any other air contaminant. However, a facility cannot use the de minimis exception if (OEPA 2021):

- A CAA requirement or specific State emission standard limits the emissions of any air pollutant or restricts the operation of a source, to less than 10 pounds per day
- The source emits radionuclides
- The source alone or in combination with similar sources has potential emissions of any air pollutant in excess of 25 tons per year
- The source emits more than one ton per year of any hazardous air pollutant or combination of air pollutants

A facility that claims the de minimis exemption must maintain detailed records that demonstrate compliance and that actual emission do not exceed di minimis thresholds, and that the emissions from the source in question, in combination with similar air contaminate sources at the same facility, if present, do not result in potential emissions of any air contaminant from the facility in excess of 25 tons during the preceding calendar year (OEPA 2021).

Permanent exemptions are another category of exemptions. However, permanent exemptions do not apply to an emissions unit subject to certain national emission standards for hazardous air pollutants (NESHAPs), certain MACT standards, or emission units subject to certain new source performance standards (NSPS).

Permit by rule (PBR) exemptions may include the following (OEPA 2021):

- Emergency electrical generators, pumps, and compressors greater than 50 horsepower but used less than 500 hours per year
- Resin injection/compression molding equipment that uses no more than 1,000 pounds of VOC in mold release agents and flatting spray per year
- Small crushing and screening plants (nonmetallic mineral processing plants)
- Remediation projects for soil vapor extraction with a total combined emission rate of less than 15 pounds per day and lasting less than 18 months
- Remediation projects for soil-liquid extraction with total combined emission rates of less than 15 pounds per day and lasting less than 18 months
- Auto body refinishing shops using no more than three thousand gallons of all VOC-containing material per calendar year
- Gas stations with Stage I vapor controls
- Gas stations with Stage I and II vapor controls
- Natural gas-fired boilers and heaters with a maximum rated heat input capacity of greater than 10 million BTUs per hour and less than or equal to 100 million BTUs per hour
- Small printing facilities emitting no more than 10 tons of VOCs, no more than 5 tons of a single HAP, and no more than 10 tons of combined HAPs per calendar year
- Mid-size printing facilities emitting no more than 25 tons of VOCs, no more than 5 tons of a single HAP, and no more than 12.5 tons of combined HAPs per rolling 12 month period

Commonly used exemptions include the following (OEPA 2021):

- Boilers, heaters, furnaces, and dryers rated less than 10 million BTUs per hour that burn only natural gas, liquefied petroleum gas, or distilled oil
- Fossil fuel or wood-fired boilers and heaters rated less than one million BTUs per hour
- Equipment used exclusively for the mixing and blending of materials at ambient temperature to make water-borne adhesives, coating, or binders
- Bench scale laboratory equipment, fume hoods, and paint sample preparation booths
- Resin injection molding equipment using less than one million pounds of resin annually
- Storage tanks for inorganic liquids or pressurized gases
- Storage tanks for organic liquids of less than 19,815 gallons capacity and equipped with submerging filling

- Storage tanks with low vapor pressure organic liquids
- Acid storage tanks of 7,500 gallons of capacity or less
- Solvent recycling units (stills) less than 20 gallons in capacity
- Unheated solvent cleaning tanks with a surface area of less than 10 square feet and not using chlorinated solvents
- Grinding, machining, abrasive blasting, wood working, and pneumatic conveying operations controlled by an internally vented dust collector under 4,000 CFM
- Aluminum die casting machines
- Gas stations equipped with Stage I vapor controls
- Maintenance welding
- Refrigerant reclaiming and recycling machines
- Small natural gas compressor engines are used for maintenance activities
- Emergency electrical generators, compressors, and pumps rate at 50 horsepower or less
- Mobile treatment units or vacuum trucks used to contain and/or prevent further migration of a hazardous material spill during an emergency response

Each PBR provision contains qualifying criteria, emission limits, conditions for operation, and requirements for record-keeping and reporting. Many of these requirements are similar or identical to those found in air pollution limits issued under a permit. Facilities that desire to operate under PDR must submit a notification to the applicable regulatory authority.

Given the rather extensive operating conditions and circumstances of operating with or without a permit, conducting an inventory of all air emission sources and contacting the local air permitting authority should always be the first step.

7.3.4.5 Air Pollutant Capture from Stationary Sources

The approach to addressing air pollution is to capture the pollutant(s) as near as possible to the source. According to USEPA (2008a), there are two types of air pollutant sources; (1) mobile sources, such as automobiles, trucks, buses, farm machinery, airplanes, and (2) stationary sources, such as industrial manufacturing facilities, chemical production facilities, pharmaceutical companies, and refineries. According to USEPA (2008a and 2008b), 90% of air contamination is attributable to motor vehicle exhaust. However, as described in Chapter 8, there have been significant improvements in air quality have been achieved in the last few decades through many programs and initiatives to reduce the amount and type of contaminants from motor vehicle exhaust that include the following (USEPA 2008b):

- Increasing fuel efficiency.
- Decreasing the amount of emissions.
- Installing air pollution control devices on vehicles (such as catalytic converters)
- Increasing public awareness (e.g., driving less and not refueling during critical periods).
- Reformulating fuels (e.g., eliminating the use of lead as an additive).
- Developing more efficient engines.
- Requiring inspections and routine maintenance, if necessary (e.g., vehicle emissions inspections).

Stationary sources are divided into two categories that include (1) gaseous contaminants and (2) particulate matter. Removal of contaminants from the gaseous phase is accomplished through various technologies that include (USEPA 2008b):

- Contact condenser
- Surface condenser
- Thermal incinerator
- Catalytic incinerator

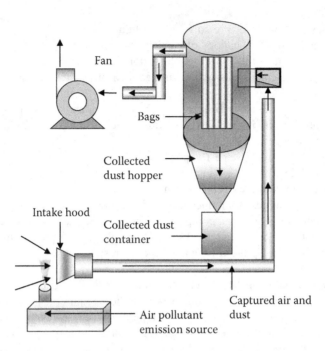

FIGURE 7.4   Generalized schematic of a dust collector.

Removal of particulates is usually accomplished with the installation of a dust collector (also referred to as a baghouse), or a similar device called a wet scrubber. The process of collecting dust (particulates) using a baghouse involves capturing the particulate matter at or very near the emission source by placing the source under a vacuum. The particulates are captured using a filter, such as a fabric before the air stream is exhausted into the atmosphere. Capture efficiencies can be very high and may routinely exceed 95%. Figure 7.4 shows a schematic of a dust collector. Figure 7.5 shows a photograph of a dust collector. Figure 7.6 is an example of duct work that was not functioning properly and was leaking emissions.

7.3.4.6   Compliance Testing of Air Emissions

Compliance testing of air emissions is commonly conducted using two methods. One is commonly referred to as Stack Testing and the other is conducting visual observation and is commonly referred to as Opacity readings or Method 9 reading. Each is described in greater detail below.

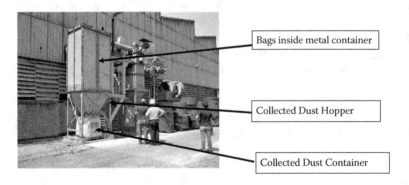

FIGURE 7.5   Photograph of a dust collector (Photograph by Daniel T. Rogers).

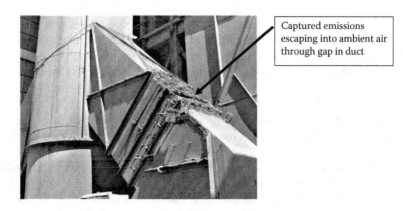

Captured emissions escaping into ambient air through gap in duct

FIGURE 7.6   Example of captured emissions leaking into ambient air (Photograph by Daniel T. Rogers).

7.3.4.6.1   Stack Testing

A **Stack Test** is a quantitative test that measures the amount of a specific regulated pollutant, pollutants, or surrogates being emitted; demonstrates the capture efficiency of a capture system; or determines the destructive or removal efficiency of a control device used to reduce emissions at facilities subject to the requirements of the CAA. Stack testing is an important tool used to determine a facility's compliance status with emission limits, or capture or control efficiencies as outlined in the applicable facility permit (USEPA 2021g, 2021n).

Performance tests are conducted by independent testing companies contracted by a facility that is required to perform source testing. Performance testing is typically required to be conducted in the facility air permit, such as a Title V permit, and is typically mandated to be conducted every 5 years but can vary based on geography, whether the facility is in a nonattainment area, or based on the pollutant of concern. In most instances, the regulatory authority must be notified of the scheduled performance test at least 30 days in advance so that a representative from the regulatory authority can be present if desired. In addition, the following information may be required to be submitted to the regulatory authority (Minnesota Pollution Control Agency 2021, USEPA 2021n):

- A detailed test plan
- The company that will be conducting the test
- Conduct a pretest meeting
- A complete and detailed test report

A stack test typically involves the use of several test methods to measure gas flow in the stack and the concentration of one or more pollutants in the emissions. Individual tests are generally 1 hour in length and typically three 1-hour tests are conducted. In addition, performance testing will only be accepted by the regulatory authority as long as the production process is operating at 95% or greater of its rated capacity. The test itself involves the collection of a representative sample from a properly located sampling port and analyzing the sample either in the field or in a laboratory using USEPA-approved methods and quality assurance and quality control measures. Therefore, before conducting a performance test and knowing that the results of the performance test will be compared to the facility permit, the following items should be considered (USEPA 2021n):

- Understand your permit
- Record data properly
- Ensure that the testing company is well qualified
- Establish testing procedures that are well understood by all those involved
- Ensure that facility management is aware of the test

- Ensure that operations are at levels required to be considered a valid test
- Be aware of special circumstances that might influence testing results, such as weather conditions
- Ensure that the test is conducted using applicable and required safety measures

7.3.4.6.2   Method 9 – Visual Opacity

Many stationary air emission sources discharge visible emissions into the atmosphere. These emissions are typically in the shape of a plume. USEPA has developed a qualitative test termed Visual Opacity or Method 9 which involves the determination of plume opacity by a qualified observer. The method includes procedures for the training and certification of observers and procedures to be used in the field for the determination of plume opacity. The appearance of a plume as viewed by an observer depends upon a number of variables some of which may be controllable and some of which may not be controllable in the field. Variables that can be controlled to an extent to which they no longer exert a significant influence upon the plume appearance include (USEPA 2021o):

- Angle of the observer with respect to the plume
- Angle of the observer with respect to the sun
- Point of observation of attached and detached steam plume
- Angle of the observer with respect to a plume emitted from a rectangular stack with a large length to width ratio

Other variables which may not be controllable in the field are luminescence and color contrast between the plume and the background against which the plume is viewed. These variables affect the ability of the observer to accurately assign opacity values to the observed plume. In general, a plume is most visible and presents the greatest apparent opacity when viewed against a contrasting background, and the least apparent opacity is often observed under conditions where there is no or little contrast with the background (USEPA 2021o).

General procedures for conducting a Method 9 Opacity Test include the following (USEPA 2021o):

- Position: The qualified observer shall stand at a distance sufficient to provide a clear view of the emission with the sun oriented in the 140-degree sector behind the observer. The observer should conduct observations from a position so that the line of sight is approximately perpendicular to the plume direction. The observer's line of sight should not include more than one plume at a time when multiple stacks are involved.
- Field records: The observer shall record the name of the facility, emission location, type of facility, observer's name and affiliation, a sketch of the observer's position relative to the source, approximate wind direction, estimated wind speed, description of sky conditions (presence and color of clouds), and plume background. These observations are to be recorded on a field data sheet at the time the opacity readings are initiated and completed.
- Observations: Opacity observations shall be made at the point of greatest opacity in that portion of the plume where condensed water vapor is not present. The observer shall not look continuously at the plume, but instead shall observe the plume momentarily at 15-second intervals.
- Attached steam plumes: When condensed water vapor is present within the plume as it emerges from the emission outlet, opacity observations shall be made beyond the point in the plume at which condensed water vapor is no longer visible. The observer shall record the approximate distance from the emission outlet to the point in the plume at which the observations are conducted.
- Detached steam plumes: When condensed water vapor in the plumes condenses and becomes visible at a distinct distance from the emission outlet, the opacity of emissions should be evaluated at the emission outlet prior to the condensation of water vapor and the formation of a steam plume.

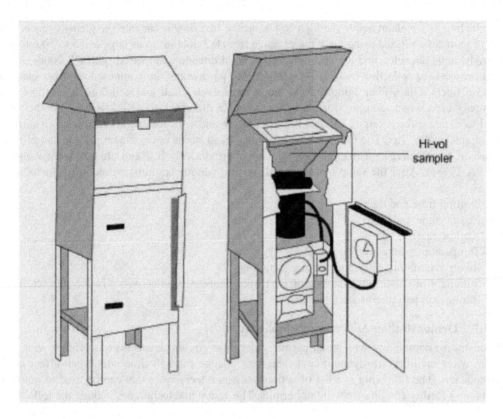

FIGURE 7.7 Outdoor particulate air sampler.

- Recording observations: Opacity observations shall be recorded to the nearest 5% at 15-second intervals on an observational record sheet. A minimum of 24 observations shall be recorded. Each momentary observation recorded shall be deemed to represent the average opacity of emissions for a 15-second period.

7.3.4.6.3 Outdoor Air Sampling

Outdoor air sampling for particulates is conducted using high-volume air sampling equipment as depicted in Figure 7.7 (USEPA 1998b). The ambient air sampling device shown pumps air from all directions across a 7-inch by 9-inch exposed filter inside the shelter at flow rates ranging from 39 to 60 cubic feet of air per minute. The roof design depicted in Figure 7.8 of the shelter is a standard

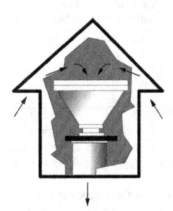

FIGURE 7.8 Air flow pattern through outdoor particulate sample.

design acting as a plenum above the filter to permit the free flow of air into the plenum space. The sizes of particles collected using the sampler range from 0.3 microns to as large as 25 to 50 microns in aerodynamic diameter, and are dependent upon wind direction and speed (USEPA 2008c).

Filters used for collecting outdoor air sampling for particulates are composed of either glass or cellulose fibers. Glass fiber samples have been extensively used and cellulose fiber filters are commonly used when sampling for heavy metals. One drawback to cellulose fibers is their potential to increase the potential for adsorption of water and enhance artifact formation of nitrates and sulfates on the filter. The filters are kept in a clean and sterile environment prior to sampling at a constant temperature and relative humidity of approximately 3% and are precisely weighed (USEPA 1998b). After the sample is collected, several parameters must be recorded, including:

- Starting time and date
- Ending time and date
- Airflow rate
- Temperature ranges during sampling
- Relative humidity difference during sampling
- Summary of conditions that may affect results (construction activities in the area, etc.)
- Barometric pressure at start and end

7.3.4.7   Demonstrating Air Permit Compliance

Demonstrating compliance with an air permit under most circumstances is not as simple as it may appear. Most facilities are dynamic and therefore, change through time which can affect compliance status. The following is a list of action items and techniques that can be used to assist in maintaining facility compliance with air permits. The action and techniques include the following:

- Regularly conducting detailed air emission inventories.
- Being aware that any facility modification that has the potential to increase emissions should be regularly evaluated
- Evaluate every permit condition to ensure that the facility does not accept a permit condition it cannot maintain
- Fully understand that every permit condition is clearly defined
- Fully understand that every permit condition is properly documented

7.3.5   Mobile Sources

Up to this point, the discussion has focused on stationary sources of air pollution. Let's now turn to the other portion of the CAA and discuss mobile sources. A **mobile source** is defined as all on-road vehicles such as automobiles and trucks and off-road vehicles such as trains, ships, aircraft, and farm equipment (USEPA 2021p). Today, mobile sources are responsible for (USEPA 2021p):

- More than 50% of smog-forming VOC emissions
- More than 50% nitrogen oxides (NOx)
- More than 50% of the other toxic air emissions in the United States

The total vehicle miles traveled in the United States increased by 178% between 1978 and 2005 and continue to increase at a rate of two to three percent per year. Currently, there are approximately 210 million cars and light trucks in the United States. During the past 35 years, Americans are driving more vans, trucks, and Sport utility vehicles (SUVs) which typically pollute the air three to five times more than small cars (USEPA 2021p).

The CAA takes a comprehensive approach to reduce pollution from these types of mobile sources by requiring manufacturers to build cleaning operating engines, refiners to produce cleaner-burning fuels, and certain areas with air pollution problems to adopt and run passenger vehicle inspection and maintenance programs. In addition, operating cleaner buses that use propane instead of gasoline, hybrid buses, and electric buses further reduce harmful emissions, especially in heavily populated cities and urban areas where exposure to harmful air pollutants is at its greatest (USEPA 2007a, 2021c, 2021f). Lastly, USEPA has issued a series of regulations affecting diesel trucks and nonroad vehicles such as lawnmowers, garden equipment, recreational vehicles, and boats so that as people buy new vehicles and equipment, overall emissions per vehicle will be reduced. Of course, overall emissions depend on a combination of number of mobile sources and amounts of emitted pollutants per source.

The challenge is that as our population increases and so does the number of mobile sources which in large part offsets any efficiencies gained through improved technology. However, since 1970, new cars purchased today are over 90% more efficient. These improvements are from improved engine efficiency, changes in refueling operations, changes in the composition of fuels, and improved catalytic converters. Improvements must continue because the number of cars in the world exceeded 1 billion in 2010 and is expected to reach 2.5 billion by 2040 (USEPA 2021p).

One of USEPA's earliest accomplishments after the CAA was passed in 1970 was the elimination of leaded gasoline. In the mid-1970s USEPA began its lead phase-out effort by proposing to limit the amount of lead that could be used in gasoline. By the summer of 1974, leaded gasoline was widely unavailable in the United States. This effort was followed by even stronger restrictions in the 1980s and by 1996, leaded gasoline was finally banned as a result of the Clean Air Act (USEPA 2021c).

However, the news hasn't always been positive. The CAA revisions in 1990 required the use of oxygenated gasoline in certain areas of the United States that were prone to excessive smog and carbon monoxide formation as a result of the incomplete combustion of fuel in automobiles (USEPA 2021c). The chemical of choice to add to gasoline was Methyl Tertiary Butyl Ether (MTBE). MTBE is a volatile organic compound (VOC) that is formed by a chemical reaction between methanol and isobutylene and was used to raise the oxygen level in gasoline so that it burned at a higher temperature and resulted in more complete combustion and less pollution emitted by the automobile. MTBE is a chemical that is toxic, mobile, and persistent with a high CRF value. Shortly after MTBE was used as an additive to gasoline, it started showing up in groundwater some of which was used as a source of drinking water. As a result of groundwater impacts, congress mandated the elimination of MTBE as a gasoline additive in 2005 but the damage was done. Numerous groundwater aquifers in the United States were contaminated with MTBE and can no longer be used as a source of drinking water and are being cleaned up at a cost of hundreds of millions of dollars and will continue for perhaps decades (USEPA 2006).

7.3.6 Acid Rain Reduction

Acid rain reduction is discussed here because its part of the CAA and in Chapter 5 because acid rain also affects climate change. The CAA Amendments of 1990 established what is called an allowance market system is known today as the Acid Rain Program. The Acid Rain Program is a market-based initiative created by the USEPA with the purpose of reducing overall atmospheric levels of the principal acid rain-causing chemicals, which are sulfur dioxide (SOx) and nitrogen oxides (NOx). The program is an implementation of emission trading that primarily targets coal-burning power plants, allowing them to buy and sell emission credits called "allowances" according to individual plant needs and costs (USEPA 2008d).

Phase I of the program required that significant reductions be achieved by January 1, 1995, largely requiring 110 electric power generating plants to reduce sulfur dioxide emission rates by 2.5 pounds per million BTU. Each of the electric power generating plants was identified in the statute by name, address, and quantity of emissions allowances in tons of allowable sulfur dioxide emissions per year. New generating plants built after 1978 were required to limit sulfur dioxide to an emission rate of approximately 0.6 pounds per million BTUs(USEPA 2008d). As an incentive for reducing emissions, for each ton of sulfur reduced below the applicable emissions limit, owners of the generating unit received an emissions credit that they could use at another unit or sell. This legitimized a market for sulfur dioxide emission credits and was administered by the Chicago Board of Trade. Power generating units that installed flue-gas desulfurization equipment (e.g., scrubbers) or other qualified equipment, which reduced sulfur emission by 90% qualified for an extension of two years, provided that they already owned allowances to cover their total actual emissions for each year of the extension period (USEPA 2008d, 2015).

Phase II of the program required that all-fossil-fired unit over 75 million megawatts in size were required to limit emissions of sulfur dioxide to 1.2 pounds per million BTUs generated by January 1, 2000. Thereafter, they were required to obtain an emissions credit for each ton of sulfur emitted and were subject to a mandatory fine of $2,000 for each ton emitted in excess of the credits held. USEPA distributes credits equivalent to nearly 9 million tons each year (USEPA 2015).

To reduce nitrogen oxide (NOx) emissions, many electric power generating plants installed low NOx burner retrofits and reduced NOx emissions by approximately 50%. This technology was readily available, so installation of NOx control was considerably less expensive and easy to install and comply with the required CAA amendments. Since the program was initiated, SOx emissions have been reduced by 40% and acid rain levels have dropped by 65% when compared to 1976 levels. USEPA has estimated that the cost of continued compliance ranges from 1 to 2 billion US dollars per year (USEPA 2015).

7.3.7 GREENHOUSE GASES AND CLIMATE CHANGE

Greenhouse Gases are briefly discussed here because they part of the CAA. However, greenhouse gases are discussed in great detail in the previous chapter. In 2009, USEPA concluded that under section 202(a) of the CAA greenhouse gases threaten both the public health and the public welfare and that greenhouse gas emissions from motor vehicles have contributed to that threat (USEPA 2013). This action by USEPA had two distinct findings which are (USEPA 2021q):

1. The **Endangerment Finding** in which the Administrator concluded that the mix of atmospheric concentrations of six key, well-mixed greenhouse gases threatens both the public health and the public welfare of current and future generations. These six greenhouse gases are considered by USEPA to be the compounds that constitute the air pollutants that threaten the public health and welfare. The compounds include the following (USEPA 2021q):
   a. Carbon dioxide (CO_2)
   b. Methane (CH_4)
   c. Nitrous oxide (N_2O)
   d. Hydrofluorocarbons (HFCs)
   e. Perfluorocarbons (PFCs)
   f. Sulfur hexafluoride (SF_6)
2. The **Cause or Contribute Finding** in which the Administrator concluded that the combined greenhouse gas emissions from new motor vehicles and motor vehicle engines contribute to the atmospheric concentrations of the six listed greenhouse gases and hence to the threat of climate change.

The USEPA issued these endangerment findings in response to a 2007 United States Supreme Court case of Massachusetts v. EPA, when the court ruled that greenhouse gases are air pollutants according to the Clean Air Act (USEPA 2021q).

In 2010 and in response to the Endangerment Finding, USEPA required industry to track facility emissions and report GHG emissions if the facilities if the sources of GHG emissions were greater than 25,000 metric tons of carbon dioxide equivalent to the newly formed USEPA Greenhouse Gas Reporting Program under 40 CFR Part 98 (USEPA 2013).

Greenhouse gases originate from both stationary sources and mobile sources. An example of a stationary source of greenhouse gas emissions would be a coal-fired power plant. An example of a mobile source would be an automobile. Greenhouse gas emissions from mobile sources are now greater than any other source passing electrical generation as of December 2018 (USEPA 2021r). USEPA and the National Highway Traffic Safety Administration (NHTSA) have established economy standards for light-duty vehicles, commercial trucks and buses, aircraft, and federal fleets that reduce the generation of greenhouse gas emissions over time (USEPA 2021r).

Individual opportunities to lower air pollution include the following (USEPA 2021r):

- Conserve energy – turn off appliances and lights when you leave a room
- Recycle paper, plastic, glass bottles, cardboard, and aluminum cans
- Keep woodstoves and fireplaces well maintained. Consider replacing old wood-burning stoves with an EPA-certified model
- Plant deciduous trees in locations around your home to provide shade in the summer and light into your home during the winter
- Purchase green electricity, if possible, from your utility company
- Wash clothes with warm and cold water instead of hot water
- Lower the thermostat in your home by even just a couple degrees
- Lower the thermostat on your hot water heater
- Connect your outside lights to a timer
- Use solar lighting when and where possible
- Use low-VOC or water-based paints, stains, finishes, and paint strippers
- Test your home for radon
- Do not smoke in your home
- Purchase "Energy Star" products
- Choose vehicles to purchase that are low-polluting
- Choose consumer products that have less packaging and that are recyclable.
- Shop with a reusable canvas bag instead of using paper or plastic bags
- Purchase rechargeable batteries
- Keep vehicle tires properly inflated
- Fill gas tanks in the evening during the summer months
- Avoid spilling gasoline or topping off the tank while refueling
- Attempt to limit or avoid drive-thru lines
- Use public transportation
- Walk
- Ride a bike
- Get regular engine tune-ups and regular maintenance checks
- Use an energy-conserving motor oil
- Request flexible work hours
- Request to work from home, if possible
- Report vehicles with excess exhaust
- Join a carpool
- Check daily air quality forecasts
- Remove indoor asthma triggers
- Avoid outdoor asthma triggers
- Minimize sun exposure

7.3.8 Upper Atmospheric Ozone

Ozone in the stratosphere, a layer of the atmosphere 10–30 miles above the Earth, serves as a shield, protecting life on the surface of the Earth from ultraviolet radiation. The 1990 amendments to the CAA required USEPA to established a program for phasing out production and use of ozone-destroying compounds termed ODS, which were used as aerosol propellants in consumer products such as hairspray and deodorants, and as coolants in refrigerators and air conditioners (USEPA 2021s).

Specific ODS compounds include the following (USEPA 2021s):

- Chlorofluorocarbons (CFCs)
- Hydrochlorofluorocarbons (HFCs)
- Halons
- Methyl bromide
- Carbon tetrachloride
- Methyl Chloroform

Scientists have been monitoring ozone levels in the upper atmosphere since the 1970s and in 1987, 190 countries, including the United States and other industrialized Nations, signed the Montreal Protocol which called for an elimination of chemicals that destroy the ozone layer.

In 1996, production of ODS in the United States ceased for many of the compounds capable of the most harm such as CFCs, halons, and methyl chloroform. Scientists estimate that it will take another 60 years for the upper atmosphere to recover. The largest observed ozone hole attributed to ODS was estimated to be 28.3 million square kilometers in September 1998 and has been shrinking ever since (NASA 2021).

In addition to phasing out ODS compounds, the CAA includes other steps to protect the ozone layer. The CAA encourages the development of ozone-friendly substitutes, and because of this, many products have been reformulated. For instance, aerosol propellants and refrigerators no longer use ODS compounds. However, there are a few examples where alternatives for ODS compounds have not been developed. These include some applications in the medical field and an effective substitute for methyl bromide which is a pesticide that is used by farmers in the United States (USEPA 2021s).

7.3.9 ENFORCEMENT

Enforcing environmental laws is a central part of USEPA's strategic plan to protect human health and the environment and violations of the CAA are no different. To ensure that environmental regulations are taken seriously by the regulated community, the USEPA has established an Enforcement Policy and has published several guidance documents that provide detail so that enforcement measures are applied equally and consistently (USEPA 2021s, 2021t). In addition, USEPA has a top priority to protect whole communities disproportionately affected by pollution through what is termed the environmental justice network (USEPA 2021s).

In large part, the reason why the United States has been so successful at improving the environment and protecting human health has been because USEPA has an effective enforcement program. Purposeful acts that harm the environment have been greatly reduced in large part because of education and the threat of enforcement. No longer is there a mentality that "It's not a violation until I get caught". In fact, it may be a good time to rethink enforcement policies that have not changed in 30 years.

There are generally two levels of enforcement penalties in the United States which are civil and criminal (USEPA 2021u). The criminal enforcement program at USEPA was established in 1982 and was granted full law enforcement authority by Congress in 1988 and employs special agents and investigators, forensic scientists and technicians, lawyers, and support staff. Civil enforcement arises most often through an environmental violation. Civil enforcement does not take into consideration what the responsible party knew about the law or regulation that was violated. Environmental criminal liability is triggered through some level of intent (USEPA 2021g). A simple explanation of the difference between a civil or criminal violation is that a criminal act is characterized and a "knowing violation" where a person or company is aware of the facts that create the violation. A conscious and informed action brought about the violation. In contrast, a civil violation may be caused by an accident or mistake (USEPA 2021s, 2021t, 2021u).

Types of enforcement results from a civil violation include settlements, civil penalties, injunctive relief, and supplemental environmental projects (SEPs) (USEPA 2021s). In addition, USEPA has published penalty guidance for specific compounds, compound groups, or types of activities that may lead to a violation of the CAA and include the following (USEPA 2021u):

- Permit violations
- Vinyl chloride
- Asbestos
- VOC's
- Volatile HAPs
- Residential wood heaters
- Stratospheric Ozone

To evaluate the level of seriousness of an environmental violation, one must answer three basic questions:

1. Who first discovered the environmental violation? For the most part, environmental regulations are based on self-reporting. It is generally a worse situation if the regulatory authority discovers a violation, especially one that should have been self-reported.
2. Did the violation cause harm? It is generally a worse situation if the violation caused harm to human health or the environment. A permit violation or exceedance is generally considered a situation that has the potential to cause harm.
3. Was the violation a purposeful act? It will always be a worse situation if the violation was a purposeful act.

To evaluate the level of a potential fine associated with a violation of the CAA, the following are generally considered (USEPA 2021u):

- How long did the violation occur? Under many circumstances, fines are determined on a per-day violation that occurred and can range from $7,500 per day to as much as $37,500 per day.
- What was the economic benefit of the violation considered from a point of view of delayed costs and avoided costs? In addition, compelling public concerns are considered.
- Did the violation cause harm to human health and the environment? To evaluate the level of risk, the toxicity of the pollutant, sensitivity of the environment, pollutant mass released, and length of time of the violation are each considered.
- Did the violation continue to occur after discovery?
- Size of the violator
- Litigation risk
- Ability to pay
- Offsetting penalties such as those paid to State and Local governmental agencies or citizen groups
- Did more than one violation occur?

Settlement of violations under the CAA often includes the following (USEPA 2021u):

- Payment of a fine. As stated above, fines are accessed on how many days the violation occurred and the level of seriousness of the violation itself, such as, did the violation caused harm to human health and the environment.

- Permit modification. Usually, when a permit modification is justified, a stricter emission standard is applied.
- Supplemental environmental project (SEP). If a facility must upgrade equipment or invest capital expense in order to return to compliance, a portion of that expense may be used to offset the fine amount. However, the amount of offset is generally capped at 50%.

Certainly, circumstances surrounding any violation are considered. However, any violation must be taken as a serious matter and, in the end, may determine whether any facility continues to operate.

7.4 SUMMARY OF THE CLEAN AIR ACT

Between 1970 and 1990, USEPA established regulations for only six pollutants and these were termed "Criteria Pollutant." During this time period, USEPA regulated only one chemical compound at a time with limited success at actually reducing pollution. The 1990 CAA amendments took a completely different approach to address air pollution. The 1990 Amendments required USEPA to regulate a list of chemical compounds labeled Toxic Air Pollutants which numbered approximately 188 different chemicals.

In addition, the Amendments required USEPA to take steps to reduce pollution by requiring air pollution sources to install controls or change production processes. These requirements, which were a different direction from regulations of the day ultimately made excellent sense to regulate by category rather than by compound, since many individual sources of air pollution most often emitted a mixture of chemical compounds rather than just one. Developing controls and process changes for industrial source categories resulted in major reductions in releases of multiple air pollutants to the air at a time (USEPA 2007a, 2007b).

USEPA has published regulations covering a wide range of industrial categories, including chemical plants, incinerators, dry cleaners, and manufacturers of wood furniture just to name a few. Harmful air toxins from large industrial sources have been reduced by nearly 70%. This number is even more significant when considering that the industry has grown significantly. These regulations apply to mostly "major" sources of air pollution but also to small smaller sources called "Area" sources. In some cases, USEPA does prescribe a specific control technology but sets a performance level based on a technology or other practices already used by the better-controlled and lower-emitting sources in an industry (USEPA 2021c).

The CAA has greatly influenced mobile sources of air pollution, especially automobiles which, according to USEPA has achieved a 90% reduction in air pollutants since the CAA was enacted in 1970. However, as demonstrated by the example of MTBE, efforts to reduce air pollution have had detrimental and harmful collateral effects in causing groundwater pollution.

The CAA has also reduced ODS emissions and the continued destruction of the ozone layer. However, the USEPA was not alone in this endeavor. Nearly 200 other countries also participated in efforts to eliminate ODS use and restore the ozone layer (2021c). We shall explore in later chapters the future role of global efforts in addressing pollution issues on Earth.

One last item that the CAA has introduced into the United States environmental arsenal, is the single operating permit for air which is termed the Title V permit. Environmental regulations are very complex and streamlining the process under one permit has made it much easier for industry and the regulatory authority as well. As we have discovered in this chapter and will discuss in later chapters, opportunities exist to further improve environmental regulations to increase the efficiency of implementing and understanding environmental regulations and ultimately better protecting human health and the environment.

The CAA has made a major impact in improving the air quality throughout the United States, especially with respect to automobile exhaust, ozone, particulate matter, and other criteria pollutants. According to USEPA (2018b), compared to emissions in 1980, carbon monoxide emissions have

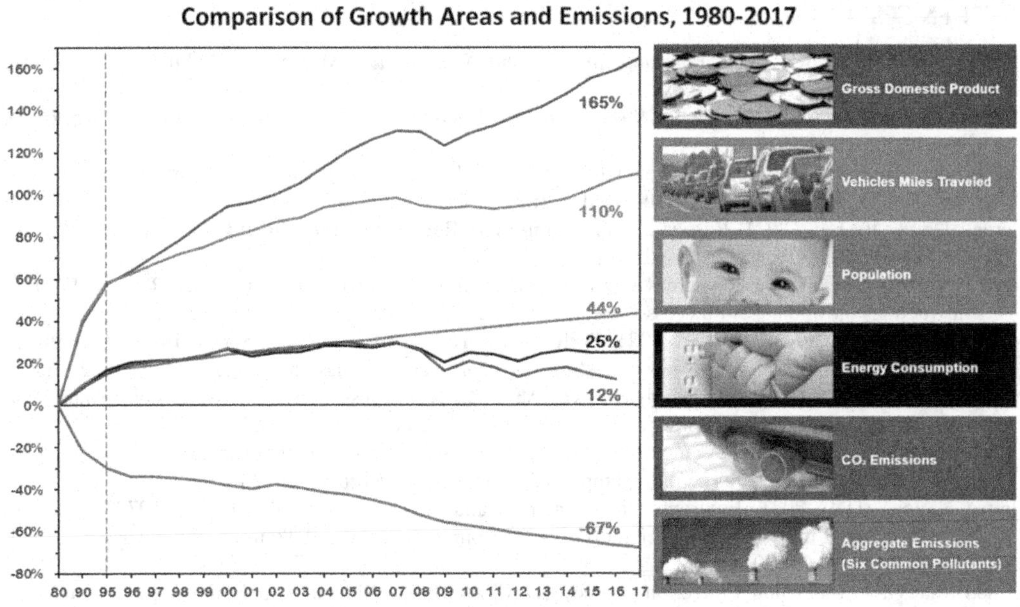

FIGURE 7.9 Reduction of air emissions since 1980.

United States Environmental Protection Agency (USEPA). 2021v. Air Quality – 2018 National Summary. https://www.epa.gov/air-trends/air-quality-national-summary (accessed February 15, 2021).

been reduced by 72%, lead by 99%, nitrogen oxides by 61%, volatile organic compounds by 54%, particulate matter (PM10) by 61% and, sulfur dioxide by 89% (see Figure 7.9) (USEPA 2021v).

Improvements on exhaust from diesel trucks and buses have been slowed and are especially noticeable in urban areas. In addition, other sources of particulate matter are of high concern in urban areas as well and need improvement. This need is in part the result of an increasing population and that urban areas continue to grow as population centers that will inevitably increase automobile congestion and emissions if not addressed as quickly as possible.

As we shall see in the next chapter, even though the United States is in need of improving urban air, many locations in the rest of the world are far worse when it comes to urban air pollution. Actions by some countries (i.e., the United Kingdom) include a carbon tax placed on vehicles driven in urban areas with high pollution levels to discourage unnecessary travel.

Questions and Exercises for Discussion

1. You are requested to review a Title V permit. One item that catches your eye is a statement that reads "Opacity readings should be recorded from the highest emitting stack." Provide an explanation as to why this language can be misinterpreted and what actions do you recommend to be undertaken?

2. You are participating in an environmental audit at an industrial manufacturing facility. You discover that the facility has installed new transformers and because of the new transformers, the facility has increased its production output by 20%. List what actions should be undertaken to evaluate whether the facility is in non-compliance with its Title V air permit.

3. You are reviewing a Title V permit for an industrial manufacturing company and one of the permit conditions states that "the facility shall not emit more than 3.1 grams of XYZ chemical per kilogram of XYZ used." What question(s) should you ask the facility engineer?

REFERENCES

Bates, C. G. and Ciment, J. 2013. *Encyclopedia of Global Social Issues*. M.E. Sharpe Publishers. New York. 1450 p.

Brownwell, F. W. and Zeugin, Z. B. 1991. *Clean Air Handbook*. Government Institutes, Inc. Rockville, Maryland. 318 pages.

California Environmental Protection Agency (CalEPA). 2021. Air Emission Background Information. https://www.arb.ca.gov/ei/general.htm (accessed February 15, 2021).

Cornell University Law. 2021. Resource Conservation and Recovery Act (RCRA) Overview. https://www.law.cornell.edu/wex/rcra (accessed February 15, 2021).

Iowa Department of Natural Resources (IDNR). 2021. Title V Forms and Instructions. Title V Forms / Instructions (iowadnr.gov) (accessed February 15, 2021).

Minnesota Pollution Control Agency. 2021. Performance Testing for Stationary Source Emissions. https://www.pca.state.mn.us/air/performance-testing-stationary-source-emissions (accessed July 31, 2021).

National Aeronautics and Space Administration (NASA). 2021. Largest-Ever Ozone Hole Over Antarctica. https://visibleearth.nasa.gov/view.php?id=54991 (accessed February 15, 2021).

Ohio Environmental Protection Agency (Ohio EPA). 2021. Air Permit Exemptions. https://www.ohio.gov/portal/41/sb/publications/airpermit/exemptions.pdf (accessed February 15, 2021).

United Nations (UN). 2021. Environmental Principles and Concepts Derived from the 1972 Stockholm Conference and the 1992 Rio Declaration on Environmental Law and Training. www.UNEP.org (accessed February 15, 2021).

United States Environmental Protection Agency (USEPA). 1998a. Potential to Emit: A guide for Small Businesses. EPA-456/B-98-003. *Office of Air Quality*. Research Triangle Park. North Carolina. 56 p.

United States Environmental Protection Agency (USEPA). 1998b. *Guidance for Using Continuous Air Monitors*. United States Environmental Protection Agency Office of Air Quality Planning and Standards. Washington, D.C.

United States Environmental Protection Agency (USEPA). 2006. Methyl Tertiary Butyl Ether (MTBE). https://archive.epa.gov/mtbe/html/faq.html (accessed February 15, 2021).

United States Environmental Protection Agency (USEPA). 2007a. Guide to the Clean Air Act. *Office of Air Quality Planning and Standards*. Research Triangle Park, North Carolina. 24 p.

United States Environmental Protection Agency (USEPA). 2007b. The Plain English Guide To The Clean Air Act. *Office of Air Quality Planning and Standards*. Research Triangle Park, North Carolina. 26 p.

United States Environmental Protection Agency (USEPA) 2008a. EPA's Report on the Environment. EPA/600/R-07/045F. Washington, D.C.

United States Environmental Protection Agency (USEPA). 2008b. Latest Findings on National Air Quality: Status and Trends Through 2006. *USEPA*. EPA454/R-07-007. Research Triangle Park, NC.

United States Environmental Protection Agency (USEPA). 2008c. *List of Designated Reference and Equivalent Methods for Air Sampling*. United States Environmental Protection Agency National Exposure Research Laboratory. Washington, D.C.

United States Environmental Protection Agency (USEPA). 2008d. Acid Rain. https://www.epa.gov/acidrain (accessed February 15, 2021).

United States Environmental Protection Agency (USEPA). 2013. USEPA Greenhouse Gas Reporting Program. https://epa.gov/green-house-gas-reporting (accessed February 15, 2021).

United States Environmental Protection Agency. 2015. Acid Rain Emission Reductions. USEPA Archive. https://www.epa.gov/acid-rain-reductions (accessed March 22, 2017).

United States Environmental Protection Agency (USEPA). 2019. AP-42: Compilation of Air Emission Factors. AP-42: Compilation of Air Emissions Factors | Air Emissions Factors and Quantification | US EPA (accessed December 29, 2019).

United States Environmental Protection Agency (USEPA). 2021a. National Environmental Policy Act. www.epa.gov/NEPA (accessed February 15, 2021).

United States Environmental Protection Agency (USEPA). 2021b. Summary of Environmental Regulations of the United States. https://epa.gov/summary-environmental-laws (accessed February 15, 2021).

United States Environmental Protection Agency (USEPA). 2021c. Summary of the Clean Air Act. https://www.epa.gov/laws-regulations/summary-clean-air-act (accessed February15, 2021)

United States Environmental Protection Agency (USEPA). 2021d. NAAQS Table. https://www.epa.gov/criteria-air-pollutants/naaqs-table (accessed February 15, 2021).

United States Environmental Protections Agency (USEPA). 2021e. Non-attainment areas of the United States. www.eps.gov/nonattainmentmap (accessed March 2, 2017).

United States Environmental Protection Agency (USEPA). 2021f. Hazardous Air Pollutants (HAPs). https://www.epa.gov/haps (accessed February 15, 2021).

United States Environmental Protection Agency (USEPA). 2021g. Area Sources of Urban Air Toxics. https://www.epa.gov/urban-air-toxics/area-sources-urban-air-toxics (accessed February 15, 2021).

United States Environmental Protection Agency (USEPA). 2021h. National Emission Standards for Hazardous Air Pollutants (NESHAPs). https://www.epa.gov/NESHAPs (accessed February 15, 2021).

United States Environmental Protection Agency (USEPA). 2021i. Title V Operating Permits. https://www.epa.gov.title-v-operating-permits (accessed February 15, 2021).

United States Environmental Protection Agency (USEPA). 2021j. Minor Source and Synthetic Minor Source Permits. https://www.epa.gov/tribal-air/true-minor-source-and-synthetic-minor-source-permits. (accessed February 15, 2021).

United States Environmental Protection Agency (USEPA). 2021k. *New Source Review (NSR) Permitting*. New Source Review (NSR) Permitting I US EPA (accessed February 15, 2021).

United States Environmental Protection Agency (USEPA). 2021l. New Source Review and PSD Basic Information. https://www.epa.gov/nsr/prevention-significant-deterioration-basic-information (accessed February 15, 2021).

United States Environmental Protection Agency (USEPA). 2021m. True Minor Source Permits. https://www.epa.gov/tribal-air/true-minor-source-and-synthetic-minor-source-permits. (accessed February 15, 2021).

United States Environmental Protection Agency (USEPA). 2021n. *Clean Air Act National Stack Testing Guidance*. Clean Air Act National Stack Testing Guidance I Compliance I US EPA (accessed January 19, 2021).

United States Environmental Protection Agency (USEPA). 2021o. Method 9 – Visual Opacity. https://www.epa.gov/emc/method-9-visual-opacity (accessed February 15, 2021).

United States Environmental Protection Agency (USEPA). 2021p. Improved Vehicle Emissions Guide. https://www.epa.gov/vehicleemissions (accessed February 15, 2021).

United States Environmental Protection Agency (USEPA). 2021q. Endangerment and Cause or Contribute Findings for Greenhouse Gases. Section 202(a) of the Clean Air Act. https://www.epa.gov.climatechange/endangerment (accessed February 15, 2021).

United States Environmental Protection Agency (USEPA). 2021r. Regulations for Greenhouse Gas (GHG) Emissions. https://www.epa.gov/regulations-emissions-vehicles-and-engines/regulations-greenhouse-gas-ghg-emissions (accessed February 15, 2021).

United States Environmental Protection Agency (USEPA). 2021s. *Enforcement Policy*. Guidance and Publications. https://www.epa.gov/enforcement/enforement-policy-guidance-publications (accessed February 15, 2021).

United States Environmental Protection Agency (USEPA). 2021t. *Enforcement Policy*. Guidance and Publications. https://www.epa.gov/enforcement/criminal-enforement-overview (accessed February 15, 2021).

United States Environmental Protection Agency (USEPA). 2021u. Basic Information on Enforcement. https://www.epa.gov/enforcement/basic-information-enforement (accessed March 31, 2021)).

United States Environmental Protection Agency (USEPA). 2021v. Air Quality – 2018 National Summary. https://www.epa.gov/air-trends/air-quality-national-summary (accessed March 31, 2021).

8 Protecting the Water through Environmental Regulations

8.1 INTRODUCTION

By the 1960s, water pollution in the United States had become so severe and degraded that pollution and the effects of pollution could no longer be ignored. At that time, there were several indications that surface waters of the United States were heavily impacted and polluted through human activities. Some of these were as follows (United States Environmental Protection Agency [USEPA] 2021a):

- Pollution in the Chesapeake Bay killed fish that resulted in an estimated loss to the fishing industry of $3 million annually in 1968.
- Bacteria levels in the Hudson River were 170 times greater than what was considered the safe limit as measured in 1969.
- An estimated 26 million fish were killed in Lake Thonotosassa, Florida, due to wastewater discharges from four food processing plants in 1969.
- A floating oil slick on the Cuyahoga River, just southeast of Cleveland, Ohio, caught fire causing significant damage to two railroad trestles. The cause of the fire was never determined but investigations concluded that discharges of volatile organic compounds to the River with low flashpoints provided enough fuel that a passing train, for instance, could have provided an ignition source to start the fire.
- The Department of Health, Education and Welfare's Bureau of Water Hygiene reported that 30% of drinking water samples had detected chemicals exceeding recommended Public Health Service limits.
- A study by the FDA in 1971 concluded that over 70% of samples of Swordfish had mercury at concentrations that were considered unfit for human consumption.

In the late 1960s, it was not an issue of sustainability but had become a survival issue at many locations within the United States because water pollution had become so severe that it put public health at significant risk.

Even though we must remind ourselves of the factors that influence the effectiveness of environmental regulations. We must also evaluate why did water pollution in the United States get so severe before the action took place to address the issue. As a reminder, those factors that influence the effectiveness of environmental regulations include (Bates and Ciment 2013, United Nations 2021):

- Political will
- Environmental tragedies, harm to the environment, and lessons learned
- Cost
- Geography and climate
- Enforcement
- Incentives
- Lack of overriding social factors such as war, political structure, greed, poverty, hunger, lack of infrastructure, educational awareness, and corruption

DOI: 10.1201/9781003175810-8

Factors that may have influenced a delay in enacting strong environmental regulations to improve water quality in the United States prior to the passage of the Clean Water Act (CWA) in the early 1970s were likely influenced by the following:

- Instances of water pollution were isolated and did not reflect water quality degradation nationwide.
- A belief system that the Earth was so large that widespread pollution was not possible.
- A belief system that the Earth would filter out pollution given enough time.
- Addressing water pollution would be too expensive and was not necessary.

These speculative factors turned out to be false. The reality was that pollution was widespread, and the Earth was not so large to dilute water pollution to safe levels. In addition, while the Earth can adsorb some pollutant load, the amount and types of pollution released exceeded the capacity that nature could handle. One other factor that was clear even in the 1960s is that certain types of pollutants did not degrade and bioaccumulated within many species and even increased in concentrations in some species. Examples include halogenated volatile organic compounds such as tetrachloroethene and trichloroethene, polychlorinated biphenyls (PCB), and the insecticide dichlorodiphenyltrichloroethane (DDT). These are synthetic compounds that do not degrade and may last for several decades to centuries in the environment once released (Rogers 2020).

Another factor that seems obvious now is the United States did not have a robust library of scientific research that documented the scale and degree of water pollution on a national scale. This made it difficult to assess how severe the issues were and what actions were required to minimize and reverse the effects of water pollution.

Along with our example situations, each of the above-listed factors either influences the effectiveness of environmental regulations in a positive way or undermines environmental regulations. While reading this chapter, keep in mind the three basic environmental fundamentals, which are as follows:

- Protect the air
- Protect the water
- Protect the land
- Protect living organisms and cultural and historical sites

This chapter concentrates on the CWA and the Safe Drinking Water Act (SDWA). There are other water-related environmental regulations that we will discuss in a later chapter when we address protecting living organisms and cultural and historical sites.

8.2 CLEAN WATER ACT

By the time the CWA was passed in 1972, the USEPA estimated that nearly two-thirds of the United States lakes, rivers, and coastal waters had become unsafe for fishing or swimming due to untreated sewerage being dumped into open water. To address these concerns, USEPA has divided the CWA into four subgroups which include the following (USEPA 2021a):

- Point source discharges to surface water
- Wetland protection
- Stormwater runoff
- Spill prevention

We will describe each separately in the following sections.

8.2.1   POINT SOURCE SURFACE WATER DISCHARGE

The CWA authorized the USEPA to regulate what are called **Point Sources** that discharge pollutants in the waters of the United States through what was termed the National Pollutant Discharge Elimination System (NPDES) permit program. The term "Point Sources" is defined as those locations where wastewater is generated that originated from a variety of locations and activities such as municipal and industrial operations, including treated wastewater, process water, cooling water, and stormwater runoff from drainage systems. A total of 46 of the 50 States actually implement the program under authorization from Congress and oversight of the USEPA. USEPA directly implements the program in Idaho, New Mexico, Maine, and New Hampshire as well as Native American lands (USEPA 2021a).

Under the CWA, the USEPA has implemented pollution control programs such as establishing wastewater treatment standards for certain industries and water quality standards for contaminants in the surface waters of the United States. The CWA made it unlawful to discharge any pollutant from a point source into navigable waters unless a permit was obtained through the NPDES program. Individual homes that are connected to a municipal system, have a septic system or do not discharge to surface water, do not require a permit under the program (USEPA 2021b).

USEPA has compiled water quality criteria for human health, aquatic life, and organoleptic effects, such as taste and odor for approximately 150 pollutants (USEPA 2021c). USEPA developed these criteria to provide guidance to the States and Native American Tribes to use to establish water quality standards and provide a basis for controlling discharges or releases of pollutants to the environment. Table 8.1 presents select compounds on the current list while the entire list can be viewed at https://www.epa.gov/wqc/national-recommended-water-quality-criteria-human-health-criteria-table (USEPA 2021c). USEPA has established concentrations of each listed pollutant for "water + organism" and "organism" only. The concentrations recommended by USEPA are those levels that are not expected to cause adverse health effects to humans (USEPA 2021c). A water quality standard defines the water quality goals of a water body, or portion thereof, by designating the use or uses to be made of the water body and by setting the minimum criteria that protect the designated uses that include the following (USEPA 2010):

- Propagation of fish, shellfish, and wildlife
- Recreation in or on the water
- Consideration as a drinking water source
- Agricultural
- Industrial
- Navigation
- Other uses

There are two basic types of NPDES permits which are individual and general permits. These permit types share the same components but are used under different circumstances and involve different permit issuance processes (USEPA 2010).

An individual permit is a permit that is specifically tailored to an individual facility. Most industrial businesses or locations have an individual permit. After submitting the required information to the regulatory authority, the agency develops a permit for that facility on the basis of information from the permit application and other sources, which may include the following (USEPA 2010):

- Previous permit requirements
- Discharge monitoring reports
- Technology and water quality standards
- Total maximum daily loads
- Ambient water quality data
- Other source materials and reports

TABLE 8.1

Recommended USEPA Water Quality Standards for Human Health

Compound	CAS No.	Water + Organism (µg/L)	Organism Only (µg/L)	Compound	CAS No.	Water + Organism	Organism Only
Acenaphthene	83329	70	90	Methoxychlor	72435	0.02	0.02
Acrolein	107028	3	400	Methyl bromide	74839	100	10,000
Acrylonitrile	107131	0.061	7	Methylene chloride	75092	20	1,000
Aldrin	309002	7.7E-7	7.7E-7	Nickel	7440020	610	4,600
Alpha-Hexachlorocyclo-hexane	319846	0.00036	0.00039	Nitrates	1479755	10,000	Under review
Alpha-Endosulfan	959988	20	30	Nitrobenzene	98953	10	600
Anthracene	120127	300	400	Nitrosamines	None	0.0008	1.24
Antimony	7440360	5.6	640	Nitroso-dibutylamine	924163	0.0063	0.22
Arsenic	7440382	0.018	0.14	Nitroso-diethylamine	55185	0.0008	1.24
Asbestos	1332214	7E+7 fibers/L	Under review	Nitroso-pyrrolidine	930552	0.016	34
Barium	7440393	1,000	Under review	N-Nitroso-dimethylamine	62759	0.00069	3.0
Benzene	71432	0.58	16	N-Nitrosodi-n-propylamine	621647	0.005	0.51
Benzidine	92875	0.00014	0.11	N-Nitroso-diphenylamine	86306	3.3	6.0
Benzo(a)anthracene	56553	0.0012	0.0013	Pentachloro-benzene	608935	0.1	0.1
Benzo(a)pyrene	50328	0.00012	0.00013	Pentachloro-phenol	87865	0.03	0.03
Benzo(b) fluoranthene	205992	0.0012	0.0013	pH	None	5–9	Under review
Benzo(k) fluoranthene	207089	0.012	0.013	Phenol	108952	4,000	300,000
Beryllium	744047	4	4	PCBs	See	Arochlor	6.4E-5
6.4E-5							
Beta-hexachlorocyclo-hexane	319857	0.008	0.14	Pyrene	129000	20	30
Beta-endosulfan	3321365	20	40	Selenium	7782492	170	4,200
Bis(2-chloro-1-methylethyl)ether	108601	200	4,000	Dissolved Solids	None	250,000	Under review
Bis(2-chloroethyl) ether	111444	0.03	2.2	Salinity	None	25,000	Under review
Bis(2-ethylhexyl) phthalate	117817	0.32	0.37	Tetrachloroethene	127184	10	29
Bis(chloroethyl) ether	542881	0.00015	0.017	Thallium	7440280	0.24	0.47
Bromoform	75252	7	120	Toluene	108883	57	520
Butylbenzyl phthalate	85678	0.10	0.10	Toxaphene	8001352	0.00070	0.00071
Cadmium	7440439	5	5	Trichloroethylene	79016	0.6	7

TABLE 8.1 (Continued)

Recommended USEPA Water Quality Standards for Human Health

Compound	CAS No.	Water + Organism (μg/L)	Organism Only (μg/L)	Compound	CAS No.	Water + Organism	Organism Only
Carbon tetrachloride	56235	0.4	5	Vinyl Chloride	75014	0.022	1.6
Chlordane	57749	0.00031	0.00032	Zinc	7440666	7,400	26,000
Chlorobenzene	108907	100	800	1,1,1-Trichloroethane	71556	10,000	200,000

Source: United States Environmental Protection Agency (USEPA). 2021c. National Recommended Water Quality Criteria-Human Health Criteria Table. https://www.epa.gov/wqc/national-recommended-water-quality-criteria-human-health-criteria-table (accessed February 20, 2021).

μg/L, microgram per liter.

The permit authority then issues the permit to the facility for a defined period of not more than 5 years, which then the permitted facility must reapply usually no later than within 6 months prior to the expiration date of the permit or other designated time frame defined within the permit (USEPA 2010).

A general permit covers multiple facilities in a specific category of discharges or of sludge use or disposal practices. General permits are typically for sewer districts, urbanized areas, cities, or towns. The regulation also allows a general permit to cover any other appropriate division or combination of such boundaries. For instance, USEPA has issued general permits to cover multiple states, territories, and Native American Tribes where USEPA is the permitting authority (USEPA 2010).

The major components of an NPDES permit include the following (USEPA 2010):

- Cover Page: Contains the name and location of the permittee, a statement authorizing the discharge, and a listing of the specific locations for which the discharge is authorized.
- Effluent Limitations: The primary mechanism for controlling discharges of pollutants to receiving water are the effluent limitations. Effluent limitations typically include a list of each pollutant in the effluent and an acceptable concentration for each pollutant. In addition, the permit typically include daily and or monthly maximums of loading to the receiving waters.
- Monitoring and Reporting Requirements: Monitoring and reporting requirement are used to characterize
  o Waste streams
  o Receiving waters
  o Wastewater treatment efficiency
  o Compliance with permit conditions
- Special conditions: Special conditions are developed to supplement numeric effluent limitations. Examples may include
  o Additional monitoring activities
  o Additional or special studies or sampling
  o Best management practices
  o Compliance schedules
- Standard Conditions: Standard conditions are those conditions that apply to all NPDES permits and delineate the legal, administrative, and procedural requirements of the NPDES permit.

8.2.2 STORMWATER

Stormwater runoff occurs when precipitation from rain or snowmelt flows over the ground. Impervious surfaces such as driveways, sidewalks, parking lots, streets, and roofs of houses and buildings prevent some or all stormwater from naturally soaking into the ground. These discharges may contain pollutants of various types and concentrations depending upon the chemistry of the rainwater, potential contaminants or surfaces the rainwater comes into contact with, and for how long the contact exists. Therefore, USEPA requires that municipalities, industry, and construction sites obtain a permit that outlines what controls are necessary to prevent or minimize the potential for pollutants to contact pollutants and adversely affect surface water quality (CalEPA 2001, USEPA 2009).

The NPDES Stormwater Program, existing as a separate set of regulations, was put into place in the CWA amendments of 1987 and 1990, regulates discharges from municipal storm sewer systems, construction activities, industrial activities, and those designated by USEPA due to water quality impact. The stormwater regulation went into effect in two phases (USEPA 2009).

Phase I required permits for the following five categories (USEPA 2010):

1. Discharges permitted before February 4, 1987
2. Discharges associated with industrial activity
3. Discharges from Large Municipal Separate Storm Sewer Systems (MS4s) (systems serving a population of 250,000 or more
4. Discharges from medium MS4s (system serving a population of greater than 100,000, but less than 250,000)
5. Discharged evaluated by the permitting authority to be a significant source of pollution or which contributes to a violation of a water quality standard

Phase II required smaller jurisdictions of less than a population of 100,000 to come into compliance by 2008, and to implement USEPA's six minimum control measures that included the following (USEPA 2010):

- Public information and education
- Public involvement and participation
- Illicit discharge detection
- Construction site stormwater runoff control
- Post-construction stormwater management
- Pollution prevention/housekeeping

Phase II also required numerous small MS4s, construction sites of one to five acres, and industrial facilities owned or operated by small MS4s which were previously exempted in Phase I to also obtain a stormwater permit (USEPA 2010).

The CWA and related regulations define the specific industrial and municipal stormwater sources that must apply and obtain an NPDES permit. The CWA also recognizes that other sources, such as commercial properties, may need to be regulated on a case-by-case or category-by-category basis based on additional information or localized conditions (USEPA 2010).

The authority to regulate other sources based on the localized adverse impact of stormwater on water quality through the NPDES permits is commonly referred to as "Residual Designation" authority. In 2008, USEPA issued records of decisions requiring additional NPDES stormwater permits in specific areas within the Charles River watershed in Massachusetts and the Long Creek watershed in Maine. In July 2013, the Conservation Law Foundation, Natural Resources Defense Council, and American Rivers filed a petition seeking USEPA to render a decision that commercial, industrial, and institutional sites contribute to violations of water quality standards and therefore, required NPDES permits (USEPA 2009).

A stormwater permit is termed a Stormwater Pollution Prevention Plan (SWPPP). An outline of a typical SWPPP is outlined below (Wisconsin Department of Natural Resources 2021):

1. Overview
   a. General facility information includes name, location, property size, operations, physical characteristics, and site map.
   b. Statement of Introduction: This states the citation and reference to applicable stormwater regulations that apply to the specific facility.
   c. Statement of Objectives: States the primary goal of the stormwater permit which is to improve the quality of surface waters by reducing the amount of pollutants potentially contained in stormwater runoff.
2. Stormwater Pollution Prevention Team: This section identifies specific individuals at the facility who will be responsible for developing, implementing, maintaining, and revising the SWPPP.
3. Potential Sources of Pollution: This section includes a comprehensive assessment of all potential sources of stormwater sources at the facility and includes the following:
   a. Site map: A site is a critical item included in any SWPPP. The map or maps typically include the following:
      i. Facility property boundaries
      ii. A depiction of the storm drain collection and disposal system, including all known surface and subsurface conveyances that are clearly labeled.
      iii. Any secondary containment structures
      iv. The location of all outfalls, which should also be numbered for reference purposes, that discharge channelized flow to surface water, groundwater, or wetlands
      v. The drainage area boundary for each stormwater outfall
      vi. The surface area in acres draining to each outfall, including the percentage that is impervious defined as those areas that are paved, roofed, or highly compacted soil, and the percentage that is previously defined as those areas that are grass and woods
      vii. Any structural stormwater controls
      viii. Name and location of receiving waters
      ix. Location of activities and materials that have the potential to contaminate stormwater shall also be identified. These locations typically refer to any outside storage of materials, drums, and other containers (even if empty), machines, waste storage locations, maintenance shops, used oil storage, loading and unloading locations, onsite vehicle storage, and parking (e.g., forklifts).
   b. Summary of Sampling Data: Not every facility will have sampling data, especially if applying for a first-time permit and are not renewing a permit. If renewing a permit, historical data is required to be provided. Sampling data will vary from site to site since operations and chemical use may vary greatly depending on operations. However, typical historical data that may be available may include the following:
      i. Heavy metals
      ii. Total solids
      iii. Total dissolved solids
      iv. Oil and grease
      v. Total petroleum hydrocarbons
      vi. Volatile organic compounds
      vii. Semi-volatile organic compounds

       viii. Polychlorinated biphenyls
        ix. pH
         x. Temperature
        xi. Turbidity
       xii. Other compounds

   c. Inventory of Potential Sources of Contamination: The following have been identified as potential sources of stormwater contamination:
         i. Outdoor manufacturing areas
        ii. Rooftops contaminated from industrial activity or emission deposition from a pollution control device
       iii. Areas of significant soil erosion, discoloration, or stressed vegetation
       iv. Industrial plant storage yards
        v. Onsite access roads or rail lines and spurs
       vi. Material handling locations (storage, loading, unloading, transportation, or conveyance of any raw material, finished product, intermediate product, by-product, solid or liquid water
       vii. Shipping and receiving locations
      viii. Manufacturing buildings
       ix. Residual treatment, storage, and disposal locations
        x. Storage areas (including tank farms) for raw materials, finished or intermediate materials
       xi. Refuse sites
      xii. Disposal locations
      xiii. Areas containing residual pollutants from past industrial activity, including previous areas of leaks or spills
      xiv. Vehicle maintenance and cleaning areas
       xv. Any other area having the capability or possibility of contaminating stormwater runoff

4. Other Plans Incorporated by Reference: These may include the following:
   a. Preparedness, Prevention, and Contingency Plan
   b. Spill Control and Countermeasures Plan
   c. National Pollution Discharge Elimination System Permit
   d. Toxic Organic Management Plan
   e. Occupational Safety and Health Administration Emergency Action Plan
   f. Preventative Maintenance Plan

5. Best Management Practices: These items describe stormwater management controls labeled best management practices that would be implemented to reduce the amount of pollutants in stormwater discharged from the facility and would typically include the following:
   a. Source Elimination (e.g., removing the risk by relocation indoors thereby eliminating the chance that rainwater can come into contact with the source
   b. Source Area Controls for those areas that can't be eliminated including:
         i. Erosion control measures
        ii. Improved housekeeping
       iii. Preventative maintenance
       iv. Quarterly visual comprehensive inspections and documentation
        v. Spill prevention and response procedures
       vi. Employee training
      vii. Bulk Storage management improvements
   c. Residual Pollutants (e.g., capping with an imperious layer)
   d. Stormwater Treatment may be necessary and may include the following:

> > i. Preventative measures (e.g., redirecting stormwater)
> > ii. Diversions
> > iii. Containment
> > iv. Other controls
> e. Facility Monitoring includes the following:
> > i. Visual inspections
> > ii. Sampling and analysis

6. Record Keeping: All reports and records pertaining to the permit coverage shall be retained 5 years from the date of the permit. The records are to be maintained onsite and shall be made available upon request of the permit authority.
7. Certification of the SWPPP: To be made by the plan preparer and an authorized representative of the facility in charge.

8.2.3 WETLANDS

Placement of dredged or fill material into wetlands, lakes, streams, rivers, estuaries, and certain other waters are also regulated under the CWA in Section 404. The goal of Section 404 is to avoid and minimize losses of wetland and other waters and to compensate for unavoidable loss through mitigation and restoration. Section 404 is jointly implemented by USEPA and the Army Corps of Engineers (United States Army Corps of Engineers 2021).

In addition, property development may also be subject to wetlands regulations if wetlands are present. **Wetlands** are defined by the United States Army Corps of Engineers (USACE) and the USEPA as those areas that are inundated or saturated by surface or groundwater as a frequency and duration sufficient to support, and under normal circumstances, do support a prevalence of vegetation typically adapted for life in saturated soil conditions. In order for an area to be classified as wetland, a wetland evaluation must typically be conducted by a wetland professional to determine the presence of hydrophytic vegetation hydric soils, and wetland hydrology indicators are present and if so, the area is typically considered a wetland (United States Army Corps of Engineers 2021). If a proposed development has the potential to destroy or otherwise influence or impact any portion of a wetland, a permit will be required from either the USACE or perhaps even the applicable State agency where the property is located (United States Army Corps of Engineers 2021).

8.2.4 POLLUTION PREVENTION AND SPILL PREVENTION CONTROL AND COUNTERMEASURES

As part of the pollution prevention/housekeeping requirements of the CWA, USEPA requires facilities that store oil or other petroleum products prepare a Spill Prevention Control and Countermeasures (SPCC) Plan (USEPA 2012). The purpose of an SPCC plan is to prevent the release of oil or petroleum products into a navigable waterway of the United States. It is not the purpose of an SPCC Plan that describes actions to be conducted to clean-up a spill once it occurs. This would be a separate plan commonly termed a Spill Response Plan (SRP) (USEPA 2012).

SPCC requirements are classified into three tiers that include small facilities, Mid-size facilities, and large facilities. What determines which class a specific facility would be included is based on the storage capacity of petroleum products. Regardless of the classification, all SPCC plans must be prepared in accordance with the oil pollution guidelines in the Federal Code of Regulations (CFR) 40 CFR Part 112 (USEPA 2012).

Tier I requirements include the following (USEPA 2012):

- Above-ground petroleum product storage capacity of 10,000 gallons or less.
- Storage is calculated by adding up all above-ground storage containers of petroleum products at a capacity of 55 gallons or more. This includes transformers if they contain oil and are not the dry type and contain more than 55 gallons capacity.

- No single discharge of petroleum to navigable waters or adjoining shorelines each exceeding 42 gallons within any 12-month period, 3 years before the certification date.
- No two discharges of petroleum to navigable water or adjoining shorelines each exceeding 42 gallons.
- Having little or no piping and very few oil transfer stations.

If the above criteria are met, an owner or operator may complete and self-certify a plan template instead of a full professional engineer-certified plan.

Tier II requirements include the following (USEPA 2012):

- Above-ground petroleum product storage capacity of 10,000 gallons or less.
- Storage is calculated by adding up all above-ground storage containers of petroleum products at a capacity of 55 gallons or more. This includes transformers if they contain oil and are not the dry type and contain more than 55 gallons capacity.
- No individual above-ground petroleum storage container is greater than 5,000 gallons.
- Meets all other Tier I conditions.

If the above Tier II criteria are met, an owner or operator may self-certify a spill plan in accordance with requirements outlined in 40 CFR, Part 112.7, in lieu of a professional engineer-certified plan.

Tier III requirements include the following (USEPA 2012):

- Above-ground petroleum storage capacity greater than 10,000 gallons
- Meets all other Tier I and II conditions

A licensed professional engineer must prepare and certify the plan for Tier III sites.

There are 12 basic requirements of any SPCC plan which are (USEPA 2012) as follows:

1. Description of any spills in the past 12 months, including
   a. Corrective action measures undertaken
   b. Plans to prevent reoccurrence
2. Layout of facility, including diagrams markings locations and contents of
   a. All oil and petroleum storage containers and capacities, including transformers, if applicable
   b. All underground or buried tanks
   c. All transfer station, if any
   d. All connecting piping
3. Predictions of the direction, rate of flow, and total quantity of oil or petroleum that could be discharged
4. A complete explanation of spill containment and any diversionary structures or equipment including
   a. Dikes
   b. Berms
   c. Retaining walls
   d. Curbing
   e. Culverts, gutters, or other drainage systems
   f. Spill diversion/retention ponds
   g. Double-walled tanks with interstitial monitors
   h. Absorbent materials
   i. Tools (i.e., shovels and other implements)
   j. Personal protective clothing (i.e., boots, shields, gloves, etc.)
   k. Location and contents of spill kits

5. Facility maintenance of containment area drainage that includes
   a. Stormwater in berms and dikes
   b. Dike drainage practices
   c. Management of areas that are not contained
6. Bulk storage practices that include
   a. Verification that tank material and construction are compatible with material stored
   b. Secondary containment means such as double-walled tanks or physical containment with a capacity equal to the largest capacity container plus an additional 10%
   c. Procedure to ensure that drainage of secondary containment areas does not release oil or petroleum products.
7. Transfer protocol method that includes
   a. Methods to limit corrosion of buried piping, if any
   b. Methods to inspect and maintain above-ground valves and piping
   c. Procedures, barriers, and methods that warn vehicles to avoid damaging above-ground piping and storage areas
8. Tank truck loading and unloading practices that include
   a. Documentation that loading and unloading procedures meet Department of Transportation (DOT) requirements
   b. Loading and unloading area containment capacity is calculated and sufficient
   c. Containment methods for unloading and loading areas are satisfactory
   d. Methods to prevent vehicle departure before transfer lines are disconnected
9. Inspection and documentation methods that include
   a. Method to ensure the plan is being implemented properly
   b. Retain records for at least three years
10. Site security that includes
    a. Restricting access to oil handling and storage areas
    b. Methods to secure tank valves, pumps, and loading and unloading connections when in standby status
11. SPCC training programs that include
    a. Operation and maintenance of equipment
    b. Overview of applicable environmental regulations and requirements
    c. Designation of an SPCC Plan coordinator
    d. Training schedule
    e. Personal training records
12. A "Certification of Applicability of Substantial Harm Criteria" form. If the potential exists that a release could cause substantial harm then a facility must prepare a site-specific response plan to the applicable regulatory authority.

8.2.5 Summary of the CWA

The CWA established the basic structure for regulating discharges of pollutants into the water of the United States. It also established water quality standards for surface waters. The CWA made it unlawful to discharge any pollutant from a point source into navigable waters unless a permit was obtained. USEPA has estimated that cleaning up and preventing polluted discharges to surface water, mainly through the construction of publically owned treatment works (POTWs), has cost over 56 billion dollars since the CWA was passed in 1972 (USEPA 2021b). There is no doubt that the CWA has been one of many great success stories of environmental regulations in the United States thus far. However, much work remains especially from nonpoint sources of water pollution (USEPA 2021b).

Nonpoint source water pollution occurs when atmospheric deposition, rainfall, or snowmelt flows over land or through the ground, dissolves or carries pollutants, and deposits them into rivers, lakes or streams, and coastal waters or introduces them into groundwater (USEPA 2021b). Regulating NPS of pollution is much more difficult to regulate because the pollution is generally, by the nature of its definition, diffuse and widespread throughout the environment. According to USEPA (2021b), NPS pollution includes the following:

- Fertilizers including phosphorus and nitrogen
- Herbicides
- Insecticides
- Oil
- Grease
- Salt from roads and irrigation
- Bacteria from livestock and septic tanks

One of the best examples that demonstrates the need for NPS regulations under the CWA is the Gulf of Mexico hypoxic zone. It is commonly referred to as the Gulf of Mexico Dead Zone because the oxygen levels within the zone are too low to support marine life. Sometimes it is even referred to as Red Tide. The Dead Zone was first documented as early as the 1970s and originally occurred every few years (Carlisle 2000). Now it occurs every year and in some years is so huge that it is roughly equivalent in size of the State of Georgia which is nearly 100,000 square kilometers and has also appeared as far away as the Florida coast (NOAA 2018). NOAA now releases routine bulletins to inform boaters and beach visitors of the presence of what they term, Gulf of Mexico Harmful Algal Bloom (NOAA 2018). The main cause has been traced to NPS of pollution, mainly rural and agricultural land in the Mississippi River Basin and other drainage basins that discharge to the Gulf of Mexico that carry dissolved fertilizers. Once in the Gulf of Mexico, the fertilizers provide nutrients favorable for algae growth and deplete the oxygen supply in the water resulting in massive (in the millions or more) marine life deaths (Malakoff 1998, Carlisle 2000).

8.3 SAFE DRINKING WATER ACT

Almost immediately after the CWA was passed, the need to establish water quality standards for drinking water under the jurisdiction of the USEPA was a priority to protect human health and provide consistency and coordination with the CWA (USEPA 2021d).

8.3.1 INTRODUCTION

The SDWA of 1974 was established to protect the quality of drinking water in the United States. This law focuses on all waters actually or potentially designed for use as drinking water from above-ground or below-ground sources, which includes groundwater (USEPA 2021d). In the United States, there are approximately 170,000 community water systems (USEPA 2021d). USEPA defines a community water system as a water system that has 15 or more service connections or serves 25 people per day for 60 or more days per year (USEPA 2021d). Drinking water standards apply to water systems differently based on their type and use. The types of water systems include the following (USEPA 2021d):

- Community water system: A community water system is a water system that serves the same people year-round. Examples include residences that include homes, apartments, and condominiums in cities, small towns, and mobile home parks.
- Noncommunity water system: A noncommunity water system is a public water system that serves the public but does not serve the same people year-round. There are two types of noncommunity systems:

1. Nontransient Noncommunity water system: This type of water system serves the same people more than six months a year, but not year-round. An example would be a school with its own water supply.
2. Transient noncommunity water system: This type of water system serves the public but not the same individuals for more than six months a year. An example of this type of water system may be a rest area or campground.

The SDWA authorizes the USEPA to establish minimum standards to protect tap water and requires all owners or operators of public water systems to comply with these standards termed Primary Drinking Water Standards. **National Primary Drinking Water Standards** (NPDWS) are defined as those drinking water standards that are based on health-related criteria established by USEPA (2021e). The 1996 amendments to the SDWA required USEPA to consider a detailed risk assessment, cost assessment, and best available peer-reviewed science when revising the standards (USEPA 2021d). USEPA and State governments, which can be approved by USEPA to implement these rules on behalf of USEPA, also encourage compliance with an additional set of standards termed Secondary Standards. Secondary Standards are commonly referred to as nuisance-related compounds (USEPA 2021d). Currently, there are approximately 170,000 public water supplies in the United States that are subject to the SDWA. Private wells are not covered under the Act. The SDWA also does not apply to bottled water. Bottled water is regulated by the FDA under the Federal Food, Drug, and Cosmetic Act (USEPA 2021d).

8.3.2 Maximum Contaminant Levels

NPDWS regulations include both mandatory levels, termed Maximum Contaminant Levels or MCLs, which are enforceable and nonenforceable health goals termed Maximum Contaminant Level Goals or MCLGs for each contaminant listed. MCLs have additional significance because they can be used under the Superfund Law, which we will discuss later in the chapter, as "Applicable or Relevant and Appropriate Requirements" in setting cleanup requirements at contaminated sites on the National Priority List (NPL). NPDWS are organized into six groups of compounds, which include the following (USEPA 2021d, 2021e, 2021f):

- Microorganisms
- Disinfectants
- Disinfection byproducts
- Inorganic chemicals
- Organic chemicals
- Radionuclides

Table 8.2 lists the current MCLs or MCLGs for the chemicals regulated under the primary drinking water standards. Table 8.3 lists the current MCLs or MCLGs for secondary standards. The complete lists of each are at https://www.epa.gov/dwstandardsregulations. (USEPA 2021g).

8.3.3 National Secondary Drinking Water Regulations

National Secondary Drinking Water Regulations (NSDWRs) are nonenforceable guidelines regulating contaminants that may cause cosmetic effects, such as skin or tooth discoloration, or aesthetic effects, such as taste, odor, or color, in drinking water. USEPA recommends secondary standards for water systems but does not require systems to comply with the standards. However, states may choose to adopt the secondary standards as enforceable (USEPA 2021g).

Table 8.3 lists the current MCLs or MCLGs for secondary standards. The complete lists of each are at https://www.epa.gov/dwstandardsregulations (USEPA 2021h).

TABLE 8.2
MCLGs for Select Parameters

Contaminant	MCLG	Sources of Contaminant
Cryptosporidium	Zero	Human and animal fecal waste
Legionella	Zero	Present naturally in water
Total Coliform (*E. coli*)	Zero	*E. coli* only originates from human and animal fecal waste
Total Trihalomethanes	0.080 Mg/L	Byproduct of disinfection
Chlorine (as Cl_2)	4.0 mg/L	Byproduct of disinfection
Arsenic	0.10 mg/L	Naturally occurring, farming, and industry
Asbestos	7E+7 fibers/L	Naturally occurring, decay of water mains
Barium	2 mg/L	Naturally occurring and drilling waste
Cadmium	0.005 mg/L	Naturally occurring, water pipe corrosion
Chromium (total)	0.1 mg/L	Naturally occurring, industrial discharges
Copper	1.3 mg/L	Naturally occurring, household plumbing
Cyanide	0.2 mg/L	Fertilizer plants, industrial discharges
Lead	0.015 mg/L	Naturally occurring, water pipe corrosion
Mercury	0.002 mg/l	Naturally occurring
Nitrate (N)	10 mg/L	Naturally occurring. fertilizer
Selenium	0.05 mg/L	Naturally occurring
Acrylamide	Zero	Sewerage treatment chemical
Alachlor	Zero	Herbicide, agricultural runoff
Atrazine	0.003 mg/L	Herbicide, agricultural runoff
Benzene	0.005 mg/L	Gasoline, industrial discharges
Benzo(a)pyrene (PAHs)	0.0002	Oil and industrial discharges
Carbon tetrachloride	0.005	Industrial discharges
Chlordane	0.002	Banned termiticide
Chlorobenzene	0.1 mg/L	Industrial discharges
2,4-D	0.07	Herbicide, farming runoff
Dalapon	0.2 mg/L	Herbicide, farming runoff
1,2-Dibromo-3-chloropropane	Zero	Agricultural runoff
o-Dichlorobenzene	0.6 mg/L	Industrial discharges
1,2-Dichloroethane	0.005 mg/L	Industrial discharges
Cis-1,2-Dichloroethylene	0.07 mg/L	Industrial discharges
Trans-1,2-Dichloroethylene	0.1 mg/L	Industrial discharges
Dichloromethane	0.005 mg/L	Industrial discharges
1,2-Dichloropropane	0.005 mg/L	Industrial discharges
Di(2-ethylhexyl) phthalate	0.006 mg/L	Industrial discharges
Dioxin	3.0E-8 mg/L	Waste incineration and other combustion
Diquat	0.02 mg/L	Herbicide, agricultural runoff
Endrin	0.002 mg/L	Insecticide, agricultural runoff
Trichloroethylene	0.005 mg/L	Industrial discharges
Ethylbenzene	0.7 mg/L	Gasoline and industrial discharges
Tetrachloroethylene	0.005 mg/L	Industrial discharges

Source: United States Environmental Protection Agency (USEPA). 2021d. Drinking Water Contaminants – Standards and Regulations. https://www.epa.gov/dwstandardsregulations (accessed February 20, 2021).
mg/L, milligram per liter.

TABLE 8.3
Secondary Drinking Water Standards

Contaminant	Secondary MCL	Effect
Aluminum	0.05–0.2 mg/L	Colored water
Chloride	250 mg/L	Salty taste
Color	15 color units	Visible tint
Copper	1.0 mg/L	Metallic taste, blue-green staining
Corrosivity	Noncorrosive	Metallic taste, corroded pipes/fixtures, staining
Fluoride	2.0 mg/L	Tooth discoloration
Foaming agents	0.5 mg/L	Frothy, cloudy, bitter taste, odor
Iron	0.3 mg/L	Rusty color, sediment, metallic taste, reddish or orange staining
Manganese	0.05 mg/L	Black to brown color, staining, bitter metallic taste
Odor	3 TON (threshold odor number)	Rotten egg odor, musty or chemical odor
pH	6.5–8.5	Low pH: bitter taste, corrosionHigh pH: slippery feel, soda taste, deposits
Silver	0.1 mg/L	Skin discoloration, graying of the white portion of the eye
Sulfate	250 mg/L	Salty taste
Total Dissolved Solids (TDS)	500 mg/L	Hardness, deposits, colored water, staining, salty taste
Zinc	5 mg/L	Metallic taste

Source: United States Environmental Protection Agency. 2021e. Drinking Water Standards. https://www.epa.gov/safedrinkingwater (accessed February 20, 2021).
mg/L, milligram per liter.

The benefits include the following (USEPA 2021c):

- Reduction of contaminants at the point of consumption or exposure
- Cost savings due to extending the useful life of water mains and service lines
- Energy savings from transporting water more easily through the smoother, less corroded distribution system
- Reduced water losses through leak and broken water main prevention

8.3.4 LEAD AND COPPER RULE

Lead and copper enter drinking water primarily through plumbing materials. For instance, water pipping in the United States is typically copper. In addition, lead was historically used as a component of the material that was used to connect copper piping together. In 1991, the USEPA published a regulation to control lead and copper in drinking water. The treatment technique for the rule requires water systems to be monitored at customer taps. If lead concentrations exceed an action level of 0.015 milligrams per liter (mg/L) or if copper concentrations exceed 1.3 mg/L in more than 10% of customer taps sampled, the system must undertake a number of additional actions to control corrosion (USEPA 1992).

Lead is rarely present in significant quantities in naturally occurring sources of water, such as lakes, streams, rivers, or groundwater. Lead from pipes, faucets, and fixtures can dissolve into water or can sometimes enter as flakes or small particles. To keep lead from entering the water, USEPA requires some systems, including those that are having difficulty controlling lead, to treat water using certain chemicals that keep the lead in place by reducing corrosion. When corrosion control alone is not sufficient to control lead exposure, USEPA requires systems to educate the public who may be using the affected water system about the risks and USEPA requires that the water system be upgraded to remove lead from the system (USEPA 2021i).

8.3.5 CORROSION CONTROL

Corrosion control is perhaps the single most cost-effective method a water distribution system can use to treat iron, copper, zinc, and sometimes lead (USEPA 2021i).

Corrosion control is most often a function of monitoring and controlling the pH of the water supply. A pH that is less than 7 may lead to an increase of corrosion and lead to leach of metals from the pipes and water mains. This leads to an increase in concentration of some heavy metals in the water that include copper, iron, zinc, and lead. Lead may increase in concentration in older water systems since lead was more widely used as solder and pipe connections historically. This resulted in the enactment of the Lead and Copper Rule by USEPA in 1991 in an effort to reduce the amount of lead and copper in drinking water from plumbing and also lowered the MCL for lead in drinking water from 50 µg/L to 15 µg/L (USEPA 1992).

Corrosion control is not used for treatment to remove heavy metals from the water supply but has a secondary benefit of preventing the leaching of heavy metals from the water distribution system. Therefore, corrosion control should be viewed as not necessarily increasing the concentration of heavy metals not already present from the source (USEPA 2021i).

Applying USEPA regulations for safe drinking water is complex due to numerous factors that can influence water quality as evidenced by the situation that occurred in Flint, Michigan, when the water supply was switched to the Flint River as its source and ultimately resulted in higher lead concentrations in the drinking water supply (USEPA 2016).

8.3.6 SUMMARY OF THE SDWA

The availability of clean water is crucial to sustain life. Water quality for drinking water and all potable use in the United States is considered good. but shows signs of degradation in both surface water and groundwater sources. The United States has invested significantly in providing clean water through effective environmental regulations, with very few exceptions. Future regulatory protections will require replacing old and degraded service lines, many of which contain lead. Some water service lines also leak and break, especially during the colder months in the northern portion of the United States.

8.4   SUMMARY OF WATER REGULATIONS

The CWA and DWA have been tremendously successful at improving water quality within the United States within the last 50 years. This success has set the example throughout the world. However, there are still an estimated 2.8 billion people who do not have access to clean water. In addition, water sources in the United States show signs of degradation of both surface water and groundwater. In addition, overuse has become a significant issue, especially in the western portion of the United States where water is scarce. The scarcity of water in the western states can be traced to an increasing population, agriculture, and polluted water from anthropogenic sources.

Water regulations can only go so far to protect water. To protect water in the future will require coordinated sustainability measures that go well beyond environmental regulations. This is where sustainability measures must be employed to address issues with water scarcity and pollution from sources that are in need of further regulations. These other areas include improved land-use practices, modifying agricultural practices, wetland restoration, addressing nonpoint sources of pollution, conservation measures, population increases, and others. Many of these efforts are discussed in greater detail later in this book.

Questions and Exercises for Discussion

1. You are a member of a team conducting a wastewater discharge review of an industrial facility. You notice that the facility discharges to a small stream that flows adjacent to the facility. You notice a sheen on the water surface at a downstream location of the facility. What set of actions should you conduct?

2. You are a member of a team conducting a stormwater review of an industrial facility. You notice that there are several 55-gallon drums stored outside, next to a building on a paved surface that are labeled used oil, and appeared to be full. You also notice that there is no roof over the drums and one drum has an open lid, and a storm drain is less than 5 meters away from the drums. What set of actions do you feel are appropriate and why?

3. You are a facility environmental engineer at a manufacturing facility. You review stormwater monitoring data and notice that a stormwater sample had a zinc concentration of 0.30 milligrams per liter. The facility permit states that the discharge limit for zinc is 0.11 milligrams per liter. You also notice that a rainwater sample collected during the same stormwater event had a zinc concentration of 0.20 milligrams per liter. What set of actions would you undertake?

4. You are reviewing analytical data of drinking water samples collected from an elementary school and notice a sample concentration of 400 microgram per liter in one of several samples. What set of actions would you undertake?

REFERENCES

Bates, C. G. and Ciment, J. 2013. *Encyclopedia of Global Social Issues*. M.E. Sharpe Publishers. New York. 1450 p.

California Environmental Protection Agency (CalEPA). 2001. *Storm Water Sampling Guidance Document*. California Stormwater Task Force: Sacramento, California. 30 p.

Carlisle, E. 2000. *The Gulf of Mexico Dead Zone and Red Tides. The Louisiana Environment*. Tulane University. New Orleans. Louisiana. 6 p.

Malakoff, D. 1998. Death by Suffocation in the Gulf of Mexico. *Journal of Science*. Vol. 281. pp. 190–192.

National Oceanic and Atmospheric Administration (NOAA).2018Ocean Pollution. https://www.noaa.gov/resource-collections/ocean-pollution (accessed August 12, 2021).

Rogers, D. T. 2020. *Urban Watersheds: Geology, Contamination, Environmental Regulations, and Sustainability*. CRC Press. Boca Raton, FL. 606 p.

United States Army Corps of Engineers. 2021. Overview of Wetlands Regulation in the United States. https://www.saw.usace.army.mil/wetlands (accessed February 15, 2021).

United Nations. 2021. The United Nations Environment Programme. https://www.unenvironment.org/environment-you (accessed August 12, 2021).

United States Environmental Protection Agency (USEPA). 1992. Lead and Copper Rule. https://www.epa.gov/dwreginfo/lead-and-copper-rule (accessed February 20, 2021).

United States Environmental Protection Agency (USEPA). 2009. Industrial Stormwater Monitoring and Sampling Guide. EPA 832-B-09-003. Washington D.C. 49 p.

United States Environmental Protection Agency (USEPA). 2010. National Pollution Discharge Elimination System Permit Writer's Manual. United Stated Environmental Protection Agency. Office of Water. Washington D.C. 320 p.

United States Environmental Protection Agency (USEPA). 2012. Spill Prevention, Control and Countermeasures Plan (SPCC) Program. United States Environmental Protection Agency (USEPA). Office of Emergency Management. Washington D.C. https://www.epa.gov/emergencies (accessed February 20, 2021).

United States Environmental Protection Agency (USEPA). 2016. Flint Drinking Water Documents. https://www.epa.gov/flint/flint-drinking-water-documents (accessed February 20, 2021).

United States Environmental Protection Agency (USEPA). 2021a. Summary of the Clean Water Act. https://epa.gov/laws-regulations/summary-clean-water-act (accessed February 20, 2021).

United States Environmental Protections Agency (USEPA). 2021b. Basic Information about Nonpoint Source (NPS) Pollution. https://www.epa.gov/nps/basic-inofmration-about-nonpoint-source-nps-pollution (accessed February 20, 2021).

United States Environmental Protection Agency (USEPA). 2021c. National Recommended Water Quality Criteria-Human Health Criteria Table. https://www.epa.gov/wqc/national-recommended-water-quality-criteria-human-health-criteria-table (accessed February 20, 2021).

United States Environmental Protection Agency (USEPA). 2021d. Summary of the Safe Drinking Water Act. https://www.epa.gov/laws-regulations/summary-safe-drinking-water-act (accessed February20, 2021).

United States Environmental Protection Agency (USEPA). 2021e. Drinking Water Standards. https://www.epa.gov/safedrinkingwater (accessed February 20, 2021).

United States Environmental Protection Agency (USEPA). 2021f. National Recommended Water Quality Criteria-Human Health Criteria Table. https://www.epa.gov/wqc/national-recommended-water-quality-criteria-human-health-criteria-table (accessed February 20, 2021).

United States Environmental Protection Agency (USEPA). 2021g. Drinking Water Contaminants – Standards and Regulations. https://www.epa.gov/dwstandardsregulations (accessed February 20, 2021).

United States Environmental Protection Agency (USEPA). 2021h. Drinking Water Standards. https://www.epa.gov/safedrinkingwater (accessed February 20, 2021).

United States Environmental Protection Agency (USEPA). 2021i. Understanding the Lead and Copper Rule. https://www.epa.gov/ground-water-and-drinking-water/lead-and-copper-101 (accessed February 20, 2021).

Wisconsin Department of Natural Resources. 2021. Sample Stormwater Pollution Prevention Plan (SWPPP). https://www.dnr.wi.gov/topic/stormwater/documents/sampleSWPPP.pdf (accessed February 20, 2021).

9 Protecting the Land through Environmental Regulations

9.1 INTRODUCTION TO LAND REGULATIONS

This chapter addresses environmental regulations that are focused on protecting the land. In the United States, the Resource, Conservation, and Recovery Act (RCRA) and the Comprehensive Environmental Response, Compensation and Liability Act (CERCLA) form the foundation of environmental regulations that protect the land.

9.2 RESOURCE, CONSERVATION, AND RECOVERY ACT

The Resource Conservation and Recovery Act was enacted in 1976 and is the principal law in the United States addressing land-based disposal of solid and hazardous waste (USEPA 2021a). It was enacted to address problems that the United States faced from its growing volume of municipal and industrial waste.

In very general terms, RCRA defines a waste as anything that is discarded (USEPA 2021a). RCRA recognizes that a solid waste may not be a solid. Under RCRA, many solid wastes are liquid, semi-solid, or contained gaseous material (USEPA 2021a). RCRA set goals for

- Protecting human health and the environment from the hazards of waste disposal.
- Energy conservation and natural resources.
- Reducing the amount of waste generated, through source reduction and recycling.
- Ensuring the management of waste in an environmentally sound manner.

RCRA also established standards for the treatment, storage, and disposal of hazardous waste in the United States. In the United States, RCRA is responsible for (USEPA 2021a)

- Managing approximately 2.5 billion tons of solid, industrial, and hazardous waste
- Overseeing 6,600 facilities with over 20,000 process units in the full permitting universe
- Working to address more than 3,700 existing contaminated facilities in need of cleanup
- Reviewing a possible 2,000 additional facilities that may need cleanup in the future
- Providing nearly $100 million dollars in grant funding to assist states to implement authorized waste programs
- Providing incentives and opportunities to reduce or avoid greenhouse gas emissions through material and land management practices

It is important to note and highlight some of RCRA's accomplishments nationwide to understand how RCRA is currently viewed and what the future made hold. The major accomplishments of RCRA include (USEPA 2021a)

- Developing a comprehensive nationwide system to manage waste from its point of generation to its final resting place which is commonly referred to as "cradle to grave"
- Establishing the framework for states to implement effective municipal solid waste and nonhazardous waste management programs

DOI: 10.1201/9781003175810-9

- Preventing contamination from adversely impacting communities and becoming future Superfund sites.
- Restoring 18 million acres of contaminated lands, nearly equal to the size of South Carolina
- Creating partnership and award programs to encourage companies to modify manufacturing practices to generate less waste and reuse materials safely
- Enhancing perceptions of wastes as valuable commodities that can be part of new products through USEPA's sustainable materials management efforts
- Creating and enhancing the nation's recycling infrastructure and increasing municipal solid waste recycling/composting rate from less than 7% to nearly 35%

RCRA divides those directly affected by the regulation into three categories that include (USEPA 2021a)

1. Generators: Generators must determine whether they are creating any wastes that would be classified as hazardous. If so, they must obtain a tracking number and manifest, ensure proper storage and labeling of the waste, and keep records.
2. Transporters: Transporters ensure that they comply with USEPA and USDOT requirements for transportation of hazardous materials (HAZMAT) and must ensure proper packaging, labeling, reporting, and record keeping.
3. Treatment, Storage, and Disposal Facilities (TSD): Treatment, Storage, and Disposal Facilities have by far the most complex and stringent requirements. They must obtain permits, which require inspections and monitoring, as well as comply with the manifest system.

There are three main components of RCRA called subtitles. The first describes and defines a nonhazardous waste, the second defines and describes a hazardous waste, and the third addresses petroleum underground storage tanks. In addition to the three subtitles, RCRA also includes methods for analytical testing of solid wastes under RCRA and is called Test Methods for Evaluating Solid Waste (SW-846). This is significant because what the SW-846 provision has created is a consistent set of methods for fully characterizing any solid waste (Cornell University Law 2021 and USEPA 2021b).

The three main components of RCRA and SW-846 are described in greater detail in the following sections.

9.2.1 SUBTITLE D – NONHAZARDOUS WASTE

Regulations under Subtitle D addresses Non-hazardous waste which essentially banned open dumping of waste and set minimum federal criteria for the operation of municipal waste and industrial waste landfills, including (USEPA 2021a)

- Design criteria
- Locations restrictions
- Financial assurance
- Corrective action or cleanup
- Closure requirements

States play a lead role in implementing these regulations and may set more stringent requirements but cannot set less stringent requirements. States may not set less stringent requirements. USEPA approves State programs and if a State does not have an approved program, then the waste facilities must demonstrate that they meet the requirements.

RCRA defines a solid waste in broad terms and includes, solids, sludges, liquid, semisolids, or contained gaseous material. In defining waste, USEPA focuses on the actual waste, not materials that were still part of the manufacturing process. USEPA intended to exclude recycling but also wanted to prevent fraudulent recycling efforts. Thus, USEPA uses the following 5-factor test to determine the definition of waste (USEPA 2021a):

1. Whether the material is typically discarded on an industry-wide basis.
2. Whether the material replaces raw material when it is recycled and the degree to which its composition is similar to that of the raw material.
3. The relation of the recovery practice to the principal activity of the facility.
4. If the material is handled prior to reclamation in a secure manner that minimizes loss and prevents releases to the environment.
5. Other factors, such as the length of time the material is accumulated.

Nonhazardous solid wastes include certain hazardous wastes which are exempted from the Subtitle C regulations, such as hazardous wastes from households and from conditionally exempt small quantity generators, which we will define later. Oil and gas exploration and production wastes, such as drilling cutting, produced water, and drilling fluids are categorized as "special wastes" and are also exempt from Subtitle C. Subtitle D also includes garbage (e.g., food containers, coffee grounds), nonrecycled household appliances, residue from incinerated automobile tires, refuse such as metal scrap, construction materials, and sludge from industrial and municipal wastewater facilities and drinking water treatment plants (USEPA 2021a).

9.2.2 Subtitle C – Hazardous Waste

Regulations under Subtitle C address Hazardous Waste and were enacted to ensure that hazardous waste is managed safely from the moment it is generated to its final disposal. This is termed "cradle-to-grave." Subtitle C regulations set criteria for hazardous waste generators, transporters, and treatment, storage, and disposal facilities. This also includes permitting requirements, enforcement, and corrective action or cleanup. An important definition central to RCRA and its enforcement is what defines a hazardous waste. From a universal perspective, a hazardous waste is defined as a waste that could pose a threat to human or public health or the environment if it were to be released. From this basic definition, the potential list of hazardous wastes is large and diverse and indeed this is a true statement. Therefore, USEPA also has the following definition of hazardous waste (USEPA 2021a, 2021c):

A solid waste, or combination of solid waste, which because of its quantity, concentration, or physical, chemical, or infectious characteristics may (a) cause, or significantly contribute to, an increase in mortality or an increase in serious irreversible, or incapacitating reversible, illness; or (b) pose a substantial present or potential hazard to human health or the environment when improperly treated, stored, transported, or disposed of, or otherwise managed.

This definition is broad as well. However, it does provide a general indication of which wastes USEPA intended to regulate as hazardous, but it obviously does not provide the clear scientific distinctions necessary for waste generators to determine whether their wastes pose a significant threat to warrant regulation or not in most cases. Therefore, USEPA developed more specific criteria for defining hazardous waste and decided on two definitions; a statutory definition and a regulatory definition. The statutory definition is presented above and serves as a general guideline in what would become the regulatory definition which we discuss in much greater detail below. Hazardous wastes can be solids, liquids, and contained gases. They can be by-products of manufacturing processes, discarded used materials, or discarded unused commercial products, such as residual cleaning fluids (solvents), cleaning products (bleach or ammonia), paints, pigments, pesticides, or a multitude of other items (USEPA 2021a). RCRA divides the regulatory definition of a hazardous waste into three major categories (USEPA 2021a):

1. Characteristic Waste
2. Listed Waste
3. Universal Waste

There are a couple of other categories of hazardous wastes or wastes that are sometimes managed as hazardous and they include used oil and wastes that are considered hazardous because of what is called the "contained-in rule" or the "derived-from policy." In addition, there are what is called "Land Disposal Restrictions" (LDRs) which must be met for hazardous wastes. We will discuss each in the sections below along with manifesting hazardous wastes in its own section.

9.2.2.1   Characteristic of Hazardous Waste

Under RCRA (USEPA 2021a), a characteristic waste is materials that are known or tested to exhibit one or more of the following four hazardous traits:

- Ignitibility
- Reactivity
- Corrosivity
- Toxicity

Ignitibility is a trait of wastes that can create fires under certain conditions, undergo spontaneous combustions, or have a flashpoint of less than 60°C (140°F). Examples include used oil and solvents. Test methods to determine whether a waste is ignitable and therefore a hazardous waste include the (1) Pensky-Martens Closed-Cup Method, (2) the Staflash-Closed-Cup Method, and (3) Ignitibility of Solids, which are each described in SW-846 Sections 1010, 1020, and 1030, respectively (USEPA 2005, 2021d).

Corrosivity is a trait of materials, including solids, that are acids or bases, or that produce acidic and alkaline solutions. Aqueous wastes with a pH of less than or equal to 2.0 or great than or equal to 12.5 are corrosive and therefore hazardous wastes. A liquid may also be corrosive if it is able to corrode metal containers, such as storage tanks, drums, or barrels. Spent battery acid is an example. Test methods to determine whether a waste exhibits the characteristic of corrosivity are pH Electronic Measurement and Corrosivity Towards Steel in SW-846 Methods 9040 and 1110, respectively (USEPA 2005, 2021a).

Reactivity is a trait of wastes that are unstable under normal conditions. They can cause explosions or release toxic fumes, gases, and vapors when heated, compressed, or mixed with water. Examples include lithium-sulfur batteries and unused explosives. Wastes are evaluated for reactivity using criteria set forth in the hazardous waste regulations in SW-846 (USEPA 2021e).

Toxicity is a trait of wastes that are harmful or fatal when ingested or absorbed (e.g., wastes containing mercury, lead, DDT, PCBs, etc.). When toxic wastes are disposed, the toxic constituents may leach from the waste and impact groundwater. The characteristic of toxicity contains eight subsections described below. A waste is a toxic hazardous waste if it is identified as being toxic by any of the eight subsections that include (USEPA 2021d)

1. TCLP: Toxic as defined through the application of a laboratory test procedure called Toxicity Characteristic Leaching Procedure or TCLP Test. The analytical method is described in USEPA SW-846 Method 1311. The TCLP identifies wastes (as hazardous) that may leach hazardous concentrations of toxic substances into the environment. The result of the TCLP test is compared to the Regulatory Level (RL) (USEPA 2021e). Table 9.1 lists some common chemical compounds with their RLs.

TABLE 9.1

Select Maximum Concentrations of Contaminants for Toxicity Characteristic

Contaminant	Hazardous Waste Code	Regulatory Level (RL) (mg/L)	Contaminant	Hazardous Waste Code	Regulatory Level (RL) (mg/L)
Arsenic (As)	D004	5.0	Barium (Ba)	D005	100.0
Benzene	D018	0.5	Cadmium (Cd)	D006	1.0
Carbon Tetrachloride	D019	0.5	Chlordane	D020	0.03
Chlorobenzene	D021	100.0	Chloroform	D022	6.0
Chromium (Cr)	D007	5.0	o-Cresol	D024	200.0
m-Cresol	D024	200.0	p-Cresol	D025	200.0
Cresol (total)	D026	200.0	2,4-D	D016	10.0
1,4-Dichlorobenzene	D027	7.5	1,2-Dichloroethane	D028	0.5
1,-Dichloroethylene	D09	0.7	2,4-Dichlorotoluene	D030	0.13
Endrin	D012	0.02	Heptachlor	D031	0.008
Hexachlorobenzene	D032	0.13	Hexachlorobutadiene	D033	0.5
Hexachloroethane	Do34	3.0	Lead (Pb)	D008	5.0
Lindane	D013	0.4	Mercury (Hg)	D009	0.2
Methoxychlor	D014	10.0	Methyl ethyl ketone	D035	200.0
Nitrobenzene	D036	2.0	Pentachlorophenol	D037	100.0
Pyridine	D038	5.0	Selenium (Se)	D010	1.0
Silver (Ag)	D011	5.0	Tetrachloroethylene	D039	0.7
Toxaphene	D015	0.5	Trichloroethylene	D040	0.5
2,4,5-Trichlorophenol	D041	400.0	2,4,6-Trichlrophenol	D42	2.0
2,4,5-TP (Silvex)	D017	1.0	Vinyl chloride	D043	0.2

Source: United States Environmental Protection Agency (USEPA). 2021d. Resource Conservation and Recovery Act Listed and Characteristic Hazardous Wastes. Office of Solid Waste. Washington, D.C. https://www.epa.gov/hazarodus-wastes (accessed March 24, 2021).

Note: mg/L, milligram per liter.

2. Total and WET: Toxic as defined through the application of laboratory test procedures called the "total digestion" and the "Waste Extract Test," commonly referred to as the WET. The results of each of these laboratory tests are compared to their Soluble Threshold Limit Concentrations (STLCs) (USEPA 2021e).

3. Acute Oral Toxicity: Toxic because the waste either is an acutely toxic substance or contains an acutely toxic substance if ingested. A waste identified as being toxic would have an acute oral LD50 less than 2,500 mg/kg or simply calculated LD50. We will define and explain LD50 when we discuss health risk assessments in a later chapter.

4. Acute Dermal Toxicity: Toxic because the waste either is an acutely toxic substance or contains an acutely toxic substance if inhaled. A waste is identified as being toxic would have a dermal LC50 less than 4,300 mg/kg or simply calculated LC50. Again, we will define and explain LC50 when we discuss health risk assessments in a later chapter.

5. Acute Inhalation Toxicity: Toxic because the waste either is an acutely toxic substance or contains an acutely toxic substance if inhaled. A waste is as identified as being toxic if it has a dermal LC50 less than 10,000 mg/kg. Again, we will define and explain LC50 when we discuss health risk assessments in a later chapter. USEPA test method 3810 in SW-846 termed Headspace is the recommended analytical test to evaluate inhalation toxicity.

6. Acute Aquatic Toxicity: Toxic because the waste is toxic to fish. A waste is aquatically toxic if it produces an LC50 less than 500 mg/L when tested using the Static Acute Bioassay Procedures for Hazardous Waste Samples (USEPA 2021e).
7. Carcinogenicity: Toxic because it contains one or more carcinogenic substances. A waste is identified as being toxic if it contains any of the specified carcinogens at a concentration of greater than or equal to 0.001% by weight.
8. Experience or Testing: A waste may be toxic, and therefore a hazardous waste, even if it is not identified as toxic by any of the seven criteria listed above. At the present time, only wastes containing ethylene glycol (e.g., spent antifreeze) have been identified as toxic in this subsection.

9.2.2.2 Listed Hazardous Waste

Under RCRA (USEPA 2021d), a hazardous waste can also be what is called a "Listed Waste." A listed waste is a waste that appears on one of four RCRA hazardous waste lists described as follows (USEPA 2021d):

F-Listed Wastes: F-Listed Wastes identify wastes from many common manufacturing and industrial processes, such as solvents that have been used for cleaning or degreasing. Since the processes producing these wastes occur in many different industry sectors, the F-listed wastes are known as wastes from nonspecific sources. Nonspecific means that the waste does not originate from one specific industry or one specific industrial or manufacturing process.

K-Listed Wastes: K-Listed Wastes include certain wastes from specific industries, such as petroleum refining or pesticide manufacturing. Also, certain sludges and wastewaters from treatment and production in these specific industries are examples of source-specific wastes.

P-Listed Wastes: P-Listed Wastes include specific commercial chemical products that have not been used, but that will be (or have been) discarded. Industrial chemicals, pesticides, and pharmaceuticals are examples of commercial chemical products that appear on the P-List and become hazardous waste when discarded.

M-Listed Wastes: M-Listed Wastes include certain wastes known to contain mercury, such as fluorescent lamps, mercury switches, and the products that house mercury switches, and mercury-containing novelties. These types of wastes may also be included as types of Universal Wastes which we will discuss later in this chapter. Therefore, clarification from the applicable regulatory authority may be required.

Now that a waste has been characterized as hazardous it must be disposed of in timely fashion. Typically, waste containers such as a 55-gallon drum that are full and have been characterized as hazardous must be affixed with an appropriate label that identifies it as hazardous. Figure 9.1 is an example of an appropriate label. For some wastes, it may take a few weeks or even longer before enough waste is accumulated to fill a container. The maximum accumulation time allowed ranges between 90 and 270 and is dependent upon (USEPA 2021d):

- How much waste is accumulated in any month?
- The generator status of the facility (i.e., small quantity generator, large quantity generator, etc.).
- Whether the waste is acutely hazardous or extremely hazardous.

Typically, the "clock" begins on the first date on which any amount of hazardous waste begins to accumulate during that month (USEPA 2021e).

9.2.2.3 Universal Hazardous Waste

A Universal Waste is a category of waste materials designated as hazardous waste but containing materials that are very common. Universal wastes are widely produced by households and businesses alike and typically include the following (USEPA 2021d):

FIGURE 9.1 Hazardous waste label.

Source: United States Environmental Protection Agency (USEPA). 2021d. Resource Conservation and Recovery Act Listed and Characteristic Hazardous Wastes. Office of Solid Waste. Washington, D.C. https://www.epa.gov/hazarodus-wastes (accessed March 24, 2021).

- Televisions
- Computers
- Cell phones
- Other electronic devices such as VCRs, portable DVD players, etc.
- Fluorescent tubes and bulbs, high-intensity discharge lamps, sodium vapor lamps, and electric lamps that contain added mercury, as well as other lamps that exhibit a characteristic of a hazardous waste (e.g., typically lead).
- Pesticides
- Mercury-containing equipment including thermostats, switches, mercury thermometers, pressure and vacuum gauges, dilators and weighted tubing, mercury gas flow regulators, dental amalgams, counterweights, dampers, mercury-added novelties such as jewelry, ornaments, and footwear. Of special note of clarification is that mercury-containing equipment may also be considered as a listed M-waste. In these cases, clarification from the applicable regulatory authority may be required.
- Cathode Ray Tubes (CRTs) which are typically the glass picture tubes removed from devices such as televisions and computer monitors
- Nonempty aerosol cans

9.2.2.4 Used Oil

Some States, such as California, regulate used oil as a hazardous waste even if they do not exhibit any characteristic of a hazardous waste. The term **Used Oil** is a defined term meaning any oil that has been refined from crude oil, or any synthetic oil that has been used and, as a result of use, is contaminated with physical or chemical impurities. Other materials that contain or are contaminated with used oil are also commonly considered a hazardous waste as well and are subject to Used Oil regulations under RCRA 40 CFR Part 279.22 (USEPA 2021f).

In general, used oil must be kept in approved containers with secondary containment that does not leak and must be properly labeled as "Used Oil" (USEPA 2021f).

9.2.2.5   Contained-In Policy and Derived-From Rule

Materials can be hazardous even if they are not specifically listed or don't exhibit any characteristic of a hazardous waste. This is often confusing so let's try and make sense of a couple of hazardous waste concepts to understand if a waste is classified as a hazardous waste in a step-by-step approach. For example, let's discuss what is considered "the contained-in rule" (USEPA 2021a).

The contained-in rule or contained-in policy under RCRA empowers hazardous waste with the power to transform solids, otherwise not hazardous, into a hazardous waste. USEPA separates the contained-in rule from the mixture or derived-from rule as separate and distinct. This power to render a solid waste into a hazardous waste stems from USEPA's general policy that "once a hazardous waste always a hazardous waste." Once a hazardous waste is hazardous, it is presumed to be forever hazardous regardless of changes in its form or its combination with other substances. This rule provides that any mixture of a listed hazardous waste and a solid waste is itself, in totality, a hazardous waste regardless of the concentration of the hazardous constituents in the waste even if it no longer exhibits any toxicity (USEPA 2021a).

The derived-in rule differs slightly in concept from the contained-in rule but the outcome is the same. To best explain the derived-from rule is through an example. Let's say an ash is created through incineration of a listed hazardous waste. Since the ash is created from a listed hazardous waste, the ash itself is considered a listed hazardous waste. Thus, the ash produced by burning the listed hazardous waste bears the same hazardous waste code and regulatory status as the originally listed waste, regardless of the ash's actual physical and chemical properties (USEPA 2021a).

USEPA's basic position on termination of a material's status as a listed hazardous waste is simple. Once a listed hazardous waste has been classified as a hazardous waste it generally remains a hazardous waste forever. There are two explicit bases for termination of this status. First, an unlisted waste that is hazardous only because of a characteristic, ceases to be a hazardous waste if it no longer exhibits a hazardous waste characteristic. Second, a listed waste which is a waste containing a listed waste, or a waste derived from a listed waste ceases to be a hazardous waste only if it has been "delisted" from classification. Although a characteristic waste can lose its status as a hazardous waste without action by USEPA, in most cases a listed waste will remain hazardous until USPEA affirmatively grants a petition to reclassify the material as nonhazardous (USEPA 2021a). Some States, such as California have a process by which a listed hazardous waste that impacts a media such as soil can be delisted if it is demonstrated that the listed waste is present in insignificant concentrations based on a risk-based evaluation (California Department of Toxic Substance and Control [CalDTSC] 2021).

By adopting the mixture and derived-from rules, USEPA established incentives for generators to keep waste streams separate and to limit the amount of hazardous waste generated since the cost for treatment and disposal is much higher for a hazardous waste compared to a nonhazardous waste (2021a).

9.2.2.6   Land Disposal Restrictions

The 1984 amendments to RCRA called the Hazardous and Solid Waste Amendments (HSWA) required USEPA to establish treatment standards termed **Land Disposal Restrictions** (LDRs) for all listed and characteristic hazardous wastes destined for land disposal. For wastes that are restricted, the amendments to RCRA required USEPA to set concentration levels or methods of treatment, both of which are called Treatment Standards, that substantially diminish the toxicity of the wastes or reduce the likelihood that hazardous constituents contained within the wastes will migrate from the disposal site and potentially impact human health or the environment in an adverse way (USEPA 1991). The major reason for established LDRs was for the prevention of groundwater impacts, usually through the leaching of contaminants from hazardous wastes from landfills and potentially impacting potable water supplies.

LDR provisions focus on eliminating or greatly reducing the probability of impacting groundwater through proper treatment of hazardous or toxic constituents in hazardous waste before land disposal. The processes may include (United States Environmental Protection Agency USEPA 1991):

- Incineration, which destroys organic hazardous compounds.
- Stabilization or immobilization, which binds toxic metals to the mass itself making it less likely to migrate. An example would be mixing the toxic material with concrete, mortar, a bentonite clay, or other material to render the waste immobile.

There are three main sections of the LDR program and they include (United States Environmental Protection Agency USEPA 1991):

1. Disposal Prohibition: This requires waste-specific treatment standards to be met before disposal of the waste on land. A facility can meet such standards by either:
   a. Treating the hazardous chemicals in the waste to meet appropriate levels. Many such methods are approved by USEPA for use, excluding dilution or
   b. Treating hazardous waste with a USEPA-approved technology. Once the waste is appropriately treated, it can then be land disposed.
2. Dilution Prohibition: Ensures that wastes are treated properly and that diluting the waste to meet LDRs is not permitted. This is because of the assumption that dilution does not reduce the overall toxicity of a hazardous waste.
3. Store Prohibition: Hazardous waste cannot be stored indefinitely.

From the moment a hazardous waste is generated, it is subject to LDRs. If a hazardous waste generator produces more than 220 pounds (or 2.2 pounds of acute hazardous waste) of hazardous waste in a calendar month, they must identify the type and nature of the waste, and also determine the course of applicable treatment of the waste before land disposal. Generators that produce less than these amounts and are conditionally exempt small quantity generators (CESQG) of hazardous waste are exempt from LDR requirements, as are waste pesticides and container residues which are disposed of by farmers on their own land, and newly-listed wastes which USEPA has not yet established treatment standards (USEPA 2021e, Ohio EPA 2014).

9.2.2.7 Analytical Methods

Much of what defines a hazardous waste is dependent upon what chemical compounds are contained within the waste. Therefore, for consistency, USEPA developed a manual describing acceptable analytical, sampling methods, and quality control for assuring that wastes were properly characterized. The document USEPA published was entitled Test Methods for Evaluating Solid Waste SW-846 or simply SW-846. It was issued in 1980 and has changed through time as technology is advanced in analytical procedures. Currently, the SW-846 manual is over 3,500 pages (USEPA 2021e). SW-846 has become one of the most important documents for not only assessing waste but also in conducting investigations at potential hazardous waste sites and in conducting environmental investigations in general.

9.2.2.8 Hazardous Waste Manifests

RCRA requires that generators of hazardous waste and owners or operators of hazardous waste treatment, storage, or disposal facilities use what is called the Uniform Hazardous Waste Manifest

(USEPA Form 8700-22) and if necessary, the continuation sheet (EPA Form 8700-22A) for both interstate and intrastate transportation. General Instructions for completing a manifest are provided at https://www.epa.gov/hwgenerators/uniform-hazardous-waste-manifest-instructions-sample-form-and-continuation-sheet. The return receipt of the manifest from the disposal facility must be returned to the generator within 45 days from when the waste was signed and shipped from the generator facility (USEPA 2021a, 2021g).

9.2.3 Subtitle I – Petroleum Underground Storage Tanks

The operation of Underground Storage Tanks (USTs) became subject to RCRA with the enactment of the Hazardous and Solid Waste Amendments of 1984 which amended RCRA by inserting Subtitle I. It was estimated at the time that there were approximately 2 million USTs in the United States. Currently, the estimate is 560,000. The UST regulations under Subtitle I cover tanks storing petroleum or a listed hazardous waste and define the types of tanks permitted. The UST requirements under Subtitle I set standards for (USEPA 2021a):

- Groundwater monitoring
- Double-walled tanks
- Release detection, prevention, and correction
- Spill control
- Overfill control

Up until the middle portion of the 1980s, UST was largely unregulated, except for those that contained hazardous wastes. It was not until the catastrophic failure of USTs and large-scale groundwater contamination occurred in Provincetown, Massachusetts, Northglenn, Colorado, and Dover-Warpole, Massachusetts did Congress recognize the need for UST regulation. Subtitle I imposed new obligations and potential liabilities on a large and diverse number of entities and businesses including (USEPA 2021a):

• Service stations	• Petroleum bulk tanks
• Petroleum bulk plants	• Municipal warehouses
• Shopping centers	• School bus yards
• Farms	• Factories
• Automobile dealerships	• Delivery services

The UST regulations apply to owners and operators of regulated UST systems. An **operator** is defined as any person in control of, or having responsibility for, the daily operation of a UST system. The definition includes current operators only, and implies that one must actively manage a tank system to be considered an operator (USEPA 2021a).

An **owner** is defined as any person who owns a UST system, unless that system was taken out of service prior to November 8, 1984. For UST systems no longer in use at that date, the owner is the person who owned the system immediately prior to its discontinuation. A person is defined broadly to include an individual, trust, firm, joint-stock company, federal agency, corporation, state, municipality, commission, state political subdivision, any interstate body, or the government of the United States (USEPA 2021a). The UST regulations impose joint and several liability on UST owners and operators and do not place primary responsibility on one over the other (USEPA 2021a).

When USTs are currently being operated, it is not difficult to identify at least one person responsible for compliance with the UST regulations. By definition, there is no operator of an abandoned or out-of-service UST system. If a UST was in use after November 8, 1984, and the UST remains in the ground, the current property owner may fall within the statutory definition as an owner. If the UST was abandoned before November 8, 1984, by a previous property owner, it may be impossible to locate records identifying the owner of the UST immediately before dis-continuation of use because UST registration was not historically required and UST closures were not reported because that was not required (USEPA 2021a).

In amending Subtitle I of RCRA, USEPA required every owner of a UST system to notify the designated state or local agency of the existence of the UST system, age, size, type, locations, product stored, and uses for each UST. Initial notification forms were due by May 8, 1986. In addition to the reporting requirement for operating USTs, owners of USTs taken out of service after January 1974 but still in the ground were required to notify a designated state agency of the existence of the UST by May 8, 1986, as well. This notice also required a UST owner to specify, to the extent known, the date that the UST was taken out of service, the age of the UST when it was taken out of service, size, type, product stored, location, and the type and quantity of substances that remained stored in the UST (USEPA 2021a).

USTs brought into service after December 22, 1988, must meet design and construction requirements. RCRA specified that owners or operators notify the designated state agency within 30 days of the existence of the new UST. The notification form for new USTs requires the owners or operators to certify compliance with the following requirements (USEPA 2021a):

1. That the UST system was properly installed by a state-certified UST contractor
2. That cathodic protection was installed for steel tanks and associated piping
3. That financial responsibility requirement we met
4. That release detection was installed and meets regulatory standards
5. That spill and overfill prevention measures were installed and meet regulatory standards

9.2.4 SUMMARY OF THE RESOURCE CONSERVATION AND RECOVERY ACT

RCRA, along with the CAA, SWA, and the SDWA form the basic fundamentals of environmental protection in the United States with the exception of cleaning up legacy sites which we will address next. RCRA was created essentially to deal with huge increases in the amount of waste that was being created throughout the United States as a direct result of our industrialization and population increase. The success of RCRA can't be understated. RCRA has changed our behavior in many ways and has involved everyone in participating in taking care of our environment and not just an industry group for almost all of us now recycle much of our discarded materials that were once considered garbage is now transformed into new products that we use and includes used tires, wood and pulp products, plastic, glass, and especially scrap metal which is now considered a commodity and has grown into an industry of itself. Looking back, there are some challenges that RCRA has failed to address proactively, and one of the most significant is municipal or household wastes. We have discussed at length what industry faces in characterizing and disposing of wastes properly in order to protect human health and the environment, but have not addressed with the same vigor our everyday household garbage which is generally exempted from RCRA but contains many of the same components and chemicals as many hazardous wastes but is not regulated.

Figure 9.2 shows some common household products that typically become hazardous wastes if they were generated by the industry as opposed to households. This to many scientists does not make sense at all when a chemical that is known to be harmful to human health and the en-vironment and is a hazardous waste but is not fully regulated as such. The Earth does not care how or by whom the material is released into the environment, it's toxic to the Earth just the same.

Next, we will discuss the Comprehensive Environmental Response Compensation and Liability Act which focuses on addressing cleaning up contaminated sites.

FIGURE 9.2   Household items that can become contaminants.

Source: United States Geological Survey (USGS). 2006. Volatile Organic Compounds in the Nation's Ground Water and Drinking-Water Supply Wells. USGS Circular 1292. Reston, VA.

9.3   MEDICAL WASTE

Medical waste is a subset of waste primarily generated at health care facilities, such as hospitals, physicians' offices and clinics, blood banks, and veterinary hospitals and clinics, medical research centers, and laboratories. **Medical waste** is defined as healthcare waste that may contain or is contaminated with blood, body fluids, or other potentially infectious materials (USEPA 2020a). Concern surrounding the potential health hazards of medical waste grew suddenly after medical wastes washed up along beaches and the coastline of the eastern United States in the 1980s (USEPA 2020a).

USEPA was the agency responsible for regulating medical waste from 1988 when medical waste first became regulated under the Medical Waste Tracking Act of 1988 took effect. However, the Act expired in 1991. Currently, medical waste is regulated by each State.

Historically, treatment and disposal of medicals wastes typically involved incineration. However, because of concerns over lowered air quality through the incineration of medical wastes (USEPA 2020a). Alternatives for treatment and disposal of medical wastes include (USEPA 2020a).

- Thermal treatment, such as microwaves
- Steam sterilization, such as an autoclave
- Electropyrolysis
- Chemical systems

In general, each State requires that medical waste be carefully collected in appropriately labeled containers, are treated, and then disposed of under a manifest, as a regulated medical waste (USEPA 2020a). There are several different types of containers depending on the type of medical waste.

9.4   LOW-LEVEL RADIOACTIVE WASTE

Low-level radioactive wastes are often associated with medical facilities such as X-ray and other machines at medical and dental facilities. Low-level radioactive wastes are regulated under the Low-level Radioactive Waste Policy Act (LLRWPA) of 1980. As amended in 1985. Under the Act, each state is responsible for the management and disposal requirements of low-level radioactive wastes (USEPA 2020b). The LLRWPA does not regulate high-level radioactive wastes, which is the responsibility of the Department of Energy. Typically, high-level radioactive wastes are associated with nuclear power plants and those generated by or associated with the United States Department of Defense (Nuclear Regulatory Commission [NRC] 2021).

Each facility planning on installing a low-level radioactive source must obtain a license from the NRC or State. The facility must be designed, constructed, and operated to meet safety standards depending on the source (NRC 2021). Disposal of low-level radioactive sources is strictly regulated. Disposal options are limited and licensed transporters are generally few. Therefore, advance planning is key. Manifests are also complex and comprehensive. An example manifest is available for review at https:///nrc.gov/cdn/legacy/waste/llw-disposal.ml20178a433-nureg-br-0204-rev3-form-541-example-filled.jpg (3542×2729) (nrc.gov).

9.5 COMPREHENSIVE ENVIRONMENTAL RESPONSE, COMPENSATION, AND LIABILITY ACT

CERCLA of 1980 is commonly referred to as Superfund or Polluters Pay Law. CERCLA authorized USEPA to recover natural resource damages caused by hazardous substances released into the environment. Created out of CERCLA was the Toxic Substance and Disease Registry (ATSDR) (USEPA 2021h).

9.5.1 INTRODUCTION

CERCLA is significant for addressing poor past practices that impacted the environment through the introduction of contaminants. CERCLA was fundamental in establishing guidelines for investigating sites of environmental contamination, conducting risk assessment, establishing clean-up criteria, pollution prevention, and funding research to develop remedial technologies.

9.5.2 POTENTIAL RESPONSIBLE PARTIES

USEPA may identify responsible parties for contamination from releases of hazardous substances to the environment (polluters) and either compel them to investigate and clean-up sites, or it may undertake the investigation and cleanup themselves using the Superfund (a trust fund) and recover all response costs and even more in some instances from polluters by referring to the Department of Justice for cost recovery proceedings. As of 2016, there were 1,328 sites listed, 391 sites delisted, and 55 new sites proposed for a total of 1,774 Superfund sites. Figure 9.3 is a map showing the distribution of CERCLA sites in the United States (USEPA 2021h).

Approximately 70% of CERCLA investigation and cleanup activities have been paid for by responsible parties (PRPs). The exceptions occur when the responsible party cannot be found or is unable to pay for the cleanup. The majority of the funding came from a tax on the petroleum and chemical industries, but since 2001, most of the funding now comes from taxpayers (USEPA 2021h).

USEPA's enforcement program under CERCLA is based on the "**polluters pay principle**," which stipulates that the party responsible for the pollution pays for cleaning up the pollution. Over the past 35 years, USEPA has recovered over $35 billion from potentially responsible parties for cleanup at CERCLA sites. This principle is one of many highlights of USEPA environmental protection that has been copied worldwide (USEPA 2021j).

9.5.3 CERCLA HISTORY

CERCLA was enacted in 1980 in response to threats from sites of environmental contamination, including three sites, in particular, Times Beach in Missouri, Love Canal in New York, and the Valley of Drums in Kentucky. As part of CERCLA, USEPA establish a Hazardous Ranking System (HRS) to prioritize each Superfund Site related to its threat to human health and the environment. In 1982, USEPA published its first list of Superfund sites and titled it the National Priority List (NPL) (USEPA 2021h).

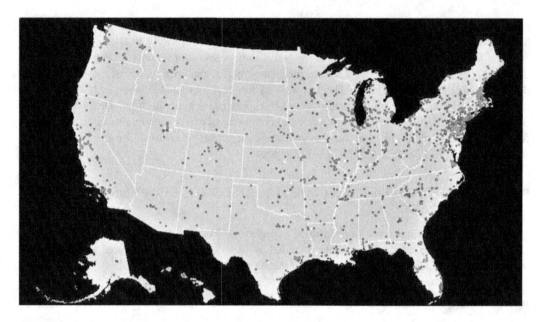

FIGURE 9.3   CERCLA or superfund sites in the United States.

Source: United States Environmental Protection Agency (USEPA). 2021i. CERCLA Sites in the United States. https://www.epa.gov/CERCLA-sites (accessed March 24, 2021).

Other prominent sites that have shaped not only Superfund, but have had significant contributions to environmental science include

- Toms Rivers and the Reich Farm Superfund site in New Jersey
- Wells G and H Superfund site in Woburn Massachusetts
- PG&E Hinckley Site in California

A brief discussion of each of the six sites mentioned above is covered in the following sections.

9.5.3.1   Times Beach

Times Beach is now a ghost town located approximately 17 miles southwest of St. Louis, Missouri. Times Beach was a town of more than 2,000 people before all were evacuated in 1983 because of dioxin contamination. During the 1960s, a pharmaceutical and chemical company was located in Times Beach and produced some thick oily wastes that contained dioxin. Approximately 18,500 gallons of the dioxin-containing oily waste ended up with a waste oil contractor who mixed the material with other sources of waste oil and sprayed the material at three local horse arenas and 23 miles of roads in Times Beach as a dust suppressant (USEPA 2021k).

The Center for Disease Control (CDC) initially became involved in August 1971 when local citizens reported the deaths of numerous horses that used one or more of the horse arenas sprayed with the dust suppressant. The CDC discovered an old tank containing approximately 4,300 gallons of oily waste that contained dioxin at a concentration of 340 ppm. In 1974, the CDC collected soil samples from the area and detected trichlorophenol, PCBs, and dioxin. USEPA did not become heavily involved until 1979 when they discovered 90 drums of oily waste at the site with dioxin concentrations as high as 2,000 ppm. In May and June 1982, USEPA collected soil samples from the area and detected dioxin as high as 0.3 ppm in soil. The CDC recommended cleanup level for dioxin at the time was as low as 0.001 ppm. On December 23, 1982, the CDC publicly recommended that Times Beach not be inhabited. Subsequently, 800 residential properties and 30 businesses were abandoned. In 1997, the cleanup was completed at a cost of nearly 200 million dollars (Sun 1983, USEPA 2021k).

9.5.3.2 Love Canal

Love Canal is a neighborhood within Niagara Falls, New York. The site that became the epicenter of contamination was a 29-acre parcel that was used as a landfill. Originally planned as a canal, the site was abandoned at the turn of the 20th century and was used as a dump by the City of Niagara Falls. In 1942, the Niagara Power and Development Company granted permission to Hooker Chemical to dump wastes in the canal. In 1948, the City of Niagara ended the disposal of refuse into the canal and the Hooker Chemical Company became the sole user and owner of the site. Hooker Chemical ceased dumping in 1952 and capped the dump with clay. During its lifespan of 5 years, an estimated total of 21,800 tons of chemicals mostly consisting of sludges, caustics, alkalines, chlorinated hydrocarbons, and volatile organic hydrocarbons were disposed in the landfill, many in 55-gallon drums (Beck 1979, USEPA 2021l).

By the early 1950s, the City of Niagara Falls was in the midst of a population increase and purchased the site from Hooker Chemical to build a school. In 1955, 99th Street School opened with 400 children enrolled. The area surrounding the dump continued to develop in the 1950s with a total of 800 private residences built in the surrounding area. Soon after development, residents began to complain of odors and black substances in their yards and public playgrounds (USEPA 2021l).

It took until 1977, when the New York State Department of Health and Environmental Conservation conducted a large scientific study of the air, groundwater, and soil in the area and detected a number of organic compounds in the basements of several residences adjacent to the former dump. Compounds detected included, benzene, chlorinated hydrocarbons, PCBs, and dioxin (Beck 1979 and USEPA 2021l). By 1978, Love Canal became a national media event and then on August 7, 1978, then-President Jimmy Carter announced that Love Canal was a federal health emergency. In total 950 families were evacuated. This was the first time that emergency funds were used for a situation that was not a natural disaster. When CERCLA was passed, Love Canal was placed first on the list of Superfund sites. USEPA finished the cleanup of Love Canal in 2004 at an estimated cost of $400 million dollars (USEPA, 2004, 2021l).

9.5.3.3 Valley of Drums

The Valley of Drums Site is a 23-acre site in northern Kentucky, near Louisville. The site became a collection site for wastes dating back into the 1960s. It initially caught the attention of local government officials when some drums that were onsite site caught fire and took firefighters more than a week to extinguish. However, no action was taken at the time since no laws existed at the time to take any further action. In 1978, a Kentucky Department of Natural Resources and Environmental Protection (KDNREP) investigation of the property estimated that over 100,000 drums of waste were located onsite and that an estimated 27,000 drums were buried and the remaining containers were discharged into onsite pits and trenches (USEPA 2021m). Figure 9.4 shows some of the drums discovered.

The KDNREP investigation revealed that many of the drums onsite were deteriorated and their contents spilled onto the ground and were washed into a nearby creek. Analysis of collected samples detected the presence of several heavy metals, PCBs, and 140 other chemical substances. USEPA initiated an emergency cleanup in 1979 but quickly realized that the magnitude of the project was beyond what they were prepared for and that was used by members of Congress as an additional reason why CERCLA was needed (USEPA 2021m).

The Valley of Drums site was placed on the Superfund List in 1983 when a large-scale cleanup of the site began and ended in 1990. In 1996, the site was removed from the Superfund List. However, more buried drums were discovered during an inspection of the site in 2008 outside the cleanup area. Currently, USEPA is monitoring the site and has authorized an additional $400,000 in funds to remove and properly disposed of more drums discovered buried drums (USEPA 2021m).

FIGURE 9.4   Photograph of valley of drums superfund site, circa 1979.

Source: United States Environmental Protection Agency (USEPA). 2021g. Superfund Site Profile: Valley of Drums, Kentucky. https://www.epa.gov/valley-of-drums-profile (accessed March 24, 2021).

9.5.3.4   Wells G&H Superfund Site in Woburn Massachusetts

Groundwater extraction Wells G&H was developed by the City of Woburn, Massachusetts in 1964 and 1967, respectively. The wells were capable of pumping two million gallons of water per day and were initially intended to supplement an existing water system that supplied water to the residents of Woburn. In 1979, the Massachusetts Department of Environmental Protection (MDEP) conducted analytical testing of water from the two wells and discovered the presence of several chlorinated volatile organic compounds in the samples that included (USEPA 1989a):

- 1,1,1-trichloroethane (TCA)
- *Trans*-1,2-dichloroethene (trans DCE)
- Tetrachloroethene (PCE)
- Trichloroethene (TCE)

Concentrations ranged from 1 to 400 ppb. MDEP immediately shut the wells down. In 1981, USEPA conducted a hydrogeologic investigation of a 10-square mile area surrounding the wells to evaluate the source of the contamination and to define the extent of groundwater contamination. Five facilities were identified as sources of contamination and the site was placed on the Superfund list in December 1982 (USEPA 1989a). Remedial response actions were first initiated by USEPA in 1983 and are still being conducted more than 35 years later. USEPA has been maintaining close community relations since 1986. Community involvement has been significant and was even the subject of a movie titled, "A Civil Action."

9.5.3.5   Ciba-Geigy and Reich Farm Site in Toms River New Jersey

Ciba-Geigy Superfund site is located in Toms River, New Jersey. The site is approximately 1,400 acres in size, of which 320 acres were developed and were formerly used to manufacture synthetic organic pigments. From 1952 to 1966, the facility discharged treated effluent into Toms River, which discharged directly into the Atlantic Ocean. Beginning in 1966, the facility discharged effluent directly into the Atlantic Ocean through a pipeline. In addition, the facility disposed of solid wastes onsite at several locations (USEPA 1989b).

During the 1970s and early 1980s, the facility conducted several investigations at the direction of the New Jersey Department of Environmental Protection (NJDEP). USEPA eventually listed the site as a Superfund site in 1982. In 1984, NJDEP discovered that the facility was illegally disposing of drums containing hazardous waste onsite which lead to a criminal investigation. As a result of the criminal investigation, indictments of several company officials were charged with illegal disposal of chemical waste, filing false documents, making false and misleading statements, and illegal disposal of hazardous waste (1989b).

In 1985, USEPA began conducting a remedial investigation of the site to fully define the extent of contamination in soil and groundwater both onsite and offsite. The investigation was completed in 1988 (USEPA 1989b). Remedial activities included excavation of drums, contaminated soil, and groundwater treatment that continues to this day more than 30 years later.

The Reich Farm Superfund Site in Toms River, New Jersey, is in close proximity to the Ciba-Geigy Superfund Site and was designated a Superfund Site in 1983. According to USEPA (1988), a three-acre portion of the farm was leased for the temporary storage of used 55-gallons drums in 1971. However, it was subsequently discovered that wastes from the drums may have been dumped into trenches and that this activity contaminated groundwater (USEPA 1988).

Compounds detected in groundwater included (USEPA 1988)

- Styrene-Acrylonitrile Trimer
- Tetrachloroethene (PCE)
- Trichloroethene (TCE)
- 1,1,1-trichloroethane (TCA)
- Chlorobenzene
- 1,2-dichloroethene (DCE)
- Other volatile organic compounds

Contaminants listed above and many others not listed contaminated groundwater, which residents of Toms River used as a source of drinking water. This led to a significant toxicological risk evaluation that is still ongoing (USEPA 2021n).

9.5.3.6 PG&E Hinkley Site in California

The PG&E Site was located 2 miles from the town of Hinkley, California, and 12 miles west of Barstow, California. Between 1952 and 1966, PG&E used hexavalent chromium as corrosion protection in its cooling tower water. The wastewater from this process was discharged to unlined ponds onsite which subsequently contaminated groundwater. The resulting groundwater contaminant plume of hexavalent chromium migrated from the site and was estimated to be eight miles long and 2 miles wide and impacted drinking water in Hinkley (California Regional Water Quality Control Board 2021). In 1993, citizens of Hinkley filed suit against PG&E and settled in 1996 for 333 million dollars (San Francisco Chronicle 1996). This site was also the subject of a movie titled "*Erin Brockovich*." Remediation at the site is ongoing.

Although this site is not actually a federal Superfund Site, it does follow a similar profile to the five other sites briefly described above. Similarities include a significant toxicological risk study, large groundwater contaminant plume, and a high remedial cost.

9.5.3.7 Summary of CERCLA History

The six sites highlighted above have been often referred to as the sites that influenced the formation and passage of CERCLA or have demonstrated a lack of knowledge of how contaminants behave in the subsurface geological environment. In addition, four of the six sites involve agriculture or agricultural land.

These examples have provided scientific information that has highlighted the following:

- The need for groundwater protection zones
- The need, development, and procedures for conducting toxicological risk evaluations
- The development of groundwater modeling
- The difficulty in remediating contaminated groundwater
- The enormous cost of remediating contaminated groundwater
- The toxicity, mobility, and persistence of chlorinated VOCs
- The toxicity, mobility, and persistence of hexavalent chromium

9.5.4   SITE SCORING

USEPA developed the Hazard Ranking System (HRS) to score potential sites of environmental contamination. A score of 28.5 was usually high enough to be placed on the NPL or Superfund List. The HRS is a numerical process whereby it assigns integers to factors that relate to risk based on site conditions. The factors are grouped into three categories (USEPA 2021h):

1. Likelihood that a site has released or has the potential to release hazardous substances into the environment
2. Characteristics of the waste, including toxicity, quantity, etc.
3. Potential human exposure pathways

There are four pathways for the potential human exposure category that includes the following (USEPA 2021h):

1. Surface water migration (i.e., drinking water sources, sensitive habitats such as a wetland, human food chain, etc.)
2. Soil exposure (i.e., dermal contact, dust, etc.)
3. Groundwater migration (i.e., drinking water)
4. Air migration

After scores are calculated for one or all the pathways, they are combined using a root-mean-square equation to determine the overall site score. It should be noted that some sites score high for just one pathway if the potential exposure is severe enough and the toxicity of the contaminant or contaminants of concern is also high.

The identification of a site on the NPL is intended primarily to guide USEPA in the following (USEPA 2021h):

- Determining which sites warrant further investigation to assess the nature and extent of the risks to human health and the environment
- Identifying what CERCLA-financed remedial actions may be appropriate
- Notifying the public of the site which USEPA believes warrant further investigation
- Notifying PRPs that USEPA may initiate a CERCLA-financed remedial action

A Potential Responsible Party (PRP) is a possible polluter who may eventually be liable under CERCLA for contamination or misuse of a particular property or resource. Four classes of PRPs may be liable for contamination at a Superfund Site and they include the following (USEPA 2021h):

1. The current owner or operator of the site
2. The owner or operator of a site at the time that disposal of a hazardous substance, pollutant, or contaminant occurred

3. A person who arranged for the disposal of a hazardous substance, pollutant, or contaminant at a site
4. A person who transported a hazardous, pollutant, or contaminant to a site, who also has selected that site for the disposal of the hazardous substances, pollutants, or contaminants

Inclusion of a site on the Superfund List does not itself require PRPs to initiate action to clean up a site, nor does it assign liability to any one entity or person. The Superfund List serves primarily informational purposes, notifying the government and the public of those sites or releases that appear to warrant a remedial action. Under Superfund, the key difference in the authority to address hazardous substances and pollutants or contaminants is that the cleanup of pollutants or contaminants, which are not hazardous substances, cannot be compelled by unilateral administrative order (USEPA 2021h).

An important provision of CERCLA is that if a PRP spends money to clean up a CERCLA site, that party may sue other PRPs in a contribution action. CERCLA liability has generally been judicially established as joint and several among PRPs to the government for cleanup costs in that each PRP is hypothetically responsible for all costs, but CERCLA liability is allocable among PRPs in contribution based on comparative fault. An "orphan share" is the share of costs at a Superfund site that is attributed to a PRP that is unidentifiable or insolvent. USEPA as a matter of long-standing policy attempts to treat all PRPs equitably and fairly (USEPA 2021h).

9.5.5 REMEDIATION CRITERIA

Most states have passed their own versions of CERCLA so that sites that do not qualify to become a Superfund site but have had a release of hazardous substances above cleanup levels get addressed. Cleanup levels established at the state level vary from state to state depending on various factors. However, much is based on USEPA exposure scenarios and toxicological data. States have established cleanup levels for different types of land use, typically residential and non-residential (industrial or commercial), and by media or exposure pathway including the following:

* Groundwater or drinking water
* Soil
* Groundwater – surface water interface
* Inhalation to indoor air
* Inhalation to outdoor or ambient air

Table 9.2 lists the generic residential cleanup levels for select chemical compounds for the state of Michigan (Michigan Department of Environmental Quality 2016). The exposure pathways listed in the table are residential direct contact, commercial-industrial direct contact, and construction worker direct contact. The complete list can be viewed at https://www.michigan.gov/documents/deq/deq-rrd-chem-CleanupCriteriaTSD_527410_7.pdf (Michigan Department of Environmental Quality 2016). The values listed under the Groundwater-surface water interface in Table 9.2 are values within a zone where groundwater becomes surface water before mixing with surface water or exposure to the atmosphere. The values listed for air are presented in micrograms per kilogram in soil immediately beneath a building for indoor air and immediately beneath the groundwater surface if an obstruction is not present (Michigan Department of Environmental Quality 2016).

The values listed for indoor and ambient air should be considered screening values since they are not actual air samples collected from a typical breathing zone. In Michigan, groundwater is defined as any water encountered beneath the surface of the ground regardless of quantity or if it occurs in a naturally occurring formation.

As a comparison, Table 9.3 presents soil cleanup criteria for New Jersey (New Jersey Department of Environmental Protection [NJDEP] 2008, 2015, 2017), Illinois Environmental Protection Agency

TABLE 9.2

Generic Residential Cleanup Levels for Select Compounds

Contaminant	CAS No.	Soil (µg/kg)	Ground-water (µg/L)	Surface Water Interface (µg/L)	Indoor Air (µg/kg)	Ambient Air (µg/kg)
Volatile Organic Compounds						
Benzene	71432	100	5	4,000	1,600	13,000
Bromobenzene	108861	550	18	NA	3.10 E+5	4.50 E+5
Bromochloromethane	74755	1,600	80	ID	1,200	9,100
Bromodichloromethane	75274	1,600	80	ID	1,200	9,100
Bromoform	75252	1,600	80	ID	1.50 E+5	9.0 E+5
Bromomethane	74839	200	10	35	860	11,000
n-Butylbenzene	104518	2.60 E+5	80	44,000	5.4 E+7	2.90 E+7
Sec-Butylbenzene	135988	1,600	80	ID	1.0 E+5	1.0 E+5
Tert-Butylbenzene	98066	78,000	80	ID	3.1 E+8	9.70 E+9
Carbon tetrachloride	56235	100	5	900	190	3,500
Chlorobenzene	108907	2,000	100	25	1.3 E+5	7.70 E+5
Chloroethane	75003	8,600	430	1,100	2.9 E +6	3.0 E+7
Chloroform	67663	1,600	80	350	7,200	45,000
Chloromethane	74873	5,200	260	ID	2,300	40,000
2-Chlorotoluene	95498	3,300	150	ID	2.70 E+5	1.20 E+6
4-Chlorotoluene	106434	900	150	360	4.3 E+5	9.60 E+6
1,2-Dibromo-3-chloropropane	96128	100	.2	ID	260	260
1,2-Dibromoethane	106934	20	80	ID	1.1 E+5	2.60 E+6
Dibromomethane	74953	1,600	80	ID	7,000	7,000
1,2-Dichlorobenzene	95501	14,000	600	13	1.1 E+7	3.90 E+7
1,3-Dichlorobenzene	541731	170	6.6	28	26,000	79,000
1,4-Dichlorobenzene	106467	1,700	75	17	19,000	77,000
Dichlorodifluoromethane	75718	95,000	80	ID	5.3 E+7	5.50 E+8
1,1-Dichloroethane	75353	18,000	880	740	2.30 E+5	2.10 E+6
1,2-Dichloroethane	10762	100	5	360	2,100	6,200
1,1-Dichloroethene	75354	140	7	130	62	1,100
Cis-1,2-Dichloroethene	156592	1,400	70	620	22,000	1.80 E+5
Trans-1,2-Dichloroethene	156592	2,000	100	1,500	23,000	2.80 E+5
1,2-Dichloroporopane	78875	100	5	230	4,000	25,000
1,4-Dioxane	123911	1,700	85	2,800	NLV	NLV
Ethylbenzene	100414	1,500	74	18	87,000	7.20 E+5
Hexachlorobutadiene	87673	26,000	50	ID	1.30 E+5	1.30 E+5
Isopropylbenzene	98828	91,000	800	28	4.0 E+5	1.70 E+6
Methylene chloride	75092	100	5	1,500	45,000	2.10 E+5
Methyl-*tert*-butyl ether (MTBE)	163404	800	40	7,100	9.9 E+6	2.50 E+7
1,1,1,2-Tetrachloroethane	630206	1,500	8.5	78	36,000	54,000
1,1,2,2-Tetrachloroethane	79345	170	8.5	35	4,300	10,000
Tetrachloroethene (PCE)	127184	100	5	60	11,000	1.70 E+5
Toluene	108883	16,000	790	270	3.3 E+5	2.80 E+6
1,2,4-Trichlorobenzene	120821	4,200	70	99	9.6 E+6	2.80 E+7
1,2,3,-Trichlorobenzene	87616	4,200	70	99	9.6 E+6	2.80 E+7
1,1,1-Trichloroethane (TCA)	75556	4,000	220	89	2.50 E+5	3.80 E+6
1,1,2-Trichloroethane	79005	100	5.0	330	4,600	17,000

Source: Michigan Department of Environmental Quality. 2016. Cleanup Criteria and Screening Levels Development and Application. https://www.michigan.gov/documents/deq/deq-rrd-chem-CleanupCriteriaTSD_527410_7.pdf (accessed March 24, 2021).

Note: µg/kg, microgram per kilogram; µg/L, microgram per liter; NA, Not available; ID, Insufficient data; NLV, Not likely to volatilize.

TABLE 9.3

Soil Cleanup Levels for New Jersey, Illinois, New York, California, and USEPA

Contaminant	CAS ID No.	New Jersey[1] (mg/kg)	Illinois[2] (mg/kg)	California[3] (mg/kg)	USEPA[4] (mg/kg)
Acetone	67641	1,000	25	20	2.9
Benzene	71432	3	0.03	5	0.0023
Bromobenzene	108861	NL	NL	NL	0.042
Bromochloromethane	74755	11	NL	NL	0.021
Bromodichloromethane	75274	NL	0.6	NL	0.00036
Bromoform	75252	86	0.8	NL	0.00087
Bromomethane	74839	79	NL	NL	0.0019
Methyl ethyl ketone	789833	1,000	17	NL	1.2
Tert-Butylbenzene	98066	NL	NL	NL	1.6
Carbon tetrachloride	56235	2	0.07	NL	0.00018
Chlorobenzene	108907	37	1	NL	0.053
Chloroethane	75003	NL	NL	NL	0.081
Chloroform	67663	19	0.6	NL	0.000061
Chloromethane	74873	520	NL	NL	0.049
2-Chlorotoluene	95498	NL	NL	NL	0.23
4-Chlorotoluene	106434	NL	NL	NL	0.24
1,2-Dibromo-3-chloropropane	96128	NL	NL	NL	0.081
1,2-Dibromoethane	106934	NL	NL	NL	0.0000021
Dibromomethane	74953	NL	NL	NL	0.0021
1,2-Dichlorobenzene	95501	5,100	17	NL	0.03
1,3-Dichlorobenzene	541731	5,100	NL	NL	0.00082
1,4-Dichlorobenzene	106467	570	2	NL	0.00046
Dichlorodifluoromethane	75718	NL	NL	NL	0.03
1,1-Dichloroethane	75353	8	0.06	NL	0.00078
1,2-Dichloroethane	10762	6	0.02	NL	0.000048
1,1-Dichloroethene	75354	8	0.06	5	0.00078
Cis-1,2-Dichloroethene	156592	4	0.4	5	0.011
Trans-1,2-Dichloroethene	156592	4	0.7	NL	0.11
1,2-Dichloroporopane	78875	10	0.03	NL	0.00015
2,2-Dichloropropane	594207	NL	NL	NL	0.013
1,3-Dichloropropane	142289	NL	2NL	NL	0.13
1,4-Dioxane	123911	NL	NL	10	0.000094
Ethylbenzene	100414	1,000	13	5	0.0017
Hexachlorobutadiene	87673	1	NL	5	0.00027
Isopropylbenzene	98828	NL	NL	NL	0.084
Methylene chloride	75092	49	0.02	10	0.0029
Methyl-tert butyl ether (MTBE)	1634044	NL	0.32	NL	0.0032
1,1,1,2-Tetrachloroethane	630206	170	NL	NL	0.00022
1,1,2,2-Tetrachloroethane	79345	34	NL	NL	0.00003
Tetrachloroethene (PCE)	127184	4	0.06	5	0.0051
Toluene	108883	1,000	12	5	0.76
1,2,4-Trichlorobenzene	120821	68	5	NL	0.0034
1,2,3,-Trichlorobenzene	87616	NL	NL	NL	0.021
1,1,1-Trichloroethane (TCA)	75556	210	2	NL	2.8

(Continued)

TABLE 9.3 (Continued)

Soil Cleanup Levels for New Jersey, Illinois, New York, California, and USEPA

Contaminant	CAS ID No.	New Jersey[1] (mg/kg)	Illinois[2] (mg/kg)	California[3] (mg/kg)	USEPA[4] (mg/kg)
Trichloroethene (TCE)	75694	23	0.06	5	0.00018
Trichlorofluoromethane	96184	NL	NL	NL	3.3

Notes

[1] New Jersey Department of Environmental Protection (NJDEP). 2008. Introduction to Site-Specific Impact to Ground Water Soil Remediation Standards Guidance Document. Trenton, NJ. 10 p. https://www.nj.gov/dep/srp/guidance/rs/igw (accessed March 24, 2021); New Jersey Department of Environmental Protection (NJDEP). 2015. Site Remediation Program: Remediation Standards. http://www.nj.gov/dep/srp/guidance (accessed March 24, 2021); and New Jersey Department of Environmental Protection (NJDEP). 2017. Soil Cleanup Criteria. https://www.nj.gov.srp/guidance/scc (accessed March 24, 2021).

[2] Illinois Environmental Protection Agency (IEPA). 2017. Tiered Approach to Corrective Action Objectives (TACO). Springfield, IL. 284 p.

[3] California Department of Toxic Substances and Control (DTSC). 2013. Chemical Look-Up Table. Technical Memorandum. Sacramento, CA. 5 p.; and California Environmental Protection Agency (CalEPA). 2005. Use of Human Health Screening Levels (CHHSLs) in Evaluation of Contaminated Properties. Sacramento, CA. 65 p.

[4] United States Environmental Protection Agency (USEPA). 2021o. Regional Screening Levels. https://www.epa.gov/RSLs (accessed March 24, 2021).

mg/kg, milligram per kilogram; NL, Not listed

[IEPA] 2017, California Department of Toxic Substances (DTSC) Control Chemical Look-Up Table (California Department of Toxic Substances and Control DTSC 2013 and California Environmental Protection Agency CalEPA 2005), and USEPA Regional Soil Screening Levels (RSLs) (USEPA 2021o).

Criteria presented in Table 9.3 generally correspond to a carcinogenic risk of 1 in 100,000 and a Hazard Quotient of 1 for residential properties. The values presented for New Jersey are not for the protection of groundwater. Soil remediation values for New Jersey protective of groundwater are calculated by NJDEP on a site-by-site basis. The lessons from examining Tables 9.2 and 9.3 are the following:

- Cleanup or remediation criteria vary substantially from by media which include soil, groundwater, surface water, indoor air, and outdoor air
- Cleanup or remediation criteria vary substantially based on land use which typically includes residential, commercial, industrial, construction worker, or recreational
- Cleanup or remediation criteria vary between each state and with USEPA

9.5.6 SUMMARY OF COMPREHENSIVE ENVIRONMENTAL RESPONSE, COMPENSATION, AND LIABILITY ACT

CERCLA was originally designed as a mechanism to fund clean-up of "orphaned" sites in the United States using the "Polluters Pay Principle." However, CERCLA has meant much more than just a funding mechanism. CERCLA also was a catalyst in developing investigative techniques and protocols, risk assessment evaluations, remediation technologies, and remediation target values for hundreds of chemical compounds in different media. Over its more than 35-year history, USEPA has recovered over $35 billion to fund cleanup at thousands of CERCLA sites in the United States (USEPA 2021p). In addition, the "Polluters Pay Principle," is used worldwide as the preferred method for protecting human health and the environment at difficult and complex sites of contamination.

9.6   SUMMARY OF LAND REGULATIONS

As important the CAA is to the air and the CWA is to water, RCRA and CERLA form the foundation of environmental regulation that relates to the land. RCRA and CERLA play a critical role in creating a more sustainable environment, especially in urban areas. Put in simplistic terms, CERCLA deals with Anthropogenic impacts from our past, and RCRA deals with preventing Anthropogenic impacts in the future. Both have now become part of our environmental fabric and have been adopted by most countries of the world.

We will now turn our attention to environmental regulations that focus on protecting living organisms and cultural and historical sites.

• Lead	0.8 mg/L
• Cadmium	0.5 mg/L
• Chromium	8.0 mg/L
• Selenium	0.5 mg/L
• Silver	0.1 mg/L

Questions and Exercises for Discussion

1. **An auto repair shop uses mineral spirits as a parts washer solvent. The solvent does not contain any halogenated or listed solvents and its flashpoint is greater than 140°F. When the solvent becomes dirty, it is distilled. The distilled solvent is placed back into use. The residual solids leftover from the distillation process are the waste that must be characterized. The waste was transferred to an appropriate container and labeled as "Parts Washer Waste, Analysis Pending." A representative sample of the waste was collected and sent to a certified laboratory under chain-of-custody for waste characterization. The laboratory conducted the required tests and a portion of the results are as follows:**

   How should the waste be classified and how should it be disposed?

2. **Auto Body Shop. The exhaust filters in the paint spray booth became saturated with overspray from the painting operation. Since the body shop uses many different types of paint and primers, it's difficult to determine whether the filters are hazardous or not without laboratory testing. A representative filter is removed and sampled. The remaining filters are placed into a container and are label "Waste Filters Pending Analysis." The sample is sent to a certified laboratory under the chain of custody for analysis. The laboratory conducted the required tests and a portion of the results are as follows:**

• Lead	9.1 mg/L
• Chromium	0.4 mg/L
• Barium	0.8 mg/L
• Methyl ethyl ketone	10 mg/L

How should the waste be classified and how should it be disposed?

3. **General Manufacturing Facility.** The facility receives large steel components which the facility re-manufactures. The process requires that the components be dismantled, and the surfaces are prepared for re-finishing. The metal components are placed into a sandblasting machine and are cleaned under high-pressure blasting sand. After a few weeks of use, the blasting media (sand) is no longer effective and must be replaced. A representative sample of the sand is collected and the used sand from the operation is placed in a metal 55-gallon drum and labeled sand "Waste Sand from Sand Blasting – Analysis Pending." The sample collected for waste characterization is sent to a certified laboratory under the chain of custody for analysis. The laboratory conducted the required tests and a portion of the results are as follows:

• Arsenic	0.5 mg/L
• Cadmium	3.5 mg/L
• Chromium	25 mg/L
• Lead	40 mg/L

4. **You are conducting a solid waste inspection of a facility and notice that a drum labeled "garbage only" contains paint residue and spent aerosol cans. What actions should be undertaken?**

REFERENCES

Beck, E. C. 1979. The Love Canal Tragedy. Journal of the Environmental Protection Agency. Washington, D.C. https://www.epa.gov/epa/aboutepa/love-canal-tragedy.html (accessed March 24, 2021).

California Department of Toxic Substances and Control (DTSC). 2013. *Chemical Look-Up Table.* Technical Memorandum. Sacramento, CA. 5 p.

California Department of Toxic Substance and Control (CalDTSC). 2021. Hazardous Waste. https://www.dtsc.ccelern.csus.edu/wasteclass (accessed March 24, 2021).

California Environmental Protection Agency (CalEPA). 2005. *Use of Human Health Screening Levels (CHHSLs) in Evaluation of Contaminated Properties.* Sacramento, CA. 65 p.

California Regional Water Quality Control Board. 2021. PG & E Hinkley Chromium Cleanup. https://www.waterboards.co.gov/lahontan/water_issues/projects/pge/ (accessed May 24, 2021).

Cornell University Law. 2021. Resource Conservation and Recovery Act (RCRA) Overview. https://www.law.cornell.edu/wex/rcra (accessed March 24, 2021).

Illinois Environmental Protection Agency (IEPA). 2017. *Tiered Approach to Corrective Action Objectives (TACO).* Springfield: IL. 284 p.

Michigan Department of Environmental Quality. 2016. Cleanup Criteria and Screening Levels Development and Application. https://www.michigan.gov/documents/deq/deq-rrd-chem-CleanupCriteriaTSD_527410_7.pdf (accessed March 24, 2021).

New Jersey Department of Environmental Protection (NJDEP). 2008. Introduction to Site-Specific Impact to Ground Water Soil Remediation Standards Guidance Document. Trenton, New Jersey. 10 p. https://www.nj.gov/dep/srp/guidance/rs/igw (accessed March 24, 2021).

New Jersey Department of Environmental Protection (NJDEP). 2015. Site Remediation Program: Remediation Standards. http://www.nj.gov/dep/srp/guidance (accessed May 24, 2021).

New Jersey Department of Environmental Protection (NJDEP). 2017. Soil Cleanup Criteria. https://www.nj.gov.srp/guidance/scc (accessed March 24, 2021).

Nuclear Regulatory Commission (NRC). 2021. Low-Level Waste Disposal. Low-Level Waste Disposal | NRC.gov (accessed November 9, 2021).

Ohio Environmental Protection Agency (Ohio EPA). 2014. *Land Disposal Restrictions (An Overview). Division of Materials and Waste Management.* Ohio EPA. Columbus, OH. 3 p.

San Francisco Chronicle. 1996. PG&E to Pay 333 Million in Pollution Suit. https://www.sfgate.com/article/PG-E-to-Pay-333-million-in-pollution-suit-3303933.php (accessed March 24, 2021).

Sun, M. 1983. Missouri's Costly Dioxin Lesson. *Science.* Vol. 219. pp. 367–369.

United States Environmental Protection Agency (USEPA). 1988. Superfund Record of Decision: Reich Farms, Toms River, New Jersey. Office of Emergency and Remedial Response, Washington, D.C. EPA/ROD/R-02-88/070. 108 p.

United States Environmental Protection Agency (USEPA). 1989a. Superfund Record of Decision: Wells G & H, Woburn, Massachusetts. Office of Emergency and Remedial Response, Washington, D.C. EPA/ROD/R-01-89/036. 85 p.

United States Environmental Protection Agency (USEPA). 1989b. Superfund Record of Decision: Ciba-Geigy, Toms River, New Jersey. Office of Emergency and Remedial Response, Washington, D.C. EPA/ROD/R-02-89/076. 120 p.

United States Environmental Protection Agency (USEPA). 1991. Land Disposal Restrictions: Summary of Requirements. USEPA Office of Solid Waste and Emergency Response. Washington, D.C. 86 p.

United States Environmental Protection Agency (USEPA). 2004. EPA Removes Love Canal from Superfund List. https://www.epa.gov./admpress/love-canal (accessed March 24, 2021).

United States Environmental Protection Agency (USEPA). 2005. Introduction to Hazardous Waste Identification (40CFR Parts 261). USEPA Office of Solid Waste and Emergency Response (5305 W). EPA530-K-05-012. Washington, D.C. 30 p.

United States Environmental Protection Agency (USEPA). 2020a. Medical Waste. https://www.epa.gov/medicalwaste (accessed December 30, 2020).

United States Environmental Protection Agency (USEPA). 2020b. Radiation Regulations and Laws. https://www.epa.gov/radiationregulationsandlaws (accessed December 30, 2020).

United States Environmental Protection Agency. 2021a. Summary of the Resource Conservation and Recovery Act. https://www.epa.gov/rcra/summary (accessed March 24, 2021).

United States Environmental Protection Agency (USEPA). 2021b. Criteria for the Definition of Solid Waste and Solid and Hazardous Waste Exclusions. https://.www.epa.gov/criteria-definition-solid-waste (accessed March 24, 2021).

United States Environmental Protection Agency (USEPA). 2021c. Defining Hazardous Waste. https://www.epa.gov/hw/defining-hazardous-waste-listed-characteristic-and-mixed-radiological-wastes (accessed March 24, 2021).

United States Environmental Protection Agency (USEPA). 2021d. Resource Conservation and Recovery Act Listed and Characteristic Hazardous Wastes. Office of Solid Waste. Washington, D.C. https://www.epa.gov/hazarodus-wastes (accessed March 24, 2021).

United States Environmental Protection Agency (USEPA). 2021e. *USEPA Office of Solid Waste. SW-846 Test Methods Manual.* USEPA. Washington, D.C. 1357 p. https://www.USEPA.gov/sw-846 (accessed March 24, 2021).

United States Environmental Protection Agency (USEPA). 2021f. Used Oil Regulations Under the Resource Conservation and Recovery Act. Office of Solid Waste. Washington, D.C. https://www.epa.gov.used-oil (accessed March 24, 2021).

United States Environmental Protection Agency (USEPA). 2021g. Hazardous Waste Manifests. https://www.epa.gov/hwgenerators/uniform-hazardous-waste-manifest-instructions-sample-form-and-continuation-sheet (accessed March 24, 2021).

United States Environmental Protections Agency (USEPA). 2021h. CERCLA. https://www.epa.gov.superfund (accessed March 24, 2021).

United States Environmental Protection Agency (USEPA). 2021i. CERCLA Sites in the United States. https://www.epa.gov/CERCLA-sites (accessed March 24, 2021).

United States Environmental Protection Agency (USEPA). 2021j. Laws and Regulations. https://www.epa.gov/laws-regulations/regulations (accessed March 24, 2021).

United States Environmental Protection Agency (USEPA). 2021k. Times Beach, Missouri Archive. https://www.epa.gove/times-beach (accessed March 24, 2021).

United States Environmental Protection Agency (USEPA). 2021l. Love Canal Tragedy. https://www.epa.gov/history/love-canal (accessed March 24, 2021).

United States Environmental Protection Agency (USEPA). 2021m. Superfund Site Profile: Valley of Drums, Kentucky. https://www.epa.gov/valley-of-drums-profile (accessed March 24, 2021).

United States Environmental Protection Agency (USEPA). 2021n. EPA to Update Community on Toms
    River, New Jersey Superfund Site. https://archive.epa.gov/epa/newsreleases/epa-update-community-
    toms-river-nj-superfund (accessed March 24, 2021).
United States Environmental Protection Agency (USEPA). 2021o. Regional Screening Levels. https://www.
    epa.gov/RSLs (accessed March 24, 2021).
United States Environmental Protection Agency (USEPA). 2021p. Superfund Enforcement: 35 years of
    Protecting Communities and the Environment. https://www.epa.gov/enforcement/superfund-enforcement-
    35-years-protecting-communities-and-environment (accessed March 24, 2021).
United States Geological Survey (USGS). 2006. Volatile Organic Compounds in the Nation's Ground Water
    and Drinking-Water Supply Wells. USGS Circular 1292. Reston, VA.

10 Protecting Living Organisms and Cultural and Historic Sites

10.1 INTRODUCTION

This chapter describes environmental regulations that relate to the protection of living organisms and cultural and historically significant sites and have sustainability concepts or benefits imbedded within them. Many of the benefits are obvious and others have additional narrative to explain the connection between the specific regulations and sustainability. While reading this chapter, keep in mind the following basic environmental fundamentals, which are referred to numerous times throughout this book:

- Protect the air
- Protect the water
- Protect the land
- Protecting living organisms and historic and cultural sites

10.2 ENVIRONMENTAL LAWS OF THE UNITED STATES WITH SUSTAINABILITY IMPLICATIONS

Laws that are commonly attributed to protecting human health and the environment are given in the following list (USEPA 2021a). Those that appear in **Bold** and have not been described earlier in this book, have significance to sustainability, protecting living organisms, and Cultural and historically significant sites. These include:

- Rivers and Harbors Act of 1899
- **Antiquities Act of 1906**
- Atomic Energy Act of 1946
- Atomic Energy Act of 1954
- **Brownfield Revitalization Act of 2002**
- **Clean Air Act (CAA) of 1970**
- **Clean Water Act (CWA) of 1972**
- **Coastal Zone Management Act of1972**
- **Comprehensive Environmental Response, Compensation, and Liability Act (CERCLA) of 1980**
- Emergency Planning and Community Right-to-Know Act of 1986
- **Endangered Species Act of 1973**
- Energy Policy Act of 1992
- **Energy Policy Act of 2005**
- Federal Power Act of 1935
- Federal Feed, Drug, and Cosmetic Act (FFDCA) of 1938
- Federal Insecticide, Fungicide, and Rodenticide Act (FIFRA) of 1947
- Fish and Wildlife Coordination Act of 1968
- **Food Quality Protection Act of 1996**
- **Fisheries Conservation and Management Act of 1976**
- **Global Climate Protection Act of 1987**

- Hazardous Materials Transportation Act (HMTA) of 1975
- Insecticide Act of 1910
- Lacey Act of 1900
- Low-Level Radioactive Waste Policy Act of 1980
- **Marine Protection Act of 1972**
- **Marine Mammal Protection Act of 2015**
- Medical Waste Tracking Act of 1980
- **Migratory Bird Treaty Act 1916**
- Mineral Leasing Act 1920
- National Environmental Policy Act of 1969
- **National Forest Management Act of 1976**
- **National Historic Preservation Act of 1966**
- **National Parks Act of 1980**
- Noise Control Act 1974
- Nuclear Waste Policy Act of 1982
- **Ocean Dumping Act of 1988**
- Occupational Safety and Health Act (OSHA) of 1970
- **Oil Spill Prevention Act of 1990**
- **Pollution Prevention Act (PPA) of 1990**
- Refuse Act of 1899
- **Resource, Conservation, and Recovery Act (RCRA) of 1976**
- Rivers and Harbors Act of 1899
- Safe Drinking Water Act (SDWA) of 1974
- **Superfund Amendments and Reauthorization Act (SARA) of 1986**
- **Surface Mining Control and Reclamation Act of 1977**
- **Toxic Substance Control Act (TSCA) of 1976**
- **Wild and Scenic Rivers Act of 1968**

The number of environmental laws and regulations is a very long list and encompasses hundreds of thousands of pages of rules. We shall discuss each Act separately in the later sections and will go into substantial detail for those that focus on addressing pollution.

10.2.1 Toxic Substance Control Act

The Toxic Substance Control Act (TSCA) authorizes USEPA to screen existing and new chemicals used in manufacturing and commerce to identify dangerous products or uses that should be subject to federal control. TSCA was first passed in 1976 and has been amended in 1986, 1988, 1990 (twice), 1992, 2007, 2008, 2010, and 2016. Both naturally occurring and synthetic chemicals are subject to TSCA, with the exception of chemicals already regulated under other federal laws that include food, drugs, cosmetics, firearms, ammunition, pesticides, herbicides, and tobacco (Schierow 2013).

TSCA has significance when examining it from a sustainability perspective. We have already learned that some chemicals are very expensive to remediate once they are released into the environment, such as PCBs, PFAS, PIP, chlorinated hydrocarbons, 1,4 Dioxane, dioxin, DDT, and many others. In addition, many of these chemicals have significantly harmed the environment and other life forms, such as DDT.

This is what makes TSCA powerful and significant in pollution prevention because TSCA has the potential to remove these chemicals from use before it's too late. Which may represent that TSCA is more proactive than the majority of other environmental regulations.

10.2.1.1   Restricting Chemicals

The purpose of TSCA was to protect the public from unreasonable risk of injury to health or the environment by regulating the manufacture and sale of chemicals. TSCA does not address wastes produced as byproducts of manufacturing. Instead TSCA attempted to exert direct government control over which types of chemicals could and could not be used in actual use and production. For example, the use of chlorofluorocarbons in manufacturing is now strictly prohibited in all manufacturing processes in the United States, even if no chlorofluorocarbons are released into the atmosphere. The types of chemicals regulated under TSCA are in two broad groups: New and Existing. New chemicals are defined as any chemical substance which is not included in the chemical substance list compiled and published under TSCA, Section 8(b). This list included all of the chemical substances manufactured or imported into the United States prior to 1979, which were grandfathered into TSCA and were considered safe and included over 62,000 chemicals (USEPA 2021b).

When first passed, TSCA limited the manufacture, processing, commercial distribution, use, and disposal of chemical substances including polychlorinated biphenyls (PCBs), asbestos, radon, and lead-based paint. At first USEPA could only limit the commercial use of new chemicals, and even then only immediately after the chemicals were introduced in commerce. Once a chemical substance established itself in the commercial world, USEPA could only restrict new uses of the substance. Thus, USEPA's influence was limited under the 1976 version of TSCA. This created much criticism of TSCA from environmental groups through the years that focused on the "after the fact focus" of TSCA in that it failed to protect individuals before substances were available in products. In addition, TSCA was difficult to implement because the number of substances, totaling approximately 62,000 that were grandfathered, remained on the market. Lastly, fully assessing the risks posed by a new chemical was expensive and time consuming (Markell 2014). To put things into perspective, as of 2014, the list of chemicals in use in the United States is greater than 84,000 and the number of chemicals that USEPA has what they consider to be fully evaluated numbers only 250 (United States Environmental Protection Agency 2015). Even though TSCA gives the authority to USEPA to test existing chemicals, USEPA has had difficulty in obtaining the data needed from industry to determine their risks and the cost to conduct the tests themselves has been evaluated as too costly (Markell 2014).

USEPA has been successful in restricting only 9 chemicals in its history of over 40 years since promulgated in 1976 and none since 1984. Those substances currently restricted include (USEPA 2021b):

- Polychlorinated biphenyls (PCBs)
- Chlorofluorocarbons (CFC's)
- Dioxin
- Asbestos
- Hexavalent chromium
- Four nitrite compounds that are either:
  - Mixed mono and diamides of an organic acid
  - Triethanolanime salt of a substituted organic acid
  - Triethanolanime salt of tricarboxylic acid
  - Tricarboxylic acid

Even though many critics seem to believe that TSCA has failed in many respects, it still represents a potentially powerful tool for pollution prevention and can and should play a vital role in our society (Markell 2014).

In June 2016, TSCA was amended significantly and gave USEPA the authority and responsibility to proactively evaluate the risks of all chemical substances, which will be conducted in a multi-step process that is outlined as the following (National Law Review 2016):

- Developing a screening process to identify high-priority chemical substances
- Designating chemical substances as high or low priority'
- Evaluating the risks of high priority chemicals
- Determining whether any chemical presents unreasonable risk
- Issuing rules to restrict the use of substances that present an unreasonable risk.

The 2016 amendments define high priority substances as those chemicals that may present an unreasonable risk of injury to health or the environment because of a potential hazard and a potential route of exposure including an unreasonable risk to a potentially exposed or susceptible subpopulation that includes but not limited to any of the following (National Law Review 2016):

• Infants	• Children	• Pregnant women
• Workers	• Ederly	

Accordingly, USEPA has stated that its intent is to accord preference to those substances with persistence and bioaccumulative tendencies and known human carcinogens with high acute and chronic toxicity (USEPA 2021b). In the near-term, USEPA has identified 10 chemicals as high priority and are currently conducting risk evaluations. They include the following (USEPA 2021b):

• 1,4-Dioxane	• 1-Bromopropane
• Asbestos	• Carbon Tetrachloride
• Cyclic Aliphatic Bromide Cluster	• Methylene Chloride
• N-methylpyrolidone	• Pigment Wiolet 29
• Tetrachloroethylene	• Trichloroethylene

An analysis of the given list shows that it includes four of substances that belong to the same chemical group called chlorinated or halogenated volatile organic compounds (CVOC's or HVOC's respectively), and are also known as chlorinated solvents. Those include Carbon Tetrachloride, Methylene Chloride, Tetrachloroethylene, and Trichloroethylene. In addition, 1,4-Dioxane is associated with the breakdown of Tetrachloroethylene and Trichlorethylene. Therefore, of the 10 chemicals USEPA has initially placed in its high priority list, half belong or are associated with the same group of compounds, CVOC's or chlorinated solvents.

10.2.1.2 Regulating PCBs under TSCA

PCBs are a group of synthetic organic chemical compounds consisting of carbon, hydrogen, and chlorine atoms. According to USEPA (2021b), PCBs are still released into the environment from sources that include:

- Poorly maintained hazardous wastes sites that contain PCBs
- Illegal or improper dumping of PCB wastes
- Leaks or releases from electrical transformers containing PCBs
- Disposal of PCB-containing consumer products into municipal or other landfills not designed to handle hazardous waste
- Burning some wastes in municipal and industrial incinerators

According to USEPA (2021b), PCBs are not defined as a hazardous waste under RCRA. PCBs are considered TSCA wastes and are regulated under Title 40 CFR Part 761 Section 261.8.

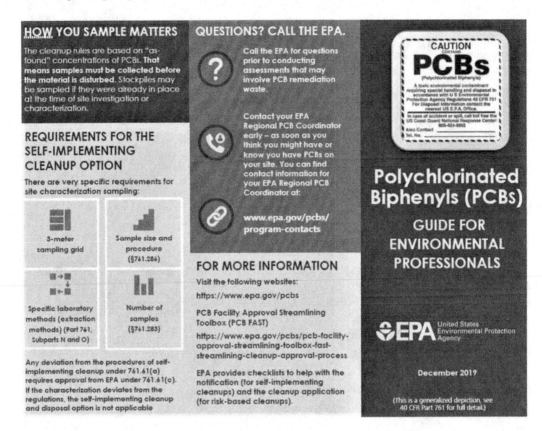

FIGURE 10.1 PCB guide (From United States Environmental Protection Agency. Summary of the Toxic Substance Control Act. https://epa.gov/laws-regulations/summary-toxic-substances-control-act (accessed March 24, 2021), 2021b).

Cleanup of PCBs that have been released into the environment vary widely depending on the location of where the PCBs were detected or reside in the environment. For instance, if PCBs are detected in river or lake sediments, which is often the case, clean up requirements may be very low because PCBs tend to bioaccumulate and do not readily degrade and potentially enter the food chain. In general, if the source of PCBs detected originated at a concentration greater than 50 parts per million (ppm), it would be considered a PCB waste and TSCA regulations apply. Cleaning up PCBs that originate from a source greater than 50 ppm requires approval from USEPA (USEPA 2021b). Figures 10.1 and 10.2 provide a flow chart and guidance when dealing with PCBs.

10.2.2 SUPERFUND AMENDMENTS AND REAUTHORIZATION ACT (SARA) AND EMERGENCY PLANNING AND COMMUNITY RIGHT-TO-KNOW ACT

Superfund Amendments and Reauthorization Act (SARA) of 1986 revised CERCLA largely in response to a 1984 tragedy in Bhopal, India, where thousands of residents were killed or injured by exposure to a chemical, methyl isocyanate (MIC), gas that leaked from a Union Carbide plant (USEPA 2021c). In response to continuing community concerns regarding hazardous materials and the Bhopal, India, tragedy, Title III of SARA was developed and is known at the Emergency Planning and Community Right-to-Know Act (EPCRA).

SARA is significant to sustainability in that SARA provides for exchange of information and engineering and safety controls that are put into place before a potential release occurs. The

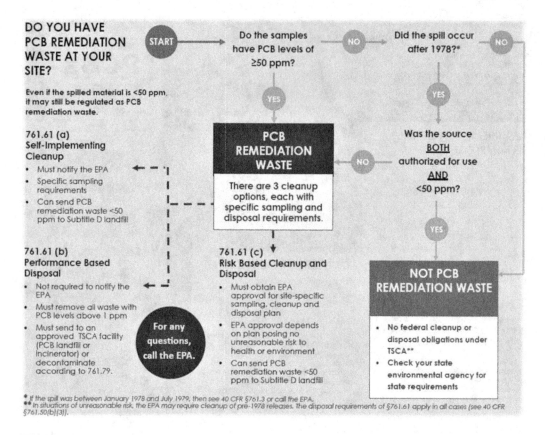

FIGURE 10.2 Flowchart for PCB wastes (From United States Environmental Protection Agency. Summary of the Toxic Substance Control Act. https://epa.gov/laws-regulations/summary-toxic-substances-control-act (accessed March 24, 2021), 2021b).

objective then is to educate the public of risk and provide for engineering cand safety controls to minimize or eliminate harm to human health and the environment if a release occurs.

During the early morning hours of December 3, 1984, a Union Carbide plant in a village just south of Bhopal, India, released approximately 40 tons of MIC into the air. MIC is used in manufacturing pesticides and is considered a lethal chemical. The gas quickly and silently diffused near the ground surface and in the end killed thousands of people and injured tens of thousands of residents (Eckerman 2005). Title III of SARA commonly called EPCRA was a direct result of the tragedy in India.

There are many examples of incidents in the United States that involve hazardous substances and the purpose of EPCRA was to prevent or minimize the potential for an incident like what occurred in India from ever happening in the United States (USEPA 2021c). The purpose of EPCRA includes:

1. To encourage and support emergency planning for responding to chemical accidents, and
2. To provide local governments and the public with information about chemical hazards in their communities.

To facilitate cooperation between industry, interested citizens, environmental and other public-interest groups and organizations, and the government at all levels, EPCRA establishes an ongoing forum at the local level called the Local Emergency Planning Committee (LEPC). LEPCs are governed by the State Emergency Response Commission (SERC) in each state. Under EPCRA, SERCs and LEPCs are responsible for (USEPA 2021d):

1. Preparing emergency plans to protect the public from chemical accidents
2. Establishing procedures to war and, if necessary, evacuate the public in case of an emergency
3. Providing the public citizens and local governments with information about hazardous chemicals and accidental releases of chemical in their communities
4. Assisting in the preparation of public reports on annual release to toxic chemicals into the air, water and soil

EPRCRA does not place limits on which chemicals can be stored, used, released, disposed, or transferred at a facility. It only requires a facility to document, notify, and report information (USEPA 2021d). Emergency planning under EPCRA is found in Sections 301–303 and are to ensure that state and local communities are prepared to respond to a potential chemical accident. As a first step, each state had to establish a SERC. In turn the SERC designated local emergency planning districts. For each district, the SERC appoints, supervisors and coordinates the activities of the LEPCs. The LEPC in turn, must develop and emergency response plan for its district and review it annually. The membership of the LEPC includes representatives from public and private organizations as well as a representative from every facility subject to EPCRA emergency planning requirements which we will discuss later in this section (USEPA 2021d).

The plan developed by the LEPC must have the following information (USEPA 2021d):

- Identify affected facilities and transportation routes
- Describe emergency notification and response procedures
- Designate community and facility emergency coordinators
- Describe methods to determine the occurrence and extent of a release
- Identify available response equipment and personnel
- Outline evacuation plans
- Describe training and practice programs and schedules
- Contain methods and schedules for exercising the plan

USEPA has established the requirements of when a facility is subject to EPCRA emergency planning reporting. USEPA has published a list of what it terms "extremely hazardous substances (EHS)." For each EHS, the list includes its name, the Chemical Abstract Service number of the substance, and a number that is called the Threshold Planning Quantity (TPQ) and is expressed in pounds. If a facility has within its boundaries an amount of an EHS equal to or greater than the TPQ, the facility is subject to EPCRA and must notify both the SERC and the LEPC. The facility must also appoint an emergency response coordinator to work with the LEPC in developing and implementing the local emergency plan at the facility (USEPA 2017aa).

Emergency Release Notification requirements are described in Section 304 of EPCRA and defines if a facility is subject to release reporting requirements even if it is not subject to the provisions described in Section 301–303. Section 304 applies to a facility which stores, produces, or uses a **Hazardous Chemical** defined as any chemical which is a physical hazard or health hazard, and releases a reportable quantity (RQ) of a substance contained in either the list of extremely hazardous substances or the list of CERCLA hazardous substances (USEPA 2021d).

Under EPCRA, the RQ is the determining factor in whether a release must be reported. This is a number that is expressed in pounds that is assigned to each chemical in USEPA's list of extremely hazardous substances or the CERCLA list of hazardous substances. If the amount of a chemical released to the environment exceeds its corresponding RQ, the facility must immediately report the release to the appropriate LEPC and SERC and provide written follow-up as soon as practicable. In addition to immediate notification, the facility is required to provide a written follow-up report that includes an update to the original notification, any additional information in the response action conducted, known, anticipated, or health risk incurred, and any information on medical care needed by any exposed person or population (USEPA 2021d).

Community Right-to-Know Reporting Requirements are described in Section 311 and 312. The purpose of these requirements is to increase community awareness of chemical hazards and to assist in emergency planning, when and if ever needed. Section 311 and 312 apply to any facility that is required by the OSHA under its Hazard Communication Standard to prepare or have available a Safety Data Sheet (SDS) for a hazardous chemical or that has onsite, for any one day in a calendar year, an amount of a hazardous chemical equal to or greater that the following threshold limits established by USEPA (USEPA 2021d):

- 10,000 pounds (4,500 kg) for hazardous chemicals
- Lessor of 500 pounds (230 kg) or the threshold planning quantity (YPQ) for extremely hazardous substances.

If a facility is subject to reporting under these sections, it must submit information the SERC and LEPC and the local fire department with jurisdiction over the facility under two categories that include SDS reporting and inventory reporting. SDS reporting requirements specifically provide information to the local community about mixtures and chemicals present at a facility and their associated hazards. For all substances whose onsite quantities exceed the threshold values, the facility must submit the following (USEPA 2021d):

- A copy of the SDS for each above-threshold chemical onsite or a list of the chemicals grouped into categories
- Submit any change within three months, an SDS or list of additional chemical which meet the reporting criteria

Inventory reporting is designed to provide information on the amounts, location and storage condition of hazardous chemicals and mixtures containing hazardous chemicals present at any facility. The inventory has two forms. The first is call Tier I. The Tier I form contains aggregate information for applicable hazard categories and must be submitted yearly by March 1. The Tier II form contains more detailed information, including specific names of each chemical. This form is submitted upon request of the agencies authorized to receive the Tier I form. It can also be submitted yearly in lieu of the Tier I form (USEPA 2017aa). Toxic Chemical Release Inventory Reporting is described in Section 313 of EPCRA. Under this section, the USEPA is required to establish a Toxics Release Inventory (TRI) which is an inventory of routine toxic chemical emission from certain facilities. The original requirements under this section have been expanded by the passage of the Pollution Prevention Act of 1990 and will be described further in a later section. The TRI must now include information on source reduction, recycling and treatment. To obtain this data, EPCRA requires each required facility to submit a Toxic Chemical Release Inventory Form (Form R) to the USEPA and designated state officials each year on July 1 (USEPA 2021d).

A facility must file a Form R if the facility (USEPA 2021d):

- Has more than 10 full time employees
- Is in a specified Standard Industrial Classification Code
- Manufactures more than 25,000 lb/year of a listed toxic chemical
- Processes more than 25,000 lb/year of a listed toxic chemical
- Otherwise uses more than 10,000 lb/year of a listed toxic chemical

- Manufacture, processes or otherwise uses a listed persistent bio-accumulative toxic (PBT) chemical above the respective PBT's reporting threshold. PBT reporting thresholds can vary anywhere from 0.1 grams for dioxin compounds to 100 pounds (45 kg) for lead.

USEPA has updated some aspects of the Form R rules in 1999, 2001, and in 2006, which related to adding chemicals or adjusting thresholds for chemical on either the list of extremely hazardous substances or the list of CERCLA hazardous substances. Therefore, companies should review the lists routinely for changes and modifications (USEPA 2021d).

Table 10.1 is the partial list of Extremely Hazardous Substances defined in Section 302 of EPCRA. The full list can be viewed at https://www.epa.gov/epcra/final-rule-extremely-hazardous-substance-list-and-threshold-planning-quantities-emergency (USEPA 2015).

10.2.3 Pollution Prevention Act

The Pollution Prevent Act (PPA) of 1990 created a national policy to prevent or reduce pollution at the source whenever possible. As discussed in the previous section, it also expanded the Toxics Release Inventory under SARA and EPCRA. The PPA focused on reducing and preventing pollution and not just for industry but also the government itself and the general public to reduce waste and encouraged and emphasized recycling efforts (USEPA 2021e). The PPA directly applies to any sustainability initiative or program in that the PPA is specifically designed to minimize or eliminate the potential release of hazardous substances to the environment.

Up to this point in the rapid enactments of environmental laws beginning just 20 years before the PPA, the focus was on compliance. Now, with the PPA, there was a shift in focus from compliance to prevention and reducing amounts of wastes generated. However, as we discussed earlier in this chapter, it was in 1971 when public service announcements first were aired on television drawing attention to waste disposal and recycling. Why did it take so long? The answer is of course, like many other aspects of human interaction with nature, is complex and has much to do with how and to what extent humans impact the natural environment which we are still learning. The PPA is also considered the first real legislative effort in the United States to take up the subject of sustainability. Pollution prevention as we shall see in subsequent chapters is addressed as a subset of sustainability. Emphasis was placed on preventing pollution from being produced, another words, preventing pollution at the source was preferred through technological innovation and advancements in manufacturing. If it could not be prevented, then the focus should be on recycling in an environmentally safe manner. The PPA goes on to also state that disposal of waste or other types of releases into the environment should be conducted only as a last resort (USEPA 2021e).

10.2.4 Brownfield Revitalization Act

The Brownfield Revitalization Act of 2006 amended CERCLA provided funding to assess and cleanup brownfields, clarified CERCLA liability protections, and provided funds to enhance state and Native American Tribal response programs. USEPA defines a **Brownfield** as a property, the expansion, redevelopment, or reuse of which may be complicated by the presence or potential presence of a hazardous substance, pollutant, or contaminant (USEPA 2021f and 2021g). USEPA estimates that there are more than 450,000 brownfield sites in the United States.

The Brownfield Revitalization Act was enacted in large part so that these properties, many of which are impacted with hazardous substances, can be investigation, cleaned up, and be re-developed (USEPA 2021f and 2021g).

However, CERCLA had to be revised because under CERCLA, United States courts have ruled that a buyer, lessor, or lender may be held responsible for remediation of hazardous substance residues, even if a prior owner caused the contamination. Therefore, CERCLA was amended to include what is termed "All Appropriate Inquiry" which outlined procedures for the performance of a Phase I

TABLE 10.1

Partial List of Extremely Hazardous Substances

Compound	Compound	Compound
Acetone cyanohydrin	Acetone thiosemicarbazide	Acrolein
Acrylamide	Acrylonitrile	Acryloyl chloride
Adiponitrile	Aldicarb	Aldrin
Allyl alcohol	Allylamine	Aluminum phosphide
Aminopterin	Amiton	Amiton oxalate
Ammonia	Amphetamine	Aniline
Aniline, 2,4,6-trimethyl	Antimony pentafluoride	Antimycin A
ANTU (Alpha-Naphthlthiourea)	Arsenic pentoxide	Arsenous oxide
Arsenous trichloride	Arsine	Azidoazide azide
Azinphos-ethyl	Azinphos-methyl	Benzal chloride
Benzenamine, 3-(trifluoromethyl)-	Benzenearsonic acid	Benzimidazole, 4,5-dichloro-2-(trifluoromethyl)-
Benzothrochloride	Benzyl chloride	Benzyl cyanide
Bicyclo(2.2.1)heptane-2-carbonitrile	Bis(chloromethyl) ketone	Bitoscanate
Boron trichloride	Boron trifluoride	Boron trifluoride w/ methyl ether
Bromadiolone	Bromine	Cadmium oxide
Cadmium stearate	Calcium arsenate	Camphechlor
Cantharidin	Carbachol chloride	Carbamic acid
Carbofuran	Carbon disulfide	Carbophenothion
Chlordane	Chlorfenvinfos	Chlorine
Chlorine trifluoride	Chlormethos	Chlormequat chloride
Chloracetic acid	2-chlorothanol	Chloroethyl chloroformate
Chloroform	Chloromethyl methyl ether	Chlorophacinone
Chloroxuron	Chlorthiophos	Chromin chloride
Cobalt carbonyl	Cobalt, (2,2-(1,2-ethanediylbis)	Colchicine
Coumaphos	Cresol, -o	Crimidine
Crotonaldehyde	Crotonaldehyde, (E)	Cyanide
Cyanogen bromide	Cyanigen iodide	Cyanophos
Cyanuric fluoride	Cycloheximide	Cyclohexylamine
Decaborane	Demeton	Demeton-S-methyl
Dialifor	Diborane	Dichloroethyl ether
Dichloromethylphenylsilane	Dichlorvos	Dicrotophos
Diepoxybutane	Diethyl chlorophosphate	Digitoxin
Diglycidyl ether	Digoxin	Dimefox
Dimethoate	Dimethyl mercury	Dimethyl phosphorochloridothioate
Dimethyl-p-phenylenediamine	Dimethyl-p-phenylebediamine	Dimethylcadmium
Dimethyldichlorosilane	Dimethylhydrazine	Dimetilan
Dinitrocresol	Dinoseb	Dineterb
Dioxathion	Diphacinone	Disulfoton
Dithiazanine iodide	Dithobiuret	Endosulfan
Endoothion	Endrin	Epichlorohydrin
EPN	Ergocalciferol	Ergotamine tartrate
Ethanesulfonyl chloride, 2-chloro-	Ethanol, 1,2-dichloro-. Acetate	Ethion
Athoprophos	Ethylbis(2-chloroethyl)amine	Ethylene oxide
Ethylemediamine	Ethyleneimine	Ethylthiocyanate
Fenamiohos	Fenitrothion	Fensulfothion

TABLE 10.1 (Continued)
Partial List of Extremely Hazardous Substances

Compound	Compound	Compound
Fluenetil	Fluorine	Fluoroacetamide
Fluoroacetic acid	Floroacetyl chloride	Fluoroantimonic acid
Fluorouracil	Fonofos	Formaldehyde

Source: United States Environmental Protection Agency, USEPA 2015. List of Lists. EPA 55-B-15-001. Office of Solid Waste and Emergency Response. Washington, D.C. 125p.

Environmental Site Assessment (ESA). By conducting a Phase I ESA, a new property owner may create a safe harbor or protection from liability, known as the "Innocent Landowner Defense" for new purchasers or lenders for transactions involving properties. In the United States, Phase I ESAs have become the standard type of environmental investigation employed when initially investigating a property (USEPA 2021f). The BRA has significant sustainability features in that it was purposely designed to recycle impacted and abandoned properties and return them to productive re-use.

Standards for conducting Phase I ESAs were originally published by the American Society for Testing Materials (ASTM) in 1993 and were revised in 1997, 2000, and 2005 (American Society for Testing Materials 2005). On November 1, 2006, the USEPA published federal standards for conducting Phase I ESAs termed "Standards and Practices for All Appropriate Inquiries" (AAI) by amending the Comprehensive, Environmental Response, Compensation and Liability Act (CERCLA) of 1980, commonly known as the Brownfield Revitalization Act. As stated earlier, the Phase I Environmental Site Assessment (ESA) is typically the first environmental investigation conducted at a specific property (Rogers 2020).

A Phase I ESA is conducted with the objective of qualitatively evaluating the environmental condition and potential environmental risk of a property or site. A **property** is defined here as a parcel of land with a specific and unique legal description. A **site** is defined here as a parcel of land including more than one property or easement and typically refers to an area of contamination potentially affecting more than one property. As the first environmental investigation, the Phase I ESA is often regarded as the most important activity because all subsequent decisions concerning the property are, in part, based on the results of the Phase I ESA (Rogers 2019). Therefore, great care, scrutiny, scientific inquiry, and objectivity should be exercised while conducting the Phase I ESA. According to United States Environmental Protection Agency (2005) requirements, an environmental professional, such as a geologist, or environmental scientist must conduct the Phase I ESA. General requirements for conducting a Phase I ESA include (Rogers 2020):

- Extensive review of current and historical written records, operations, and reports
- Extensive site inspection
- Interviews with knowledgeable and key onsite personnel
- Assessment of potential environmental risks from offsite properties
- Data gap or data failure analysis

A Phase I ESA is typically a non-invasive assessment, performed without sampling or analysis. In some instances, limited sampling may be conducted on a case-by-case basis if in the professional judgment of the person conducting the assessment–or due to other requests or mitigating factors–sampling is justified. In most cases, collecting and analyzing samples is usually deferred to the Phase II investigation, but if it does occur, the sampling conducted during a Phase I ESA may include the following (Rogers 2020):

- Sediment
- Surface water
- Waste material
- Lead-based paint
- Mold

- Drinking water
- Groundwater
- Soil
- Suspect asbestos-containing material
- Radon

The environmental professional conducting the Phase I ESA must perform extensive research and review all available written records. These researching activities apply not only to the property in question, but also to the surrounding properties within a radius of up to one-mile of the investigated property or site. The research/review process typically involves (Rogers 2020):

- Title search and environmental liens
- Historical chemical ordering documents
- Historical aerial photographs
- Engineering diagrams
- Previous environmental incident reports
- Environmental compliance documents
- Fire insurance maps
- USGS topographic maps
- USGS geologic maps
- Native American Tribal records
- Building permits

- Historical operations documents
- Historical photographs
- Engineering reports
- Historical environmental reports
- Safety data sheets (SDS's)
- Hazardous substance inventories
- Soil Conservation Service maps
- USGS investigations reports
- Environmental agency maps
- Local governmental records
- Construction diagrams and blueprints

The site inspection consists of a walk-through of the property or site. Items to evaluate and document during the site inspection include (Rogers 2019):

- Interviews with key personnel
- Areas absent of vegetation
- Above-ground storage tanks (ASTs)
- Chemical storage areas
- The 'back 40" (the rear of the facility)
- Storage sheds
- Special labeling
- Recent excavations or land disturbance
- General topography
- Mold
- Insects
- Sumps
- Broken concrete
- Weather conditions
- Nearest water body
- Potential asbestos-containing materials
- Utilities
- Offsite inspection
- Potential contaminant migration pathways

- Stressed vegetation
- Stained soil or pavement
- Underground storage tanks (USTs)
- Back doors
- Signage
- Refuse storage and containers
- Evidence of fill or mounding
- Depressions in the land surface
- Wetlands
- Animal scat
- Pits and trenches
- Floor and roof drains
- Areas not inspected or inaccessible
- Recent precipitation events
- Evidence of wells or borings
- Potential septic tanks
- Onsite and offsite dumping or fill
- Soil type(s)
- Potential ecological and human receptor pathways

Once the environmental professional has completed the data collection and site inspection portion of the Phase I ESA, an evaluation of whether there is evidence of an existing release, a past release, or a material threat of a release of any hazardous substance or petroleum is conducted. If a product has been released and made its way into structures on the property or into the ground, groundwater, or surface water of the property, then this situation is termed a Recognized Environmental Condition (REC). **Recognized Environmental Condition (REC)** is defined as the presence or likely presence of any hazardous substance or petroleum product on a property or site under conditions that may materially affect or threaten the environmental condition of the property or site, human health, or the environment (American Society for Testing Methods [ASTM] 2005 and United States Environmental Protection Agency, 2005). If a REC is discovered, further investigation will likely be recommended to evaluate its potential significance. Many sites have more than one REC, and many of these RECs may require further evaluation. REC's are intended to exclude de minimis conditions generally not presenting a threat to human health or the environment and typically would not be the subject of an enforcement action if brought to the attention of appropriate governmental agencies. However, there may be impacts encountered whose perceived severity falls between a de minimus condition and a REC. In these situations, the term **environmental concern** is applied. Items listed as environmental concerns become RECs if left unattended or lead to a release (Rogers 2020).

Historical aerial photographs are effective sources of information, and often help with environmental investigations. In urban areas especially, aerial photographs from several sources are readily available. These sources include (Rogers 2020):

- Private local companies specializing in aerial photography
- Private national companies specializing in aerial photography
- Local and state historical societies
- Local and state agencies
- Utility companies
- Local companies
- Federal agencies such as USEPA, USGS, Soil Conservation Service, National Forest Service, Bureau of Land Management, National Park Service.

10.2.5 LACEY ACT OF 1900

The Lacey Act of 1900 is a conservation law that prohibits trade in wildlife, fish, and plants that have been illegally taken, possessed, transported, or sold. Essentially, the Lacey Act made poaching illegal. From a sustainability perspective, this is significant because the Lacey Act has the benefit of protecting and regulating wildlife. The Act also regulated the introduction of birds and other animals to places where they have never existed. This is now commonly referred to as invasive species. The Lacey Act is still active and has been amended several times. The latest occurring in 2008 (United States Fish and Wildlife Service 2021a).

10.2.6 ANTIQUITIES ACT OF 1906

The Antiquities Act of 1906 gave the President of the United States the power to create National Monuments with the purpose to protect historic landmarks, historic and prehistoric structures, and other objects of historic or scientific interest. The Antiquities Act was passed as a direct result of concerns over theft from destruction of archaeological sites and was designed to provide an expeditious means to protect federal lands and resources. President Theodore Roosevelt used the authority in 1906 to establish Devil's Tower in Wyoming as the first National Monument and used the Act 36 times during his presidency which is the most for any President. Sixteen of the 19 President's since have used this authority and have created 151 monuments. Some worthy of note include (United States National Park Service 2021a):

- Grand Canyon National Park
- Grand Teton National Park
- Zion National Park
- Olympic National Park
- Statue of Liberty

Since it has been enacted, hundreds of millions of acres of land has been set aside and protected, including marine environments (United States National Park Service 2021a). This is significant from a sustainability perspective to not to protect treasured national landmarks and culturally significant structures, but also to protect vast tracks of land and wilderness.

10.2.7 MIGRATORY BIRD TREATY ACT 1916

The Migratory Bird Treaty Act of 1916 is an environmental treaty between the United States and Canada. The Act makes it unlawful to pursue, hunt, take, capture, kill, or sell birds that appear on the protected list. The Act does not discriminate between live or dead birds and also grants full protection to any bird parts including feather, eggs, and nests. Currently, there are over 800 species of birds on the list (United States Fish and Wildlife Service 2021b). The migratory Bird Treaty Act is one of the first legislative measures acknowledging that countries must cooperate in protecting wildlife in order to maintain sustainable wildlife populations.

10.2.8 NATIONAL PARKS ACT OF 1916

The National Parks Act of 1916 is also known as the Organic Act and established the National Park Service so that management of all the national parks would be conducted under one authority (United States National Park Service 2021b). These areas include a diverse variety of land of national importance and include (United States National Park Service 2021b):

• National Parks	• National Monuments
• National Memorials	• National Military Parks
• National Battlefields	• National Historic Sites
• National Parkways	• National Recreation Areas
• National Seashores	• National Scenic Riverways
• National Scenic Trails	• National Cemeteries
• National Heritage Areas	• Nationals Scenic Rivers

From the time Yellowstone National Park was created in 1872, each national park or other lands deemed of national importance was managed separately. The passage of the National Parks Act of 1916 combined management of all these protection lands and locations to the newly created National Parks Service.

The NPA, considered by many to be USA's best idea, has significant sustainability implications. The NPA has also provided scientists with areas to conduct research in wildlife behavior and examining sustainable ecological restorations methods and effectiveness.

10.2.9 NATIONAL HISTORIC PRESERVATION ACT OF 1966

The National Historic Preservation Act of 1966 was intended to preserve historical and archaeological sites in the United States. The Act created the National Register of Historic Places. Although the Act was not officially passed until 1966, many landmarks in the United States have been protected, such as George Washington's home, Mount Vernon and Thomas Jefferson's home as well. Protection of these historic places did not seem appropriate under the Antiquities Act of 1906. Supporters and historians desired a more specifically tailored Act that was specific to historical places (United States Advisory Council on Historic Preservation 2021).

10.2.10 WILD AND SCENIC RIVERS ACT OF 1968

The Wild and Scenic Rivers Act of 1968 was passed to protect rivers in the United States because they possessed scenic, recreational, geologic, fish and wildlife, historic, cultural, other similar values. Rivers, or portions or sections of a river, that obtain the designation are preserved in their free-flowing condition and are not dammed or otherwise impeded. The Act essentially vetoes the licensing of new hydroelectric power plants. Currently, there are over 200 rivers, totaling more than 12,500 miles of river in 38 states and Puerto Rico that have been designated under the Act. By comparison, more than 75,000 large dams are located across the United States that have modified an estimated 600,00 miles of rivers in the United States (United States National Park Service 2021c). Designation as a wild and scenic river is not the same as a national park designation, and generally does not confer the same level of protection as a wilderness area designation. However, it does protect the free-flowing nature of a river in federal or non-federal areas and does not alter property rights (United States National Park Service 2021c).

10.2.11 FISH AND WILDLIFE COORDINATION ACT OF 1968

The Fish and Wildlife Coordination Act of 1968 is an amended Act that was first passed in 1934 with the purpose of protecting fish and wildlife when federal action resulted in the control or modification of a natural stream or body of water. The Act is implemented by the Department of Interior and is authorized to conduct the following to preserve fish and wildlife (United States Fish and Wildlife Service 2021c):

- Developing, protecting, rearing, and stocking all species of wildlife, resources thereof, and their habitat
- Controlling losses from disease and other causes
- Minimizing damages from overabundant species
- Providing public shooting and fishing areas
- Carrying out other necessary measures
- Conduct surveys and investigations of the wildlife of the public domain
- Accept donations of land and other contributions from private sources

10.2.12 COASTAL ZONE MANAGEMENT ACT OF 1972

The Coastal Zone Management Act of 1972 was established to preserve, protect, develop, restore, or enhance coastal zones of the United States. This Act was in response and a recognition that coastal areas of the United States were under significant stress due to development pressures.

The Act established the National Coastal Zone Management Program (NCZMP) and the National Estuarine Research Reserve System (NERRS). The State of Washington was the first to adopt the program in 1976. The NCZMP is administered by the National Oceanic and Atmospheric Administration (NOAA). The program is designed to establish a basis for protecting, restoring, and establishing a responsibility in preserving and developing coastal communities and resources where they are under the highest pressures. The area does not just include the coastal areas of the Atlantic, Pacific or Artic Oceans alone but also includes the Great Lakes and island territories to ensure a healthy and thriving ecosystem for future generations (NOAA 2021a).

10.2.13   MARINE PROTECTION ACT OF 1972

The Marine Protection Act of 1972 was the first Act to specifically manage natural resources through an ecosystem approach. The ecosystem-based management approach is an integrated management approach that recognizes the full array of interactions within an ecosystem, including humans, rather than considering single issues, species, or ecosystem services in isolation. The ecosystems approach has been integrated into other marine regulations including the Magnuson-Stevens Fishery Conservation and Management Act and the Marine Mammal Protection Act of 2015. Originally the Marine Protection Act of 1972 set aside marine mammals as untouchable and unapproachable. It is viewed as the first animal rights legislation in the United States. The Act was amended through the Magnuson-Stevens Fishery Conservation and Management Act and Marine Mammal Protection Act of 2015.

The Act prohibited the act of hunting, killing, capture, or harassment of any marine mammal, including the Polar Bear. There are some exceptions and include capture of marine mammals for scientific study and Alaska Native subsistence (NOAA 2021b).

10.2.14   ENDANGERED SPECIES ACT OF 1973

The Endangered Species Act of 1973 was designed to protect animal species from extinction from a consequence of human interference from economic growth and development un-tempered by adequate concern and conservation. The Act was established in an attempt to reverse a trend of those identified endangered species from extinction at whatever the cost (United States Fish and Wildlife Service 2021a). The Act is administered by the Fish and Wildlife Service (FWS) and the National Oceanic and Atmospheric Administration (NOAA). Listing status is a sliding criteria that indicates the degree to which a certain species is threated with extinction. The listing status used in the United States is as follows (United States Fish and Wildlife Service 2021d):

- E = Endangered. Endangered is the highest of alerts meaning the closest to extinction in the United States and generally means any species that is in danger of extinction throughout all or a significant portion of its natural habitat range.
- T = Threatened. Threatened is any species which is likely to become endangered within the foreseeable future throughout all or a significant portion of its natural habitat range.
- C = Candidate. Candidate is a species that is under consideration due to a drop in population or because of loss of a significant reduction in its natural habitat.

The first list of endangered species in the United States included 14 mammals, 36 birds, 6 reptiles and amphibians, and 22 fish. According to the United States Fish and Wildlife Service (FWS) there are currently a total of 711 animal species on the list and 941 plant species for a total of 1,652 on the Endangered Species List just in the United States (United States Fish and Wildlife Service 2021d). Some of the animal species in the United States that are or were on the Endangered Species list include the following (United States Fish and Wildlife Service 2021d):

• Bald Eagle – removed in 2007	• Whooping Crane
• California Condor	• Kirkland's Warbler
• Peregrine Falcon – removed in 1999	• Gray Wolf
• Mexican Wolf	• Red Wolf
• Grey Whale	• Grizzly Bear – removed in 2007
• California Southern Sea Otter	• Florida Key Deer
• Hawaiian Goose	• Virginia Big-Eared Bat
• Black-Footed Ferret	• Florida Grasshopper Sparrow

Worldwide, there are thousands of species of animals and plants that are in danger of extinction because of direct human involvement that includes development, habitat loss and destruction, poaching, pollution and other factors. The World Wildlife Fund maintains a general list of those animals that are in danger and also includes an additional category different than that of the United States by the addition of another category called Critically Endangered. Some of the animals currently listed by the World Wildlife Fund as critically endangered or endangered include (World Wildlife Fund 2021):

• Amur Leopard	• Black Rhino	• Bornean Orangutan
• Cross River Gorilla	• Eastern Lowland Gorilla	• Javan Rhino
• Malayan Tiger	• Sumatran Orangutan	• Orangutan
• South China Tiger	• Sumatran Elephant	• Mountain Gorilla
• Sumatran Rhino	• Sumatran Tiger	• Amur Tiger
• Asian Elephant	• Western Lowland Gorilla	• African Elephant
• Bluefin Tuna	• Blue Whale	• Bonobo
• Chimpanzee	• Borneo Pygmy Elephant	• Galapagos Penguin
• Hectors Dolphin	• Ganges River Dolphin	• Indian Elephant
• Indochinese Tiger	• North Atlantic Wright Whale	• Red Panda
• Sea Lion	• Snow Leopard	• Sri Lanka Elephant
• Tiger	• Black Spider Monkey	• African Giraffe

The given list is just to name a few of the more familiar animals that are now considered threatened with extinction. Unless major efforts from the world community are conducted immediately, the loss of these animals and countless others are likely inevitable. The following list includes some North American species that have gone extinct just since 1500 AD and include (International Union on the Conservation of Nature [IUCN] 2021):

• Great Auk – 1852	• Passenger Pigeon – 1914
• California Golden Bear – 1922	• Ivory-Billed Woodpecker – 1942
• Labrador Duck – 1878	• Eastern Cougar – 2011 declared
• Cascade Mountain Wolf – 1940	• Eastern Elk – 1887
• Mexican Grizzly Bear – 1964	• Newfoundland Wolf – 1911
• Sea Mink – 1860	• Caribbean Monk Seal – 1952
• South Rocky Mountain Wolf – 1935	• Southern California Kit Fox – 1903
• Carolina Parakeet – 1918	• Dusky Seaside Sparrow – 1987
• Heath Hen – 1932	• Eskimo Curlew – 1981
• Bachman's Warbler – 1988	• Tacoma Pocket Gopher – 1970

According to the World Wildlife Fund (2021) and the International Union for Conservation of Nature (2021), hundreds of species of animals are now extinct directly due to humans in several different ways including, hunting, habitat destruction, poisoning, development, and pollution. According to the United Nations, more than 1 million species of animals and plants are threatened with extinction as a direct result of human activities (United Nations 2021). As another measure of animal life on Earth, the WWF estimates that animal life on Earth has been reduced by more than 70% during the last 50 years (WWF 2020).

10.2.15 Fisheries Conservation and Management Act of 1976

The Fisheries Conservation and Management Act of 1976 is also known as the Magnuson Fishery Conservation and Management Act and the Sustainable Fisheries Act. The Act was established to (United States Fish and Wildlife Service 2021e):

- Establish national fish hatchery conservation and management standards
- Provide for the sustained participation of fishery dependent communities and minimize economic impacts to those communities
- Modify operation of established Fishery Management Councils
- Mandate action to identify overfished species
- Mandate action to rebuild fishery populations to those species that have been over fished
- Establish guidelines to identify essential fish habitat
- Establish a fishing capacity reduction program
- Conduct research on fishery management, conservation, and economic characteristics of fisheries

10.2.16 National Forest Management Act of 1976

The National Forest Management Act of 1976 was enacted to establish guidelines for the management of renewable resources on national forest land. The Act specifically addresses timber harvesting within National Forests. The purpose of the Act was to protect National Forests from permanent damage from excessive logging and clear cutting practices (United States Forest Service 2021). Throughout the history of the United States there has been numerous legal battles and conflicts between environmental groups and the timber industry. The demand for timber for home building, farming, and other uses in the United States has always seemed to be high and has placed significant pressure on the environment through habitat destruction.

One of the more contentious conflicts was over the Northern Spotted Owl in the Pacific Northwest. The Northern Spotted Owl experienced a decrease in population levels that threatened its extinction. A major study was undertaken to evaluate the causes of the population decrease and the results indicated that there were three major threats to the owl population decrease that involved timber harvesting that included (1) variability of birth and death rates through time, (2) loss of genetic variation, and (3) random catastrophes.

The passage of the Act required the National Forest Service to conduct an inventory of all National Forests and to use a systematic and interdisciplinary approach to resource management which also included monitoring biological effects of timber harvesting (United States Forest Service 2021). Although deforestation in the United Sates has not been as much as we will discover in other countries of the world in the next chapter, the decrease is still significant. In 1620, approximately 4.1 million square kilometers of the United States was forest. The low point was in 1926 at 2.9 million square kilometers. Since 1926, the amount of forested land in the United States is 3.0 million square kilometers (United States Department of Agriculture 2014). Figure 10.3 shows the amount of forest in the United States in 1620 and Figure 10.4 shows the amount in 1926.

10.2.17 Surface Mining Control and Reclamation Act of 1977

The Surface Mining Control and Reclamation Act of 1977 was enacted for the purpose of addressing the environmental effects of coal mining. The Act created two programs that included:

- Regulating active coal mines
- Reclaiming abandoned mine lands

FIGURE 10.3 Amount of forest in the United States in 1620. (From United States Department of Agriculture. U.S. Forest Resource Facts and Historical Trends. https://www.fia.fs.fed.us/library/brochures/docs/2012/ForestFacts_1952-2012_English.pdf (accessed March 31, 2014).

FIGURE 10.4 Amount of forest in the United States in 1620. (From United States Department of Agriculture. U.S. Forest Resource Facts and Historical Trends. https://www.fia.fs.fed.us/library/brochures/docs/2012/ForestFacts_1952-2012_English.pdf (accessed March 31, 2014).

The Act also created the Office of Surface Mining within the United States Department of Interior to ensure consistency and fair application of the requirements under the Act (United States Office of Surface Mining and Reclamation 2021).

The Act was developed because of environmental concerns with the effects of surface coal mining or strip mining. Coal has been mined in the United States since the 1740s but surface or strip mining did not become widespread until the 1930s and by the mid 1970s accounted for more than 605 of coal mining. The Act established the following (United States Office of Surface Mining and Reclamation 2021):

- Uniform Standards of Performance. This provision standardized what was required nationwide and eliminated confusion and differences in regulations from state to state.
- Permitting. The Act required that each proposed mine obtain a permit that described:
  - Pre-mining environmental conditions and land use
  - Proposed mining activities and area
  - Post mining reclamation
  - How the mine will achieve Act requirements
  - How the land will be used following completion of mining activities
- Bonding. The Act required that mining companies post a bond sufficient to cover the cost of reclaiming the mined land
- Inspection and Enforcement. The Act permitted inspections and enforcement of violations by the federal government.
- Land Restrictions. The Act prohibited mining activities within the National Parks and wilderness areas. It also allowed for citizen challenges to proposed mining developments.

The Act also created a reclamation fund to assist in funding reclamation for mines closed before the passage of the act. Some funds ae used to pay for emergencies such as landslides, land subsidence, fires, and the conduct high priority cleanups (United States Office of Surface Mining and Reclamation 2021).

10.2.18   NUCLEAR WASTE POLICY ACT OF 1982

The Nuclear Waste Policy Act of 1982 established a program with the intention of providing a safe and permanent location for the disposal of highly radioactive wastes. During the 40 years before enactment of the law, there was no regulatory framework in place for disposal of nuclear wastes generated in the United States (Nuclear Regulatory Commission [NRC] 2021).

Some of the waste that was generated had a half-life of more than a million years and was temporarily stored in various locations and types of containers. Most of the wastes generated were as a result of manufacturing nuclear weapons. Approximately 77 million gallons of nuclear waste had been generated and was being stored in South Carolina, Washington and Idaho. In addition, there were 82 nuclear power plants that produced electricity and also generated nuclear waste in the United States in 1982. The nuclear waste at the nuclear power plants consisted of spent fuel rods which were stored in water at the reactor sites and many of the plants were in danger of running out of storage space (NRC 2021).

The Nuclear Waste Policy Act created a timetable and procedure for establishing a permanent repository for high-level radioactive waste by the mid-1990s and provide temporary storage for those sites that were running out of space. The USEPA was directed to establish public health and safety standards for potential releases of radioactive materials and the NRC was required to promulgate regulations that covered construction, operation, and closure of repository location. Generators of nuclear waste were to be charge a fee to cover the costs of disposal of nuclear wastes (NRC 2021). Another provision of the Act required the Secretary of Energy to issue guidelines for selecting sites of two permanent underground nuclear waste repositories. The Department of

Energy (DOE) studied sites in Washington near the Hanford Nuclear Reservation, in Nevada near the nuclear testing site, and sites in Utah, Texas, Louisiana, and Mississippi. Other sites were studied along the east coast of the United States but were quickly ruled out as possible storage locations (NRC 2021).

In 2002, Yucca Mountain was selected as the only site to be characterized as the permanent repository for all the nation's nuclear waste. However, the State of Nevada used its veto power and rejected the recommendation and challenge the recommendation in court. The United States Court of Appeals upheld the State of Nevada's appeal, ruling that USEPA's 10,000 year compliance period for site selection was too short and was not consistent with the National Academy of Sciences recommended compliance period of 1 million years.

USEPA subsequently revised the standard to 1 million years. At issue is selecting a disposal site that will safely contain the nation's nuclear waste for 1 million years. A license application selecting the Yucca Mountain location was submitted in 2008 and is still under review by the NRC. As of 2010, several lawsuits have been filed in federal courts to contest the legality of the selection process and proposed location of Yucca Mountain as the selected permanent repository site (NRC 2021 and USEPA 2021h). In the meantime, nuclear waste is being stored at various locations in temporary containers.

10.2.19   GLOBAL CLIMATE PROTECTION ACT OF 1987

The Global Climate Protection Act of 1987 delegated responsibility to the USEPA to develop and propose a coordinated national policy on global climate change. In a report to congress in 1991, USEPA stated that in order to address global climate change the following elements must be addressed (USEPA 1991):

- Scientific research on climate change to reduce uncertainties with respect to operation of the climate system and to improve understanding of the impacts of human activities on future climate conditions and responses
- Assessment of environmental, social, and economic impacts of climate change
- Evaluation of policy options and practices to limit, mitigate, or adapt to climate change, including assessment of effectiveness and social and economic impacts
- Development and implementation of feasible and cost-effective polies and practices

USEPA stated in its report to Congress in 1991 that the nature of human existence on the Earth has changed dramatically during the last century. The world population has increased threefold. However, this was in 1991, now in 2021 (30 years later) the human population is estimated at over 7 billion people as opposed to just over 5 billion in 1991 (United States Census Bureau 2021). USEPA (1991) further states that industrial production has increased by a factor of 50 and the world economy has increased by a factor of 20. In all, USEPA concludes that human activity is changing the features of Earth and the chemistry of the atmosphere. Since the Act was passed in 1987, USEPA has supported and produced hundreds of reports and academic journal articles that have been valuable in building our understanding of causes, impacts, and potential solutions to climate change. The CAA gave USEPA the authority to regulate emissions from power plants and other large sources of greenhouse gases, and now requires facilities that are large emitters of greenhouse gases to report emissions to USEPA (USEPA 2021i). USEPA has developed and implemented voluntary programs to cut greenhouse gas emissions through the following (USEPA 2021i):

- Natural Gas STAR, which is a program to limit methane emissions
- Coalbed Methane Outreach Program to encourage mine owners and operators to capture methane rather than allow its escape in the atmosphere

- Environmental stewardship partnership programs to address the most potent greenhouse gases emitted from the aluminum, semiconductor, refrigerant, power, and magnesium industries
- Energy Star Program, which is a voluntary program to assist businesses and individuals protect the climate through energy efficiency

Currently, reductions of greenhouse gas emissions are voluntary and Congress has yet to pass legislation to cut greenhouse gas emissions. USEPA (1991) stated that in order to effect real change in reducing the effects of global climate change from greenhouse gas emissions that it must be through international cooperation. Thus far, efforts to reduce greenhouse gas emissions under an enforceable international treaty have failed, in part, due to language that limits greenhouse gas emissions on developed countries such as the United States but does not limit greenhouse gas emission on developing nations, such as China and India (USEPA 2021i).

10.2.20   Ocean Dumping Act of 1988

The Ocean Dumping Act of 1988 amended the Marine Protection, Research and Sanctuaries Act of 1972 to prohibit the dumping of sewage sludge and industrial waste into the ocean after 1991 (USEPA 2021j). Before 1972, the ocean was in effect, a dumping ground that included sewerage, sludges, heavy metals, organic chemical wastes, industrial waste, and radioactive waste (USEPA 2021j). The Act prohibits the dumping of most any waste into the ocean including (USEPA 2021j):

- High-level radioactive waste
- Radiological, chemical, and biological warfare agents
- Persistent inert synthetic or natural materials which may float or remain in suspension in the ocean
- Sewerage sludge
- Medical wastes
- Industrial wastes containing the following:
  - Organohalogen compounds
  - Mercury and mercury compounds
  - Cadmium and cadmium compounds
  - Oil of any kind or in any form
  - Known carcinogens, mutagens, or teratogens

Passage of the Act in 1988 was motivated in part by the garbage barges that would dump garbage from New York City into the ocean.

10.2.21   Oil Spill Prevention Act of 1990

The Oil Spill Prevention Act of 1990 was enacted in large part due to the March 24, 1989 Exxon Valdez oil spill in Prince Williams Sound off the Alaskan shore and spilled 11 million gallons of crude oil. At the time, it was the largest oil spill in the history of the United States. Cleanup of the oil was mainly achieved through mechanical means and included (USEPA 2021k):

- Applying oil booms to concentrate the oil and pump the oil into containers aboard recovery ships
- Washing shorelines
- Physically recovering oil from shorelines and beaches
- Excavating shorelines and beaches

Burning the oil and applying a dispersant were both attempted but had little or no effect. Billions of dollars were spent recovering oil and cleaning tens of miles of shoreline and saving and cleaning affected wildlife (USEPA 2021k and NOAA 2021c). Immediate effects of the spill included estimated deaths from 100,000 to as high as 250,000 seabirds, nearly 3,000 sea otters, 300 harbor seals, nearly 250 Bald Eagles, 22 orcas, and an unknown number of salmon and herring (NOAA 2021c). The Oil Spill Prevention Act resulted in instrumental changes in oil production, transportation, and distribution that included the following (USEPA 2021k):

- Establishing a fund for future cleanup of spills
- Increasing spill response preparedness
- Increasing reliability in mapping and location of ships
- Improving ship designs
- Instituting measures and controls to minimize spills during loading and unloading
- Increasing financial assurance requirements

Another oil spill worthy of note is the Deepwater Horizon oil spill in 2010. On April 20, 2010, the offshore drill rig Deepwater Horizon was drilling a well 41 miles off the southeast coast of Louisiana in water that was approximately 5,000 feet deep when an explosion occurred that lead to the destruction of the drilling rig and spilled millions of gallons of oil into the Gulf of Mexico.

An accurate estimate on the amount of oil released from the Deepwater Horizon oil spill in 2010 vary widely from just a few million gallons to much more. Cleanup however, required billions of dollars. The majority of the cleanup has been completed but monitoring continues (NOAA 2021d).

10.2.22 Energy Policy Act of 1992 and 2005

The Energy Policy Act of 1992 was enacted to sets goals and mandates to increase clean energy use, improve overall energy efficiency and lesson the nation's dependence on imported energy. The Act set standards for many different sectors of the economy including establishing standards for buildings, utilities, equipment standards, renewable energy, and alternative fuels (National Energy Institute 2005). The Act was amended in 2005 to provide tax incentives for alternative energy development such as wind energy. The Act also made it easier for a new technology to expand which was oil fracking by exempting waste fluids generated from the Clean Air Act, Clean Water Act, and the Safe Drinking Water Act, and CERCLA (National Energy Institute 2005).

An unforeseen disadvantage to wind turbines is that they kill birds, especially raptors, migratory birds, and bats. Some estimates range from 250,000 to several million birds are killed annually in the United States and this will likely increase as more wind turbines are installed and old ones are replaced with wind turbines that are much larger and kill even more birds because of increased reach and spin speed of the turbine (Eveleth 2013 and Curry, 2021). However, vertical wind turbines are much more bird friendly because they are more compact and can be installed in urban areas.

10.2.23 Food Quality Protection Act of 1996

The Food Quality Protection Act (FQPA) was enacted to standardize the way USEPA would manage the use of pesticides and amended FIFRA and FFDCA. It mandated a health-based standard for pesticides used in foods, provided for special protection for babies and infants, streamlined the approval process of safe pesticides, established incentives for the creation of safer pesticides, and required pesticide registrations remain current. FQPA established a new safety standard that must be applied to all food commodities, that being "reasonable certainty of no harm." In addition, USEPA was required to consider this new risk standard as it applies to babies and infants. The FQPA required that re-testing of all existing pesticide tolerance levels be conducted within 10 years and that USEPA must account for "aggregate risk" and "cumulative exposure" to pesticides with similar mechanisms of toxicity (USEPA 2021l).

As a result of the FQPA, USEPA banned methyl parathion and azinphos methyl because of the risks posed by these two pesticides to children. In 2000, USEPA banned an additional pesticide, chloropyrifis that was common in agricultural, household cleaners, and commercial pest control products because of concerns on children's health (USEPA 2021l). FQPA further raised debate over using clinical studies of pesticides on humans. In 2004, the National Academy of Science released a report supporting the use of clinical studies under very strict regulations, citing that the benefit outweighed the risk to the individual. In 2005, USEPA adopted the Human Studies Regulation that allows human studies, with the exception of pregnant women or children and mandate a strict ethical code (USEPA 2021l).

10.2.24 MARINE MAMMAL PROTECTION ACT OF 2015

The Marine Mammal Protection Act of 2015 amended the earlier 1972 Act in that is required that nations exporting fish and fish products to the United States are required to demonstrate that harvesting fish does not also kill or injure marine mammals in excess of standards established under the Act. The Act essentially leveled the playing field for domestic and international fishing interests. To comply, other nations that sold fish or fish products to the United States could adopt marine mammal conservation techniques acceptable or in compliance with United States standards or demonstrate compliance using or developing other acceptable methods or standards (NOAA 2021e).

10.3 SUMMARY OF UNITED STATES ENVIRONMENTAL REGULATIONS THAT PROTECT LIVING ORGANISMS, CULTURAL, AND HISTORICALLY SIGNIFICANT SITES AND THEIR SUSTAINABILITY RELEVANCE

After reading this chapter, you should have a much greater understanding of just how much environmental regulation we have in the United States and why. It should also be apparent that staying in compliance is a complex undertaking requiring much skill and expertise. There is much we can be proud of when we examine the improvements of our air quality, water quality, addressing sites of environmental contamination, which now number in the tens of thousands of sites across the United States. Laws and regulations originated from many circumstances because of unintentional actions that ended up making quite a mess and causing significant harm to human health, the environment, or both. Some of the incidences that we have touched upon include the Exxon Valdez spill in 1990, Deepwater Horizon in 2010, Love Canal, Bhopal India in 1984, Valley of Drums, and Toms River to the killing and extinction of animal species that also nearly included the American Bison, California Condor, Grizzly Bear, and numerous others. However, this is a reactionary approach not a proactive approach. When we discuss sustainability later in this book, we will discuss the need to move beyond reactive measures to proactive sustainability measures to improve our environment in the future. If there are delays in enacting sustainability measures, it may be too late.

The amount and complexity of environmental laws and regulations in the United States is due to the fact that we are an industrialized society and consume enormous amounts of energy and goods and our hunger for those goods and energy continues to rise as we continue to improve our standard of living. The United States produces large volumes of waste of all types and as our population continues to increase, the amount of waste produced will also continue to increase.

These central facts combined with the dynamic and innate complexities of the natural world add an almost infinite number of negative outcomes in a world with over 7 billion people all wanting to improve their standard of living. The fact is, if we do not have effective and comprehensive environmental regulation, we simply threaten our existence and numerous other species as we are slowly devoured in our own garbage.

Our advances in science have created many technologies that have improved our standard of living and in some rare circumstances have improved our relationship with nature. But our technological advancements have also created many unintended negative side effects that question the benefit of some technologies. An example involves the creation of synthetic compounds that nature has difficulty breaking down. Some of these compounds include polychlorinated biphenyls, chlorinated hydrocarbons, and many pesticides, including DDT which began the environmental movement in the United States in earnest over 50 years ago. This is to say nothing about plastics which we all should know about since we commonly use and discard huge amounts of plastic in our everyday life.

Although efforts are underway to reduce our energy needs and our veracious hunger for consuming nature, one thing seems certain, our thirst for consumer goods and energy needs will continue to grow, and as our population increases our energy needs will increase even faster. This means that environmental regulation will have to grow with us in volume and complexity as we consume more and more of Earth's resources and replace it with our waste. The next chapter will focus on environmental regulations and pollution worldwide. We will focus on industrialized nations and a few developing nations with large populations including India and China. We will not be examining countries with what are considered extreme hardships that include Somalia, Syria, Afghanistan, Iran and others. We will evaluate how countries environmental laws compare to the United States. Some findings might be rather surprising in that it will become apparent that a few countries are more advanced at protecting human health and the environment than the United States. Surprising still might be that impression that some countries either lack environmental laws or the will to enforce the environmental laws they have due to other overriding social or economic challenges. As we shall see in the next chapter, it is not enough just to have environmental regulations. Many other factors are required to have an effective environmental regulatory framework that actually accomplishes that ultimate goal of protecting human health and the environment.

Questions and Exercises for Discussion

1. **What are the weaknesses of TSCA according to environmental groups?**
2. **How can TSCA be improved?**
3. **What incident occurred that influence the creations of Superfund Amendments and Reauthorization Act (SARA) and Emergency Planning and Community Right-to-Know Act?**
4. **Of the regulations covered in this chapter, which three are the most important to you and why?**
5. **Can you think of an environmental regulation that is needed to be improved or created and why?**

REFERENCES

American Society for Testing Materials. 2005. *Standard Practice for Environmental Site Assessments.* E1527-05. ASTM, West Philadelphia, PA.

Curry, A. 2021. *Will Newer Wind Turbines Mean Fewer Bird Deaths.* The National Geographic. https://www.nationalgeographic.com/news/energy/2014/04/140427altamont-pass (accessed April 5, 2021).

Eckerman, I. 2005. *The Bhopal Saga – Causes and Consequences of the World's Largest Industrial Disaster.* Universities Press. London, United Kingdom. 284p.

Eveleth, R. How Many Birds to Wind Turbines Really Kill?. 2021. Smithsonian Magazine, Washington, D.C. https://www.smithsoniation.com/smart-new/how-many-birds-do-wind-turbines-kill (accessed April 5, 2021).

International Union on the Conservation of Nature (IUCN). 2021. Extinct Species in North America since 1500ad. https://www.iucn.org (accessed March 31, 2021).

Markell, D. 2014. An Overview of TSCA, Its History and Key Underlying Assumptions, and Its Place in Environmental Regulation. Volume 32. No 448. *New Directions in Environmental Law. Washington University Journal of Law and Policy.* p. 333–376.

National Energy Institute. 2005. Summary of the Energy Policy Act of 1992 and 2005. https://www.nei.org/summary-of-energy-policy-act (accessed March 31, 2021).

National Law Review. 2016. What's New About the Revised TSCA – Toxic Substance Control Act. https://www.natlawreview.com/article/tsca (accessed March 31, 2021).

National Oceanic and Atmospheric Administration (NOAA). 2021a. Coastal Zone Management Act. https://www.noaa.gov/coastal-zone-management-act.html (accessed March 31, 2021).

National Oceanic and Atmospheric Administration (NOAA). 2021b. Marine Mammal Protection Act. https://noaa.gov/marine-mammal-protection-act (accessed March 31, 2021).

National Oceanic and Atmospheric Administration (NOAA). 2021c. Exxon Valdez Oil Spill. Office of Response and Restoration. https://www.noaa.gov/exxon-valdez-oil-spill-restoration (accessed March 31, 2021).

National Oceanic and Atmospheric Administration (NOAA) 2021d. Deepwater Horizon Oil Spill. Office of Response and Restoration. https://noaa.gov/deepwater-horizon-oil-spill (accessed March 21, 2021).

National Oceanic and Atmospheric Administration (NOAA) 2021e. Summary of the Marine Mammal Protection Act of 2015. https://noaa.gov/Marine-mammal-protection-act-summary (accessed March 31, 2021).

Nuclear Regulatory Commission. 2021. Nuclear Waste Policy Act of 1982. https://www.nrc.gov/NWPA (accessed March 31, 2021).

Rogers, D. T. 2019. *Environmental Compliance and Sustainability: Global Challenges and Perspectives.* CRC Press. Boca Raton, FL. 583p.

Rogers, D. T. 2020. *Urban Watersheds: Geology, Contamination, Environmental Regulations, and Sustainability.* CRC Press. Boca Raton, FL. 606p.

Schierow, L. J. 2013. *The Toxic Substance Control Act (TSCA): A summary of the Act and Its Major Requirements.* Congressional Research Service (CRS) Report to Congress, Washington, D.C. 16p.

United Nations. 2021. *Intergovernmental Science-Policy Platform on Biodiversity and Ecosystems Services (IPBES) Summary for Policymakers of the Methodological Assessment Report of the IPBES Scenarios and Models of Biodiversity and Ecosystems Services.* https://www.ipbes.net/system/tdf/downloads/pdf/spm_deliverable_3c_scenarios_20161124.pdf?file=1&type=node&id=15245 (accessed March 31, 2021).

United States Advisory Council on Historic Preservation. 2021. *National History Preservation Act.* https://www.achp.gov (accessed March 31, 2021).

United States Census Bureau. 2021. *United States Population through Time and World Populations Clock.* https://www.UScensus.gov/USpopluation (accessed March 31, 2021).

United States Department of Agriculture. 2014. *United States Forest Resource Facts and Historical Trends.* https://www.fia.fs.fed.us/library/brochures/docs/2012/ForestFacts_1952–2012_English.pdf (accessed March 31, 2021).

United States Environmental Protection Agency (USEPA). 1991. *U.S. Efforts to Address Global Climate Change: Report to Congress.* United States Department of State, Washington, D.C. 87p.

United States Environmental Protection Agency (USEPA). 2005. Standards and Practice for All Appropriate Inquiries. *40 Code of Federal Regulation (CFR), Part 312.* US Government Printing Office, Washington, DC.

United States Environmental Protection Agency, USEPA. 2015. *List of Lists. EPA 55-B-15-001.* Office of Solid Waste and Emergency Response, Washington, D.C. 125p.

United States Environmental Protection Agency. 2021a. *Summary of Environmental Regulations of the United States.* https://epa.gov/summary-environmental-laws (accessed March 31, 2021).

United States Environmental Protection Agency. 2021b. *Summary of the Toxic Substance Control Act.* https://epa.gov/laws-regulations/summary-toxic-substances-control-act (accessed March 31, 2021).

United States Environmental Protection Agency (USEPA). 2021c. *Superfund Amendments and Reauthorization Act (SARA).* https://www.epa.gov/SARA (accessed March 31, 2021).

United Stated Environmental Protection Agency (USEPA). 2021d. *Emergency Planning and Community-Right-to-Know Act (EPCRA).* https://www.epa.gov/EPCRA (accessed March 31, 2021).

United States Fish and Wildlife Service. 2021e. Fishery Conservation and Management Act of 1976. https://www.fws.gov/FCMA (accessed August 19, 2021).

United States Environmental Protection Agency (USEPA). 2021e. *Pollution Prevention Act.* https://www.epa.gov/pollution-prevention-act (accessed March 31, 2021).

United States Environmental Protection Agency (USEPA). 2021f. *Innocent Landowner Defense.* https://www.epa.gov/enforcement/innocent-landowner (accessed March 21, 2021).

United States Environmental Protection Agency (USEPA). 2021g. *Brownfields.* https://www.epa.gov/brownfields/all-appropriate-inquiry (accessed March 21, 2021).

United States Environmental Protection Agency (USEPA). 2021h. *Summary of the Nuclear Waste Policy Act.* https://www.epa.gov/laws-regulations/summary-nuclear-waste-policy-act (accessed March 31, 2021).

United States Environmental Protection Agency (USEPA). 2021i. *Global Climate Change Act.* https://www.epa.gov/GCCA (accessed March 31, 2021).

United States Environmental Protection Agency (USEPA). 2021j. *Summary of the Oil Pollution Act of 1990.* https://epa.gov/oil-pollution-act (accessed March 31, 2021).

United States Environmental Protection Agency (USEPA). 2021k. *USEPA Exxon Valdez Spill Profile.* https://www.epa.gov/emergency-response/exxon-valdez-spill-profile (accessed March 31, 2021).

United States Environmental Protection Agency (USEPA). 2021l. *Summary of the Food Quality Protection Act.* https://www.epa.gov/fqpa-summary (accessed March 31, 2021).

United States Fish and Wildlife Service. 2021a. *Summary of the Lacey Act.* https://.fws.gov/us-conservation-laws/lacey-act-html (accessed March 31, 2021).

United States Fish and Wildlife Service. 2021b. *Summary of the Migratory Bird Treaty Act.* https://www.nps.gov/migratory-bird-act (accessed March 31, 2021).

United States Fish and Wildlife Service. 2021c. *Fish and Wildlife Coordination Act.* https://www.fws.gov/fish-and-wildlife-coordination-act.html (accessed March 31, 2021).

United States Fish and Wildlife Service 2021d. *Endangered Species Act.* https://www.fws.gov/endangered/laws-policies/ (accessed March 31, 2021).

United States Fish and Wildlife Service. 2021e. *Fishery Conservation and Management Act of 1976.* https://www.fws.gov/FCMA (accessed March 31, 2021).

United States Forest Service. 2021. *Forest Conservation and Management Act.* https://www.nfs.gov/FCMA (accessed March 31, 2021).

United States National Park Service. 2021a. *Summary of the Antiquities Act.* https://www.nps.gov/history/antiquities-act-html (accessed March 31, 2021).

United States National Park Service (NPS). 2021b. *History of the National Park Service.* https://www.nps.gov/history (accessed March 21, 2021).

United States National Park Service. 2021c. *Wild and Scenic Rivers Act.* https://www.nps.gov/wild-and-scenic-rivers-act-html (accessed March 31, 2021).

World Wildlife Fund (WWF). 2020. Living Planet Report – 2020. *Bending the curve of Biodiversity Loss.* Almond, R. E. A., Grooten, M. and Peterson, T. (Eds). WWF, Gland, Switzerland. 159p.

United States Office of Surface Mining and Reclamation. 2021. *Surface Mining Control and Reclamation Act.* https://www.osmra.gov (accessed March 31, 2021).

World Wildlife Fund (WWF). 2021. Species List of Endangered, *Vulnerable, and Threatened Animals.* https://www.wwf.org/species/endangered-list (accessed March 31, 2021).

11 An Analysis of Global Pollution Assessments and Standards

11.1 INTRODUCTION

This chapter and the next four chapters will explore the environmental regulations and pollution of many countries of the World, including the oceans and Antarctica. The chapters are organized by continent and then by country. We will start with a summary of global pollution and assessments which will set the stage for addressing the question about how bad is pollution globally. We then will move on to examining North America and South America, then Europe, Asia, Africa, Oceania and then conclude with Antarctica and the Oceans of the world. Table 11.1 lists each country by continent that will be examined. To keep the scope of this book manageable, we have to limit the number of countries to be evaluated. In addition, we will focus on regulations that primarily address pollution. We will concentrate on industrialized nations and a few developing nations with large populations including India and China. This is because many who will be interested in environmental regulations in countries other than the United States will likely be interested in the regulations of the countries chosen to be highlighted. The countries that will be evaluated account for 70% of the human population and 75% of the land area on Earth.

To widen our prospective and grasp a more comprehensive and educated view, we will also evaluate countries that are important or unique to all humans on Earth. These countries will provide us with special insight into unique issues, situations, and problem-solving challenges due to a host of complex circumstances that some countries are now facing. These countries have special challenges that, the combination of which, are unique and will require us to explore potential reasons behind some of the challenges faced by these nations as a result of social concerns, political unrest or lack of action, economic, geographic, geological or other factors that result in poor or excellent environmental performance. Some of these countries include Australia, New Zealand, Tanzania, Egypt, Brazil, Argentina, Korea, Peru, Malaysia, and Chile.

We will focus our discussion of each country on the fundamentals, which include:

- Protect the air
- Protect the water
- Protect the land
- Protect living organisms and historic and cultural sites

To address each of the listed fundamentals, we will examine of how each country addresses air pollution, water pollution, solid and hazardous waste, and remediation of polluted sites. We will examine how most countries address legacy pollution issues through a discussion of the framework that each country has developed to investigate and remediate historical sites of environmental contamination. This means that we must also examine the significant historical sites of contamination in each country because the experiences and lessons learned from these sites have influenced and uniquely shaped each country's choices in developing environmental regulations perhaps more than most other factors.

We will compare and contrast how these countries environmental laws differ and see how they compare to the United States. Some findings might be rather surprising in that it will become apparent

DOI: 10.1201/9781003175810-11

TABLE 11.1
Countries and Areas in Which Environmental Regulations
Are Examined

North America	Europe and the European Union
Canada	Austria
Mexico	Belgium
	Bulgaria
South America	Croatia
Argentina	Cyprus
Brazil	Czech Republic
Peru	Denmark
Chile	Estonia
Asia	Finland
China	France
India	Germany
Japan	Greece
Saudi Arabia	Hungary
South Korea	Ireland
Indonesia	Italy
Malaysia	Latvia
	Lithuania
	Luxemburg
Oceania	Malta
Australia	Netherlands
New Zealand	Norway
Africa	Poland
Egypt	Portugal
Kenya	Romania
Tanzania	Russia
South Africa	Slovakia
	Slovenia
Antarctica	Spain
	Sweden
The Oceans	Switzerland
	Turkey
	United Kingdom

that some countries are more advanced compared to the United States in how they have chosen to organize and implement their environmental regulations and where they focus internal resources. In addition, the governments of some countries have different attitudes and relationships with industry. The regulatory agencies of some countries have chosen to form working partnerships with industry with the intent to work more closely together to solve complex environmental issues rather than an advisory type approach which much too often results in polarizing points of view. The result of working together saves time, effort, and resources and in the long run, has resulted in the fact that some counties are better at protecting human health and the environment than the United States.

Surprising still might be that some countries either lack key environmental laws or the will to enforce the environmental laws they have due to other social or economic challenges or the political realities some countries face. One overriding difficulty many developing countries face is lack of infrastructure, training, and education to properly implement their own environmental regulations. Another overriding factor is climate and climate change which in some instances has greatly increased the negative effects of pollution on the environment and humans.

Lastly, we will examine Antarctica and the world's oceans. Antarctica is a continent that has no permanent human settlements. As a first thought you might ask yourself why should we examine Antarctica, there should not be any significant environmental issues in Antarctica correct? To most of you the answer will be humbling and perhaps even disturbing. We shall have to wade through the information and data presented in this chapter to begin to understand and appreciate the depth and magnitude of the challenges that are confronting humans from an environmental perspective and that are very evident even in Antarctica. We will also briefly discuss the world's oceans which we will learn has been the great dumping ground for human waste for centuries.

To start our journey of environmental regulations and pollution of the world, lets first look at some global assessments that have been implemented by the United Nations (UN) and the World Health Organization (WHO) that describe the current status of pollution and human health on Earth.

11.2 GLOBAL ASSESSMENTS AND STANDARDS

Pollution of the planet encompasses numerous aspects of the land, air, and water. However, there are some of what are called sinks or final resting places of pollution and the largest sink on Earth are the oceans. In fact, in the geological sciences, there is a common reference that states "whatever is created eventually ends up in the ocean." Other sinks include the land and in some instances the atmosphere.

International organizations such as the WHO and the UN have begun to asses, on a global scale, the quality of fresh water and the air on Earth. A global assessment of the land has not yet been completed. Both have published guidelines for assessing, improving, or maintaining acceptable air and water quality standards as far back as 1983 (World Health Organization 2013 and United Nations 2016a and 2016b). According to World Health Organization (2013, 2015a, and 2015b) and the United Nations (2016a and 2016b), approximately 2.8 billion humans lack access to basic sanitation and improved drinking water. While this number is improving, it still represents nearly 30% of the planet's human population.

In addition, according to WHO (2021a), 9 out of 10 people on Earth breath air that is considered unacceptable because it contains high levels of pollution. The United Nations estimates that most land pollution is caused by agricultural activities, such as grazing, use of pesticides, fertilizers, irrigation, plowing, and confined animal facilities (United Nations 2016c).

Significant effort was made to include data tables for air pollution emission standards, water quality, remediation goals, and others for every chemical compound for every country evaluated. However, this was just not possible to keep this book manageable. Therefore, only select tables listing specific chemicals are provided and website addresses are provided for those not included.

11.2.1 AIR

According the World Health Organization, 9 out of 10 people on Earth breath air containing high levels of pollution. Updated estimates indicate that nearly 7 million deaths occur each year from breathing unhealthy air (WHO 2021a). Most of the deaths occur in low to middle income countries in Asia and Africa. The most problematic pollutant is fine particulate matter which are particles less than 2.5 microns and penetrate deep into the lungs cause diseases that include (WHO 2021a):

• Stroke	• Heart disease	• Lung cancer
• Pulmonary diseases	• Respiratory infections	• Pneumonia

The major sources of air pollution world-wide include (WHO 2021a):

- Inefficient use of energy in households
- Industry
- Agriculture
- Transportation
- Use of coal

The inefficient use of energy in households is directly related to access to clean cooking fuels which is the case for an estimated population of 3 billion people world-wide (WHO 2021a). The most significant types of air pollutants world-wide include (WHO 2021a):

• Particulate matter (PM10)	• Fine Particulate Matter (PM2.5)
• Carbon monoxide (CO)	• Ozone (O_3)
• Sulfate	• Nitrates
• Black carbon	• Lead
• Volatile organic compounds (VOCs)	• Benzo(a)pyrene
• Benzene	

The World Health Organization in 2017 has ranked the top ten most air-polluted cities in the world as the following (WHO 2021a):

11. Cairo, Egypt	12. Delhi, India
13. Beijing, China	14. Moscow, Russia
15. Istanbul, Turkey	16. Guangzhou, China
17. Shanghai, China	18. Buenos Aires, Argentina
19. Paris, France	20. Los Angeles, United States

The World Health Organization published the ambient air quality guidelines in 2005, which was updated in 2021 for ozone, particulate matter, sulphur dioxide, nitrogen dioxide, and carbon monoxide (see Table 11.2).

The guideline limits listed earlier are not enforceable and are based on human health risk posed by air pollutants. WHO examined 37 air pollutants but has only set guidelines for the five listed in Table 11.2. WHO includes guidance for countries in the form of statistical modelling related to exposure and risk for those compounds that are of concern but did not establish a guideline. This is due to many factors that relate to exposure, risk, geography, and other factors at specific locations (World Health Organization 2014).

Many cities throughout the world exceed WHO guidelines for air quality. Figure 11.1 shows the major cities of the world that exceed the WHO guidelines for Particulate Matter (PM2.5) on an annual basis (World Health Organization 2016a). As you can see from examining Figure 11.1, most cities in the United States, Australia, Canada, and Europe do not exceed the WHO guideline. However, many cities in southern Asia, central Africa, and Mexico City exceed the WHO guideline. The most air polluted city in the United States ranks tenth worst in the world and is Los Angeles, California. Many cities in China also exceed current WHO standards (World Health Organization 2016a)

The United States Department of State has an air quality monitoring program for many large cities of the World that is updated daily and is termed an Air Quality Index (AQI) (United States

TABLE 11.2
WHO Ambient Air Guidelines

Parameter	Units of Measure	Guideline Limit
Ozone (O$_3$)	Peak Season (ug/M^3)	60
Particulate Matter (PM10)	Yearly (ug/M^3)	15
Particulate Matter (PM 2.5)	Yearly (ug/M^3)	5
Sulphur dioxide (SO$_2$)	24 hour (ug/M^3)	40
Nitrogen dioxide (NO$_2$)	Yearly (ug/M^3)	10
Carbon Monoxide (CO)	8 hour (ug/M^3)	4

Source: World Health Organization. Ambient Air Pollution Database May 2021a. https://www.who.org/ambient-air-database. (accessed October 31, 2021).

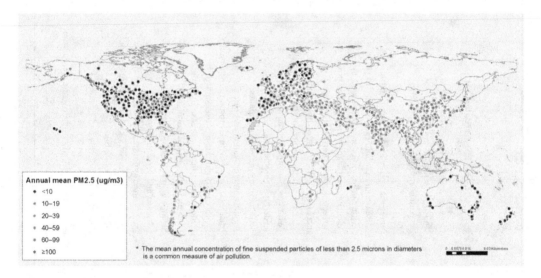

FIGURE 11.1 Annual mean PM2.5 at select cities world-wide (From World Health Organization. World Health Organization Global Urban Ambient Air Pollution Database. http://www.who.int/phe/health_topics/outdoorair/database.htm. (accessed March 31, 2021).

Department of State 2021). The U.S. Department of State air quality guidelines for the Air Quality Index for fine particulate matter (PM2.5) are listed in Table 11.3.

11.2.2 WATER

Global driving forces for issues related to water include climate change, increasing water scarcity, population growth, demographic changes, and urbanization. These driving forces are expected to place additional stress on fresh water supplies and sanitation systems and services. As climate change scenarios become increasingly reliable, existing infrastructure will need to be adapted and planning of new systems and services will need to be updated. Extreme weather conditions are also reflected in the increased frequency and intensity of natural disasters and will become the new normal (World Health Organization 2013). Figure 11.2 shows the climate regions of the world.

According to the WHO (2015a and 2015b), the most prevalent water quality problem globally is eutrophication, which is a result of high-nutrient loads (mainly phosphorus and nitrogen). The most

TABLE 11.3
Air Quality Index

AQI for PM2.5	Value (ug/m³)	Health Effects Statement	Cautionary Statement
Good	0–50	Little or no risk	None
Moderate	51–100	Unusually sensitive individuals may experience respiratory symptoms	Active children and adults, and people with respiratory disease, such as asthma, should limit prolonged outdoor exertion
Sensitive Groups	101–150	Increased aggravation of heart and lung disease and premature mortality in persons with cardiopulmonary disease and the elderly; significant increase in respiratory effects in the general population	Active children and adults, and people with respiratory disease, such as asthma, should avoid prolonged outdoor exertion; everyone else, especially children, should limit prolonged outdoor exertion
Unhealthy	151–200	Increased aggravation of heart and lung disease and premature mortality in persons with cardiopulmonary disease and the elderly; significant increase in respiratory effects in the general population	Active children and adults, and people with respiratory disease, such as asthma, should avoid all outdoor exertion; everyone else, especially children, should limit outdoor exertion
Very Unhealthy	201–300	Significant aggravation of heart and lung disease and premature mortality in person with cardio-pulmonary disease and the elderly; significant increase in respiratory effects in the general population	Active children and adults, an people with respiratory disease, such as asthma, should avoid all outdoor exertion; everyone else, especially children, should limit outdoor exertion
Hazardous	301–500	Serious aggravation of heart and lung disease and premature mortality in person with cardio-pulmonary disease and the elderly; serious risk of respiratory effects in the general population	Everyone should avoid all outdoor exertion
Beyond Index	501 and greater	Extremely high levels and health risk	Take immediate steps to reduce exposure

Source: United States Department of State. United States Department of State Air Quality Monitoring Program. http://www.stateair.net/web/post/1/5.html. (accessed March 31, 2021.)
Note: ug/m³ = micrograms per cubic meter.

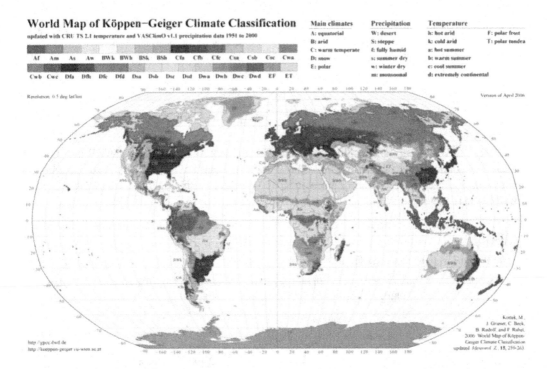

FIGURE 11.2 Climate regions of the world (From National Oceanic and Atmospheric Administration National Oceanic and Atmospheric Administration (NOAA 2019). Koppen-Geiger Climate Changes. https://sos.noaa.gov/datasets/koppen-geiger-climate-changes. (accessed August 31, 2021.)

significant sources of water pollution on Earth is agriculture. Other sources include, domestic sewerage, industrial effluents and atmospheric inputs from fossil fueled burning and wildfires. A significant issue related to water is sanitation and the spread of disease through untreated water with human excrement and from livestock and wildlife called zoonotic pathogens (World Health Organization 2013). As stated earlier, the World Health Organization (2013 an 2015a and 2015b) and the United Nations (2016a and 2016b), estimate that 2.8 billion people do not have access to proper sanitation, while (Baum et al. 2013) estimates that the actual number could be as high as 4.1 billion because most evaluations do not account for all biological disease-causing pathogens.

The WHO lists the top ten most water polluted cities in the World as the following (World Health Organization 2021b):

1. Sumgavit, Azerbaijan	2. Linfen, China
3. Tianjing, China	4. Sukinda, India
5. Vapi, India	6. La Oroya, Peru
7. Dzerzhinsk, Russia	8. Norilsk, Russia
9. Chernobyl, Russia	10. Kabwa, Zambia

Of the countries we will evaluate, only one city is located in a country that we will not evaluate, that being Zambia. Of the countries we will evaluate, four are located in the former Soviet Union, two are located in China, two in India, and one in Peru. Contaminants that have affected the most water polluted cities in the World include (WHO 2021c):

- Volatile organic compounds (VOCs)
- Oils and fuel
- Radioactive materials
- Garbage
- Zoonotic pathogens

- Biological contaminants including *e coli*
- Pesticides
- Fertilizers
- Plastic
- Other pathogens

In 2011, the WHO published minimum standards for water quality. The WHO Water Quality Standards are intended as a benchmark for those countries that have not established standards of their own and are considered minimum standards (World Health Organization (WHO) 2015a). Table 11.4 lists the WHO Water Quality Standards.

TABLE 11.4
WHO Water Quality Standards

Parameter	Standard (ug/l)	Parameter	Standard (ug/l)	Parameter	Standard (ug/l)
Acrylamide	0.5	Alachlor	20	Aldicarb	10
Aldrin and Dieldrin	0.03	Antimony	20	Arsenic	10
Atrazine	100	Barium	700	Benzene	10
Benzo(a)pyrene	0.7	Boron	2,400	Bromate	10
Bromodichloromethane	60	Bromoform	100	Cadmium	3
Carbon Tetrachloride	4	Carbofuran	7	Chlorate	700
Chlordane	0.2	Chlorine	5,000	Chlorite	700
Chloroform	300	Chlorotoluron	30	Chlorpyrifos	30
Chromium	50	Copper	2,000	Cyanzine	0.6
2,4-D	30	2,4-DB	90	DDT	1
Dibromoacetonrile	20	1,2,3-Chloropropane	1	1,2-Dibromoethane	0.4
Dibromochloromethane	100	Dichloroacetate	50	Dichloroacetonitrile	20
1,2-Dichlorobenzene	1,000	1,2-Dichloroethane	30	1,2-Dichloroethene	5-
1,4-Dichlorobenzene	300	Dichloromethane	20	Dichloroprop	20
1,2-Dichloroporpane	40	Dimethoate	6	1,4-Dioxane	50
1,3-Dichloropropene	20	Edetic Acid	600	Endrin	0.6
Di(ethylhexyl)phthalate	8	Epichlorohydrin	0.4	Ethylbenzene	300
Fenoprop	9	Fluoride	1,500	Hydroxyatrazine	200
Hexachlorobutadiene	0.6	Isoproturon	9	Lead	10
Lindane	2	MCPA	2	Mecoprop	10
Mercury	6	Methoxychlor	20	Metolachor	10
Microcystin-LR	1	Molinate	6	Monochloroacetate	20
Nickel	70	Nitrate (as NO_3^-)	50,000	Nitrite (as NO_2^-)	3,000
Nitrilotriacetate Acid	200	Pendimethalin	20	Pentachlorophenol	9
N-Nitrodimethylamine	0.1	Selenium	40	Simazine	2
Sodium dichloroisocyanurate	40,000	Styrene	20	2,4,5-T	9
Terbuthylazine	7	Tetrachloroethene	40	Simazine	2
Toluene	700	Trichloroacetate	200	Trichloroethene	20
Trifluralin	0.02	Trihalomethanes	1	Uranium	30
Vinyl chloride	0.3	Xylenes	500		

Source: WHO. Guidelines for Drinking-water Quality. Geneva, Switzerland. 516 pages. 2015a.
Note: ug/l = microgram per liter.

The WHO guidelines for drinking water are not mandatory limits. They are intended to provide a scientific point of departure for development of national or regional numerical drinking water standards for those countries that do not have standards or may have their own standards but are incomplete or, in some cases, may not be satisfactory. The WHO guidelines were established using the following criteria (World Health Organization (WHO) 2015a):

- The probability of exposure
- Chemical concentration likely to present risk of observable health risks
- Evidence of health effects and relative ease of control of the different sources of exposure

The WHO lists many other compounds that they do not provide numerical values in the guideline due to mitigating factors such as geography, environmental factors, social, cultural, economic, dietary, and other conditions affecting potential exposure. Therefore, WHO recommends that it may be necessary to undertake a drinking water assessment before establishing standards. Chemical compounds and microbes not listed in Table 7.4 but, in many instances, should be evaluated on a case by case basis after a drinking water assessment is completed, include the following (World Health Organization (WHO) 2015a and 2015b):

- *E coli* and other microbes
- Chlorothalonil
- Diazinon
- Fenamiphos
- Hexachlorocyclohexanes
- Phorate
- Quintozene
- Tributyltin oxide
- Ammonia
- Beryllium
- Bromochloroacetate
- *Bacillus thuringiensis*
- 2-Chlorophenol
- Cyanogen chloride
- 1,1-Dichloroethene
- 1,3-Dichlorobenzene
- 1,3-Dichloropropane
- Di(2-ethylhexyl)adipate
- Formaldehyde
- Hexachlorobenzene
- Inorganic tin
- Malathion
- Methyl parathion
- Monobromoacetate
- Nitrobenzene
- Permethrin
- Petroleum products
- Propanil
- Sodium
- 1,1,1-Trichloroethane
- Trichloroactonitrile

- Amitraz
- Cypermethrin
- Dinoseb
- Formothion
- Methamidophos
- Propoxur
- Toxaphene
- Trichlorfon
- Asbestos
- Bromide
- Carbaryl
- Chlorine dioxide
- Chloropicrin
- Dialkyltins
- 1,1-Dichloroethane
- 2,4-Dichlorophenol
- Diflubenzuron
- Fenitrothion
- Glyphosate
- Heptachlor
- Iodine
- Manganese
- MTBE
- Monochlorobenzene
- Navaluron
- pH
- Pirimiphos-methyl
- Pyriproxfen
- Spinosad
- TDS
- Trichlorobenzenes

- Chlorobenzilate
- Deltramethrin
- Ethylene thiourea
- MCPB
- Oxamyl
- Pyridate
- Triazophos
- Aluminum
- Bentazone
- Chlroide
- Chloral hydrate
- Chloroacetones
- Cyanide
- Dibromoacetate
- Dichloramine
- Diquat
- Endosulfan
- Fluoranthene
- Hardness
- Hydrogen sulfide
- Iron
- Methoprene
- Molybdenum
- MX
- Parathion
- 2-Phenylphenol
- Potassium
- Silver
- Sulfate
- Trichloramine
- Zinc

11.2.3 LAND

Land pollution is defined as a degradation or even destruction of the Earth's surface, soil, rock or subsurface as a result of anthropogenic activities. The expansion of housing developments, businesses, industry, infrastructure, and agriculture all in response to an increasing population over the last several decades accounts for humans modifying over 50% of the Earth's topsoil (United Nations 2016c). Land also becomes contaminated from the migration of contaminants through the soil column or when contaminants change media such those carried by the atmosphere perhaps hundreds or even thousands of miles and are deposited on the land or water perhaps in a different country.

The surface of the Earth is approximately 510,100,000 square kilometers (Coble et al. 1987). Approximately 70% is covered by water or 361,800,000 square kilometers and 30% is land or 148,300,000 square kilometers. The amount of land on Earth that is farmed is estimated at 40% of 59,320,000 square kilometers (United Nations 2016c). This is a huge number especially since the amount of urbanized land on Earth is estimated to be only 2.7% and is home to roughly half the current human population (United Nations 2016c). The amount of land currently used for farming becomes an even larger value since 10.4% of Earth's land or 15,600,000 square kilometers is covered by ice and is not farmable, 20% or 29,660,000 square kilometers is mountainous, and another 20% or 29,660,000 square kilometers is desert and has little of no topsoil (Coble et al. 1987). When factoring out those land areas not suitable for farming, only 7% of land remains available on Earth.

The WHO and USEPA (2012) have not established guidelines for soil or groundwater. However, most countries that have not developed standards for the remediation of land that in-cludes, soil, sediment, or groundwater commonly refer to a set of standards established by the Dutch in 1987 and were revised in 1994, 2001 and 2015 and are commonly referred to as the Dutch Intervention Values (DIVs) (Lijzen 2001 and Netherlands Ministry of Infrastructure and the Environment 2015).

The DIVs are based on potential risks to human health and ecosystems and are technically evaluated. The values presented usually have two integers for each chemical listed. One is a target level which is considered optimal and likely will result in no or limited health or ecological effect and the other is a level or concentration above which some intervention is necessary to lower risk posed by site specific exposure pathways.

The DIVs have been used world-wide as a benchmark for cleanup of contaminated sites in many areas of the world where cleanup values have not been established. In addition, even if there are established values, the Dutch Intervention Values are used for comparison purposes. Therefore, in many instances, the Dutch Intervention Values represent the benchmark for cleanup standards world-wide.

Table 11.5 presents a partial list of Dutch Intervention Values. A complete list can be viewed at http://www.sanaterre.com/guidelines/dutch.html. (Lijzen 2001 and Netherlands Ministry of Infrastructure and the Environment 2015)

11.2.4 SUMMARY OF GLOBAL ASSESSMENTS AND STANDARDS

It may not be clearly noticeable at this point, but there is a general lack of enforcement of in-ternational standards for dealing with pollution. As we shall see during our journey through the environmental regulations of select countries of the world, many countries struggle with meeting existing standards, in part, because enforcement is either absent or substandard. This is significant because, as we already know, pollution does not respect political boundaries.

We will now turn our attention from global standards to those of select countries. We shall begin with North America. Since the United States was covered in its own chapter we will start with a short introduction of North America and then describe the environmental regulations and challenges dealing with pollution in Canada and Mexico. We will now discuss country-specific challenges and briefly discuss pollution of Antarctica and the Oceans.

TABLE 11.5
Dutch Intervention Values for Select Compounds for Soil and Groundwater

Compound	Soil		Groundwater	
	Target Value mg/kg	Intervention Value mg/kg	Target Value ug/l	Intervention Value ug/l
Arsenic (As)	29.0	55.0	10	60
Barium (Ba)	160	625	50	625
Cadmium (Cd)	0.8	12	0.4	6
Chromium (Cr)	100.0	380	1	30
Copper (Cu)	36.0	190	15	75
Nickel (Ni)	35.0	210	15	75
Lead (Pb)	85.0	530	15	75
Mercury (Hg)	0.3	10.0	0.05	0.3
Silver (Ag)	None	15	None	40
Selenium (Se)	0.7	100	None	40
Zinc (Zn)	140	720	65	800
Chloride	None	None	100,000	None
Cyanide	None	20	5	1,500
Cyanide (complex)	None	50	10	1,500
Thiocyanate	None	20	None	1,500
Benzene	None	1.1	0.2	30
Ethyl benzene	None	110	4	150
Toluene	None	320	7	1,000
Xylenes (sum)	None	17	0.2	70
Styrene (vinylbenzene)	None	86	6	300
Phenol	None	14	0.2	2.000
Cresols (sum)	None	13	0.2	200
Naphthalene	None	None	0.01	70
Phenanthrene	None	None	0.003	5
Anthracene	None	None	0.0007	5
Vinyl chloride	None	0.1	0.01	5
Dichloromethane	None	3.9	0.01	1,000
Trichloroethene (TCE)	None	2.5	24	500
Tetrachloroethene (PCE)	None	8.8	0.01	40
Pentachlorophenol	None	12	0.04	3
Polychlorinated biphenyl (PCB) (sum)	None	1	0.01	0.01
Monochloroanilines (sum)	None	50	None	30
Dioxin (sum)	None	0.00018	None	None
Chloronaphthalene (sum)	None	23	None	6
Chlordane	None	4	0.00002	0.2
DDT (sum)	None	1.7	None	None
DDE (sum)	None	2.3	None	None
DDD (sum)	None	34	None	None
DDT/DDE/DDD sum	None	None	0.000004	None
Aldrin	None	0.32	0.000009	None
Dieldrin	None	None	0.0001	None

(Continued)

TABLE 11.5 *(Continued)*
Dutch Intervention Values for Select Compounds for Soil and Groundwater

Compound	Soil		Groundwater	
	Target Value mg/kg	Intervention Value mg/kg	Target Value ug/l	Intervention Value ug/l
Lindane	None	1.2	0.009	None
Atrazine	None	7.1	0.00029	150
Carbofuran	None	0.017	0.0009	100
Asbestos	None	100	None	None

Source: Netherlands Ministry of Infrastructure and the Environment 2015. Dutch Pollutant Standards. https://www.government.nl/ministry-of-infrastructure-and-the-environment. or at http://www.sanaterre.com/guidelines/dutch.html. (accessed March 31, 2021), 2015.)
Notes:
mg/kg = milligram per kilogram.
ug/l = microgram per liter.

Questions and Exercises for Discussion

1. Which areas of the world have the poorest air quality and why?
2. Which areas of the world have improved air quality and why?
3. How do the WHO Ambient Air Guidelines listed in Table 11.2 differ from the United States NAAQS Criteria Pollutants listed in Table 7.1?
4. Why do you think the WHO water quality are not mandatory?
5. How do the WHO Water Quality Standards differ from the Primary Dinking Water Standards mentioned in Chapter 8 for the same compounds listed in each standard?
6. How do the DIV's listed in Table 11.5 differ from USEPA limits listed in Table 9.4 for soil only?

REFERENCES

Baum, R., Luh, J., and Bartman, J. 2013. Sanitation: A Global Estimate of Sewerage Connections without Treatment and the Resulting Impact. *Journal of Environmental Science and Technology.* Vol. 47. no. 4. pp. 1994–2000.

Coble, C. R., Murray, E. G., and Rice, D. R. 1987. Earth Science. *Prentice-Hall Publishers.* Englewood Cliffs: New Jersey. 502p.

International Union of Geological Sciences (IUGS). 2016. *The Anthropocene Epoch: Adding Humans to the Chart of Geologic Time. International Geological Congress. Cape Town. South Africa.* https://www.35igc.org/verso/5/scientific-programme (accessed March 31, 2021).

Lijzen, J. P., Baars, A. J., Otte, P. F., Rikken, M. G., Swartjes, F. A., Verbruggen, E. M., and van Wezel, A. P. 2001. *Technical Evaluation of the Intervention Values for Soil/Sediment and Groundwater.* Institute of Public Health and the Environment: The Netherlands. 147p.

National Oceanic and Atmospheric Administration (NOAA). 2019. *Koppen-Geiger Climate Regions.* https://sos.noaa.gov/datasets/koppen-geiger-climate-changes (accessed August 20, 2021).

Netherlands Ministry of Infrastructure and the Environment. 2015. *Dutch Pollutant Standards.* https://www.government.nl/ministry-of-infrastructure-and-the-environment or at http://www.sanaterre.com/guidelines/dutch.html (accessed March 31, 2021).

United Nations. 2016a. *Global Drinking Water Quality Index and Development and Sensitivity Analysis Report.* UNEP Water Programme Office: Burlington, Ontario, Canada. 58p.

United Nations. 2016b. World Air Pollution Status. United Nations News Center. New York, *New York.* https://www.un.org/sustainabledevelopment/2016/09 (accessed March 31, 2021).

United Nations. 2016c. Human Development Report. *United Nations Development Programme.* http://www. hdr.undr.org/sites/2017 (accessed March 31, 2021).

United States Department of State. 2021a. *United States Department of State Air Quality Monitoring Program.* http://www.stateair.net/web/post/1/5.html (accessed March 31, 2021).

World Health Organization and United States Environmental Protection Agency. 2012. *Animal Waste, Water Quality and Human Health.* IWA Publishing: London. United Kingdom. 489p.

World Health Organization. 2013. *Water Quality and Health Strategy 2013–2020.* http://www.who.int/water_sanitation_health/dwq/en/ (accessed March 31, 2021).

World Health Organization. 2014. *Ambient Air Pollution Database May 2014.* https://:www.who.org/ambient-air-database (accessed March 31, 2021).

World Health Organization (WHO). 2015a. *Guidelines for Drinking-water Quality.* Geneva, Switzerland. 516p.

World Health Organization 2015b. *Progress on Drinking Water and Sanitation.* World Health Organization: New York, New York. 90p.

World Health Organization. 2016a. *World Health Organization Global Urban Ambient Air Pollution Database.* http://www.who.int/phe/health_topics/outdoorair/database.htm (accessed March 31, 2021).

World Health Organization. 2016b. *Air Pollution in China: Sources and Amounts.* https://www.who.org/china-air-pollution (accessed March 31, 2021)

World Health Organization. 2021a. Nine Out of Ten People Breathe Unhealthy Air. Geneva. Switzerland. https://www.who.int/news-room/detail/02-02-2018 (accessed March 31, 2021).

World Health Organization (WHO). 2021b. *World Health Organization Global Urban Ambient Air Pollution Database.* http://www.who.int/phe/health_topics/outdoorair/database.htm (accessed October 31, 2021).

World Health Organization. 2021c. United Nations Global Analysis and Assessment of Sanitation and Drinking Water: Indonesia. https://www.who.org/drinkingwater/Indonesia (accessed March 31, 2021).

12 Summary of Environmental Regulations of North and South America

12.1 INTRODUCTION

This chapter will explore the environmental regulations and pollution of North America and South America. The chapter is organized by continent and then by country. We will start with North America and review the environmental regulations of Canada and Mexico. We have already reviewed the environmental regulations of the United States. We will then make our way to South America and review, Argentina, Brazil, Chile, and Peru. To keep the scope of this book manageable, we have to limit the number of countries to be evaluated. In addition, we will focus on regulations that primarily address pollution. We will concentrate on industrialized nations and a few developing nations with large populations. This is because many who will be interested in environmental regulations in countries other than the United States will likely be interested in the regulations of the countries chosen to be highlighted. The countries that will be evaluated account for 70% of the human population and 75% of the land area on Earth.

To widen our perspective and grasp a more comprehensive and educated view, we will also evaluate countries that are important or unique to all humans on Earth. These countries will provide us with special insight into unique issues, situations, and problem-solving challenges due to a host of complex circumstances that some countries are now facing. These countries have special challenges that, the combination of which, are unique and will require us to explore potential reasons behind some of the challenges faced by these nations as a result of social concerns, political unrest or lack of action, economic, geographic, geological or other factors that result in poor or excellent environmental performance. For this chapter, these countries include Peru and Chile.

We will focus our discussion of each country on the fundamentals, which include the following:

- Protect the air
- Protect the water
- Protect the land
- Protect living organisms and cultural and historic sites

To address each of the above-listed fundamentals, we will examine how each country addresses air pollution, water pollution, solid and hazardous waste, and remediation of polluted sites. We will examine how most countries address legacy pollution issues through a discussion of the framework that each country has developed to investigate and remediate historical sites of environmental contamination. This means that we must also examine the significant historical sites of contamination in each country because the experiences and lessons learned from these sites have influenced and uniquely shaped each country's choices in developing environmental regulations perhaps more than most other factors.

We will compare and contrast how these countries' environmental laws differ and see how they compare to the United States. Some findings might be rather surprising in that it will become apparent that some countries are more advanced compared to the United States in how they have chosen to organize and implement their environmental regulations and where they focus internal

DOI: 10.1201/9781003175810-12

resources. In addition, the governments of some countries have different attitudes and relationships with industry. The regulatory agencies of some countries have chosen to form working partnerships with industry with the intent to work more closely together to solve complex environmental issues rather than an advisory type approach which much too often results in polarizing points of view. The result of working together saves time, effort, and resources and in the long run, has resulted in the fact that some counties are better at protecting human health and the environment than the United States.

Surprising still might be that some countries either lack key environmental laws or the will to enforce the environmental laws they have due to other social or economic challenges or the political realities some countries face. One overriding difficulty many developing countries face is lack of infrastructure, training, and education to properly implement their own environmental regulations. Another overriding factor is climate and climate change which in some instances has greatly increased the negative effects of pollution on the environment and humans.

12.2   NORTH AMERICA

North America is a continent entirely within the Northern Hemisphere and almost entirely within the Western Hemisphere. North America covers an area of approximately 24,709,000 square kilometers (9,540,000 square miles) and covers 4.8% of the total surface area of Earth and 16.5% of Earth's land surface (Marsh and Kaufman 2015).

North America is the third largest continent on Earth, after Asia and Africa, and is fourth in population after Asia, Africa, and Europe. As of 2019, the estimated population of North America was 575 million people in 23 independent states, or about 7.5% of the total human population on Earth (United Nations 2021a). The United States at 327 million, Mexico at 130 million, and Canada at 33.4 million, make up the majority (approximately 85%) of the human population of North America. Estimates of human habitation of North America range from approximately 40,000 to 17,000 years ago. Together with South America, this age range of human habitation is rather recent compared with other continents excluding Antarctica (United Nations 2021a).

The North American Free Trade Agreement (NAFTA) addresses the issues of environmental protection of the three participating countries being, the United States, Canada, and Mexico. However, it leaves establishing environmental rules and standards to the three participating countries, with two exceptions (United States Department of State 2021a):

1. They are to comply with existing treaties between themselves, and
2. They are not to reduce their environmental standards as a means of prompting investment in the business.

As we shall see in the following sections, the three countries that primarily comprise North America that we will examine may have similar environmental regulations, but they differ in practice significantly due to social concerns, corruption, lack of enforcement, and physical geography.

12.2.1   CANADA

Canada is the second-largest country in the land area on Earth (9,984,670 square kilometers) with significant natural resources including timber, petroleum, and minerals. The population of Canada estimated in 2009 was 33,487,208, which is about 10% of the population of the United States. It's a significant trade partner of the United States, in fact, approximately 75% of exported goods from Canada are to the United States and the majority of the population of Canada lives within 200 kilometers of the United States–Canadian border (Canadian Government 2021a). Canada closely resembles the United States in its free-market-oriented economic system and pattern of industrial production. Canada and the United States also share several industries (Canadian government 2021a).

Since the early 1970s, provincial governments have adopted several environmental laws and regulations pertaining to the protection or enhancement of the environment. These laws and regulations require duties and obligations on individuals or companies that cannot be ignored and which, in the event of noncompliance or of an environmental occurrence, release, or incident, can potentially lead to individual or company liability not only under the environmental laws but also in civil or common law context. In addition, the same can be said for the laws that have been adopted at the federal level (Canadian Government 2021a).

Canada was given the right to self-govern from England in 1849 and officially became its own country in 1867. The Constitution Act of 1867 described which level of government (federal or provincial) can do within its jurisdiction. However, environmentally-related issues had not yet been contemplated. As a result, legislative and enforcement authority over the environment is split between the federal and provincial governments who each have adopted their own set of rules (Canadian Government 2021a). This has led to an extra level of complexity with each province being different, similar to that of the United States.

The federal government has exclusive jurisdiction over criminal law, coastal and inland fisheries, navigation, federal works, and laws for peace, order, and good government. The provinces have jurisdiction over property and civil rights as well as matters of a local and private nature. Provinces can also delegate many environmental powers to municipalities. As a result of this delegation of power, municipalities regulate matters such as noise, nuisances, pesticides, sewers, and land use planning. As an example, the provincial government of Quebec has delegated responsibility over air emissions and wastewater discharges to the sewer system and waterways within its boundary to the Montreal Metropolitan Community (Canadian Government 2021a).

12.2.1.1   Environmental Regulatory Overview of Canada

The main environmental laws at the federal level in Canada include the following (Environment Canada 2016):

*   Canadian Environmental Protection Act (CEPA) of 1999
*   Fisheries Act of 2012
*   Transportation of Dangerous Goods Act
*   Species at Risk Act
*   Canadian Environmental Assessment Act

The provinces and territories of Canada have adopted some form of the environmental legislation at the federal level. The following is a general list of each province and dominant environmental law (Environment Canada 2016):

*   Province of Ontario, there is the Environmental Protection Act, Ontario Water Resources Act, Safe Drinking Water Act, Endangered Species Act, and the Environmental Assessment Act, which constitute the majority of legislation in Ontario (Province of Ontario 2021).
*   Province of Quebec, the main environmental legislation is called the Environmental Quality Act under which a series of subsections address hazardous materials, biomedical wastes, residual materials, air quality, and contaminated sites (Province of Quebec 2021).
*   Province of Alberta, the main environmental regulations are included in the Environmental Protection and Enhancement Act (Province of Alberta 2021).
*   Province British Columbia, it's the Environmental Management Act (Province of British Columbia 2021).
*   Province of Nova Scotia, it's the Environment Act (Province of Nova Scotia 2021).
*   Province of Manitoba, it's the Environment Act (Province of Manitoba 2021).

- Province of Saskatchewan, it's the Environmental Management and Protection Act (Province of Saskatchewan 2021).
- Province of New Brunswick, it's the Clean Environment Act (Province of New Brunswick 2021).
- Provinces of Newfoundland and Labrador, it's the Environment Act (Province of Newfoundland and Labrador 2021).
- Northwestern Territory, it's the Environmental Protection Act (Northwest Territories 2021).
- Yukon Territory, it's the Environment Act (Yukon Territory 2021).

In Canada, the purpose of environmental law is to promote the protection of the environment and particularly in the context of sustainable development. In Canada, "environment" or "natural environment" is defined as the air, land, water (including groundwater), and all other external conditions or influences under which humans, animals, and plants live or are developed or have dynamic relations. A "contaminant" is defined as any solid, liquid, gas, odor, heat, sound, vibration, radiation, or any combination, resulting directly or indirectly from human activities (Canadian Environmental Protection Act 1999). Environmental laws in Canada create a prohibition against contaminating the environment and establish duties to protect the quality of the environment and obligations associated with permits and approvals that allow impacts (discharges) to the environment, but in a controlled and regulated manner in order to prevent harm to the environment (Environment Canada 2016). Failure to comply may result in sanctions, fines, and administrative orders, much the same as in the United States.

12.2.1.2 Air

Air in Canada is regulated under the Canadian Environmental Protection Act of 1999. Canada regulates stationary and mobile sources such as internal combustion engines. In addition, Canada and the United States have entered into a transboundary air agreement in 2003 to identify and address transboundary air pollution (Canadian Government 2021b, 2021c). This action was emphasized and was a priority for Canada because of the realization that much of the air pollution in Canada was generated in the United States.

As in the United States, Canadian regulations on air pollution have concentrated on particulate matter (PM), ozone (O_3), nitrogen dioxide (NO_x), carbon dioxide (CO_2), sulphur dioxide (SO_2), and volatile organic compounds (VOCs). Since 2000, each of the pollutants listed above has decreased, some greater than 50%. Currently, the air in Canada in urban locations is generally considered of good quality (WHO 2021a, United States Department of State 2021b).

As in the United States, any emissions from a stationary source likely require permitting. The first step is to file a "notice of intent," which is similar to a construction permit in the United States, if the emissions originate from an industrial source from industries such as oil and gas and other petroleum-related industry, metal melting and smelting, pulp and paper, iron and steel, potash, cement lime, chemicals, fertilizers, and manufacturing (Canadian Government 2021d).

12.2.1.3 Water

In Canada, enforceable limits for drinking water are called guidelines and were established by the Federal-Provincial-Territorial Committee and are updated as needed (Health Canada 2017). The drinking water limits for Canada are listed at https://www.healthcanada.gc.ca/waterquality (Health Canada 2017).

Comparing the Canadian drinking water criteria with that of the United States one can immediately conclude that the criteria are nearly the same as the United States. However, one size does not fit all in that there are slight differences. For example, arsenic and lead have slightly lower limits in Canada than in the United States.

According to the Canadian Government (2021e), 150 billion liters of untreated wastewater is discharged into the environment daily in Canada and represents a real threat to the environment.

Wastewater effluent is regulated under the Canadian Fisheries Act which was enacted in 2012. The Fisheries Act set mandatory minimum effluent quality standards that, if necessary, must be achieved through secondary water treatment before being discharged. The wastewater regulations apply to owners and operators of wastewater systems that collect, or are designed to collect, 100 cubic meters or more of influent per day and that discharge to surface water. Key elements of the Act require monitoring, sampling and analysis, and reporting on effluent quality and quantity (Canadian Government 2021d). Canadian federal wastewater limits are listed at https://laws-lois. justice.gc.ca.eng/regulations/SOR-2012-139/fulltext.html (Canadian Government 2021e).

Due to extreme differences in geography and activity, a risk assessment may be required to establish site-specific discharge limits depending on the receiving environment. Risk assessments often include initial characterization of the effluent and consider the characteristics of the receiving environment and mixing that occurs in an allocated mixing zone (Canadian Government 2021e). This process differs from that of the United States in that risk assessments for wastewater are rarely conducted, perhaps because there is much less wilderness in the United States compared to the northern parts of Canada. An additional water issue in Canada is stormwater. Many municipalities in Canada have difficulties dealing with stormwater combined with sewer discharges. Stormwater discharges, as in the United States, tend to overwhelm water treatment plants resulting in the mass discharge of untreated wastewater to surface water bodies (Canada Water 2021).

12.2.1.4　Solid and Hazardous Waste

In Canada, hazardous waste and hazardous recyclable materials are defined as those that are flammable, corrosive, or are inherently toxic and are regulated under the Canadian Environmental Protection Act of 1999. Canada acknowledges that many household items may be hazardous that include batteries, computers and other electrical equipment, cleaners, paints, pesticides, herbicides, and others items. Providing a separate definition for hazardous waste and hazardous recyclable material provides regulators with more flexibility in managing waste (Canadian Environmental Protection Act 1999). The Canadian Environmental Protection Act (1999) includes authority to

- Set criteria to assess the environmentally sound management of wastes and hazardous recyclable materials and refuse to permit import or export if criteria are not met
- Require exporters of hazardous waste to submit export-reduction plans
- Regulate the export and import of prescribed non-hazardous wastes for final disposal
- Control interprovincial movements of hazardous wastes and hazardous recyclable materials

The Canadian Environmental Protection Act (1999) lists nine different hazardous waste classes that include

1. Explosives	2. Gases	3. Flammable liquids
4. Flammable solids	5. Oxidizers	6. Toxics and infectious
7. Radioactive	8. Corrosives	9. Miscellaneous

The basic approach relies on the nine different hazardous wastes classes and analytical tests such as the Toxic Characteristic Leaching Procedure (TCLP), which is commonly applied in the United States to evaluate whether a waste should be classified as hazardous (Canadian Environmental Protection Act 1999).

12.2.1.5　Remediation Standards

Similar to the United States, Canada has developed remediation standards for many chemical compounds. As in the United States, the Federal government of Canada has established guidelines so that each Province or Territory could develop remediation criteria for specific chemical compounds based on toxicity, exposure assumptions, and land use. Remediation standards for the province of Ontario are listed at https://www.ontario.ca/page/soil-ground-water-and-sediment-standards-use-under-part-xv1-environmental-protection-act (Province of Ontario Canada 2011).

Perhaps the most interesting and most significant difference between the Canadian remediation goals and the United States is that Canada has established cleanup criteria for agricultural soil. The United States has not established cleanup criteria for agricultural soil. In most circumstances, agricultural land is exempt from environmental regulations under USEPA in the United States.

In the United States, circumstances, where agricultural land is included in environmental assessments, are when there is a proposed land-use change and the agricultural land is then developed as residential, commercial, or industrial. It is under this circumstance that the land is evaluated and often investigated for the presence of contamination. The detection of contamination may fall within a gray zone as to whether any detected contamination is the result of agricultural activities which may be exempt or from other activities such as open dumping, chemical spills, leaking tanks, or other activities not directly associated with agricultural activities.

12.2.1.6   Summary of Environmental Regulations of Canada

Canada's population is 10% of the United States but has a larger land area and is located north of the continental United States. These geographic and demographic factors have influenced environmental laws within Canada and also how they are applied. This is, in part, due to the northern portions of Canada's sensitivity to pollution and human habitation and that much of Canada's natural resources including oil and natural gas, forest products, and minerals are located in the northern portions of Canada. Canada has to deal with many pollution issues, namely air and water, caused by the United States since Canada shares its southern border with the United States. This is because pollution does not respect political borders and is a subject that will repeatedly appear in this chapter.

In general, the regulations of Canada are robust and have been in place long enough to be effective. Canada also has vast areas that are relatively pristine from industrial and urban development leaving much of the country with limited human-caused environmental impacts. However, having large areas of uninhabited land also places stress on providing infrastructure and pollution prevention activities by increasing the cost caused by distance, isolation, and remoteness.

12.2.2   Mexico

The official name of what we commonly refer to as Mexico is the United Mexican States (Organization for the Economic Co-operation and Development [OECD] 2015). It's the World's 8th largest country by land size at 1,972 million square kilometers with a population estimated at 120 million as of 2015, which is approximately 35% of the population of the United States and almost 4 times that of Canada. Historically, Mexico has received criticism for not instituting effective environmental laws and regulations, and then once Mexico developed laws and regulations addressing growing environmental concerns, it was again criticized for not enforcing its own laws (Organization for Economic Co-operation and Development (OECD) 2015). Over the past 20 years, that has changed. As we shall see, Mexico has made significant strides in improving protection of human health and the environment. In fact, Mexico has streamlined much of the requirements for the industry into a single operating permit, which is more than we can say the United States has accomplished. However, Mexico still has a long way to go in two major areas of the environment, that being water and air. As we shall see in the next few sections, the availability of safe drinking water and improving air quality, especially in and around Mexico City remain key challenges for Mexico.

12.2.2.1   Environmental Regulatory Overview of Mexico

Environmental regulations are present in Mexico and are patterned after the United States, specifically California. The federal government of Mexico, through the Secretariat of the Environment and

Natural Resources (SEMARNAT), has sole jurisdiction over acts that affect two or more states, acts that include hazardous waste, and procedures for the protection and control of acts that can cause environmental harm or serious emergencies to the environment. SEMARNAT's main roles include the following (Organization for Economic Co-operation and Development (OECD) 2015):

- Develop environmental policy
- Enforce environmental policy
- Assist in urban planning
- Develop rules and technical standards for the environment
- Grant or deny license
- Authorization of permits
- Decide on environmental impact studies
- Offer opinions and assist states with environmental programs

SEMARNAT enforces the law, regulations, standards, rulings, programs, and limitations issued through the National Environment Institute and the Federal Attorney Generalship of Environmental Protection (PROFEPA). Environmental Law's in Mexico address air, water, hazardous waste, pollutants, pesticides, and toxic substances. Along with environmental protection, natural resource protection, environmental impact statements, risk evaluation and determination, ecological zoning, and sanctions (Organization for Economic Co-operation and Development (OECD) 2015).

12.2.2.2 Air

In 1992, the United Nations labeled Mexico City's air as the most polluted on Earth (Mage et.al. 1996). The major air pollution sources that have been blamed for causing much of Mexico City's air pollution come from car exhaust. However, Mexico City's unique geography, located in a mountain valley, where atmospheric inversions are frequent and result in trapping air contaminants near the ground surface is also a contributing factor (Yip and Madl 2002, Marsh and Grossa 2012). Mexico City is home to more than 21 million people making it the most populated city in the western hemisphere and accounts for approximately 20% of Mexico's total population (World Population Review 2021).

Air quality in Mexico City is representative of the rest of Mexico in that automobile exhaust combined to a lesser extent of industrial emissions combined with its geography present challenges to improving overall air quality. Soon after the United Nations declared that Mexico City had some of the most polluted air and a discovery that breathing polluted air contributes to infant mortality in the urbanized world air quality regulations were initially enacted in Mexico (Mage et al. 1996 and Loomis 1999). Currently, ambient air in Mexico is regulated at the federal level by SAMARNAT. In 1993, SAMARNAT issued a series of official Mexican standards or normas (NOM), that regulate maximum concentrations of environmental contaminants in the air. These include eight criteria for pollutants (Mexico Environmental and Natural Resource Ministry (SAMARNAT) 2017):

- Carbon monoxide (CO)
- Ozone (O_3)
- Nitrox dioxide (NO_2)
- Sulfur dioxide (SO_2)
- PM2.5 (particulate matter less than 2.5 microns)
- PM10 (particulate matter less than 10 microns)
- Total suspended particulate matter (TSP)
- Lead (Pb)

The most significant air pollutants in Mexico are ozone (O_3), sulfur dioxide (SO_2), Nitrogen Oxides (NOx), volatile organic hydrocarbons (VOCs), and carbon monoxide (CO) most of which all originate from the incomplete combustion of fossil fuels (Mage et al. 1996 and United Nations 2021b). Ambient air quality standards for criteria pollutants in Mexico are listed at https://www. MexicoAirQualityStandards (United Nations 2021b).

In Mexico, there are seven general categories of industry where rules are consistently applied within each category. The seven categories include the following (Mexico Environmental and Natural Resource Ministry (SAMARNAT) 2017):

1. Federal Public Activities
2. Waterworks, general communications networks, and oil, gas, and carbon transportation activities
3. Chemical and petrochemical plants, iron and steel mills, paper factories, sugar refiners, drink manufacturers, cement producers, automotive parts makers, and the generation and transmission of electricity activities
4. Mineral and nonmineral exploration, extraction, treatment, and refining activities
5. Federal touristic development activities
6. Treatment, storage, and disposal plants of hazardous waste and activities, including nuclear waste
7. Exploitation of slow regenerating vegetation in forest and tropical jungle activities

12.2.2.3 Water

In general, Mexico has poor water quality. According to the Organization for Economic Co-operation and Development (OECD) (2015), Mexico's water resources are some of the most degraded in the world, especially in the developed cities of Mexico City, Guadalajara, and Monterrey and in the northern portion of the country which is more arid than the southern portion. Despite recent improvements and as of 2015, only 14% of Mexico's population are connected to a sewerage system that treats wastewater (Organization for Economic Co-operation and Development (OECD) 2015).

In addition, the quality of treatment in many areas is not adequate and the effluent discharge in many cases does not meet fundamental water quality standards (Organization for Economic Co-operation and Development (OECD) 2015).

Water regulations have been passed but the lack of infrastructure and economic funds for building wastewater treatment plants, associated piping, education, and enforcement of regulations is far from being realized for the majority of Mexico's population (Organization for Economic Co-operation and Development (OECD) 2015). In 1988, then-President Carlos Salinas created the Mexico National Water Commission (CONAGUA). In the beginning, CONAGUA was given the task of defining federal policies to strengthen service providers through technical assistance and financial resources. Progress has been slow and in 2015, Mexico presented a new General Water Law that proposed the transfer of several functions from both federal and state levels to newly created institutions at the watershed level. The new law also designated that each individual has the right to 50 liters of water per day. Mexico published drinking water criteria in 1996 and 1997 are listed at https://www.Mexico.nom-003-SEMARNAT-1997 (Mexico Environmental and Natural Resource Ministry (SAMARNAT) 1997).

Many of the substances with water quality limits are higher than in the United States and water quality standards only exist for a few of the more common compounds. In large part, this is due to internal struggles within Mexico to improve water quality in a nation where surface water and groundwater are often already contaminated (Organization for Economic Co-operation and Development (OECD) 2015). Stormwater regulations in Mexico do not yet exist but perhaps will be forthcoming soon. According to Mexico National Water Commission named CONAGUA (2021), 63% of the water used in Mexico was from surface water and 37% was from groundwater. Overexploitation of groundwater has resulted in significant degradation of water quality and instances of surface subsidence.

Today, CONAGUA has focused on the priorities at the national level that were set through a development plan with the aim of attaining the following (Mexico Environmental and Natural Resource Ministry 2021a):

- Improving water productivity in agriculture
- Improving access and quality to water supply and sanitation
- Supporting an integrated and sustainable water resource management in basins and aquifers
- Improving the technical, administrative, and financial development of water resources
- Consolidating user and society participation
- Identifying and reducing risk with respect to meteorological phenomena
- Evaluating the effects of climate change on the water cycle
- Creating a culture of compliance

The future challenges that Mexico faces in order to improve water quality include the following (Mexico Environmental and Natural Resource Ministry 2021a):

- Infrastructure – currently lack of infrastructure including water treatment and wastewater treatment and a reliable delivery system is such a large impediment that the government has historically not even attempted to improve its infrastructure on a large scale.
- Groundwater overexploitation – this has resulted in land subsidence in many urban areas.
- Land subsidence – primarily the result of groundwater overexploitation which has resulted in impeding the structural integrity of many urban buildings.
- Flooding – as urban areas continue to expand more land will be covered with an impervious surface that will continue to exacerbate flooding and also will have a negative effect on increased land subsidence because of reduced groundwater recharge.
- Increasing urbanization – Continued development without infrastructure construction before development continues to exacerbate the already poor distribution system, especially in large urban areas.
- Basic water quality – Currently, one of the immediate health concerns is a water-borne disease from untreated potable water that in some cases contains fecal matter.
- Intermittent supply – This has resulted in individual residences to collect rainwater from roofs during rain events and storing water in black plastic storage tanks installed on rooftops.
- Limited wastewater treatment – Continues to increase the risk of water-borne diseases, especially in larger urban areas.
- Water use for irrigation – Currently agriculture is the largest consumer of water in Mexico.
- Inefficient use of urban water – This encompasses many different items that include faulty or leaking delivery systems and open sewers.
- Limited cost recovery – Due to lack of infrastructure, some delivery systems to individual users do not account for how much was is used due to lack of water meters.

Finally, the extraction of groundwater from aquifers beneath Mexico City has resulted in significant subsidence at the surface (USGS 2021a).

The issues identified in this section may seem dire or even hopeless but, as we shall see later in this book, investment in manufacturing and slow but determined progress has made a big difference in improving water quality in Mexico over the last 20 years (Organization for Economic Co-operation and Development (OECD) 2015).

12.2.2.4   Solid and Hazardous Waste

In Mexico, the definition of hazardous waste is very similar to that of Characteristic Waste in the United States, and is defined as a substance that is explosive, ignitable, corrosive, chemically reactive, or is toxic (directly or indirectly) to plants, animals, or humans. Air, water, solid waste, and hazardous waste are all handled similar to the United States (i.e., California in particular), with one exception and that being stormwater (Basurto and Soza 2007, Rodriguez et al. 2015, and Mexico Environmental and Natural Resource Ministry 2021a). Within many Mexican states, stormwater is of little or no concern due to dry conditions, especially in the northern and west regions of the country. However, that will likely change soon.

Permits, authorizations, and documentation likely required at manufacturing facilities in Mexico include the following (Basurto and Soza 2007, Rodriguez et al. 2015, and Mexico Environmental and Natural Resource Ministry 2021a):

- Operating licensed
- Wastewater discharge Registration
- Hazardous Waste Generator's Manifest
- Monthly Log of Hazardous Waste Generation
- Ecological Waybills for the Incineration and/or Exportation of Hazardous Materials and Wastes
- Semi-annual Report on Hazardous Wastes Sent to Recycling, Treatment, or Final Disposition
- Accidental Hazardous Waste Spill Manifest
- Delivery, Transport, and Receipt of Hazardous Waste Manifests
- Environmental Impact Studies

In Mexico, the operating license is a key document/permit and must be obtained before operations begin. However, in some instances this may be difficult to obtain far in advance since many items may change that will affect the details in the license that render the license invalid even before operations begin (Mexico Environmental and Natural Resource Ministry 2021a). In addition, if any changes are made within the facility, the operating license may no longer be valid and must be amended, much as it is in the United States and as we shall see, as it is in most of the World. In the United States, for instance, with an air permit, if a modification in operations is made that increases or has the potential to increase emissions, then a permit modification is necessary, and in some circumstances, a new permit may be required, that being a construction permit. This same principle applies in most other countries of the World.

In Mexico, except in relation to the petroleum and petrochemical industries or those treatment facilities of hazardous waste, a single operating license may be obtained through a single application. The single operating license integrates the various permits, licenses, and authorizations for matters of environmental impact, water service, emissions of atmospheric contaminants, and generation and treatment of hazardous waste (Mexico Environmental and Natural Resource Ministry 2021a). Obtaining a single license that covers all environmental matters is very different than how environmental permits are handled in the United States. A single operating license or permit that covers all aspects of the potential environmental impact a facility may have upon the environment is far more manageable in most all aspects than having separate permits for every source. With the exception of Title V permits for air emissions, the United States issues permit separately.

12.2.2.5 Remediation Standards

In 2004, Mexico passed what is termed "Mexico's General Law for Prevention and Integral Management of Wastes", commonly referred to as Mexico's Waste Law. The passage of this law significantly advanced Mexico's efforts to develop a detailed regulatory framework for dealing with legacy sites and the discovery of new or unknown sites of environmental contamination (Mexico Environmental and Natural Resource Ministry 2021b). It mirrors the United States Brownfields Revitalization Act and All Appropriate Inquiry.

The Mexican Waste Law created strict liability against owners and possessors (e.g., operators) of a contaminated site. Previously, parties who caused the contamination were held responsible for clean-up. The Mexican Waste Law holds that one does not have to have caused the contamination to be held liable for its cleanup. To protect new landowners, the law mandates disclosure of known contamination by hazardous materials or wastes from owners to potential third-party buyers or tenants (Mexico Environmental and Natural Resource Ministry 2021b).

In addition, the Mexican Waste Law does not allow for the transfer of a site contaminated with hazardous materials or wastes without express authorization from Mexico's environmental ministry. Lastly, the environmental ministry or SEMARNAT will not grant approval until the contamination is remediated or until an agreed plan has been approved by all parties. The impact of the Mexican Waste Law and its regulation has been realized in situations where properties contain known or suspected contamination (Mexico Environmental and Natural Resource Ministry 2021b). As one can imagine, potential purchasers are reluctant to take the title of contaminated property without recourse to prior studies that characterize the potential contaminate issue(s) and define the responsibility and costs associated with addressing the contamination. Sellers, on the other hand, are reluctant to conduct environmental investigations because of the potential to discover the contamination that may require notice to SEMARNAT and that may cost significant time and resources to address.

The most common method to achieve cleanup objectives in Mexico is to simply excavate the contaminated media, usually soil, and dispose of the material in a landfill, if possible. The Mexico Environmental and Natural Resource Ministry (SAMARNAT) (2004, 2005, and 2015) has published remedial objectives at https://www.Mexicana.nom-147-SEMARNAT/ss-2003. https://www.Mexicana. nom-138-SEMARNAT/ss-2003. and https://www.Mexicana.nom-133-SEMARNAT/ss.2003(Mexico Environmental and Natural Resource Ministry (SAMARNAT) 2004, 2005, and 2015). Cleanup levels for soil have been divided into groups and are based on exposure scenarios and land use similar to the United States and other countries. The categories include residential, agricultural, commercial, and industrial. Occasionally, some of the groups are combined under one heading. Note that in Mexico, agricultural land is specifically noted and has applicable cleanup objectives unlike in the United States (Mexico Environmental and Natural Resource Ministry 2021b).

The objective of remediation in Mexico is much the same throughout the world and that is to either eliminate or reduce contaminant concentrations or to control levels of contamination through institutional controls or other acceptable methods such that the risk posed by the presence of the contamination no longer puts human health in danger or adversely affect the environment. Remedial efforts in Mexico should accomplish the following (Mexico Environmental and Natural Resource Ministry 2021b):

- Permanently reduce contaminant concentrations
- Reduce contaminant bioavailability, solubility or both
- Avoid contaminant dispersion in the environment
- Establish institutional controls, if necessary

There are two regulatory agencies that are responsible for overseeing compliance with the investigation and remediation of contaminated sites in Mexico and they are the Procuraduria Federal de Proteccion al Ambiente (PROFEPA), which is a regulatory division of SEMARNAT, and the Health Secretariat, which is involved in cases of risk to human health (Mexico Environmental and Natural Resource Ministry 2021b).

12.2.2.6 Summary of Environmental Regulations in Mexico

Mexico has made significant progress in environmental regulations in the last two decades but lacks significant infrastructure, such as wastewater treatment in Mexico City and other locations. Other significant impediments include corruption and a lack of political priorities on issues pertaining to the environment. Mexico also struggles with consistent enforcement of environmental regulations in parts of the country.

12.3 SOUTH AMERICA

South America forms the southern landmass of the Americas, generally considered as south of Panama. South America is composed of three basic types of terrain, the Andes Mountains that span nearly the entire western boundary, the interior Amazon River Basin, and the Guiana Highlands in the east (United Nations 2021a). South America has a very diverse climate from one of the wettest in Columbia and Chile to the driest desert in the world, the Atacama (United Nations 2021c, 2021d, and 2021e). South America is composed of 12 countries and has a total estimated population of 427 million and represents only 5.6% of the total human population on Earth. The land area of South America is 17,840,000 square kilometers, making it the 4th largest continent behind Asia, Africa, and North America. The population density of South America is 21.4 per square kilometer (United Nations 2021a). Most people live along the eastern and western coasts of the continent.

12.3.1 Argentina

Argentina is bordered by the Atlantic Ocean to the East, Andes Mountains and Chile to the West, Bolivia and Paraguay to the north, and Brazil and Uruguay to the northeast. It is the 8th largest country in the world with 2,780,400 million square kilometers of land. Argentina has a population of 42.1 million making it the 32nd most populous country. The population density of Argentine is 16.2 people per square kilometer. The largest city in Argentina is Buenos Aires, which is also the capital of the country and has a population of over 15 million (United Nations 2021c). Argentina has a diverse economy that consists of manufacturing, agriculture, forestry, fishing, construction, transport, and utilities. The gross domestic product of Argentina was over $639 billion in 2017 which translates into a per capita income of $3,500 per individual, which is well below that of many other nations (World Bank 2021a).

12.3.1.1 Environmental Regulatory Overview of Argentina

Environmental issues faced by Argentina include air pollution, water pollution, deforestation, salinity, and lack of environmental controls and enforcement (United Nations 2021c).

12.3.1.2 Air

Air pollution in Argentina is most acute in its capital city of Buenos Aires. However, there is very limited data on current and historical air pollution levels in Buenos Aires and throughout Argentina. Argentina has established ambient air quality standards but they have not been implemented for the following (Argentina Secretariat of the Environment 2021a):

- Carbon monoxide
- Nitrogen dioxide
- Sulfur dioxide
- Ozone
- Particulate matter (PM10)
- Fine particulate matter (PM2.5)

Argentina's ambient air standards are listed at http://www.fmed.uba.ar/depto/toxico/plaguicidas/ 20.284 (Argentina Secretariat of the Environment 2021a).

12.3.1.3 Water

The environmental management of water bodies under federal jurisdiction in Argentina is regulated under Law 25.688 (Argentina Secretariat for the Environment 2021b). Drinking water in urban areas of Argentina is generally regarded as safe. However, up to an estimated 20% of drinking water sources in rural areas is not considered safe (United Nations 2021c).

Argentina is struggling with the cleanup of the 64-kilometer long Matanzas-Riachuelo River which flows through Argentina's most populous city of Buenos Aires and is considered one of the most polluted rivers on Earth. The river has seen contamination for over 200 years from agriculture, livestock, tanneries, and industrial pollutants from over 15,000 industrial facilities that have discharged into the river and are located within the river drainage basin (World Bank 2013). Cleanup of the river was to begin in 2018 with assistance from the World Bank, Argentina Secretariat for the Environment and Sustainable Development, and several municipalities along and near the river at a cost of several billion $US dollars (World Bank 2021a). Cleanup is expected to continue for decades before being complete. This situation along the Matanzas-Riachuelo River is but one example of the challenges that plague the development, implementation, and enforcement of water resources management in Argentina. Management and changing water management techniques in Argentina are further hindered because multiple institutions at the national, provincial, and river basin levels have jurisdiction (World Bank 2021b).

Identified issues that must be addressed include the following (World Bank 2000, 2021c).

- Incomplete or outdated legal and regulatory framework
- Limited capacity in water management at the central and provincial levels
- Outdated procedures for water resources planning
- Lack of integrated national water resources information system
- Deficient water resources monitoring system
- Serious water pollution problems
- High risk of flooding in urban and rural areas
- Lack of flood risk reduction strategies
- Lack of appropriate incentives for conservation and pollution reduction

12.3.1.4   Solid and Hazardous Waste

Solid and hazardous waste is regulated in Argentina by Law 24.051, which addresses the generation, transport, handling, treatment, and final disposal of hazardous waste. However, specific details on waste characterization, infrastructure, and management structure on a national level, like the management of water resources, are lacking. In Argentina, a waste is hazardous if it can damage living beings for contaminating land, air, water, or the environment. Any facility that generates a hazardous waste must register with National Registry of Hazardous Waste Generators and Operators which is required to operate any business that generates a hazardous waste (Argentina Secretariat for the Environment 2021c). Just as in the United States and several countries of the world, the hazardous waste tracking system in Argentina is a "Cradle to Grave" system (Argentina Secretariat for the Environment 2021c).

12.3.1.5   Remediation

As with water and solid and hazardous waste, remediation standards on a national basis are lacking in Argentina and are plagued by several impediments that include overlapping jurisdictions, competing interests, lack of reliable cleanup standards, and lack of infrastructure (United Nations 2021c).

12.3.1.6   Summary of Environmental Regulations of Argentina

Argentina is a rather large country for its population. Argentina's significant environmental issues appear to be concentrated in and immediately around its largest city of Buenos Aires and involve air, water, and land pollution. Investigation and cleanup of one of the world's most polluted rivers will likely be a benchmark achievement or failure for Argentina's future at protecting human health and the environment. Cleanup of a 64-kilometer river in an urban environment in any country, including the United States, is a monumental task and is perhaps too much to ask for a country that is struggling with social and economic issues as well.

12.3.2 BRAZIL

Brazil occupies 48% of the continent of South America and has a total land area of 8,514,215 square kilometers. The only countries larger in land area are Russia, Canada, China, and the United States. Much of the climate of Brazil is tropical, with its southern portion being temperate. The largest river in Brazil, and the second-longest in the world, is the Amazon River with a length of 6,992 kilometers (United Nations 2021d).

Brazil has the 8th largest economy in the world with a Gross Domestic Product of $2.138 trillion dollars. Brazil has the second-largest economy in America behind the United States. Significant natural resources include gold, uranium, iron, and timber. Main trading partners with Brazil include China, the United States, Argentina, and Japan. Brazil's main industries include textiles, chemicals, cement, lumber, iron ore, tin, steel, aircraft, motor vehicles, and other machinery equipment (United Nations 2021d).

Brazil undertook an ambitious program to reduce its dependence on imported oil which accounted for 70% of its historical needs. By 2007, Brazil became self-sufficient in oil and is now one of the world's leading producers of hydroelectric power. In fact, hydroelectric power now generates 90% of Brazil's electricity needs. This was accomplished by the Itaipu Dam on the Parana River and the Tucurui Dam in Para in northern Brazil. Brazil is also expanding its use of nuclear power with two nuclear power plants currently in operation and has plans for 10 more power plants (World Bank 2021d). The population of Brazil is estimated at 207 million making it the second-highest in population in America only behind the United States. The two largest cities in Brazil include Sao Paulo, with an estimated population of 21.5 million, and Rio de Janeiro, with a population of 12.8 million. The per capita income of Brazil is approximately $15,500 or nearly one-quarter of the per capita income of the United States and about the same as China (World Bank 2021d).

12.3.2.1 Environmental Regulatory Overview of Brazil

The Amazon rainforest is the largest tropical rainforest in the world spanning 5.5 million square kilometers, 60% of which is located in Brazil. Rainforests worldwide once covered 15% of the land on Earth but due to anthropogenic activities, only 6% are left and are decreasing each year. Current environmentally related threats to the Amazon rainforest include the following (National Geographic Society 2021):

- Deforestation: Deforestation in the Amazon rainforest varies between the nine countries that are within or border the rainforest. Types of deforestation within the Amazon basin include
  - Clearing the forest by burning the forest to make room for cattle pastures for large ranches and local farmers who clear smaller, but perhaps more numerous, patches for agriculture.
  - Mining, especially for gold
  - Urbanization from expanding cities on the fringe
- Loss of biodiversity: Species lose their habitat or can no longer survive in the small fragments of forest left behind after clearing, especially if the patches of forest are not interconnected.
- Habitat degradation: Fragmentation by road building and other anthropogenic activities lead to habitat degradation that places additional negative pressures on indigenous species.
- Modified global climate: Deforestation releases CO_2 into the atmosphere and also prevents CO_2 removal and the release of oxygen into the atmosphere by the trees that are deforested.
- Interruption or loss of the water cycle: Deforestation reduces the critical water cycling services that trees provide by increasing the amount of erosion and decreasing the amount of evapotranspiration which in turn disrupts and decreases subsequent rain events.

- Social effects: Deforestation ultimately leads to the degradation of the ecosystem and contributes to the overall decline of long-term economic balance.
- Logging: Logging is another example of deforestation that destroys habitat.
- Hydroelectric plants: Damming rivers in the Amazon rainforest flood.
- Poaching: Poaching of mammals, birds, reptiles, and even insects lowers the natural population numbers in species and places additional stress on Amazon wildlife.
- Pollution: Anthropogenic activities introduce a plethora of pollutants into the rainforest ecosystem through the air, water, and land that include but are not limited to organic and inorganic compounds, heavy metals, particulate matter, suspended solids, erosion, noise, increased atmospheric carbon dioxide, decreased oxygen, and habitat loss.

12.3.2.2 Air

Air pollution in Brazil is largely due to rapid urbanization and a deforestation technique called "slash and burn" which is a method where vast tracts of forest land are cut and burned to make way for development. In addition, many smaller towns still burn garbage under a legal framework that allows prescribed burning. Efforts to improve air quality are underway, especially in Brazil's largest cities (USEPA 2021a).

Ambient air quality standards in Brazil are similar to the United States and EU and include the following compounds (Brazil Ministry of the Environment 1990):

- Particulate matter (PM10)
- Fine particulate matter (PM2.5)
- Ozone
- Sulfur dioxide
- Carbon monoxide
- Nitrogen dioxide

Brazil ambient air quality standards are located at http://www.brazilgovnews.br/standards-on-air-quality-htm (Brazil Ministry of the Environment 1990).

The National Environment Council of Brazil created the ambient air standards in 1990 and established the ambient air quality program (Brazil Ministry of Environment 2021a). Air pollution in urban areas is worst in Sao Paulo and Rio de Janeiro (United Nations 2021b).

Environmental licensing is required in Brazil for any enterprise or activity that uses or may generate pollutants or hazardous substances. In the case of an air permit, the process is similar to the United States, the European Union, China, and other countries. The following is a general outline of the permit process (Brazil Ministry of Environment 2021a):

- Initial application: The initial application is termed a Construction Permit, if it's new construction. The Construction permit requires proof of ownership, certification of use or zoning, and proposed architectural or building improvements.
- Habitation permit: Following approval of the construction permit, a habitation permit is the next step and is termed.
- Registration for the article of association.
- Business License.
- Sanitary Permit: Once the sanitary permit is obtained, other required permits for water, water treatment, and waste generation can proceed.

12.3.2.3 Water

Water resources management in Brazil is the responsibility of the National Water Agency. In general, Brazil has vast quantities of freshwater, especially in the northern portions of the country near the Amazon River. Up to 97% of the population has access to clean water. However, water pollution in Brazil from sewerage is acute and is especially so in many of the larger cities, including Rio de Janeiro and Sao Paulo (Brazil Ministry of Environment 2021a).

According to the World Health Organization (2015), only 15% of sewerage in Brazil is collected and treated before discharge. Not only does the discharge of untreated sanitary waste impact the rivers and other freshwater bodies in Brazil, but also has a noticeable and significant effect on the ocean and beaches, especially near the large cities of Sao Paulo and Rio de Janeiro. (World Health Organization 2015). Water quality standards vary from each of the 27 states in Brazil and also within many of the 389 municipalities that supply water.

Some large manufacturing facilities, as in the United States and elsewhere, have installed their own water treatment and wastewater treatment plants in Brazil. Some of these facilities require more improved water quality than the local municipality can provide and others require a more reliable source. These facilities obtain a permit to withdraw water from a water source, usually, a nearby river, treat the water before it's used, and then treat the water again before its discharged back into the river.

12.3.2.4   Solid and Hazardous Waste

Brazil amended its 1998 waste law in 2010 (Brazil Ministry of Environment 2021b) that brought together in one law a set of principles, goals, instruments, targets, and actions toward the integrated management of environmentally sound solid waste management. The Brazilian Waste Law outlines concepts and provisions similar to that of the United States and includes the following (Brazil Ministry of the Environment 2021b):

- Solid waste management
- Recycling
- Reuse
- Reduce
- Sustainability
- Pollution prevention
- Polluters pay principle
- Shared responsibility or cradle to grave concept

Hazardous wastes in Brazil follow that of most other countries and include (Brazil Ministry of the Environment 2021b)

• Explosive	• Ignitable
• Flammable	• Infectious
• Oxidizing	• Toxic
• Corrosive	• Characteristic
• Reactive	• Medical

Radioactive waste is addressed separately (Brazil Ministry of the Environment 2021b). The volumes of hazardous waste generation in Brazil from 2006 to 2015 increased by 30%. The significant challenge for Brazil recently is the management of medical waste and e-waste, both of which appear to need improvement (Macedo and Sant'ana 2015).

12.3.2.5   Remediation

Much like the rest of the world, Brazil uses risk-based standards for remediation. Remediation thus far in Brazil has generally been for its larger sites that pose the highest risk and usually involve water and river systems (Brazil Ministry of Environment 2021a).

12.3.2.6 Summary of Environmental Regulations of Brazil

Brazil is the largest and most populated country in South America and the country that has 60% of the Amazon Rainforest. Brazil's most significant environmental issues are:

- Sanitation, especially in the poor areas within its major cities
- Cleaning up river systems, which is related to sanitation
- Air pollution in larger urban areas
- Protecting the rainforest

The sanitation issue is one that plagues many countries in South America and Mexico, portions of Africa, and Asia. Unfortunately, the longer it takes to finally address the sanitation issue the harder and more costly it becomes and continues to spread disease (United Nations 2021d).

Brazil's environmental regulations seem robust but are not providing the desired results because of several factors that include

- Lack of financial resources
- Overlapping responsibilities between agencies
- Other social and economic priorities
- Lack of political will and corruption, especially with respect to addressing the larger more complex environmental issues such as protecting the rainforest and sanitation

12.3.3 CHILE

Chile is located along the extreme western boundary of South America. Chile shares land borders with Argentina, Peru, and Bolivia. Chile's western border is the Pacific Ocean. From north to south Chile extends 4,270 kilometers but averages only 170 kilometers east to west making Chile a long and narrow nation. The land area of Chile is 756,096 square kilometers. The Andes Mountains run along the entire eastern border of Chile. Chile is prone to earthquakes. In fact, Chile has experienced 28 earthquakes in the 20th century with a magnitude of 6.9 on the Richter scale or greater (USGS 2021b, United Nations 2021e). Because Chile is so long and narrow in the north-south direction and goes from sea level to nearly 7,000 meters in elevation in the Andes Mountains, the climatic regions vary significantly and often over very short distances.

The population of Chile is 18.4 million. Santiago is Chile's largest city and is the capital with a population of 6.3 million, which represents 33% of Chile's entire population. The gross domestic product of Chile was $247.03 billion in 2016. Per capita income is estimated at $15,059. Chile is considered as one of South America's leading economies (World Bank 2015).

12.3.3.1 Environmental Regulatory Overview of Chile

The main statutory environmental framework in Chile is contained in the Environmental Law (No. 19.300, 1994 and amended in 2010), which introduced the Environmental Impact Assessment System, which set forth the process for obtaining an environmental license. The 1994 law is known as the General Environmental Framework Law, which established the National Environment Commission (CONAMA), reporting directly to the president's office through the Ministry of General Secretariat of the Presidency. CONAMA coordinates government environmental policies, prepares environmental regulations, and fosters the integration of environmental concerns into other policies (United Nations 2015).

The amendments to the 1994 law enacted in 2010 created the Ministry for the Environment. The ministry for the Environment, which elevated environmental issues to a cabinet-level rank, is charged with assisting the President of the Republic of Chile in the design and implementation of policies, plans, and programs for the protection of the environment which includes (United States Library of Congress 2010)

- Biological diversity
- Renewable natural resources
- Hydraulic resources
- Sustainable development
- Environmental policy
- Regulatory framework

The Environmental Impact Assessment System requires all investment projects and productive activities to undergo a detailed process to determine the real effects of proposed operations would have on the environment (Chile Ministry of the Environment 2021a).

USEPA has been collaborating with Chile since 2004 on issues relating to (USEPA) 2017, Chile Ministry of the Environment 2021a):

- Risk management related to waste and contaminated sites
- Environmental enforcement
- Environmental forensics
- Compliance with air pollution standards
- Training
- Technical consultation
- Public participation
- Environmental education

Chile's most significant environmental issues center around deforestation and the subsequent erosion that it causes and air pollution in its capital city of Santiago. The City of Santiago experiences increased air pollution due to automobile exhaust and wood-burning, which is difficult to abate because the city is surrounded by mountains (Chile Ministry of the Environment 2021a). Many households use wood for fuel because it's more economical than other forms of energy. Other environmental issues in Chile include degraded water quality, especially in and around Santiago, and from mining operations. Unfortunately, most of Chile's mining is in the northern desert region of the country where any water quality issues are heightened because of the relative scarcity of water resources (Chile Ministry of the Environment 2021a).

12.3.3.2   Air

Air pollution in Chile ranges from good to unhealthy (WHO 2021a). Santiago usually experiences the lowest air quality in the country and is larger due to automobile exhaust and wood-burning and is exacerbated because the city is surrounded by mountains which inhibits the circulation of air during certain times of the year, especially the winter months when colder air is trapped near the surface of the ground (WHO 2021a). However, the buildup of ozone is heightened in the summer months due to increased solar radiation (Diaz-Robles et al. 2011).

Similar to the United States, areas that do not meet ambient air quality standards are designated as non-attainment areas by geographic region. When an area is designated as nonattainment, either an Atmospheric Prevention Plan (APP) or an Atmospheric Decontamination Plan (ADP) is required which is similar to the State Implementation Plan (SIP) in the United States (Chile Ministry of the Environment 2021b). These plans require that region of nonattainment lower emissions until attainment is achieved. The effectiveness of the ambient air quality is evident in that the area of Santiago has reduced its sulfur dioxide levels by 77% since 1994. Other contaminants such as ozone have been on the increase mostly due to an increasing population and more automobiles on the roads (Diaz-Robles et.al 2011). Chile's ambient air quality standards are listed at http://portal.mma.gob.cl.aire (Diaz-Robles et al. 2011, Chile Ministry of the Environment 2021b).

12.3.3.3 Water

Compared to other countries in South America, Chile has good water quality and sanitation with nearly 99% of its population with access to treated water and basic sanitation (World Health Organization 2013 and World Health Organization 2015). The northern portion of the country appears to suffer the worst water quality in Chile for two reasons (United Nations 2013):

- Water is scarce because it's a desert region.
- Most mining, especially copper, is located in the same region and requires large quantities of water that is obtained from groundwater.

The resulting water contaminated with heavy metals impacts surface water quality quite significantly partially because water is relatively scarce. In contrast, water in southern Chile is considered by some to be the purest fresh water on Earth because of its remoteness and lack of anthropogenic sources of pollution (United Nations 2013).

Unlike many other countries, water in Chile is largely privatized and has been since 1981. The advantage of privatization is that the quality of water is of better quality. However, the disadvantage is that water is expensive. In addition, it is estimated that prices will increase significantly due to an increasing population and climate change which has negatively impacted water availability through increased temperatures and a general decrease in precipitation levels since 1976 (United Nations 2013).

12.3.3.4 Solid and Hazardous Waste

Environmental regulations in Chile on solid and hazardous waste are similar to that of the United States and cover the following (Chile Ministry of the Environment 2021c):

- Solid waste
- Hazardous waste
- Storage and handling of hazardous substances
- Transporting solid and hazardous waste
- Disposal

Hazardous wastes in Chile follow that of the United States and include the following (Chile Ministry of the Environment 2021c):

• Explosive	• Ignitable	• Flammable
• Infectious	• Oxidizing	• Toxic
• Corrosive	• Characteristic	• Reactive
• Medical	• Certain residues	

Certain identified processes that produce waste are also considered hazardous and are similar to Listed Wastes in the United States. Some of these include the following (Chile Ministry of the Environment 2021c):

• Textiles	• Petroleum	• Solvents
• Used absorbents	• Refrigerants	• Propellants
• Pesticides	• Herbicides	• Adhesives and glues

12.3.3.5   Remediation

Remediation of sites of environmental contamination in Chile has focused on mining and abandoned mines. USEPA has been actively collaborating with the Chile Ministry of the Environment focusing on the following (USEPA 2017):

- Management of environmental aspects of mining
- Abandoned mine risk evaluation
- Remediation of contaminated mining sites
- Site investigation and characterization
- Conducting risk assessments
- Mapping
- Selection of remedial options
- Enforcement measures at mining sites

Additional remedial focus has been on improving air quality, especially in the urban area of Santiago through a Chilean Ministry of Environment and USEPA partnership designed to improve air quality, protect the climate, and provide public health benefits (United States Environmental Protection Agency (USEPA) 2017).

12.3.3.6   Summary of Environmental Regulations of Chile

Chile is a small country with a relatively low population located in an extremely diverse climate and geographic region. This tends to complicate regulation of the environment because there are so many different, separate and distinct, and sometimes unrelated geologic and hydrologic environments. Therefore, Chile's geology and geography heavily influence its efforts to protect and improve environmental regulations. Air quality regulations appear to be the most advanced in Chile compared with water, solid and hazardous waste, and remediation regulations. However, availability of clean water will likely be more of a pressing issue in the future due to climate change.

Chile's environmental regulations first became serious in 1994 and then were revised significantly in 2010. Together with cooperation and collaboration with the United States Environmental Protection Agency (USEPA), it does not appear that environmental matters will largely be ignored in the future. Anticipation is that future environmental regulations in Chile will be very similar to that of the United States.

12.3.4   PERU

Peru is located in western South America. Peru's west coast extends along the Pacific Ocean for 2,414 kilometers. Columbia and Ecuador are located to the north, Brazil to the east, and Bolivia and Chile to the south. Peru covers a land area of 1,279,999 square kilometers making it the 20th largest country by area in the world. Peru has an estimated population of 31 million people. The Gross Domestic Product (GDP) of Peru was $470 billion in 2017, which translates into a per capita income of $13,735. The economy of Peru has significantly improved over the last 20 years. As evidence of strong economic growth, poverty in Peru has been reduced from nearly 60% in 2004 to just below 26% in 2012 (World Bank 2021e).

The Peruvian economy is 43% service (including tourism) and 47% industry and manufacturing. Major exports include metals (copper, zinc, gold, mercury), chemicals, pharmaceuticals, machinery, textiles, and fish meal. Peru ranks 2nd in the world for copper production. Copper accounts for 60% of Peru's exports (World Bank 2021e). The capital of Peru is Lima. It is located along the western coast with the Pacific Ocean. The population of Lima and associated surrounding communities is estimated at over 12 million residents, which represents 40% of the entire population of Peru (World Bank 2021e).

The geography of Peru is rather unique in that the Andean Mountain Range, which trends north to south, divides the country from its coastal region to the west from the Amazon Basin to the east (United Nations 2005). The Andean Mountain range acts as a divide in geographical and climatic terms.

12.3.4.1 Environmental Regulatory Overview of Peru

The principal environmental issues of Peru include water pollution, soil erosion, soil pollution, and deforestation. Air pollution is significant in Lima where pollutant concentrations sometimes reach unhealthy levels (WHO 2021a). Peru is susceptible to high levels of soil erosion as deforestation occurs and is often exacerbated especially in mountainous regions with increased erosion. Water pollution and lack of sanitation are also issues, especially in a rural area where an estimated 62% of the population has access to clean water and access to sewage is as low as 22% (United Nations 2005).

The USEPA is collaborating with Peru on several environmental issues since 2009 that include air, water, waste, land degradation, remedial technologies and management, and others (USEPA 2021b).

The Peruvian Ministry of Environment is a ministry of the Cabinet of Peru created in 2008. Its function is to oversee the environmental sector with the authority to design, establish, and execute government policies concerning the environmental management and strategic development of natural resources (Peru Ministry for the Environment 2021a). Peru, by means of Law 30327, established an integrated permitting process titled "Global Environmental Certification" (IntegrAmbiente) that integrates all necessary environmental permits into one permit similar to the process in China (Peru Ministry for the Environment 2021a).

12.3.4.2 Air

The city of La Oroya located in the Andes Mountains 176 kilometers northeast of Lima is considered by many to be one of the most polluted cities in the world. The source of the pollution is a copper smelter and associated copper mine which has resulted in significantly elevated levels of lead, copper, zinc, and sulfur dioxide in the air (USEPA 2021b). Ambient air quality standards are similar to the United States and EU and are listed at http://www.temasactuales.com/assets/pdf/gratis/peraiqual.htm (Peru Ministry for the Environment 2021a).

In 2016, the World Health Organization released a study on air quality that identified Lima as the city with the worst air pollution in South America (WHO 2021a and 2021b). Given that Lima is a city of more than 12 million residents, is surrounded by mountains, and is subject to air inversions that tend to trap pollutants in the low layers of the atmosphere, it is not a big surprise that the air quality is poor.

12.3.4.3 Water

Just as with air pollution, Peru faces water pollution issues associated with its copper mining operations along with degradation of water quality due to deforestation. Access to safe water and sanitation has improved in Peru in recent years but 10% of the population remains without access to safe drinking water and nearly 80% without improved sanitation (World Health Organization 2021b). The Ministry of Housing, Construction, and Sanitation is the governing entity that formulates, approves, executes, and oversees national water and sanitation policy (Peru Ministry of Housing, Construction, and Sanitation 2021).

USEPA is collaborating with the Peruvian government to improve water quality, specifically addressing mining wastes contaminated with various metals, including mercury (USEPA 2021d). Surface water quality in particular has been gradually declining due to mining discharges to surface water that either has not been treated adequately or has not been treated at all. Other anthropogenic sources of surface water pollution are from industry, municipalities, and polluted agricultural runoff (USEPA 2021b). Wastewater discharge standards in Peru are listed at http://www.gob.pe/vivienda (Peru Ministry of Housing, Construction, and Sanitation 2021).

12.3.4.4   Solid and Hazardous Waste

Solid waste management in Peru was essentially non-existent until the Ministry of the Environment was created in 2008. Environmental concerns in Peru are not considered a national priority, including solid waste management. However, there are only nine officially recognized solid waste landfills in Peru, and five are located in and around the capital city of Lima. This means that many parts of the country still rely on informal disposal of solid waste in dumps and other poorly managed and poorly constructed locations (Peru Ministry for the Environment 2021b, Peace Corp 2016). USEPA is currently collaborating with Peru on many environmental issues including upgrading infrastructure for solid and hazardous waste management (USEPA 2021b).

12.3.4.5   Remediation

Remediation of contaminated sites in Peru is improving but still has a long way to go to be considered effective. USEPA has been assisting Peru in establishing an effective program through the following activities (USEPA 2021b):

- Management of environmental aspects of mining
- Mine risk evaluation
- Remediation of contaminated mining sites
- Site investigation and characterization
- Conducting risk assessments
- Mapping
- Selection of remedial options
- Enforcement measures at mining sites
- Managing mercury

Mercury is utilized at many gold mines in Peru to assist in extracting gold from gold-bearing ore. Releases of mercury are common along with other heavy metals including iron, manganese, lead, copper, and zinc (USEPA 2021b). These heavy metals have not only impacted the soil but have significantly degraded surface water quality and are a significant source of air pollution, especially near major mines including the city of La Oroya located in the Andes Mountains 176 kilometers northeast of Lima (USEPA 2021b).

12.3.4.6   Summary of Environmental Regulations of Peru

Although Peru does not have a significant population, the environmental concerns due to mining, especially copper and gold, and lack of infrastructure are significant. An additional item of note is that the environment has only become a national priority since 2008 when the Ministry of the Environment was established. Peru is making progress but the environmental damage in erosion, deforestation for farming and mining activities, air pollution, water quality degradation and the huge future costs of environmental remediation of legacy sites will certainly present Peru with huge economic costs well into the future.

12.4   SUMMARY AND CONCLUSION

What should be apparent after reading this chapter is that many countries have a robust set of environmental regulations and that they are very similar to those of the United States. However, many countries struggle with implementing environmental regulation to the extent that they actually work at achieving the underlying purpose of any environmental regulation which is improving human health and the environment. The reasons may vary from country to country but all have common themes that involve war, poverty, corruption, overpopulation, lack of infrastructure, political will, ineffective leadership, and other social issues.

Questions and Exercises for Discussion

1. Mexico has much environmental improvement to conduct before it can focus on sustainability? Do you agree or disagree with this comment and why?
2. In what areas can Canada improve its environmental regulations? Name at least two areas and explain why.
3. In your view, is Canada or the United States less contaminated, and why?
4. Of the countries of South America, which country is the most environmentally impaired and why?
5. Of the countries of South America, which has the most effective environmental regulations and why?
6. Of the countries of South America, which country has the least effective environmental regulation and why?
7. Of the countries of South America, provide your view as to what you would recommend to improve the air, water, land, and living organisms of the country you selected from question 6 above?
8. Now that you have read this chapter, what do you think is the most important environmental issue facing South America and why?

REFERENCES

Argentina Secretariat for the Environment. 2021a. Ambient Air Standards. *Argentina Law 20.284*. http://www.fmed.uba.ar/depto/toxicol/plaguicidas/20.284 (accessed April 20, 2021).

Argentina Secretariat for the Environment. 2021b. *Argentina Water Management Regulations Law 25.688*. https://www.mrecic.gov.ar/water (accessed April 20, 2021).

Argentina Secretariat for the Environment. 2021c. *National Definition of Waste Under Law 24.051*. http://www.mrecic.gov.ar/waste (accessed April 20, 2021).

Basurto, D. and Soza, R. 2007. Mexico's Federal Waste Regulations. *Journal of the Air and Waste Management Association*. Vol. 57. p. 7–10.

Brazil Ministry of the Environment. 1990. *CONAMA Resolution No. 3. Standards on Air Quality*. http://www.brazilgovnews.gov.br/standards-on-air-quality.htm (accessed April 20, 2021).

Brazil Ministry of the Environment. 2021a. *Ministry of the Environment*. http://www.brazilgovnews.gov.br/ministry-of-the-environment.htm (accessed April 20, 2021).

Brazil Ministry of the Environment. 2021b. *National Solid Waste Law of 2010*. http://www.wiego.gov.br/wast-law.htm (accessed April 20, 2021).

Canadian Environmental Protection Act. 1999. *Canadian Environmental Protection Act*. https://www.ec.gc.ca/Lcpe-cepa (accessed April 20, 2021).

Canadian Government. 2021a. *Canada: A Brief Overview*. https://www.cic.gc.ca/english/newcomers (accessed April 20, 2021).

Canadian Government. 2021b. *Air Quality Strategy for the Canada-US Border*. https://canada.ca/en/environment-climate-change/issues/quality-stragey-canada-united-states-border.html (accessed April 20, 2021).

Canadian Government. 2021c. *Canadian Air Quality*. https://www.canada.ca/en/environment-climate-change/services/environmental-indicators/air-quality.html (accessed April 20, 2021).

Canadian Government. 2021d. *Regulatory Framework for Air Emissions*. https://www.ec.ge.ca/doc/media_124/report_eng.pdf (accessed April 20, 2021).

Canadian Government. 2021e. *Overview of Wastewater Regulations in Canada*. http://canada.ca/en/environment-climate-change/services/wastewater/regulations.html (accessed April 20, 2021).

Canada Water. 2021. *Sustainable Stormwater Management in Canada*. https://www.watercanada.net/feature/sustainable-stormwater-management (accessed April 20, 2021).

Chile Ministry of the Environment. 2021a. *Overview and USEPA Cooperation*. http://protal.mma.gob.cl/ (accessed April 20, 2021).

Chile Ministry of the Environment. 2021b. *Chile Division of Air and Climate Change*. http://portal.mma.gob.cl.aire/ (accessed April 20, 2021).

Chile Ministry of the Environment. 2021c. *Chile Division of Solid and Hazardous Waste*. http://portal.mma.gob.cl.waste/ (accessed April 2021).

Diaz-Robles, S., Schiappacasse, L., and Cereceda-Balic, F. 2011. The Air Quality in Chile. *Journal of the Air and Waste Management Association*. Vol. 33. pp. 26–33.

Environment Canada. 2016. *Overview of Environmental Act, Regulations and Agreements*. https://www.ec.gc.ca/environmental-acts (accessed April 20, 2021).

Health Canada. 2017. *Guidelines for Canadian Drinking Water Quality. Ottawa, Canada. 18 pages*. https://www.healthcanada.gc.ca/waterquality (accessed April 20, 2021).

Loomis, D., Castillejos, M., Gold, D. R., McDonnell, W., and Borja-Aburto, V. H. 1999. Air Pollution And Infant Mortality in Mexico City. *Journal of Epidemiology*. Vol. 10. no. 2. pp. 118–123.

Macedo, A. A. and Sant'ana, L. P. 2015. Managing Emerging Hazardous Wastes in Brazil. *Unisanta Science and Technology*. ISSN 2317-1316. Unisanta Science and Technology. pp. 51–54.

Mage, D., Ozolins, G., Peterson, P., Webster, A., Orthofer, R., Vandeweerd, V., and Gwynne, M. 1996. Urban Air Pollution in Megacities of the World. *Journal of Atmospheric Environment*. Vol. 30. no. 5. pp. 681–686.

Marsh, W. B. and Kaufman, M. M. 2015. *Physical Geography*. University of Cambridge Press. Cambridge, United Kingdom. 647 p.

Marsh, W. B. and Grossa, M. 2012. *Environmental Geography*. John Wiley and Sons. New York. 426 p.

Mexico Environmental and Natural Resource Ministry (SAMARNAT). 1997. *Official National Commission on Water*. https://www.Mexico.nom-003-SEMARNAT-1997 (accessed April 20, 2021).

Mexico Environmental and Natural Resource Ministry (SAMARNAT). 2004. *Official Mexican Remediation Objective Concentrations for Heavy Metals*. https://www.Mexicana.nom-147-SEMARNAT/ss-2003 (accessed April 20, 2021).

Mexico Environmental and Natural Resource Ministry (SAMARNAT). 2005. *Official Mexican Remediation Objective Concentrations for Hydrocarbons*. https://www.Mexicana.nom-138-SEMARNAT/ss-2003 (accessed April 20, 2021).

Mexico Environmental and Natural Resource Ministry (SAMARNAT). 2015. *Official Mexican Remediation Objective Concentrations for Hydrocarbons*. https://www.Mexicana.nom-133-SEMARNAT/ss-2003 (accessed April 20, 2021).

Mexico Environmental and Natural Resource Ministry (SAMARNAT). 2016. *Official National Commission on Water*. https://www.Mexico.nom-003-SEMARNAT-2016 (accessed April 20, 2021).

Mexico Environmental and Natural Resource Ministry (SAMARNAT). 2017. Ambient Air Quality Standards for Criteria Pollutants. https://www.MexicoAirQualityStandards (accessed April 20, 2021).

Mexico Environmental and Natural Resource Ministry. 2021a. Mexico's Waste Law. https://www.gob.mx/semarnat (accessed April 20, 2021).

Mexico Environmental and Natural Resource Ministry. 2021b. Remediation of Contaminated Sites in Mexico. https://www.gob.mx/semarnat (accessed April 20, 2021).

Mexico National Water Commission (CONAGUA). 2021. *Water Use in Mexico*. https://www.gob.mx/CONAGUA (accessed April 20, 2021).

National Geographic Society. 2021. *Threats to the Amazon Rainforest*. https://www.nationalgeographic.com/environment/habitats/rainforest-threats (accessed April 20, 2021).

Organization for Economic Co-operation and Development (OECD). 2015. Summary Report of Mexico's Environmental Laws and Status. Paris, *France. 27 p.* https://www.OECD.org/Mexico/environmental (accessed April 20, 2021).

Peace Corps Peru. 2016. *Trash Talking: Solid Waste Management in Peru*. https://psu2.wordpress.com/2016/11/03/trash-talking-Peru (accessed April 20, 2021).

Peru Ministry of the Environment. 2021a. *Ambient Air Quality Standards*. http://www.temasactuales.com/assets/pdf/gratis/PERairqual.htm (accessed April 20, 2021).

Peru Ministry of the Environment. 2021b. *Solid Waste*. http://www.gob.pe/ambiento (accessed April 20, 2021).

Peru Ministry of Housing, Construction, and Sanitation. 2021. http://www.gob.pe/vivienda (accessed April 20, 2021).

Province of Alberta Canada. 2021. *Environmental Laws and Regulations*. https://www.environment.gov.ab.ca/legislation (accessed April 20, 2021).

Province of British Columbia Canada. 2021. *Environmental Laws and Regulations*. https://www.environment.gov.bc.ca/legislation (accessed April 20, 2021).

Province of Manitoba Canada. 2021. *Environmental Laws and Regulations*. https://www.environment.gov.mb.ca/legislation (accessed April 20, 2021).

Province of Newfoundland and Labrador Canada. 2021. *Environmental Laws and Regulations*. https://www.environment.gov.nf.ca/legislation (accessed April 20, 2021).

Province of New Brunswick. 2021. *Environmental Laws and Regulations.* https://www2.gnb.ca (accessed April 20, 2021).

Province Northwest Territories Canada. 2021. *Environmental Laws and Regulations.* https://www.oag-bvg.ca (accessed April 20, 2021).

Province of Nova Scotia Canada. 2021. *Environmental Laws and Regulations.* https://www.environment.gov. ns.ca/legislation (accessed April 20, 2021).

Province of Ontario Canada. Ministry of Environment, Conservation and Parks. 2011. https://www.ontario. ca/page/soil-ground-water-and-sediment-standards-use-under-part-xv1-environmental-protection-act (accessed April 20, 2021).

Province of Ontario Canada. 2021. *Environmental Laws and Regulations.* https://www.ontario.ca/ environmental-laws (accessed April 20, 2021).

Province of Quebec Canada. 2021. *Environmental Laws and Regulations.* https://legisquebec.gouv.qc.ca (accessed April 20, 2021).

Province of Saskatchewan Canada. 2021. *Environmental Laws and Regulations.* https://www,saskatchewan. ca/legislation/environmental-protection (accessed April 20, 2021).

Province Yukon Territory. 2021. *Environmental Laws and Regulations.* https://www.env.gov.yk.ca/ environment (accessed April 20, 2021).

Rodriguez, A., Castrejon-Godinez, M. L., Ortiz-Hernandez, M. L., and Sanchez-Salinas, E. 2015. Management of Solid Waste in Mexico. Fifteenth International Waste Management and Landfill Symposium. CISA Publisher. Cagliari, Italy. 7 p.

United Nations. 2005. *Republic of Peru: Country Profile.* Department of Economic and Social Affairs: New York. 18 p.

United Nations. 2013. *Climate Change Impacts on Chile Water Systems.* https://www.oecd.org/env/resources/ Chile.pdf (accessed April 20, 2021).

United Nations. 2015. *Environmental Performance Review of Chile.* http://www.oecd.org/environmental/ country-review/chile.htm (accessed April 20, 2021).

United Nations. 2021a. World Population Prospects. United Nations Department of Economic and Social Affairs. *Population Division.* https://esa.un.org/unpd/wpp/data (accessed April 20, 2021).

United Nations. 2021b. World Air Pollution Status. United Nations. News Center. New York. https://www. un.org/sustainabledevelopment/2016/09 (accessed April 20, 2021).

United Nations. 2021c. *Country Profile for Argentina.* http://www.un.org/esa/earthsummit/AR.htm (accessed April 20, 2021).

United Nations. 2021d. *Country Profile for Brazil.* http://www.un.org/esa/earthsummit/brzl-cp.htm (accessed April 20, 2021).

United Nations. 2021e. *Country Profile for Chile.* http://www.un.org/esa/earthsummit.ch-cp.htm (accessed April 20, 2021).

United States Department of State. 2021a. *Office of the United States Trade Representative.* https://ustr.gov/ nafta (accessed April 20, 2021).

United States Department of State. 2021b. *United States Department of State Air Quality Monitoring Program.* http://www.stateair.net/web/post/1/5.html. (accessed April 20, 2021).

United States Environmental Protection Agency (USEPA). 2017. *EPA Collaboration with Chile.* https:// www.epa.gov.international-cooperation/epa-collaboration-chile (accessed April 20, 2021).

United States Environmental Protection Agency (USEPA). 2021a. *EPA Collaboration with Brazil.* https:// www.epa.gov.nternational-cooporation-Brazil (accessed April 20, 2021).

United States Environmental Protection Agency (USEPA). 2021b. *EPA Collaboration with Peru.* https:// www.epa.gov.nternational-cooporation-Peru. (accessed April 20, 2021).

United States Geological Survey (USGS). 2021a. *Examples of Land Subsidence in Mexico City.* https://www. USGS.gov/edu/gallery/landsubsidence-learning.html (accessed April 20, 2021).

United States Geological Survey (USGS). 2021b. Earthquakes. https://earthquakes.usgs.gov/earthquakes/map (accessed October 21, 2018).

United States Library of Congress. 2010. *Chile: New Law Creates Ministry for the Environment.* http://www. loc.gov/law/foreign-news/article/chile-new-law (accessed April 20, 2021).

World Bank. 2000. Argentina: Water Resources Management, Policy Issues and Notes: Thematic Annexes. Volume III. World Bank. Latin America and the Caribbean Office. 190 p.

World Bank. 2013. Matanza-Riachuelo River Cleanup, *Argentina. Blacksmith Institute.* http://worldbank.org/ projects/argentine/matanza-riachuela-river (accessed April 20, 2021).

World Bank. 2015. *Economy Profile for Chile.* https://data.worldbank/country/chile (accessed April 20, 2021).

World Bank. 2021a. *The World Bank in Argentina.* https://www.worldbank.org/en.country/argentina (accessed April 20, 2021).

World Bank. 2021b. *Significant Advances in the Recovery of the Matanza-Riachuelo River Basin.* World Bank: Argentina. 6 p.

World Bank. 2021c. *Argentina: Policy for Sustainable Development in the 21st Century.* http://openknowledge.wolrdbank.org/handle/10986/14980 (accessed April 20, 2021).

World Bank. 2021d. *Global Economic Prospects: The Turning of the Tide? World Bank Corp.* Washington, D.C. 184 p.

World Bank. 2021e. *Overview of World Bank in Peru.* https://www.worldbank.org/en/country/peru/overview (accessed April 20, 2021).

World Health Organization. 2013. *Water Quality and Health Strategy 2013–2020.* http://www.who.int/water_sanitation_health/dwq/en/ (accessed April 20, 2021).

World Health Organization. 2015. *Progress on Drinking Water and Sanitation.* World Health Organization. New York. 90 p.

World Health Organization. 2021a. *World Health Organization Global Urban Ambient Air Pollution Database.* http://www.who.int/phe/health_topics/outdoorair/database.htm (accessed April 20, 2021).

World Health Organization. 2021b. *Peru: Country Profile.* http://who.int/gho/countries/per/country_profiles/per.htm (accessed April 20, 2021).

World Population Review. 2021. *Population of Mexico City.* http://worldpopulationreview.com/MexicoCity/ (accessed April 20, 2021).

Yip, M. and Madl, P. 2002. Air Pollution in Mexico City. Journal of Biophysics. Salsburg, Austria. http://www.biophysics.sbg.ac.al/mexico/air.htm (accessed April 20, 2021).

13 Summary of Environmental Regulations of Europe and Africa

13.1 INTRODUCTION

This chapter will explore the environmental regulations and pollution of Europe and Africa and is organized by continent and then by country. Overall, the total countries that will be evaluated account for 70% of the human population and 75% of the land area on Earth.

We will focus our discussion of each country on the fundamentals introduced in Chapter 1, which include:

- Protect the air
- Protect the water
- Protect the land
- Protect living organisms and cultural and historic Sites

To address each of the listed fundamentals, we will examine of how each country addresses air pollution, water pollution, solid and hazardous waste, and remediation of polluted sites. We will examine how most countries address legacy pollution issues through a discussion of the framework that each country has developed to investigate and remediate historical sites of environmental contamination. This means that we must also examine the significant historical sites of contamination in each country because the experiences and lessons learned from these sites have influenced and uniquely shaped each country's choices in developing environmental regulations perhaps more than most other factors.

13.2 EUROPE

We will discuss the environmental laws and regulations of Europe first through the 28 member European Union (EU) and then move on to other European countries that are not members of the EU but need to be covered. These other countries include Russia, Norway, Switzerland, and Turkey. We will discuss the United Kingdom (UK) as part of the EU since the UK is still operating environmentally as if it were still an EU member.

13.2.1 EUROPEAN UNION COUNTRIES ENVIRONMENTAL REGULATORY STRUCTURE AND APPROACH

The EU is a collection of 28 countries but soon to be 27 with the recent planned exit of the UK. The EU represents 4,324,773 square kilometers in area, the 7th largest in the World with a population of 507,416,607 residents as of a 2014 census (European Union 2021). The EU is considered by many to have the most extensive and strictest environmental laws of any international organization. The EU's environmental legislation addressed issues such as (European Environment Agency [EEA] 2021a):

- Acid Rain
- Thinning of the Ozone Layer
- Air Quality
- Noise Pollution
- Solid and Hazardous Waste
- Water Pollution
- Sustainable Energy

The Institute for European Environmental Policy (2021) estimated that the body of EU Environmental laws total over 500 Directives, Regulations, and Decisions.

The 1972 Paris Summit meeting of heads of state and government of the European Economic Community (EEC) is often used to pinpoint the beginning of the EU's environmental policy making and declarations (EEA 2021a).

Countries of the EU include (EEA 2021a):

• Austria	• Belgium	• Bulgaria	• Croatia
• Cyprus	• Czech Republic	• Denmark	• Estonia
• Finland	• France	• Germany	• France
• Hungary	• Ireland	• Italy	• Latvia
• Lithuania	• Luxembourg	• Malta	• Netherlands
• Poland	• Portugal	• Romania	• Slovakia
• Slovenia	• Spain	• Sweden	• United Kingdom

The EU is composed of two major institutional groups. The first group is composed of the following (EEA 2021a):

- European Commission, which initiates and implements EU law and represents the driving force and is the executive body
- Council for the EU, which provides legislative approval to laws from the Commission and represents the governments of each of the member states
- Court of Justice, which ensures compliance with the law and serves as an appellate judicial body when questions of EU law arise
- Court of Auditors, which controls the EU budget

The second group consists of the following (EEA 2021a):

- European Economic and Social Committee, which expresses the opinions of organized civil society on economic and social issues
- Committee of the Regions, which represents the local and regional authorities in Europe
- European Central Bank, which is responsible for monetary policy and managing the currency of the EU, the Euro
- European Ombudsman, which investigates EU institutions and bodies
- European Investment Bank, which assists in EU objectives by financing government projects

The EU has three forms of binding legislation as part of its institutional framework that includes (EEA 2021a):

- Directives. Directives are binding on the member states to which they are addressed regarding the results to be achieved. The member states, however, may choose how to bring about those results. A directive does not take effect until a member state passes national legislation to implement the provisions of European law, which it must do within two years of the date of passage of any specific directive. A directive normally enters into force on the date specified in the directive or on the 20th day after publication in the EU's Official Journal. Directives are the most frequently used EU laws.
- Regulations. Regulations are binding in their entirety and apply directly to all member states, similar to national laws. They are stronger and much less common than directives. In addition, regulations go into effect in the member states immediately, without national implementing legislation. Approximately 10% of EU laws are regulations. They supersede any conflicting national laws and are not allowed to be transposed into national law, even if the law is identical to the regulation. Regulations are binding the day they come into force, which like directives, is either specified in the document itself or is the 20th day after publication in the EU Official Journal.
- Decisions. Decisions are binding in the entirety on those to whom they are addressed. Decisions are individual legislative acts that differ from directives or regulations in that they are specific in nature and are used to specify detailed administrative requirements or update technical aspects of regulations or directives. They may be addressed to a certain government, an enterprise, or to even an individual.

Other forms of legislation include recommendations and opinions, which are not binding but are expected to be taken into account when decisions are made. In addition, resolutions are nonbinding statements by which the Council of Ministers expresses a political commitment to a specific objective. Often what happens is that after some time passes, usually a few years, a resolution gives way to either a directive or a regulation (EEA 2021a).

The European Commission, which is the executive of the EU consists of one commissioner from each member state and is headed by a president who serves a five-year term. Commission members are fully independent and are not permitted to take instructions from the government with which they originate (EEA 2021a).

The responsibilities of the European Commission include (EEA 2021a):

- Guardianship of treaties and the initiation of infringement proceedings against member states and others who disobey the treaties and other community law
- Negotiation of international agreements
- Adoption of technical measures to implement legislation adopted by the Council
- Matters relating to the environment, education, health, consumer affairs, the development of trans-European networks (TENS), research and development of policy, culture, and economic and monetary union
- Preparing the groundwork for incorporating former Soviet bloc countries into the EU, identifying areas where these groups can align their policies with those of the UE, and a procedure of implementation
- To initiate legislation only in areas the EU is better placed than individual member states to take effective action, thus taking the principle of "subsidiarity" into account
- Taking action when necessary against those who do not respect their treaty obligations by perhaps bring them before the European Court of Justice when compliance is not voluntary.
- Managing policies and negotiating international trade and cooperation agreements
- Acting as a mediator between conflicting interests of member states
- Sustaining, managing, and developing agricultural and regional development policies

- Developing cooperation with other European nations not members of the EU and countries in Africa, Caribbean, and the Pacific rim
- Promoting research and technological developmental programs

The EEA plays an important role in EU environmental policy by providing the Commission with information used in setting EU environmental policy. The EEA aims to support sustainable development which helps to improve Europe's environment through the provision of timely, targeted, relevant, and reliable information to policy making agents and the public. The EEA provides research and clarifies Europe's environmental issues and challenges and employs a whole range of tools to assist policymakers and the public to address environmental quality and sustainable development issues. In effect, the EEA, is separate from other institutions, and is charged with providing objective information and analysis. Other duties of the EEA is to provide objective information and analysis including a catalog of data sources and projects and prototype called the European Information and Observation Network (EINETICS), which is a computerized network for environmental data (EEA 2021a).

Current EEA members include the 28 member nations plus Bulgaria, Romania, Turkey, Iceland, Norway, and Liechtenstein. A membership agreement has been entered into with Switzerland, Albania, Bosnia and Herzegovina, Croatia, former Yugoslav Republic of Macedonia, and Serbia and Montenegro have all applied for EEA membership. EEA's priorities include (EEA 2021a):

• Air Quality	• The States of the soil
• Flora and Fauna	• Biotopes
• Land use	• Natural resources
• Waste management	• Noise emissions
• Coastal protections	• Hazardous chemicals

The need for EU environmental protection has grown through time as industrial and population growth has occurred. The Single European Act of 1987 provided the first expressed legal basis for EU environmental policy and is outlined in four basic principles that include (EEA 2021a):

- Applying the principle of "polluters pay," which is similar to the policy adopted in the United States under the passage of CERCLA in 1980
- Pollution prevention, which is similar to the Pollution Prevention Act of the United States enacted three years later
- Rectify environmental degradation at the source, when possible
- Integration of environmental protection into other EU policies

13.2.1.1 Air Pollution

In Europe, emissions of many air pollutants have decreased substantially over the past few decades. However, because of its dense population and number of mobile and stationary sources of air pollution, concentrations of many air pollutants are still too high and poor air quality persists. A significant proportion of Europe's population live in cities where the poor air quality is generally located and includes ozone, nitrogen oxides, and particulate matter. According to the EEA, approximately 90% of Europe's city inhabitants are exposed to air quality levels considered harmful to health (EEA 2021b).

Air emissions management in the EU has recently focused on greenhouse gas emissions from both mobile and stationary source, much like what is presently conducted in the United States but a bit more strict within the EU. The EU regulates greenhouse gases from mobile sources such as automobiles, diesel trucks, agricultural machinery, motorcycles, and stationary sources industry, residential and household and other sources of greenhouse gases such as tools, aircraft, and sea-going ships (EEA 2021b).

For air emissions, the EU also regulates the equivalent of Criteria Pollutants in the United States which include, ozone, sulfur dioxide, nitrogen oxides, particulate matter, carbon monoxide, and lead. European legislation on air quality is built on certain principles. Each member State is divided into zones and agglomerations. The member State is to then conduct a study within each zone or agglomeration using measurements, modeling, and other empirical techniques. Where air quality levels are exceeded or elevated, the member State is then to prepare a plan or program to ensure compliance in the future. In addition, air quality data is to be shared with the general public (EEA 2021c). The EEA regulates air much the same way as USEPA regulates air in the United States. For example, EEA regulates the same air contaminants and requires zones or agglomerations that exceed the standards and objectives to prepare a plan for obtaining attainment, which is similar to the United States where USEPA requires state regulatory agencies to prepare State Implementation Plans (SIPs) in areas of non-attainment (EEA 2021b). EEA air pollutant standards are listed at https://ec.europa/environment/air/quality/standards.htm. (EEU 2021c).

In a 2009 Directive, the EU established an Emission Trading System (ETS) as the principle method to combat climate change. It's a key instrument for delivering Europe's portion of reduction commitments under the Paris agreement of the United Nations Framework Convention on Climate Change. Since initial formation in 2005, it has evolved to become a key tool for the cost-effective reduction of industrial GHG emissions by giving companies the flexibility to make investments where they deliver the biggest gains (EEA 2021b).

The ETS works on the "cap and trade" principle on the total amount of certain GHG's, such as carbon dioxide, nitrous oxide, and perfluorinated compounds. An estimated 11,000 facilities are subject to the ETS, which include mainly heavy, energy-using sites in the power generation sector and some large manufacturing operations, and some aircraft operators. The EU ETS applies to 31 countries, those of the 28 member states and Iceland, Liechtenstein, and Norway. The overall volume of GHGs that can be emitted each year by facilities subject to the directive is set at the EU cap for the EU set at the Convention on Climate Change and the cap is reduced each year. In addition, facilities are also required to have an approved monitoring plan for monitoring and reporting annual emissions. Currently, this program cover approximately 45% of GHG emission estimated for the EU (EEA 2021b).

For industrial facilities, air permitting in the EU takes a holistic approach that differs from the United States. Air permits issued to industrial facilities are usually issued under the Industrial Emissions Directive (IED) and include conditions to prevent and control all environmental impacts from a facility taking into account its full environmental performance which are (EEA 2021d):

• Air emissions	• Water use
• Water discharge	• Land use
• Generation of solid and hazardous waste	• Use of raw materials
• Energy use and efficiency	• Noise prevention
• Accident rate and prevention	• Odors
• Site closure	• Site restoration

The EU is of the opinion that this integrated air permitting approach is best for the environment because it takes into account the whole environment and not a portion and is therefore viewed as more effective, protective, and efficient. The primary responsibility for implementation of the IED rests with each Member State. In turn, these governments assign the responsibility to a department or agency within the respective country to deliver the practical implementation activities of the directive. As an example, the UK has delegated the responsibility of implementation to the Environment Agencies within the four components parts of the country, that being, England, Wales, Scotland, and Ireland. These four agencies are empowered by legislation that transposes the IED into national law (EEA 2021d).

13.2.1.2   Water

The EU regulates water as drinking water and as all other water, that being surface water, groundwater, coastal and marine waters under the following Directives:

- Drinking Water Directive
- Water Framework Directive
- Marine Strategy Directive
- Birds and Habitats Directives
- Floods Directive

The Drinking Water Directive had its beginning in 1975 and then was culminated in 1980 in setting binding water quality targets for drinking water. The EEA drinking water quality standards are listed at https://ec.europa/drinkingwater. (EEA 2021e).

The Water Framework Directive was first developed and passed in 1975 and culminated into its final form in 1980 along with the Drinking Water Directive. The Water Framework Directive established binding targets for surface water including lakes, streams and rivers. It also established standards for fish waters, shellfish waters, bathing waters, and groundwater. At first, the priority was to target those surface water bodies that were sources of drinking water (EEA 2021f). An innovative method employed by the EEA was the concept of water management through a watershed or river basin approach. This concept has been especially challenging in the past because many river basins within the EU cross international borders, such as the Rhine River. However, from a scientific and even a practical point of view, regulating a river basin is much more effective because is it a single geographical and hydrogeological unit (EEA 2021f).

The EU also has what are termed "Additional Monitoring Parameters" that are equivalent in the United States to Secondary Drinking Water Standards. These limits are listed at https://ec.europa/drinkingwater. (EEA 2021e).

In the United States, environmental protection on a watershed wide basis is perhaps not a new concept but one that is not widely used for environmental protection. The Delaware River watershed is within 6 different states and was protected and regulated under an executive order signed by John F Kennedy during his presidency in the early 1060s (Delaware River Watershed Initiative 2021). The Colorado watershed has also been regulated, predominantly by the Bureau of Land Management within the Department of Interior, for its hydroelectric capacity in the Hoover Dam and Glen Canyon Dam and for its water diversion practices through canals and aqueduct for agricultural purposes and potable water sources, namely Las Vegas (USGS 2021). The watershed or river basin approach includes two evaluation criteria in which each river basin is evaluated, one is the general requirement for ecological protection and the other is minimum chemical standards. Within these two criteria falls protection of aquatic ecology, protection of unique and valuable habitats, protection of drinking water resources, and protection of bathing water.

A good ecological protection standard is evaluated not through a rigorous standard but on a relative scale because ecological conditions very widely within the EU. However, a good chemical protection status is defined in terms of compliance with all the quality standards established for chemical substances at the European level. These standards ensure that at least a minimum chemical quality, especially with respect to very toxic substances. Under the directive, the EEA handles groundwater differently. The EEA takes the approach for groundwater that there should not be any impacts to groundwater at all, except from naturally occurring, undisturbed sources. For this reason, the EEA has not set chemical standards for groundwater, because according to EEA, setting chemical standards then establishes the concept that there are allowed levels of pollution for groundwater (EEA 2021f and EEA 2021g).

However, a few standards have been established for groundwater and include nitrates, pesticides, and biocides, which generally originated from agricultural use and not from industry. As an additional level protection of groundwater resources, the EEA has required that there will be no direct discharges of any wastewater to groundwater (EEA 2021g). This is a different approach to that within the United States, even California. For instance, Class V injection wells are allowed in California which are essentially what is termed elsewhere in the United States as "Dry Wells." In addition, many western states have what are termed "infiltration basins" where stormwater runoff is collected and is then allowed to migrate vertically to groundwater to recharge groundwater without treatment.

Europe treats groundwater as a hidden resource which is quantitatively much more significant than surface water because, in part, it represents 97% of fresh water on Earth that is not ice (EEA 2021g). In addition, groundwater recharges many surface water sources and it mostly responsible for base flow in many river systems. Groundwater migrates or moves much more slowly than surface water, which then means that it is more susceptible to becoming contaminated from anthropogenic activities because of the longer migration time frame and hence more opportunity to be environmentally degraded. What this means in many circumstances is that pollution released many decades ago may still threaten groundwater supplies not only currently but for perhaps centuries in the future.

13.2.1.3   Solid and Hazardous Waste

Europe is densely populated and nowhere more that this becomes evident than in the EEA's approach to solid and hazardous waste. The EEA's waste management approach is heavily reliant on reuse and recycling (EEA 2021h). Over the last 20 years, European countries have shifted focus from disposal methods to prevention and recycling, and also examining methods to reduce packaging waste. The EEA has set a target of recycling at 65% of all waste by 2030 (EEA 2021h).

The EEA (2021i) regulates generators of hazardous waste through a "cradle to grave" method much like the United States. The classification of wastes into hazardous or non-hazardous waste is based on methods of generation and characteristics, much like the United States, that include whether the waste material is explosive, flammable, corrosive, toxic, infectious, a gas, oxidizer, or radioactive (EEA 2021i). Hazardous waste limits are listed at http://ec.eurpoa.eu/environment/waste/hazardous_index.html. (EEA 2021i).

13.2.1.4   Remediation

In the EU, land pollution has affected an estimated 2.5 to perhaps as high as 3.5 million sites (EEA 2021j). Cleaning up these sites will take decades and cost on average from 0.5% to 2% of GDP per year as estimated by the EEA (2021h). Therefore, Europe seems to have similar challenges with contaminated land as the United States. In the EU, contaminated land is treated as two groupings, land and groundwater. The EEA lists cleanup criteria for many chemical compounds as a target levels and intervention values. These values are commonly referred to as Dutch Intervention Values (DIV) (Netherlands Ministry of Infrastructure and the Environment 2015).

The soil intervention values indicate when the functional properties of the soil for humans, plants, and animals is considered seriously impaired or threatened. A concentration of a single chemical compound at or exceeding its respective intervention value is considered a serious case of soil or groundwater contamination if the affected volume exceeds 25 cubic meters of soil or 100 cubic meters of groundwater (Netherlands Ministry of Infrastructure and the Environment 2015). A comparison of the DIV's to cleanup levels in the United States indicated that in some instances the cleanup criterion for individual chemical compounds is lower in the EU than in the United States (i.e., vinyl chloride) (Netherlands Ministery of Infrastructure and the Environment 2015).

13.2.1.5 Summary of Environmental Regulations of the European Union

Environmental regulations of the European Union are considered by many to be the most comprehensive set of environmental regulation currently known to the World. In what started at the Paris Summit in 1972 in establishing the EU's environmental policy and basing its regulatory framework upon that of the United States Environmental Protection Agency, it has now become the system that sets the example. No more is this evident than in the field of sustainability where we shall see examples of the EU's commitment to environmental stewardship.

13.2.2 Russia

Russia is the largest country in the world encompassing much of eastern Europe and the whole northern portion of Asia. The total land area of Russia is 17,098,246 square kilometers (6,601,670 square miles). Due to its size and geographic distribution, Russia has many different climate zones that include tundra, coniferous forest, mixed and broadleaf forest, grassland, and semi-desert (Blinnikiv 2011). Russia occupies 11% of the world's land surface and stretches through 11 time zones and has 40 UNESCO biosphere reserves. The population of Russia as of 2016, was estimated at 143.4 million. Interestingly, the population of Russia in 1991 was estimated to be 148.5 million (United Nations 2021a).

According to the United Nations (2021b), the population of Russia has stabilized and has started to increase since 2015. Russia's largest city by population is its capital, Moscow with a population 11.5 million, followed by Saint Petersburg with a population of 4.8 million (United Nations 2021b). For centuries, Russia has built its economy on its vast and rich natural resources that include large regions of wilderness and timber, and mineral and energy resources and fresh water (Newell and Henry 2017).

The discussion of the environmental regulations of Russia will be more in depth than most other countries. This is because Russia is a very large country with enormous natural resources, particularly in oil and gas, forestry products, and minerals (much of which have not been fully developed), and significant biodiversity. How Russia decides to implement and enforce its environmental regulations will likely have either future positive or negative implications for the rest of the planet.

13.2.2.1 Environmental Regulatory Overview and History of Russia

Russia's relationship with the environment has historically not been considered good with many troubling concerns that have been a carry-over from its Soviet days. From the aftermath of the Chernobyl nuclear disaster (in what is the Ukraine today), the drying up of the Aral Sea, unchecked industrial emissions and discharges, and its dependence on mining, oil, and gas, Russia has generated plenty of attention from environmental organizations world-wide (Brain 2016).

According to the OECD (2021), environmental policies and regulations in Russia continue to suffer from an implementation gap that continues to this day. In addition, many regulatory agencies have undergone multiple reorganizations and continue to be fragmented that have sometimes brought many agencies to the brink of institutional paralysis.

Environmental policy in Russia strives to achieve a balance between protecting the natural world and economic development. Although this objective may sound pleasing, it has been difficult to achieve in practice or actually begin especially since these ideologies are often times in direct conflict. There are numerous environmental issues in Russia. Many of these issues have been attributed to policies of the former Soviet Union, which was a period of time when officials within the government felt that pollution control was an unnecessary hindrance to economic development and industrialization (OECD 2021). As a result, 40% of Russia's territory began experiencing symptoms of significant ecological stress by the 1990s that have been traced in large part back to a number of environmental issues that include (National Intelligence Council 2012):

- Deforestation
- Energy irresponsibility
- Unchecked disposal of hazardous wastes
- Surface mining
- Improper nuclear waste disposal

For most of the Soviet period, the task of environmental protection was fragmented and shared by more than 15 ministries, each of which was responsible for a particular economic sector but many environmental issues had overlap with other sectors that resulted in confusion over who was responsible and who had the actual power to affect change and ultimately begin to protect human health and the environment (Henry and Douhovnikoff 2008).

The Chernobyl nuclear accident in 1986 and Mikhail Gorbachev's glasnost, or policy of openness, were catalysts for change in Russia and opened the doors for environmentalism in Russia for the first time. In addition, it also gave environmental watch groups and the international community some insight and data on just how large the disregard for the environment had been during the days of the Soviet Union (Henry and Douhovnikoff 2008).

In 1988, the Russian State Committee on Environmental Protection was established and gave the Committee the authority to conduct environmental reviews for all new government projects (Russian Federal Law on Environmental Protection 2002). Environmental protection gained further popularity and importance with the collapse of the Soviet Union. One of the first laws passed by the newly formed Russian Federation was the 1991 Federal Act on the Protection of the Natural Environment. Russia also announced that it had committed itself to the principle of sustainable development in the early 1990s (Henry and Douhovnikoff 2008).

Along the lines of openness, it was announced in 1993 through a commission on the environment chaired by Aleksei Yablokov, who at the time was President Yeltsin's advisor on the environment, that the Soviet Union had disposed of 2.5 million curies of radioactive waste in the Sea of Japan starting in 1965 (Henry and Douhovnikoff 2008).

From 1991 to 2000, environmental protection officials were largely ineffective because of difficult conditions due to lack of resources, bureaucratic infighting, and lack of real authority to effect change. Intense lobbying by industrial groups eroded environmental protection over time. Other significant issues including widespread corruption, lack of funding, constant reorganization, lack of clear legal direction and laws, and pressure for economic development. Confronted with these obstacles, it was no surprise that the environmental movement in the newly formed Russian Federation failed. Since 1995, the little power that the environment authorities had gradually disappeared. In 1996, President Yeltsin reduced that status of the Ministry of Environment to Committee status and in 2000 the new president Vladimir Putin dissolved the Committee of Environment and the Federal Forest Service and passed on the responsibility to the Ministry of Natural Resources (MNR). The motivation behind this action was to encourage exploitation of Russia's natural resources in order to ignite economic development due to the Russian financial crisis of 1998 (OECD 2021 and Henry and Douhovnikoff 2008).

In Russia, environmental management relies on a set of command and control, economic, and information instruments. Russia began setting hygiene standards in 1922, at the very start of the Soviet era. Most of the environmental quality standards in Russia are a carryover from the Soviet Union. Since environmental protection in Russia was passed to the MNR in 2000 by presidential decry, the MNR operates as the executive authority and is responsible for the following (Newell and Henry 2017):

- Public policy and statutory regulation for the study, use, renewal and conservation of natural resources including hunting, hydrometeorology, environmental monitoring, and pollution control.
- Public environmental policy and statutory regulation as it pertains to production and waste management, conservation and state environmental assessments.
- Ensuring compliance with international treaties on environmental matters.
- Working with other federal agencies, states, and local governmental authorities on environmental matters

Russia is a signatory to the following select international conventions (OECD 2021):

- 1972 convention on the International Regulations for Preventing Collisions at Sea
- 1972 Convention on the Prevention of Marine Pollution by Dumping of Wastes and Other Matter
- 1973 Convention to Regulate International Trade in Endangered Species of Wild Flora and Fauna
- 1974 International Convention for the Safety of Life at Sea
- 1978 Convention for the Prevention of Pollution at Sea
- 1979 United Nations Convention on Long-range Trans-boundary Air Pollution
- 1982 United Nations Convention on the Law of the Sea
- 1985 Vienna Convention for the Protection of the Ozone Layer
- 1987 Montreal Protocol on Substances that Deplete the Ozone Layer
- 1989 Basel Convention on the Trans-boundary Movements of Hazardous Wastes and disposal
- 1992 Convention on the Protection of the Black Sea from Pollution
- 1992 Convention on Biological Diversity
- 1992 United Nations Framework Convention on Climate Change
- 1992 Convention on the Protection and Use of Trans-boundary Watercourses and International Lakes
- 1997 Kyoto Protocol on Climate Change
- 2001 Stockholm Convention on Persistent Organic Pollutants

Facilities subject to federal environmental supervision are those facilities that exceed one or more of the following (OECD 2021):

- A waste disposal site that accepts 10,000 tons of hazardous waste or more
- A facility that discharges 15 million cubic meters of wastewater per year into a water body
- A facility that emits 500 tons a year or more of polluting substances into the ambient air

Environmental laws in Russia cover 8 main topics that include (OECD 2021):

- Air emission management
- Chemical management
- Water management
- Waste management
- Safety management
- Environmental Impact Assessment
- Energy Management
- Dangerous Goods – Hazardous Materials Management

Each of the given eight main topics are similar to the United States and other countries of the world.

13.2.2.2   Air

Responsibility for issuing air emission permits in Russia is divided between two agencies. The Federal Environmental, Industrial and Nuclear Supervision Service (Rostechnadzor) is responsible for the development and approval of the procedure and methods for determining air emission limits, and issuing air emission permits for radioactive sources. The Federal Service for Supervision of Natural Resources (Rosprirodnadzor) is responsible for the development and approval of the procedure and methods for determining air emission limits from non-radioactive sources. Federal standards require that Rostechnadzor and Rosprirodnadzor must include the quantity of each polluting substance to be emitted from each source (Russian Federal Law on Environmental Protection 2002).

Rostechnadzor and Rosprirodnadzor officials have the right to enter all facilities where any air pollution sources are present to evaluate compliance with emission permits and review any relevant documents including testing and laboratory results, information logs or other related material. Rostechnadzor officials also have the right to issue violation notices and fines for sources that are not in compliance with permit terms. In 2009, the Ministry of Natural Resources and Environment set out a plan to develop three orders aimed at completing the regulatory framework for air protection that remain in the process of being developed and include (Russian Ministry of Natural Resources 2021):

1. Developing procedures and methods for determining standards for maximum permissible and temporarily agreed emissions in ambient air.
2. Creating a list of pollutants that are subject regulation
3. Providing reliable sources of information for the management of air protection

After the regulatory framework has been completed the Ministry is expected to begin work on calculating environmental damage caused by air pollution and creating an inventory of the sources of air pollution that have a measurable negative impact on air quality.

13.2.2.3   Water

Water pollution is one of Russia most significant environmental issues (World Water 2021 and Russian Ministry of Natural Resources 2021). The most significant include the following:

• A significant portion of the Russian population do not have access to safe drinking water
• Evidence of oil pollution is visible in most urban areas
• An estimated 40% of surface water sources do not meet basic environmental standards

The Federal Agency for Water Resources is responsible for enforcing laws related to water resources and is under the jurisdiction of the Ministry of Natural Resources of the Russian Federation. While water pollution from industrial sources has diminished because of a decline in manufacturing, municipal wastes and nuclear contamination are constant threats to important water supply sources in Russia (Russian Ministry of Natural Resources 2021).

Evidently, the practice of discharge of untreated wastewater and sewerage into rivers and the ocean in Russia is relatively common (United Nations 2021c). It is estimated that more than 70% of all the wastewater from municipal sources is untreated. The issue of lack of wastewater treatment from mining operations is especially acute (Yablokov 2016).

In Russia, any industrial facility that discharges wastewater is required to obtain a permit. Wastewater discharge permits are issued by Rostechnadzor for a fee. The fees for wastewater discharge are reduced for those facilities that treat the wastewater before it's discharged into a water body (Russian Ministry of Natural Resources 2021).

Facilities accepting, transporting, and treating wastewater using a centralized water supply and discharge systems are required to monitor the quality and properties of the wastewater discharged. There are also discharge limits established in individual permits that are issued for polluting substances and microorganisms (Russian Ministry of Natural Resources 2021). This is similar to the EU and NPDES permits in the United States. Specific limits on wastewater discharge are established locally on a case by case basis.

Under a Russian Presidential Decree (No. 177, 2003), and Article 63 of the Federal Law on Environmental Protection, Russia implemented a unified environmental monitoring system that includes monitoring of water bodies, internal sea waters and the Baikal lake system, along with wild animal resources, state of the subsoil resources, air quality, soil quality etc. Monitoring is organized and performed by the Federal Water Resources Agency with the purpose of providing data on the improvement of water quality (Russian Ministry of Natural Resources 2021). This is similar to research and monitoring conducted in the United States by USEPA and USGS and other United States Federal and International Organizations and even State agencies.

The main specific priorities under the program include (Russian Ministry of Natural Resources 2021):

- To provide water resources needed by the population and the economy
- To prevent and eliminate floods and other harmful effects to water
- To ensure the security of hydro-electrical installations
- To protect bodies of water from pollution
- To develop a system for monitoring and forecasting the state of water resources and to provide information in water resources
- To improve the system of state management and conservation of water resources
- To provide a regulatory and scientific and technical basis for the water supply

13.2.2.4   Solid and Hazardous Waste

In Russia, the law that addresses waste management was enacted in 1998 and is termed Industrial and Consumption Waste is Russian Federal Law No. 89-FZ, 1998 (Russian Ministry of Natural Resources 2019). Solid waste generation has increased significantly as the Russian Federation became an independent nation and as many Russian residents adopted a Western-style consumption style. The vast majority of solid waste in Russia is disposed in landfills with a small percentage being incinerated.

Russia has begun to emphasis pollution prevention initiatives through reducing and minimizing the amounts of waste, particularly industrial waste during the production process. Many facilities are required to employ best management practices to reduce the amount of industrial wastes generated. As in the United States, Russia requires each facility to keep records on the generation and disposal of industrial wastes. In addition, Russia requires industrial facilities to obtain a permit to generate hazardous wastes and sets limits on the volumes of wastes that can be generated and sets analytical limits on wastes similar to that of the United States and European Union (Russian Federal Law on Environmental Protection 2002 and Russian Ministry of Natural Resources 2021). The definition of hazardous waste in Russia is similar to the United States and the rest of the developed world and is defined as waste containing harmful substances with hazardous properties making it flammable, toxic, explosive, or reactive and also includes substances containing potentially infectious agents or those representing a direct or indirect danger to human health or the environment either alone or in combination with other substances.

Solid waste in Russia is also classified into five environmental hazard classes according to Russian Federal Law 13-FZ, 2002 (Russian Ministry of Natural Resources 2021):

- Class I – Very hazardous
- Class II – Highly hazardous
- Class III – Moderately hazardous
- Class IV – Low hazard
- Class V – Virtually harmless

Handling and transportation of hazardous waste is regulated under Russian Federal Law 89-FZ, 1998. All hazardous wastes must be properly classified n the basis of their composition and characteristics. The owner and generator of the waste is responsible for assuming the cost for proper handling of the waste.

13.2.2.5 Environmental Impact Assessment

Environmental Impact assessments are generally required under Article 32 of the Russian Federal Law No. 7-FZ for projects if there is a potential that operation of the project could have an adverse effect on the environment directly or indirectly. The procedure and nature of the work to be done in order to conduct the EIA, as well as the necessary documentation, depend on a multitude of factors similar to that of many other countries and the United States. The EIA process in Russia involves several stages that are as follows (OECD 2021 and Artic Center, 2021):

- The facility must first prepare all documentation containing a general description of the proposed project including its purpose and alternatives and submit them to the applicable regulatory authority.
- Inform the general public of the project including the location and nature of the project by using publications or announcements in the central or regional press.
- Evaluate all environmental risks
- Evaluate and analyze the state of the territory on which the project is intended to be located.
- Develop any measures that may be required to restore the environment
- Calculate the consumption of natural resources that may be necessary to the project

The legal or physical person responsible for the project must certify and approve the final version of the EIA before submitting it together with all necessary documents to the state environmental expert committee for consideration. Regulations are in effect since 2010 that require environmental audits of the progress and impact of the project to determine whether the project meets the requirements set forth in EIA and Russian environmental protections laws (OECD 2021).

13.2.2.6 Summary of Environmental Regulations and Protection in Russia

Russia is a country that has vast resources but has not performed well when it has come to environmental protection. Russia appears to have a robust set of environmental regulations but has difficulty in following their own rules. Not unlike other parts of the world, including the United States, Russia has had its share of environmental incidents that have caused huge negative impacts on human health and the environment. However, what may be different in Russia more than any other country examined in the world is that the occurrence of environmental incidents causing great harm has not resulted in change for the positive.

The Russian environmental movement which began in the late 1980s and into the early 1990s under Gorbachev's reforms have for the most part crumbled due to economic hardship and political instability largely by the current administration. Currently, the Putin administration has labeled many internal environmental groups as "anti-Russian" and have used aggressive tactics to quash openness when subject matter is focused on environmental protection (Newell and Henry 2017). This reality together with corruption, poor environmental enforcement, general lack of environmental awareness and care, indicate that environmental protection will not improve in the short term. According to Newell and Henry (2017), Russia's most significant environmental challenge is the illegal and unregulated use of its natural resources.

13.2.3 NORWAY

Norway is located on the Scandinavian Peninsula and also includes the island of Jan Mayen and the archipelago of Svalbard. Norway occupies a total area of 385,252 square kilometers (148,747 square miles) and as over January 2017, has a population of 5,258,317 (United Nations 2021a). Norway is bordered by Finland and Russia to the north-east and Sweden to the east, the Atlantic Ocean to the west, the Skagerrak to the south with Denmark beyond the strait.

Norway maintains a combination of market economy and a Nordic welfare model that includes universal health care and comprehensive social security system. Norway has extensive natural resource reserves that include petroleum, natural gas, minerals, timber, seafood, fresh water and hydropower. Norway's petroleum industry accounts for approximately 25% of the Norway's GDP which, on a per capita basis, is the world's largest producer of oil outside the Middle East (United Nations 2021a). In July 2013, the Norwegian Climate and Pollution Agency and the Norwegian Directorate for Nature Management merged into what is now called the Norwegian Environment Agency (NEA) which is under the direction of the Ministry of Climate and Environment (Norwegian Environment Agency 2021a).

13.2.3.1   Environmental Regulatory Overview of Norway

The Norwegian Environment Agency has been assigned key tasks to achieve national objectives that include (Norwegian Environment Agency 2021a):

- A stable climate and strengthened adaptability
- Biodiverse forests
- Unspoiled mountain landscapes
- Rich and varied wetlands
- An unpolluted environment
- An active outdoor lifestyle
- Well managed cultural landscapes
- Living seas and coasts
- Healthy rivers and lakes
- Effective waste management and recycling
- Clean air and less noise pollution

Norway has set goals to be achieved by 2020 that include (Norwegian Environment Agency 2021a):

- Pollution of any type will not cause injury to health or environmental damage
- Releases of substances that are hazardous to health or the environment will be eliminated
- The growth of wastes will be less than economic growth
- Wastes will be fully used for recycling and energy recovery
- To ensure safe air quality.
- Noise will be reduced by 10% of 1999 values

13.2.3.2   Air

Norwegian Environment Agency (2021a) has set mean annual air quality goals of 20 ug/m^3 for particulate matter (PM10), 8 mg/m^3 for fine particulate matter (PM2.5), and 40 mg/m^3 for Nitrogen dioxide (NO$_2$). The value for fine particulate matter (PM2.5) is lower than the value set by the EEA of 25 ug/m^3. The values for PM10 and nitrogen dioxide are the same as the EEA.

Although Norway is committed to improving its own air quality by reducing emissions, Norway has committed resources and funding for the United Nations and World Bank Pollution Management and Environmental Health (PMEH) program that focuses on improving the air quality in the major urban areas in other countries including, China, Egypt, India, Nigeria, and South Africa. In addition, Norway has provided funding to address indoor air quality by providing clean cookstoves in these same countries mentioned earlier (Norwegian Environment Agency 2021a).

13.2.3.3   Water

Water management in Norway is divided between several different ministries that include (Norwegian Environment Agency 2021a):

- Health and Care Services
- Environment
- Petroleum and Energy
- Local Government
- Regional Development

Surface water accounts for 90% of water use in Norway while groundwater accounts for the remaining 10%. Norway generally has good water quality and treats 96% of waste water before being discharged. The main sources of water contamination in Norway include agriculture, salmon farming, and the 4% of untreated wastewater. Norway also struggles with stormwater from its urban areas which tends to overwhelm its treatment plants and increases flooding potential (Norwegian Environment Agency 2021a).

13.2.3.4   Solid and Hazardous Waste

Not unlike much of the EU, Norway struggles with solid waste and limited landfill space. Norway also incinerates a portion of its solid waste for energy recovery (Norwegian Environment Agency 2021a). Norway recycles up to 97% of its plastic, which is higher than any other nation measured. The recycling effort is successful because there is an environmental tax on plastic producers and an incentive for citizens to recycle which is equivalent of up to $0.15 to $0.30 rebate per container depending on size (Norwegian Environment Agency 2021b).

In Norway, hazardous waste is defined as a waste that has the potential to cause serious pollution or involve risk of injury to people or animals. Solid and hazardous waste in Norway is handled using the EU standards (Norwegian Environment Agency 2021b).

13.2.3.5   Remediation

Norway has not ignored impacts to the environment that have resulted from oil and gas, forestry, mining, agriculture, shipping, and other activities. Given that most of Norway's inhabitants reside along its coastal areas and is located in a far northern latitude with significant cold weather climate patterns all place additional difficulties in addressing its environmental remediation efforts.

Norway follows risk assessment guidelines for evaluating and remediating sites of environmental contamination and has established remediation standards for different types of land use. Remediation standards in Norway are in some instances much lower than the United States (Norwegian Environment Agency 2021c). For example, the soil value for lead in Norway is set at 60 mg/kg, whereas an acceptable level for lead in the United States may be as high as 400 mg/kg depending on the exposure route (Norwegian Environment Agency 2021c).

Remediation standards are listed at the Norwegian Environmental Agency (2021c) at https://www.miljodirektoratet.no/old/klif/publikasjoner/andre/1691/ya1691.pdf.

13.2.3.6 Summary of Environmental Regulations of Norway

In Norway, sustainability is a fundamental principle for all development. The government's strategy is based on the principles of equitable distribution, international solidarity, the precautionary principle, polluters-pay principle, and the principle of common commitment. In part, this is achieved through an awareness and value of ecosystems and their connection with sustainability. Adequate knowledge and study of the condition of ecosystems is a necessary precondition for good nature management (Norwegian Environment Agency 2021a).

Sustainability also includes efforts to address climate change where Norway has been divided into sectors with each sector having it own climate change action plan with established targets for reducing greenhouse gases and integrating the action plan into specific sustainability targets (Norwegian Environment Agency 2021a).

13.2.4 SWITZERLAND

Switzerland is officially known as the Swiss Federation. Switzerland is located in west-central Europe. Switzerland has a land area of 41,285 square kilometers (15,940 square miles) and is a land-locked country that is located in the Alps and Swiss Plateau. As of 2016, the population of Switzerland was 8,401,201. The human density of Switzerland is 195 people per square kilometer, which is very dense considering that much of the country is uninhabitable because of the mountainous terrain. For comparison purposes, the United States has a density of 33 humans per square kilometer (United Nations 2021a). Human occupation in Switzerland dates back nearly 150,000 years. Evidence of farming dates to 5,300 BC (United Nations 2021d).

Switzerland is one of the most developed countries on Earth. Switzerland has the highest nominal wealth per adult and the eighth highest GDP per capita. Switzerland ranks near the top of many metrics of national performance including government transparency, civil liberties, quality of life, economic competitiveness, and human development. Zurich and Geneva consistently each rank among the top cities in the world with respect to quality of life (Bowers 2011).

13.2.4.1 Environmental Regulatory Overview of Switzerland

Switzerland is associated with lakes, mountains and clean air along with a good quality of life. The natural environment is an integral part of Switzerland's identity. Along these lines, Switzerland has been proactive at (Organization for Economic Co-operation and Development (OECD) 2016):

- Protecting natural resources, especially forests
- Higher density urban planning
- Reducing carbon dioxide and other greenhouse gas emissions
- Preserving water quality
- Maintaining biodiversity
- Improving air quality
- Preserving soil
- Cleaning up contaminated sites

Switzerland is under intense environmental pressure due to its population density and amount of land surface that is uninhabitable because of its mountainous nature. During the 1970s and 1980s, ambitious environmental policies were implemented that included substantial government funding in promoting environmental awareness due to recognizing the environmental decline of land areas within the country, notably its forests (European Environment Agency 2015a).

Switzerland has designed and implemented pollution abatement policies with ambitious objectives. Most of these objectives have been achieved with remarkable success and include air pollution emission rates among the lowest in Europe and very high levels of waste water infrastructure and in waste management facilities. This success was realized by means of a very active regulatory approach combined with rigorous enforcement, strong support from the public, and a considerable financial effort. The environmental policy applied in Switzerland rests on an extremely comprehensive body of federal environmental regulations. Their enactment is not without careful forethought and public input, active research, and referendums that are voted upon by the citizens on major environmental policy issues (Organization for Economic Co-operation and Development (OECD) 2016).

Although Switzerland spends approximately 1.7% on its GDP on environmental issues, which is high compared to other European countries, much work remains to be accomplished which includes (Organization for Economic Co-operation and Development (OECD) 2016):

- Meeting air pollution target for NOx, VOC's, and ozone
- Maintaining and upgrading waste water treatment infrastructure
- Upgrading solid waste management infrastructure
- Cleaning up contaminated sites
- Source identification and cleanup of non-point source water pollution
- Continuing to restore forests and natural areas

13.2.4.2   Air

Air emissions in Switzerland are regulated under the Switzerland Environmental Protection Law which was first enacted in 1983 and has been subsequently amended (Switzerland Environmental Protection Act 2021). The Environmental Protection Law set ambient air quality standards very similar to the EEA. The Environmental Protection Act established the following (Switzerland Environmental Protection Act 2021):

- Maximum emission values
- Regulations on construction and equipment
- Traffic and operating regulations
- Heat insulation in buildings requirements
- Motor fuel regulations

According to the Environmental Protection Act, Switzerland's ambient air limits must be set such that any air pollution (Switzerland Environmental Protection Act 2021):

- Does not endanger people, animals, plants, their biological communities and habitats
- Does not seriously affect the well-being of the population
- Does not damage buildings
- Does not harm soil fertility, vegetation or waters

Switzerland's air quality has improved since the mid-1980s when the Environmental Protection Act was enacted. Ambient air quality standards can be viewed at https://www.admin.ch/opc/en/classified-compilation/19830267/index.html. (Switzerland Environmental Protection Act 2021).

13.2.4.3   Water

In Switzerland, water is managed by the Federal Office of Public Health (FOPH) and the Federal Office of the Environment (FOEN). The FOPH is concerned with epidemics and infectious disease, food safety and the safety of drinking water. FOEN is concerned with long-term conservation and maintaining natural resources. Switzerland water quality standards are listed at: https://www.unece.org/fileadmin/DAM/env/water/Protocol/reports/pdf. (Switzerland Environmental Protection Act 2021).

In Switzerland, 80% of drinking water is obtained from springs while the remaining 20% is obtained from surface water sources. All potential households in Switzerland are connected to a central sewerage treatment plant or a decentralized treatment system (Switzerland Office for the Environment 2021). This represents the highest percentage of water treatment of any country that we will evaluate.

Challenges facing Switzerland with respect to water include agricultural runoff that contain fertilizers, herbicides, and pesticides. In addition, stormwater runoff is another challenge especially during very wet periods. In addition, Switzerland also has to deal with pollution originating from other countries through atmospheric deposition on land and surface water bodies (Switzerland Office for the Environment 2021).

13.2.4.4   Solid and Hazardous Waste

Solid and hazardous waste is regulated under the Switzerland Environmental Protection Act (2021). Since Switzerland is a small mountainous country, landfill space is at a premium. Therefore, Switzerland is strict on the amounts and types of wastes that are permitted and is dedicated to efforts of waste avoidance. For instance, products intended for once-only or short-term use may be prohibited because of limited landfill space. In addition, Switzerland requires any manufactures to avoid any waste generation especially hazardous waste that do not readily degrade under natural conditions and to treat any waste so that it contains as little bound carbon as possible and is insoluble (Switzerland Office for the Environment 2021).

To discourage the generation of waste, Switzerland imposes high costs on generators of solid and hazardous waste and under certain circumstances require prepaid disposal fees (Switzerland Office for the Environment 2021).

13.2.4.5   Remediation

Remediation of contaminated sites in Switzerland is regulated under the Switzerland Environmental Protection Act (2021). Switzerland has adopted the "polluters pay principle" similar to the United States. The methodology for evaluating potential contaminated sites uses the risk assessment approach similar to the EU and United States. Switzerland also requires that an environmental impact study be conducted for any new proposed development and also requires that all such reports be made public (Switzerland Office for the Environment 2021).

13.2.4.6   Summary of Environmental Regulations of Switzerland

Switzerland's environmental regulations are very comprehensive and also show a commitment from the government and residents. Switzerland has realized that it must be strict on environmental pollution matters since the country is small, mountainous, and heavily populated. Luckily, Switzerland does not have an industrial base as large as many other nearby countries but does have to deal with the challenges of pollution from agricultural sources and urbanization.

13.2.5   Turkey

Turkey is a transcontinental country in Eurasia. Portions of Turkey are located within what is considered Asia and southeastern Europe. For purposes of this book, we will treat Turkey as a part of Europe. Turkey is bordered by eight countries that include: Greece and Bulgaria to the northwest; Georgia to the northeast; Armenia, Iran, and Azerbaijani to the east, and Iraq and Syria to the south. Turkey is also bordered by seas on three sides that include: The Aegean Sea to the west; the Black Sea to the north; and the Mediterranean Sea to the south. Ankara is the capital but Istanbul is the largest city in the country. Turkey covers 783,356 square kilometers (302,455 square miles) and has a population of 79,814,871 as of 2016 that equals a human density of 102 per square kilometer (United Nations 2021a and Turkish Statistical Institute 2021).

Turkey is a charter member of the United Nations, an early member of the North Atlantic Treaty Organization (NATO), and the OECD. Turkey's growing economy and diplomatic initiatives have led to its recognition as a regional power while its location has given it geopolitical and strategic importance throughout history (Porter 2009). Turkey faces some significant environmental issues that include (Anderson 2011 and European Environment Agency 2013):

- Conservation of its biodiversity
- Air pollution
- Water pollution
- Greenhouse gases
- Land degradation

13.2.5.1 Environmental Regulatory Overview of Turkey

Current environmental laws in Turkey are similar to the EU and the United States and cover air pollution, water pollution, and land pollution and target improving human health and protecting its biodiversity. First enacted on August 11, 1983, the objective was to (Turkey Ministry of Environment and Urbanization 2021a and 2021b): *"improve the environment which is the common asset of all Turkish citizens; make better use of, and preserve the land and natural resources in rural and urban locations; prevent water, land and air pollution; by preserving vegetative and livestock assets and natural and historical richness, organize all arrangements and precautions for improving and securing health, civilization and life conditions of present and future generations in conformity with economical and social development objectives, and based on certain legal and technical principles"*

However, Turkey is facing significant economic hardships that greatly influence its ability to address its environmental issues. Therefore, Turkey faces significant challenges in improving environmental quality short term and long term if it does not conduct control measures to improve and protect human health and the environment. This is especially acute since the biodiversity of plant life in Turkey is so significant (Anderson 2011, European Environment Agency 2013, Environmental Health 2017). There are more than 3,000 endemic plant species in Turkey making it one of the richest biodiversity areas on Earth. This is due to Turkey's wide variety of habitats and unique position between three continents and three seas (Anderson 2011). In addition, Turkey has not made significant progress at remediating sites of environmental contamination and remediations standards are either lacking or not enforced (European Environment Agency 2013).

13.2.5.2 Air

Air pollution is particularly significant in Turkey especially in urban areas and is especially acute in Istanbul, Ankara, Erzurum, and Bursa and consistently ranks among the worst air in the world. In the winter months, the combustion of heating fuels combined with automobile exhaust make for especially high levels of particulate matter (European Environment Agency 2014 and Environmental Health 2017). In fact, according to the European Environment Agency, 97.2% of the urban population in Turkey is exposed to unhealthy levels of particulate matter.

Turkey is currently experiencing significant environmental deterioration of air quality due to combustion of fossil fuels, predominantly from automobiles in urban areas, and lack of enforcement and resources to employ effective air pollution control technologies at power plants and industrial facilities.

The root cause of some of the environmental issues facing Turkey are the overriding social and economic factors that Turkey is currently facing that inhibit enforcement of current environmental laws and regulations (European Environment Agency 2015b).

13.2.5.3 Water

As with air pollution, Turkey faces significant challenges with water pollution of various types including drinking water, surface water, groundwater, and the marine environment. Approximately 45,000 ships pass through the narrow Bosporus Straits every year and fears of a significant incident are real. Fears of an environmental incident have increased since the Turkish Environmental Protection Ministry has limited internal resources to oversee and inspect shipments of hazardous materials (Turkish Ministry of Environmental and Urbanization 2021b).

Turkey does not have significant surface water resources which indicates that if there is a water shortage, Turkey may not be prepared. In addition, water quality and availability of water is poor even for industry. Water quality is also significantly degraded, especially from agricultural fertilizers, pesticides and herbicides (European Environment Agency 2015b).

13.2.5.4 Solid and Hazardous Waste

Rapid growth in Turkey has resulted in the generation of significant amounts of waste materials. Turkey has not kept up the pace at establishing the required infrastructure for solid waste management. For example, in 2015 it was estimated that over 80% of the solid waste generated in Turkey was not landfilled properly and was discarded by other means (Akkoyuniu et al. 2017). In addition, there are over 2,000 open dump sites scattered throughout the country (Berkun 2011).

13.2.5.5 Remediation

As stated earlier, Turkey has environmental laws in place but currently lack enforcement abilities due to other overriding social and economic factors. Land degradation is also severely stressed from uncontrolled development leading to unchecked erosion which has also affected water quality. Turkey is also facing significant environmental deterioration of land due to overuse, lack of urban green space, poor urban planning, overdevelopment, deforestation, population growth and drought (European Environment Agency 2015b).

13.2.5.6 Summary of Environmental Regulations in Turkey

Turkey is facing some of the most significant environmental deterioration compared to other countries we have examined thus far. The main causes appear to be centered on its economic and financial hardships which have also caused a significant degree of political unrest. This indicates that if the human population at large is experiencing hardship, the natural environment is likely experiencing hardship and stress at an even larger scale and severity. As mentioned earlier, this threatens its biodiversity and ecosystems with collapse if improvements are not enacted.

13.3 AFRICA

Africa is known as the location where humans are believed to have evolved nearly 2 million years ago (Smithsonian Museum of Natural History 2017). Africa is the second largest continent and second most populous. Africa occupies approximately 30.3 million square kilometers (11.7 million square miles) and covers 6% of Earth's total surface area and 20.6% of Earth's land area (United Nations 2021a). Africa's population in total is approximately 1.22 billion and has an average age of 19.7, which is the youngest average age of any continent (United Nations 2021a).

Africa has a large diversity of ethnicities, cultures and languages and has 54 separate identified countries, nine territories, and two independent states with limited or no recognition. Some locations within Africa are developed while much of the rural areas are not. Africa has very diverse environmental climates, economics, historical ties, and governmental systems. This diversity has hindered its development and contributes to its environmental degradation and lack of environmental controls of pollution (Mwambazambi 2010 and United States Environmental Protection Agency 2021a).

Africa faces significant environmental degradation as a continent caused by many factors including (Chikanda 2009 and University of Michigan 2017):

- Over population
- Lack of pollution controls
- Urban sprawl and unchecked development
- Deforestation
- Invasive species
- Soil degradation and erosion
- Water degradation
- Lack of basic sanitation
- Political unrest and armed conflicts
- Poaching of wildlife which is significant in most Africa counties

USEPA has been involved in a collaborative effort to stabilize environmental issues in many parts of Africa. Specifically Sub-Saharan Africa relating to a growing population and industrial pollution issues that have impacted native population health and wellness, particularly vulnerable populations such as children and the elderly and economically disadvantaged. Areas of focus have centered on air quality, water quality, and reducing exposure to toxic chemicals (USEPA 2021b). USEPA has not been involved significantly or at all in collaborating with northern countries of Africa above or within the Saharan Desert for numerous reasons some of which include territorial disputes and political unrest. Specific programs where USEPA is assisting Sub-Saharan countries in Africa include (USEPA 2021b):

- Air Quality Management
- Safe Drinking water practices
- Lead paint abatement
- Reducing and managing methane emissions
- Improving environmental governance
- Improving sanitation
- Improving hazardous waste management

13.3.1 SOUTH AFRICA

South Africa is located at the southern tip of the African continent and occupies 1,220,813 square kilometers making it the 25th largest country by size. It is bordered to the north by Namibia, Botswana, Zimbabwe, and Mozambique. Swaziland and Lesotho are two countries that are located within South Africa. The Atlantic Ocean is located to the west and southwest and the Indian Ocean is located to the east and southeast. The population of South Africa as estimated in 2018 exceeds 57 million (Statistics South Africa 2021).

South Africa is largely located in a dry climatic region of Africa with most of its western region located in a semi-desert. Rainfall increases toward the east and falls primarily in summer. The southern coast of Africa. Recently, the Cape Town region of South Africa has experienced drought conditions that have now threatened its potable water supply. The central cause of the drought has been blamed on significant increases in population, which now exceeds 4 million, and on climate change (Welch 2018).

13.3.1.1 Environmental Regulatory Overview of South Africa

South African environmental law attempts to protect and conserve the environment of South Africa. South African environmental law encompasses natural resource conservation and utilization, as well as land-use planning and development, and enforcement. The South Africa National

Environmental Management Act (NEMA) of 1998 forms the framework for the protection of human health and the environment. The South Africa National Environmental Management Act covers environmental subjects that include (South Africa Government Gazette 2018):

• Sustainable development	• Intergenerational equity
• Environmental justice	• Environmental rights
• Public trust	• Precautionary principle
• Preventative principle	• Polluters-pay principle
• Local-level governance	• Common differentiated responsibility

Although some names may be different, NEMA is very similar to that which we are accustomed to in the United States and most other countries. Especially to what is described as (South Africa Department of Environmental Affairs 2021a):

- Preventive Principle which relates to treatment or capture of contaminants before they are emitted or released into the environment (i.e., installation and operation of air pollution control equipment). This is the fundamental notion of NEMA regulating generation, treatment, storage and disposal of hazardous waste, and the use of pesticides.
- Precautionary principle is designed such that lack of full scientific certainty shall not be used as a reason for postponing measures to prevent environmental degradation.
- Polluters-pay principle which is analogous to Superfund in the United States.
- Environmental justice is intended to safeguard unfair environmental discrimination toward any person or persons.
- Environmental rights does not mean to imply that the environment has rights but is intended to address the rights of the individual to an environment that is safeguarded.
- Local-level governance requires that decisions affecting local municipalities should be made by local municipalities.

NEMA defines the "environment" as the surroundings within humans exist and include (South Africa Department of Environmental Affairs 2021a and 2021b):

- The land
- The water
- The atmosphere of the Earth
- Micro-organisms
- Plant life
- Animal life
- Any part or combination of the items listed earlier and the interrelationships among and between them
- The physical, chemical, aesthetic and cultural properties and conditions that influence human health and well-being

Furthermore, NEMA goes on to also define the environment as:

The aggregate of surrounding objects, conditions and influences that impact the life and habits of man or any other organism or collection of organisms

NEMA addresses the following distinct but inter-related areas of general concern and include (South African Department of Environmental Affairs 2021a):

- Land-use planning and development
- Resource conservation and utilization
- Waste management and pollution control

One of the main purposes of NEMA was to set up a national environmental management system that would outline procedures for cooperative governance within governmental agencies such that the Act does not impose overly burdensome requirements on the private sector (South Africa National Environmental Management Act 1998). Other prominent environmental legislative Laws in South Africa include (South Africa Department of Environmental Affairs 2021a):

- Air Quality Act No. 39 of 2004
- Protected Areas Act No. 31 of 2004
- Protected Areas Amendment Act No. 31 of 2004
- Biodiversity Act No. 72 of 2005
- Clean Development Regulations Act of 2004
- Protected Areas Act No. 57 of 2004
- Environmental Conservation Act No. 50 of 2004
- Marine Living Resources Act of 2004
- Biodiversity Act: Threatened or Protected Species Amendment Act of 2011
- Environmental Impact Assessment Regulations of 2010
- Environmental Management Regulations of 2010

13.3.1.2 Air Quality Standards

South Africa Ambient Air Quality Standards were enacted in 2004 and cover compounds such as Sulphur dioxide, Nitrogen dioxide, particulate matter (PM10), Ozone, benzene, lead, and Carbon monoxide (South Africa Department of Environmental Affairs 2009a). Limits are similar to the United States and the EU and are listed at https://environment.gov.sa/legislative/acts/regulations.

13.3.1.3 Water Quality Standards

South Africa established water quality standards for drinking water and other potable uses in 1996 and does not include water that is sold as a beverage or in swimming pools (South Africa Department of Environment 1996). Water quality standards are similar to the United States and the EU and are listed at https://environment.gov.sa/legislative/acts/regulations.

13.3.1.4 Waste Management

In South Africa, solid and hazardous waste are differentiated and are regulated by the National Environmental Waste Act No 59 of 2009 (South Africa Department of Environmental Affairs 2009b). Implementation guidelines were published in 2012 (South Africa Department of Environmental Affairs 2012). Hazardous wastes are also termed Priority Wastes in South Africa. Solid and hazardous waste management in South Africa rely heavily on landfill disposal of wastes.

According to the most recent information available, South Africa produced approximately 108 million metric tons of solid waste in 2011, of which 98 million metric tons were disposed in landfills. The remaining 10 million metric tons (or just under 10%) was recorded as being recycled. The largest volumes of waste are produced by the industrial and mining sectors (South Africa Department of Environmental Affairs 2012).

In South Africa, hazardous waste is defined as (South Africa Department of Environmental Affairs 2012):

Any waste that contains organic or inorganic elements or compounds that may, owing to the inherent physical, chemical or toxicological characteristics of that waste, have a detrimental impact on health and the environment

Hazardous wastes are classified according to the following criteria (South Africa Department of Environmental Affairs 2012):

- Reactive
- Corrosive
- Flammable
- Characteristic
- Explosive
- Gases
- Radioactive
- Organic (for halogenated organic wastes)
- Infectious
- Miscellaneous dangerous substance (examples of wastes in this category may include asbestos, dry ice, and other environmentally hazardous substances that do not fall into any of the earlier-listed categories).

Classifying solid wastes by characteristic in South Africa is conducted by comparing concentration of contaminants within the waste in two ways: leachable fraction and total concentration. If the concentration of the specific substance exceeds either or both criteria then it is considered hazardous. Leachable concentration is to be determined by collecting a representative sample of the waste material and analyzing the waste using the Toxic Characteristic Leaching Procedure similar to that outlined by the USEPA. Compounds and concentration limits which are similar to the United States, EU, and Australia are provided at https://environment.gov.sa/legislative/acts/regulations (South Africa Department of Environmental Affairs 2012).

Generators of the waste must select the potential chemical contaminants that are either known to exist or may be present within the waste material. This may require knowledge of site activities, site history, or the processes which created the waste. Generators must be able to justify the chemical contaminants selected for analysis and keep records of that decision for three years. If chemical contents of a waste are unknown, then it is generally recommended to conduct a comprehensive analyses which may include (South Africa Department of Environmental Affairs 2012):

- Volatile organic compounds
- Polycyclic aromatic hydrocarbons
- Semi-volatile organic compounds
- Phenols
- Polychlorinated biphenyls
- Heavy metals

13.3.1.5   Remediation Standards

Remediation standards for South Africa were established in 2013 (South Africa Department of Environmental Affairs 2013). South Africa remediation standards are divided into categories using the following definitions:

- Contaminant is defined as any substance present in and environmental medium at concentrations that exceed natural background concentrations.
- Informal Residential means an unplanned settlement on land which has not been proclaimed as a residential and consists mainly of makeshift structure(s) not erected according to approved architectural plans.
- Remediation is the management of a contaminated site to prevent, minimize, or mitigate damage to human health or the environment.
- Soil Screening Value 1 are soil quality values that are protective of both human health and

eco-toxicological risk for multi-exposure pathways, and is inclusive of migration to a water source of either surface water of groundwater.

- Soil Screening Value 2 are soil quality values that are protective of risk to human health in the absence of a water resource or ecological exposure.
- Standard Residential means a settlement that is formally proclaimed and serviced, and generally developed with formal permanent structures including land parcels.

Remediation values are similar to the United States and EU and are listed at https://environment. gov.sa/legislative/acts/regulations. (South Africa Department of Environmental Affairs 2013).

Groundwater remediation values in South Africa follow either surface water criteria or the Dutch Intervention Values described earlier in this chapter.

The variations in the climate of South Africa allows for a wide variety of crops that range from tropical fruit to corn and tree plantations. This in turn, has led to extensive use of pesticides, herbicides, and fungicides to the point where South Africa is one of the largest importers of these types of chemicals in all of Africa (Quinn et al. 2011). Estimates on the annual total pesticide use in South Africa exceed 2,800 metric tons. However, remediation standards for most of these chemicals is lacking.

South Africa depends heavily of groundwater resources but due to drought in many areas, the resource is experiencing stress from over exploitation and many wells are now dry especially in the Cape Town region Currently dam levels are at 60% capacity and there is a water use limit placed on the population of 50 liters of water per day (South Africa Department of Environmental Affairs 2021b).

13.3.1.6 Summary of South Africa Environmental Regulations

Compared to the United States, environmental regulations in South Africa have only recently become enacted in that the National Environmental Management Act is only 20 years old and many of the detailed regulations for air, water and solid and hazardous waste are generally 10 years old. The framework of the regulations are predominantly based on USEPA. This may be true from a framework point of view but many of the details are similar to the European Union and Australia. In fact, as we shall discover when we are evaluating Australia, the solid waste classification in South Africa is nearly identical to Australia.

The environmental regulations of South Africa appear to be very robust and comprehensive, even when comparing them to the United States or the European Union. However, they are still rather recent and given the political, financial pressures, and drought in South Africa, implementation and delays in enforcement have hindered and slowed progress. Evidence of this is that recycling efforts have not seen significant progress and remain at only 10% (South Africa Department of Environmental Affairs 2012). Another very real disadvantage within South Africa is lack of sufficient infrastructure due to it geography in that it is rather isolated from other developed countries and therefore, must import technologies and equipment over a longer distance.

Finally, climate change appears to have impacted South Africa as most significantly realized with the ongoing drought and the city of Cape Town experiencing a water shortage perhaps that is more severe of any developed city on Earth at the moment. The impact is clear but the response and long-term plan to address climate change is not yet certain.

13.3.2 KENYA

Kenya is located in east central Africa on the Indian Ocean between Somalia and Tanzania. Other countries that share a border with Kenya include Ethiopia, South Sudan, and Uganda. Kenya is 582,650 square kilometers in size and has an estimated population of 48 million (Kenya National Bureau of Statistics 2021). Nairobi is the largest city in Kenya and is considered the 10th largest city in Africa with an estimated population of 6.5 million residents (Kenya National Bureau of

Statistics 2021). The name originates from the Maasai phrase Enkare Nyrobi, which translates to "cool water" and is a reference to the Nairobi River which flows through the city. Central and western Kenya are located in the Kenyan Rift Valley characterized by mountains and volcanoes, some of which are considered active. Kenya's highest Peak, Mount Kenya which exceeds 5,700 meters in elevation is located in this region (Kenya Geological Society 2021).

The population of Kenya has increased significantly over the last 50 years, especially Nairobi, where the population has increased from 0.5 million to 6.5 million. This has placed pressure on providing adequate infrastructure to support a population increase of that magnitude that quickly (Kenya National Bureau of Statistics 2021). The increase in population has also placed stress on the natural environment in Kenya and include water and air pollution from urban and industrial areas and agriculture, deforestation, pesticide and herbicide use, erosion, desertification, and poaching of wildlife (Kenya National Environment Management Authority 2021). Approximately 8% of the landmass of Kenya are contained with 22 National Parks and 28 National Reserves (Kenya Wildlife Service 2021). Kenya's economy is dominated by agriculture followed by manufacturing, much of which is agriculture related. Kenya is the banking capital of central Africa and tourism is also very significant to the economy (Kenya National Bureau of Statics 2021).

13.3.2.1   Kenya Environmental Regulations

Environmental regulations were enacted starting in 2006 and now include several Acts that cover the following environmental Areas (Kenya National Environment Management Authority 2021):

• Air quality	• Domestic water
• Industrial water discharge	• Wetlands
• Noise	• Solid waste
• Hazardous waste	• Environmental impact assessments
• Chemical regulations	• Waste transport
• Controlled substances	• Coastal protection
• Biodiversity	• Land development

13.3.2.2   Air

Air pollution in Kenya is most significant near its three largest cities, Nairobi, Mombasa, and Kisumu which account for near 30% of its population (Kenya National Bureau of Statistics 2021 and Kinny et al. 2011). Fugitive emissions are also prevalent since many roads in Kenya are dirt or gravel. Although environmental regulations in Kenya are considered robust as we shall see in this chapter, air regulations are considered weak not because of a need for additional regulation, but due to implementation and enforcement of existing laws and lack of cooperation between governmental ministries in Kenya (Barczewski 2013).

Air quality regulations were enacted in 2009 titled The Environmental Management and Co-Ordination Air Quality Regulations. The Air Quality Act of 2009 defined air pollution and meaning any change in the composition of air caused by air pollutants (Kenya Air Quality Act 2009). The Air Quality Act of 2009 sets limits for the following sources:

• Industry and other stationary sources
• Mobile sources, such as motor vehicles and required vehicle emission testing
• Occupational air quality limits
• Fugitive sources
• Particulate emissions from demolition sites

- Open burning
- Cross-border air pollution
- Exposure to hazardous substances
- Effects of stockpiling of material
- Emissions from waste incinerators

The Kenya Air Quality Act of 2009 required owners or operators of industrial sites to obtain a permit or license to emit air pollutants and required each facility to conduct a detailed study that includes:

- Initial emission assessment report
- Preliminary assessment of stationary sources
- Atmospheric impact report
- Monitoring records
- Notification of excessive emissions
- Stack emission recording and reporting
- Continuous monitoring system requirements

The Air Quality Act of 2009 also set ambient air quality tolerance limits similar to the United States and EU and are listed at https://nema.go.ke/NEMA/airqualityact (Kenya Air Quality Act 2009).

13.3.2.3 Water

Water quality regulations were enacted in 2006 with the Environmental Management and Co-Ordination Water Quality Act (Kenya Water Quality Act 2006). The Water Quality Act regulates drinking water, protects the sources of drinking water, industrial use and effluent discharge, and agricultural use, and recreational use. Water quality standards for drinking or domestic use and effluent discharge are similar to the United States and EU and are located at https://nema.go.ke/NEMA/waterqualityact (Kenya Water Quality Act of 2006).

13.3.2.4 Solid and Hazardous Waste

Management of solid and hazardous waste in Kenya is regulated by the Environmental Management and Co-Ordination Waste Management Act of 2006 (Kenya Waste Management Act 2006). The act covers topics that include:

- Solid waste
- Hazardous waste
- Industrial waste
- Pesticides and toxic substances
- Biomedical wastes
- Radioactive wastes

The Act outlines the responsibilities of the generator, transporter and disposal facility, permitting, licensing, transportation, and environmental audit procedures. It also outlines requirements for evaluation of an environmental impact assessment, training, labeling, packaging, segregation, monitoring, and classification (Kenya Waste Management Act 2006).

Hazardous waste determination in Kenya is by content and percentage of certain chemicals considered hazardous by their very nature. For example, waste that contain the following are considered hazardous in Kenya (Kenya Waste Management Act 2006):

- Radio-nuclides
- Medical waste

- Pharmaceutical, drugs or medicines
- Biocides, germicides, herbicides, insecticides, fungicides
- Wood preserving chemicals
- Organic solvent waste
- Heat treatment and tempering wastes containing cyanide
- Mineral oil waste
- PCBs
- Wastes from inks, dyes, pigments and paints
- Wastes chemicals from research, development or teaching
- Explosives
- Wastes containing metal carbonyls, beryllium, hexavalent chrome, copper, zinc, arsenic, selenium, cadmium, antimony, tellurium, mercury, thallium, lead, fluorine, phosphorus, phenol, ethers and cyanide at 1% or more by weight
- Waste at a pH of less than 2
- Waste at a pH greater than 11.5
- Wastes containing asbestos
- Halogenated organic solvents at 0.1% or more by weight
- Any congener of polychlorinated dibenzo-furan
- Any congener of polychlorinated dibenzo-p-dioxin

Other waste that are considered hazardous not listed include wastes that are (Kenya Waste Management Act 2006):

• Flammable	• Explosive	• Combustible
• Oxidizers	• Organic peroxides	• Toxic or poisonous
• Infectious	• Corrosive	• Eco-toxic
• Toxic gas	• Persistent wastes	• Leachate
• Radioactive	• Carcinogens	• Medical waste

13.3.2.5 Remediation Standards

Remediation stands in Kenya were first enacted in 2003 and was amended in 2009 and are termed the Environmental Impact and Assessment and Audit Act (Kenya Impact Assessment and Audit Act 2003). The Act requires that an environmental impact assessment be conducted before any land is developed or re-developed and must consider the following at a minimum:

- Environmental, social, cultural, economic, and legal considerations
- Identify environmental impacts and scale of impacts
- Identify and analyze alternative
- Develop an environmental management plan
- Consult with the regulatory agency on a regular basis
- Seek public comment
- Hold at least three public meetings
- Prepare a detailed report

After regulatory review and acceptance, a license will be issued. Following development, environmental audits and monitoring shall be conducted at intervals outlined in the license.

The environmental impact assessment is a detailed environmental study similar to that in the United States and usually includes extensive investigation and testing for the presence of contamination and if discovered, the nature and extent of impacts must be defined. Following the completion of testing, the assessment is required to evaluate whether there are any risks posed by the presence of contamination and lower those exposure risk to an acceptable level, if necessary (Kenya Impact Assessment and Audit Act 2003).

13.3.2.6 Summary of Environmental Regulations of Kenya

Kenya does, in fact, have a very robust set of environmental regulations in place. The regulations have not been enacted for very long, most are less than 15 years old. So it should come as no surprise that the current major obstacle is implementation and enforcement along with cooperation between ministries responsible for the environmental interpretation and enforcement. Kenya air and water environmental regulations are similar to that of United States and the European Union. However, the solid and hazardous waste regulation are unique to Kenya and are strict. This is likely due to a combination of factors that include Kenya's reliance on its National Parks and biodiversity for tourism, lack of a significant industrial base, and perhaps disincentives for additional industrial development. An example of the strict solid and hazardous waste regulations in Kenya is that Kenya considers any plastic waste to be hazardous.

13.3.3 TANZANIA

Tanzania is located in east central Africa on the Indian Ocean located along the eastern border of the country. Kenya is located immediately to the north and Mozambique, Malawi, and Zambia are located along the southern border. The Democratic Republic of Congo, Burundi and Rwanda are located immediately to the east.

Tanzania has an estimated population of slightly more than 57 million and is 947,300 square kilometers in size (Tanzania Bureau of Statistics 2021). Tanzania's highest elevation is Mount Kilimanjaro at an elevation of 5,895 meters. There are numerous national parks, conservation areas and game reserves within Tanzania including the Serengeti National Park, Ngorongoro Crater Conservation Area, and the Selous Game Reserve (Tanzania National Bureau of Statistics 2021). Lake Victoria, Africa's largest lake is located in the northwest portion of the country. Northern and central portions of Tanzania are mountainous and are part of the African Rift Valley. Most of Tanzania's population is located in the northern portion of the country and the eastern border with the Indian Ocean (Tanzania Bureau of Statistics 2021).

The largest city in Tanzania is Dar es Sallaam with an estimated population of over 5.5 million and was the former capital of the country. The capital of Tanzania is now Dodoma. Dodoma has an estimated population of greater than 2.2 million and includes outlying areas (Tanzania National Bureau of Statistics 2021). The population of Tanzania has increased significantly since 1963 when the population was estimated at 11 million. This has placed pressure on providing adequate infrastructure to support a population increase of that magnitude (Tanzania National Environmental Standards Compendium (NESC) 2018). Tanzania has the second largest economy in central Africa, second to Kenya. The economy is dominated by agriculture which employs approximately 50% of the workforce. Approximately on third of the population of Tanzania live at or below the poverty level (Tanzania Bureau of Statics 2021).

13.3.3.1 Environmental Regulatory Overview of Tanzania

Not unlike Kenya, Tanzania's increase in population has placed stress on the natural environment and include water and air pollution, deforestation, pesticide and herbicide use, erosion, desertification, and poaching of wildlife. Water-borne illnesses, such as malaria and cholera account for over half of the diseases affecting the population (Tanzania National Bureau of Statistics 2021).

13.3.3.2 Air

Air quality in Tanzania is regulated under the Environmental Management Air Quality Act of 2007. The Tanzania Air Quality Act established ambient air quality standards and emission standards for motor vehicles (Tanzania Air Quality Act 2007). In addition, Tanzania has emission standards for cement plants. Cement plants in Tanzania are numerous since the population has greatly increased over the past few decades and construction for housing and commerce has placed a large demand for building materials. The objective of the Tanzania Air Quality Act of 2007 was to:

1. Set baselines parameters on air quality
2. Enforce minimum air quality standards
3. Encourage effective and optimal air pollution control equipment and procedures
4. Ensure protection of human health and the environment

To achieve the objectives, the Tanzania Air Quality Act of 2007 conducted the following:

- Establish criterion and procedures for measuring ambient air quality
- Establish minimum ambient air quality standards
- Establish emission standards and limits for various sources of air pollution
- Prescribe stack heights
- Prescribe criteria and guidelines for air pollution controls for mobile and stationary sources of air pollutants
- Establish occupational air quality standards

The Air Quality Act requires owners or operators of air pollutant emission sources to obtain a permit and to document operations of air pollution control equipment to ensure optimal performance. Tanzania Ambient Air Quality Standards are similar to the United States and EU and are listed at https://www.parliament.go.tz/acts-list-air-quality (Tanzania Air Quality Act 2007).

13.3.3.3 Water

Water quality varies significantly in Tanzania. Urban areas generally have better water quality than rural areas. The latest estimate which is from 2015 is that 26 million people or approximately half the population lack access to at least basic water. Wastewater treatment is also a challenge in Tanzania. Of the 20 urban locations within the country, 11 provide some water treatment. Within the urban areas where wastewater treatment is possible, the range in actual hookups to sanitation range from 4% to 45% of households that are actually connected to sewers (Tanzania Ministry of Water and Irrigation 2015). Water quality in Tanzania is regulated through the Tanzania Water Resources Act (2009) and the Water Supply and Sanitation Act (2009). Tanzania Drinking Water Standards for Select Compounds are similar to the United States and EU and are located at https://www,tanzania.go.tz/egov_uploads/documents/tanzania.   (Tanzania   National   Environmental Standards Compendium (NESC) 2018).

13.3.3.4 Solid and Hazardous Waste

Solid and hazardous waste is regulated through the Tanzania Environmental Management Act of 2008, more commonly referred to as the Hazardous Waste Control and Management Regulations (2008). The Hazardous Waste Control and Management Regulations cover the following topics:

- Classification procedures
- Packaging
- Labeling

- Handling
- Transporting
- Storage
- Permitting

In Tanzania wastes are considered hazardous if they have the following (Tanzania Environmental Management Act 2008)

- Characteristics such as explosive, flammable, corrosive, reactive or have the potential to produce a leachate considered toxic
- Are generated from certain types of operations such as:
  - Medical waste
  - Pharmaceuticals, drugs or medicines
  - Biocide production
  - Wood preserving chemicals
  - Organic solvent production
  - Heat treatment and tempering operations that use cyanide
  - Used mineral oils
  - Oil and water mixtures
  - Waste containing PCBs
  - Ink, dye, pigment, paint, lacquer, and varnish operations
  - Waste from resins, latex, plasticizers, glues and adhesives
  - Waste chemical substances from research, development or teaching
  - Residues from waste treatment operations
- Waste having the following constituents:
  - Metals such as beryllium, hexavalent chromium, copper, zinc, arsenic, selenium, cadmium, antimony, tellurium, mercury, thallium, lead,
  - Inorganic cyanides
  - Asbestos
  - Phenols
  - Halogenated organic solvents
  - Ethers
  - Any congener of polychlorinated dibenzo-furan
  - Any congener of polychlorinated dibenzo-p-dioxin

13.3.3.5   Remediation Standards

Soil quality standards were established in Tanzania in 2007 with the Environmental Management Act that established maximum allowable concentrations of many compounds in soil. Groundwater standards have not been established. The intention of the Environmental Management Act of 2007 was to make clear that any intentional disposal of solid refuse or putrid solid matter onto the ground is prohibited (Tanzania Environmental Management Act 2007).

13.3.3.6   Summary of Environmental Regulations of Tanzania

The environmental regulations in Tanzania are similar to that of its neighbor Kenya. Differences between Kenya and Tanzania focus on the fact that Kenya is more developed and is further along at implementation of its regulations. Tanzania appears to struggle with providing basic sanitation and supplying water to its population. This is largely due to the lack of basic infrastructure within the country. In addition, approximately one third of the population of Tanzania that are considered living below the poverty level. Therefore, it should come as no surprise that Tanzania is also struggling with maintaining its natural environment and the health of its residents.

13.3.4 EGYPT

Egypt is located in northern Africa and has an estimated population of 99 million and a land area of 1,001,449 square kilometers (United Nations 2021a). The Mediterranean Sea forms Egypt's northern border, Libya is located to the west, Sudan to the south and Jordan and Saudi Arabia to the east. Egypt is the third most populous country in Africa behind Nigeria and Ethiopia. Approximately 95% of the population of Egypt lives along the Nile River, the Nile Delta, or along the Suez Canal. These regions are among the most densely populated areas of the world. An estimated 75% of the population of Egypt are under the age of 25 making it one of the most youthful country populations in the world (United Nations 2021a). Cairo located along the Nile River is the largest city in Egypt with an estimated metropolitan population of over 20 million residents (United Nations 2021e).

13.3.4.1 Environmental Regulatory Overview of Egypt

In Egypt, the environment is regulated by the Ministry of Environment. Environmental standards in Egypt were first enacted in 1994 and were amended in 2009 and are simply known and referred to as The Environmental Law and covers protection of the land, air, water, and marine environment (Egypt Environmental Affairs Agency 2021).

13.3.4.2 Air

Air pollution in the urban area of Egypt is considered serious and decreases life expectancy by more than two years (World Bank 2013). Air quality in Cairo is from 10 to 100 times greater than acceptable world standards. The source of much of the air pollution in Cairo is attributed to automobile exhaust from the more than 2 million cars that are on the roads daily. Excessive air pollution has also accelerated deterioration of many ancient landmarks and relics of Egypt's history. Additional sources are from industry and frequent dust storms from desert regions that surround the city (Marey et al. 2010). The Egypt Environmental Law of 2009 addresses air pollution to ensure that total pollution emitted by all sources in any area is within permissible limits (Egypt Environmental Affairs Agency 2021). Egyptian ambient air quality standards are similar to the United States and EU and are listed at https://www.eeaa.gov.eg/en-us/laws/envlaw.aspx (Egypt Environmental Affairs Agency 2021).

The Environmental Law addresses the following (Egypt Environmental Affairs Agency 2021):

- Open burning of any kind including, agricultural fields, garbage, solid waste, and other types of unauthorized burning
- Excessive motor vehicle exhaust
- Application of pesticides and herbicides
- Properly maintaining equipment and machines emitting air pollutants
- Providing appropriate and adequate ventilation
- Prohibiting smoking in public transport areas
- Limiting radioactivity in ambient air
- Regulating ozone depleting substances

To further address the air quality in the Cairo area specifically, governmental authorities have adopted various measures to address the air quality issues including (Egypt Environmental Affairs Agency 2021):

- Constructing better roads with overpasses to keep traffic moving
- Constructing satellite cities to relieve inner city congestion
- Improving mass transit
- Providing incentives for taxis and industry to convert to natural gas

However, the overall air quality has not improved and has actually worsened because of several factors that include (World Bank 2013):

- Increased population
- Increased development
- Increased agricultural activities, primarily rice farming
- Ineffectiveness of traffic pattern management
- Increased motor vehicles
- Under capacity of mass transit
- Lack of enforcement of emission standards

13.3.4.3 Water

The main source of water in Egypt is the Nile River with an estimated current extraction rate 56 billion cubic meters of freshwater every year and represents 97% of the total volume of freshwater used each year in Egypt. There are water shortages the further away from the Nile River especially in rural areas. This is not a surprise since much of Egypt is considered a desert. Major improvements in providing piped water supply to residents has been realized in that nearly 100% of residents in urban areas and 93% of residents in rural have piped water available. However, only about 50% of the population are connected to sanitary sewers (World Health Organization 2021a and 2021b). The Egyptian Environmental Law of 2009 addressed mainly the marine environment along Egypt's northern coast and the area along the Suez Canal. Drinking water is regulated by the Egyptian Ministry of Health beginning in 1995 and focused on protecting the source of drinking water, that being the Nile River (World Health Organization 2021a and 2021b). Egyptian water quality standards are similar to the United States and EU and are listed at https://www.who.int/countries/egy/en (World Health Organization 2021a).

13.3.4.4 Solid and Hazardous Waste

Solid and hazardous waste is regulated through The Environmental Law of 2009 and administrated through the Egypt Environmental Affairs Agency (2021). In Egypt, all solid and hazardous wastes are regulated and include (Egypt Environmental Affairs Agency 2021):

- No wastes shall be disposed of in unlicensed locations.
- All treatment of wastes must be conducted under an applicable permit and license.
- No wastes shall be imported.
- Appropriate health and safety measures for all those who handle or ship wastes.
- All waste generators shall be registered.

The Egyptian solid and hazardous waste regulations include (Egypt Environmental Affairs Agency 2021):

• Transportation	• Labeling	• Manifesting
• Storage	• Disposal	• Characterization
• Licensing	• Operations	• Packaging
• Treatment	• Disposal	

In Egypt, a hazardous waste is defined as (Egypt Environmental Affairs Agency 2021):

Wastes of activities and processes or their ashes that maintain their harmful properties and have no subsequent original or substitutive uses

Unlike most countries of the world, no characteristics of hazardous wastes, processes, waste streams, or constituents of wastes have been identified or defined (Ramadam and Nadim 2014).

13.3.4.5   Remediation

Egypt requires any new proposed development conduct an Environmental Impact Assessment (EIS), which must be conducted in accordance to the elements, designs, specifications, bases and pollutant loads determined by the Egyptian Environmental Affairs Agency (Egypt Environmental Affairs Agency 2021). The focus of the EIA is not on current soil or groundwater quality but only addresses future potential impacts.

13.3.4.6   Summary of Egyptian Environmental Regulations

Egypt is a very urbanized country with very dense population centers. Together with an increasing population has placed enormous stress on the health of its population (World Health Organization 2021a). Air pollution, sanitation, remediation of contaminated sites, and deterioration of ancient historical relics are of immediate and pressing concern. Near future progress on solving environmental issues do not look promising due to recent political strife and violence within the country.

13.4   SUMMARY AND CONCLUSION

Yet again, what should be apparent after reading this chapter is that many countries have a robust set of environmental regulations and that they are very similar to those of the United States. For the most part, European Union counties not only have robust environmental regulations but they are also effective. In fact, many may be more effective than their counterparts in the United States. Many countries in eastern Europe such as Russia and Turkey and others, and many countries in Africa struggle with implementing environmental regulations to the extent that they actually work at achieving the underlying purpose of any environmental regulation which is improving human health and the environment. The reasons may vary from country to country but all have common themes that involve war, poverty, corruption, over population, lack of infrastructure, political will, ineffective leadership, and other social issues.

Questions and Exercises for Discussion

1. Russia has much environmental improvement to conduct before it can focus on sustainability? Do you agree or disagree with this comment and why?
2. In what areas can the EU improve its environmental regulations? Name at least two areas and explain why.
3. In your view, what are the environmental priorities for Russia and Turkey?
4. Of the countries of Africa, which country is the most environmentally impaired and why?
5. Of the countries of Africa, which has the most effective environmental regulations and why?
6. Of the countries of Africa, which country has the least effective environmental regulation and why?
7. Of the countries of Africa, provide your view as to what you would recommend to improve the air, water, land, and living organisms of the country you selected from question 6?
8. Now that you have read this chapter, what do you think is the most important environmental issue facing Africa and Europe and why?

REFERENCES

Akkoyuniu, A., Avsar, Y., and Erguven, G. O. 2017. Hazardous Waste Management in Turkey. *Journal of Hazardous, Toxic, and Radioactive Waste.* Vol. 21. no. 4. p. 7

Anderson, Sean. 2011. *Turkey's Globally Important Biodiversity Crisis. Journal of Biological Diversity.* Volume 144. Elsevier Publishers: New York, New York. pp. 2752–2769.

Arctic Center. 2021. Environmental Impact Assessment Processes in Northwest Russia. https://www. articcenter.org/RussianEIA/process (accessed April 25, 2021).

Barczewski, B. 2013. How Well Do Environmental Regulations Work In Kenya? A Case Study. *Center for Sustainable Urban Development*. Columbia University. New York. NY. 28p.

Berkun, M., Aras, E. and Amlan, T. 2011. Solid Waste Management in Turkey. *Journal of Materials and Waste Management*. Vol. 13. no. 1. pp. 305–313.

Blinnikiv, M. 2011. *A Geography of Russia and its Neighbors*. The Guilford Press. London, United Kingdom. 425p.

Bowers, S. 2011. Swiss Top of the Rich List. The Guardian. London, *United Kingdom*. https://www. Theguardian.com/features/2011qualityoflife (accessed April 25, 2021).

Brain, S. 2016. Environmental History of Russia. *Oxford Research Encyclopedias*. Oxford University Press. Oxford, UK. 237p.

Chikanda, A. 2009. Environmental Degradation in Sub-Saharan Africa. *In Environment and Health in Sub-Saharan Africa: Managing an Emerging Crisis*. Springer Publishing. New York. NY. pp. 79–94.

Delaware River Watershed Initiative. 2021. Delaware River Watershed Initiative. https://4states/source.org/ about (accessed April 25, 2021).

Egypt Environmental Affairs Agency (EEAA). 2021. Environmental Protection Law. https://www.eeaa.gov. eg/en-us/laws/envlaw.aspx (accessed April 25, 2021).

Environmental Health. 2017. Air Pollution and Health in Turkey. *Health and Environment Alliance (HEAL)*. Brussels, Belgium. https://www.env-health.org/imf/pdf/150220 (accessed April 25, 2021).

European Environment Agency. 2013. Turkey Air Pollution Fact Sheet. http://www.eea.europe.eu/themes/air/ air-pollution-fact-sheets (accessed April 25, 2021).

European Environment Agency. 2014. Air Pollution Country Fact Sheet: Turkey. http://www.eea.europe.eu/ themes/air/air-pollution-fact-sheets (accessed April 25, 2021).

European Environment Agency. 2015a. Environmental Status of Switzerland. https://www.eea.europa.eu/ switzerland (accessed April 25, 2021).

European Environment Agency. 2015b. Turkey Country Briefing – The European Environment – State and Outlook 2015. https://eea.europe.eu/soer-2015/countries/turkey (accessed April 25, 2021).

European Environment Agency (EEA). 2021a. Overview of Environmental Regulations within the European Union. https://www.eea.europa.eu/overview (accessed April 25, 2021).

European Environment Agency (EEA). 2021b. Air Emissions Policy Context. https://www.eea.europa.eu/ themes/air/policy-context (accessed April 25, 2021).

European Environment Agency (EEA) 2021c. European Union Air Quality Standards. https://ec.europa/ environment/air/quality/standards.htm (accessed April 25, 2021).

European Environment Agency (EEA). 2021d. European Air Permitting Directives and Procedures. https://ec. europa/air/permitting (accessed April 25, 2021).

European Environment Agency (EEA). 2021e. Drinking Water Standards. https://ec.europa/drinkingwater (accessed April 25, 2021).

European Environment Agency (EEA) 2021f. Introduction to the European Union Water Framework Directive. https://ec.europa.eu/environment/water/water-framework/information (accessed April 25, 2021).

European Environment Agency (EEA). 2021g. Water Pollution Framework in the European Union. https:// www.ec.europa/waterframework (accessed April 25, 2021).

European Environment Agency (EEA). 2021h. Resource Efficiency and Waste. https://www.eea.europa.eu/ themes/waste (accessed April 25, 2021).

European Environment Agency. 2021i. Hazardous Waste. http://ec.europa/environment/waste/hazardous_ index.html (accessed April 25, 2021).

European Environment Agency (EEA). 2021j. Soil Contamination in Europe. https://www.eu.europa.soil-contamination (accessed April 25, 2021).

European Union. 2021. European Countries in Brief. https://www.eu./european-union/about-eu/countries/ member-countries_eu (accessed April 25, 2021).

Henry, L. A. and V. Douhovnikoff, V. 2008. Environmental Issues in Russia. *Journal of Annual Review of Environmental Resources*. Vol. 33. no. 1. pp. 437–460.

Institute for European Environmental Policy. 2021. Institute for European Environmental Policy. https://ieep. eu/ (accessed April 25, 2021).

Kenya Air Quality Act. 2009. National Environmental Management Authority. Nairobi, Kenya. https://www. nema.go.ke/NEMA/Airqualityact (accessed April 25, 2021).

Kenya Geological Society. 2021. Summary of the Geology and Minerals Resources of Kenya. https://www. apsea.or.ke/index/geological-society-of-kenya (accessed April 25, 2021).

Kenya Impact Assessment and Audit Act. 2003. National Environmental Management Authority. Nairobi, Kenya. https://www.nema.go.ke/NEMA/Impactassessmentandauditact (accessed April 25, 2021).

Kenya National Bureau of Statistics. 2021. *Estimated Population of Kenya in 2021*. https://www.knbs.or.ke/ (accessed April 25, 2021).

Kenya National Environment Management Authority. 2021. *Kenya State of the Environment and Outlook 2010*. https://www.nema.go.ke/downloads (accessed April 25, 2021).

Kenya Waste Management Act. 2006. National Environmental Management Authority. Nairobi, Kenya. https://www.nema.go.ke/NEMA/Wastemanagementact (accessed April 2021).

Kenya Water Quality Act. 2006. National Environmental Management Authority. Nairobi, Kenya. https://www.nema.go.ke/NEMA/Waterqualityact (accessed April 25, 2021).

Kenya Wildlife Service. 2021. *National Parks and National Wildlife Reserves in Kenya*. https://www.go.ke/national-parks (accessed April 25, 2021).

Kinny, P. L., Gatari, M., Volavka-Close, N., Ngo, N., Ndiba, P. K., Law, A., Gachanja, A., Mwaniki, S., Chillrud, S. N., and Sclar, E. 2011. Traffic Imapcts on PM2.5 Air Quality in Nairobi, Kenya. *Journal of Environmental Science and Policy*. Vol. 14. no 4. pp. 369–378.

Marey, H. S., Gille, J. C., El-Askary, H. M., Shalaby, E. A., and El-Raey, M. E. 2010. Study of the Formation of the "Black Cloud" and its Dynamics Over Cairo, Egypt Using MODIS and MISR Sensors. *Journal of Geophysical Research*. Vol. 115. no. 10. pp. 1,029–1,048.

Mwambazambi, Kalemba. 2010. Environmental Problems in Africa: A Theological Response. *Ethiopian Journal of Environmental Studies and Management*. Vol. 3. no. 2. University of South Africa. Pretoria, South Africa. pp. 54–66.

National Intelligence Council. 2012. *The Environmental Outlook in Russia*. http://www.dni.org/nic/special_russianoutlook.html (accessed April 25, 2021).

Netherlands Ministry of Infrastructure and the Environment. 2015. *Dutch Pollutant Standards*. https://www.government.nl/ministry-of-infrastructure-and-the-environment. or at http://www.sanaterre.com/guidelines/dutch.html (accessed April 25, 2021).

Newell, J. P. and Henry, L. A. 2017.The State of Environmental Protection in the Russian Federation: A Review of the Post-Soviet Era. *Journal of Eurasian Geography and Economics*. DOI: 10.1080/1538721 6.2017.1289851. (accessed August 20, 2021).

Norwegian Environment Agency. 2021a. *Ministry of Climate and Environment*. https://miljodirektoratet.no/no/Om-Miljodirektoratet?Norwegian-Environment-Agency (accessed April 25, 2021).

Norwegian Environment Agency. 2021b. *Solid and Hazardous Waste*. http://www.miljodirektoratet.no/en/legislation1/regulations/waste-regulations/chapter11/ (accessed April 25, 2021).

Norwegian Environment Agency. 2021c. *Guidelines for Risk Assessment of Contaminated Sites*. https://www.miljodirektoratet.no/old/klif/publikasjoner/andre/1691/ya1691.pdf (accessed April 25, 2021).

Organization for Economic Co-operation and Development (OECD). 2016. Summary Report of Switzerland's Environmental Laws. Paris, France. 21p. https://www.OECD.org/Switzerland/environmental (accessed April 25, 2021).

Organization for Economic Co-operation and Development (OECD). 2021. *Summary Reports of Russia*. http://www.oecd.org/russia/publicationsdocuments (accessed April 25, 2021).

Porter, M. E. 2009. Turkey's Competitiveness: National Economic Strategy and the Role of Business. *Institute for Strategy and Competitiveness*. Harvard Business School: Cambridge, MA. https://www.ics.hbs.edu (accessed April 25, 2021).

Quinn, L. P., de Vos, B. J., Fernandes-Whaley, M., Roos, C., Bouwman, H., Pieters, K. R., and Berg, J. 2011. Pesticide Use in Africa. https://www.environment.gov.sa/docs/pesticides.pdf (accessed April 2021).

Ramadam, A. R. and Nadim, A. H. 2014. Hazardous Waste in Egypt: Status and Challenges. In: *Waste Management and the Environment*. Popov, V., Itoh, H., Brebbia, C. A. and S. Kungolos (editors). WIT Press. Southhampton: United Kingdom. pp. 125–129.

Russian Federal Law on Environmental Protection. 2002. http://www.rospotrebnadzor.ru (accessed April 25, 2021).

Russian Ministry of Economic Development of the Russian Federation. 2021. Industrial Safety of Major Hazard Installations. http://en.smb.gov.ru/regulation/116fz (accessed April 25, 2021).

Russian Ministry of Natural Resources (MNR). 2021. *Ministry of Natural Resources and Environment of the Russian Federation*. https://www.mnr.ru/english (accessed April 25, 2021).

Smithsonian Museum of Natural History. 2017. Introduction to Human Evolution. *Smithsonian Institution's Human Origins Program*. Smithsonian Institution: Washington, D.C. http://humanorigins.si.edu/education/introduction-human-evolution (accessed April 25, 2021).

South Africa Department of Environment. 1996. *South African Water Quality Guidelines. Domestic Water Use. Second Edition*. Government Gazette: Pretoria, South Africa. 197p.

South Africa Department of Environmental Affairs. 2009a. *South Africa Ambient Air Quality Standards.* Government Gazette. No. 32816. Pretoria, South Africa. 12p.

South Africa Department of Environmental Affairs. 2009b. *National Environmental Waste Act No. 59 of 2009.* Pretoria, South Africa. 309p.

South Africa Department of Environmental Affairs. 2012. *National Waste Management Strategy.* South Africa Government Gazette: Pretoria, South Africa. 77p.

South Africa Department of Environmental Affairs. 2013. *National Norms and Standards for the Remediation of Contaminated Land and Soil Quality in the Republic of South Africa.* Government Gazette. Pretoria, South Africa. 15p.

South Africa Department of Environmental Affairs. 2021a. *Outline of Environmental Regulations.* https://environment.gov.sa/legislative/actsregulations (accessed April 25, 2021).

South Africa Department of Environmental Affairs. 2021b. *Cape Town Drought; Current Status.* https://www.capetowndrought.com (accessed April 25, 2021).

South Africa Government Gazette. 2018. South Africa National Environmental Management Act (NEMA). 1998. *National Environmental Management Act. Act No. 107.* Pretoria, South Africa. November 27, 1998. 37p.

Statistics South Africa. 2021. *Mid-Year 2018 Population Estimate of South Africa.* https://www.statsa.gov.sa/?=15 (accessed April 25, 2021).

Switzerland Environmental Protection Act. 2021. *Federal Act on the Protection of the Environment.* https://www.admin.ch/opc/en/classified-compilation/19830267/index.html (accessed April 25, 2021).

Switzerland Office for the Environment. 2021. *Reporting for Switzerland on the Protocol for Water and Health.* https://www.unece.org/fileadmin/DAM/env/water/Protocol/reports/pdf (accessed April 25, 2021).

Tanzania Air Quality Act. 2007. The Environmental Management. *Air Quality Regulations.* https://www.parliament.go.tz/acts-list-air-quality (accessed April 25, 2021).

Tanzania Bureau of Statistics. 2021. *Current Population of Demographics of Tanzania.* https://www.nbs.go.tz (accessed April 25, 2021).

Tanzania Environmental Management Act. 2007. Environmental Management Act. *Soil Quality Regulations.* Dar es Salaam. Tanzania. 25p.

Tanzania Environmental Management Act. 2008. Hazardous Waste Control and Management Regulations. Dar es Salaam. Tanzania. 49p.

Tanzania Ministry of Water and Irrigation. 2015. *Rural Water Supply and Sanitation Program.* https://www,tanzania.go.tz/egov_uploads/documents/tanzania (accessed April 25, 2021).

Tanzania National Environmental Standards Compendium (NESC). 2018. Drinking Water Specifications. *TZS 789:2003.* Dar es Salaam, Tanzania. 78p.

Tanzania Water Resources Act. 2009. *Tanzania Water Resources Management.* https://www.tanzania.go.tz/egov_iploads/documents/water-resources_en_sw.pdf (accessed April 25, 2021).

Tanzania Water Supply and Sanitation Act. 2009. *The Republic of Tanzania Water Supply and Sanitation Act.* Dar es Salaam, Tanzania. 47p.

Turkish Statistical Institute. 2021. *Census of the Population of Turkey 2017.* https:///www.turkstat.gov.tr (accessed April 25, 2021).

Turkey Ministry of Environment and Urbanization. 2021a. *Turkey Environmental Law.* https://csb.gov.tr/ (accessed April 25, 2021).

Turkey Ministry of Environment and Urbanization. 2021b. *Directorate of Environmental Management. Department of Marine and Coastal Management.* http://mavikart.cevre.gov.tr/en/Haberler.aspx (accessed April 25, 2021).

United Nations. 2021a. World Population Prospects. United Nations Department of Economic and Social Affairs. *Population Division.* https://esa.un.org/unpd/wpp/data (accessed April 25, 2021).

United Nations. 2021b. Population Trends of Russia. Department of Economic and Social Affairs. *Division of Population.* https://www.UN.org.unpd (accessed April 25, 2021).

United Nations. 2021c. *Outer Limits of the Continental Shelf Beyond 200 Nautical Miles (370 km) from the Baselines.* https://www.un.org/depts/los/clcs.htm (accessed April 25, 2021).

United Nations. 2021d. Human Development Report. *United Nations Development Programme.* http://www.hdr.undr.org/sites/2017 (accessed April 25, 2021).

United Nations. 2021e. *United Nations Population Programme For Egypt.* https://www.egypt.unfpa.org/em/united-nations-population-egypt-2 (accessed April 25, 2021).

United States Environmental Protection Agency (USEPA). 2021a. *Air Quality in Africa.* http://epa.gov/international/air/africa.htm (accessed April 25, 2021).

United States Environmental Protection Agency (USEPA). 2021b. *EPA Collaboration with Sub-Saharan Africa.* https://www.epa.gov.international-cooporation/epa-collaboration-sub-saharan (accessed April 25, 2021).

United States Geological Survey (USGS). 2021. *Colorado River Basin Census.* https://water.usgs.gov/watercensus/crb-fas/index.html (accessed April 25, 2021).

University of Michigan. 2017. *Environmental Issues of Africa.* https://www.globalchange.umich.edu.globalchange1 (accessed April 25, 2021).

Welch, C. 2018. *Why Cape Town is Running Out of Water and Who's Next.* https://www.news.nationalgeographic.com/2018/02/cape-town-running-out-of-water (accessed April 25, 2021).

World Bank. 2013. The Arab Republic of Egypt, Air Pollution in Greater Cairo. *Sustainable Development Department.* Middle East and North Africa Region: Washington, D. C. 150p.

World Health Organization. 2021a. *World Health Organization Country Profile – Egypt.* https://www.who.int/countries/egy/en/ (accessed April 25, 2021).

World Health Organization. 2021b. *Drinking-Water Quality Standards in the Eastern Mediterranean Region.* World Health Organization. WHO-EM/CEH/143/E: Cairo, Egypt. 44p.

World Water. 2021. *Russian Water Industry Remains at Crossroads.* http://www.waterworld.com/articles/wwi/print/volume-25 (accessed April 25, 2021).

Yablokov, A. 2016. Environmental Problems and Projections in Russia. *2016. Institute for Global Communications.* Moscow, Russia. https://www.igc.org/russia (accessed April 25, 2021).

14 Summary of Environmental Regulations of Asia and Oceania

14.1 INTRODUCTION

This chapter will explore the environmental regulations and pollution of Asia and Oceania and is organized by continent and then by country. In total, the countries that will be evaluated account for 70% of the human population and 75% of the land area on Earth.

To widen our prospective and grasp a more comprehensive and educated view, we will also evaluate countries that are important or unique to all humans on Earth. These countries will provide us with special insight into unique issues, situations, and problem-solving challenges due to a host of complex circumstances that some countries are now facing. These countries have special challenges that, the combination of which, are unique and will require us to explore potential reasons behind some of the challenges faced by these nations as a result of social concerns, political unrest or lack of action, economic, geographic, geological, or other factors that result in poor or excellent environmental performance.

We will focus our discussion of each country on the fundamentals introduced in Chapter 1, which include:

- Protect the air
- Protect the water
- Protect the land
- Protect living organisms and cultural and historic sites

14.2 ASIA

Asia is the largest and most populous continent. The land area of Asia is estimated at 43,608,000 square kilometers and represents approximately 29.4% of Earth's land surface. Asia's estimated population is greater than 4 billion and represents approximately 56% of the world's human population. China and India account for most of the population of Asia with a combined population of 2.7 billion (United Nations 2021a). Countries within the continent of Asia that we will evaluate include China, India, Japan, South Korea, Saudi Arabia, Malaysia, and Indonesia. We have already evaluated Russia, most of which lies within Asia, but was evaluated as part of Europe. This is because the western part of Russia, including its capital of Moscow, is located within the continent of Europe.

14.2.1 CHINA

China is located in southeast Asia and is the third largest country by area in the world behind Russia and Canada with a land area of 9,596,960 square kilometers. The geography of China is perhaps the most wide ranging of any country on Earth. As an example, China's lowest point is Turpan Pendi at 154 meters below mean sea level, which is one of the lowest places on Earth. In contrast, China's highest point is Mount Everest, the highest point on Earth at 8,848 meters above mean sea level. Due to the extremes in elevation differences, China's climatic range is from tropical along the south eastern portion of the country to subarctic along the western boundary. China's estimated population now exceeds 1.42 billion making it the most populous country on Earth (United Nations 2021a). Most of China's population is located along its eastern coastline near the Yellow Sea and the East China Sea. The two largest cities in China include Shanghai and China's capital, Beijing. Shanghai has an estimated population of 24.5 million and Beijing has an estimated population of 21 million (United Nations 2021a).

China's economy is now considered the largest in the World at $23.12 trillion, followed by the European Union at $19.9 trillion, and then the United States at $19.3 trillion. However, China is still a relatively poor country in terms of its standard of living at only $15,600 per person compared to the United States at $59,500 per person based on total GDP. This is staggering to imagine that China's economy is larger than that of the United States but will likely continue to grow significantly in the future. China has built its economy on relatively cheap labor. China's largest trading partner is the United States (United States Department of State 2021a, United States Department of Commerce 2018, and United States Census Bureau 2021).

14.2.1.1 Environmental Regulatory Overview of China

China faces significant environmental issues due to its huge population, rapid growth, and relatively low standard of living. Those environmental issues include air and water pollution, water shortages in the northern part of the country, deforestation, sanitation, soil and groundwater impacts, soil erosion, and desertification (USEPA 2021a). Much of the air pollution is due to its heavy dependence on coal, the rapid increase in the number of motor vehicles, especially in its urban regions, and industry (USEPA 2021a). Figure 14.4 is an example of poor air quality along the Great Wall of China near Beijing.

China is a party to many international environmental treaties, protocols, and agreements including signing the Kyoto Protocol to reduce carbon emissions. Others include the following (United Nations 2021b)

- United Nationals Convention to Combat Desertification
- Marine Dumping
- Ozone Layer Protection
- Endangered Species
- International Tropical Timber Agreement
- International Convention for the Regulation of Whaling
- Ship pollution
- Wetlands
- Convention on Biological Diversity
- Climate change
- Antarctic Environmental Protocol
- Nuclear Test Ban Treaty

USEPA and China's Ministry of Ecology and Environment (MEE) have been collaborating for several years to strengthen and expanded China's capabilities at mitigating pollution that now present significant challenges to human health and the environment in China (USEPA 2021a). Objectives of the collaboration have been to improve many of China's environmental protection programs including the following (USEPA 2021a)

- Improving air quality
- Reducing water pollution
- Preventing exposure to chemicals and toxics
- Remediating soil

- Treating hazardous waste
- Disposal of hazardous and solid waste
- Improving environmental enforcement
- Promoting environmental compliance
- Improving existing environmental laws
- Enhancing environmental education
- Chemical management

USEPA has also been collaborating with China's Ministry of Science and Technology on joint research to better access air emission sources and impacts on China to better develop mitigation practices and to enhance sustainability. Areas of research have included (USEPA 2021a):

- Soil remediation
- Groundwater remediation
- Water sustainability
- Toxicology
- Motor vehicle emissions
- Clean cooking methods
- Air pollution monitoring and assessment

14.2.1.2   Air
According to the WHO (2021a), 16 of the world's 20 cities with the worst air quality are located in China. Contributors to the poor air quality include:

- Motor vehicle exhaust
- Industry
- Combustion of coal, much of which is high in sulphur

Coal is considered the most significant source of air pollution in china. China generates 80% of its electricity from coal generating plants. China burns six million tons of coal per day (USEPA 2021a). Motor vehicles also represent a significant source of air pollution in China. It is estimated that China will have 400 million vehicles by 2030, which by any calculation, means that motor vehicle pollutants will continue to be a significant issue in the future. Just as in other countries, the main components of the air pollution in China include (USEPA 2021a):

- Particulate matter (PM10)
- Fine particulate matter (PM2.5)
- Sulphur dioxide
- Nitrogen oxide
- Nitrogen dioxide
- Ozone

China leads the world in premature deaths caused by air pollution at 1.2 million per year according to the World Health Organization (2021a) and the National Geographic Society (2021a).

China first promulgated ambient air standards in 1978 (Siddiqi and Chong-Xain 1984). Article II of the law adopted in 1978 states:

The State protects the environment and natural resources and prevents and eliminates pollution and other hazards to the public.

From passage of the law, China developed ambient air quality standards that we announced in 1982. There are three divisions in the ambient air standards in China that include (Siddiqi and Chong-Xain 1984):

- Class I: This level represents the ideal standard, at which there is no direct or indirect harmful effect of any type on humans or ecosystems, even under long-term exposure.
- Class II: This level is set at what is believed to be threshold concentrations for effects on sensitive plants, and above which chronic effects on humans may become noticeable. This level is set as the standard in areas that are urban residential, commercial, cultural, or rural.
- Class III: This is the level considered necessary to protect people from acute or chronic poisoning, and to protect animals and plants (not including sensitive plants)

During the 2008 Olympics, China instituted a mandate that shut down factories three weeks prior to the 2008 Summer Olympics and placed restrictions on the use of motor vehicles. The result was a measurable positive effect on the health of people in Beijing (Rich et al. 2010). In 2015, China formed the Clean Air Alliance of China (Clean Air Alliance of China (CAAC) 2015) to formulate recommendations to improve ambient air quality in China. China is now beginning to address the challenge of improving its air quality with the formation of the CAAC (USEPA 2021a).

14.2.1.3   Water

Water pollution in China is as important as air pollution but is less widely known. China suffers from water shortages and severe water pollution. The foundation behind China water shortage and pollution is its huge population combined with rapid industrialization and a historical lack of environmental oversight and enforcement (United Nations 2013).

Along with the challenges of water pollution, sanitation is also a significant contributor to its poor water quality. Approximately 100 million Chinese do not have access to a reliable water source and an estimated 460 million Chinese do not have access to basic sanitation. Although these estimates have improved greatly in recent decades, China still faces significant challenges (World Health Organization WHO 2015a and 2015b). According to the China Research Academy of Environmental Sciences (2013), both surface water and groundwater are heavily polluted in many areas of China. This is exacerbated by the uneven distribution of groundwater and surface water resources across China due to its physical geology. Northern portions of China account for approximately 65% of the land area but only have 30% of the water resources.

In contrast, the southern portion of China which accounts for 35% of the land area has 70% of the surface and groundwater resources. This uneven distribution has led to over extraction and exploitation of groundwater in northern China, which has led to a decrease in water levels in some groundwater supply wells of more than 100 meters in some cases (China Research Academy of Environmental Sciences 2013). Causes of water pollution in China have been identified as the following (China Research Academy of Environmental Sciences 2013):

- Discharge and infiltration of untreated urban stormwater runoff
- Discharge and infiltration of untreated industrial discharges
- Runoff and infiltration from areas of soil contamination
- Runoff and infiltration from agricultural areas
- Lack of groundwater protection legislation
- Lack of awareness

In response to the degradation of both surface and groundwater, the Chinese Ministry of Environment has launched a series of research efforts to understand both surface water and groundwater impacts so that by the year 2020 surface water and groundwater impacts will be identified and risks reduced (China Research Academy of Environmental Sciences 2013). As stated earlier in this section, China faces challenges in supplying clean drinking water to its population. Approximately 60% of China's population obtains its drinking water from surface water bodies that mainly include rivers, and 40% obtains its drinking water from groundwater. The majority of groundwater withdrawal for drinking water occurs in the northern regions of China (World Health Organization 2015b). China established drinking water criteria on July 1, 2012 by adopting the World Health Organization Guidelines for drinking water (WHO2015a).

14.2.1.4 Solid and Hazardous Waste

Efforts to regulate solid and hazardous waste in China have been plagued by numerous significant social, economic, and political challenges. In 2007, the first national census of hazardous waste in China estimated that China produced 45 million metric tons of hazardous waste but only 10 million metric tons were produced from industry (Wencong 2012). However, much information was lacking that includes (Wencong 2012):

- Types of waste
- How the wastes were generated
- How the wastes were transport
- Where the wastes were disposed

The first law addressing solid waste in China was passed in 1995 and was titled Law of the People's Republic of China on the Prevention and Control of Environmental Pollution by Solid Waste (China Ministry of Ecology and Environment [CMEE] 2021). As part of China's solid waste law, a national catalogue of hazardous waste was developed to identify and classify solid wastes as either hazardous or not hazardous. In China, as in most other countries, a waste is hazardous if the waste is any of the following (CMEE 2021):

• Corrosive	• Ignitable	• Flammable
• Medical waste	• Toxic	• Reactive
• Explosive	• Pesticide	• Herbicide

Other types of wastes considered hazardous in China include (CMEE 2021):

• Electronic wastes	• Fluorescent light bulbs
• Waste Hg-containing thermometers	• Waste mineral oils
• Waste solvents	• Waste film and photographs
• Waste Ni-Cd battery cells	• Disinfectants
• Waste paints	

Exempted from the law in China are wastes generated by households in daily life, which is similar to that of the United States (CMEE 2021). Until recently, China did not have a reliable infrastructure for disposal of hazardous wastes. Therefore, many industrial sites stored wastes until a satisfactory transport method and disposal site was established nearby that was in compliance with China's solid waste law. However, historically solid wastes were disposed of anywhere convenient including farm fields, next to ponds or lakes, or wastes were simply burned. China is currently examining and will likely revise its solid and hazardous waste regulations sometime in 2021 (CMEE 2021).

14.2.1.5   Remediation

In 2014, The China Ministry of Environmental Protection announced five standards of environmental protection (CMEP Notice No. 14 of 2014) that provided a framework for future regulatory development of soil and groundwater pollution prevention and remediation (MEP 2014). The five standards included (China Ministry of Environment Protection 2014):

- Technical guidelines for Environmental Site Investigation
- Technical Guidelines for Environmental Site Monitoring
- Technical Guidelines for Risk Assess of Contaminated Sites
- Technical Guidelines for Site Soil Remediation
- Definition of Terms of Contaminated Sites

On August 31, 2018, China passed a new law on soil pollution and control. The new law will go into effect on January 1, 2019. The law stipulates that the China Ministry of Environmental Protection establish soil remediation standards and conduct a census on the condition of soils nationwide every 10 years (China Ministry of Environmental Protection 2021). The law strengthens the responsibilities of governments and the responsibilities and liabilities of polluters. In addition, the law also require the polluters of farmland to develop rehabilitation plans, provide them the appropriate government authority, and to carry out the plans.

Until the China Ministry of Environmental Protection adopts cleanup values for pollutants of its own, the Dutch Intervention Values (DIVs) are commonly used for comparison purposes to evaluate whether soil, sediment, and groundwater require cleanup and to what extent.

14.2.1.6   Summary of Environmental Regulations in China

Environmental regulations in China are similar to that of the United States and European Union with two major exceptions that are:

- Requirement that each facility obtain one operating permit that covers all operating conditions including environmental aspects
- Many regulations are either rather new (i.e., air and water) or are currently under development (i.e., land)

China is a rapidly developing nation with significant air, water, sanitation, and land pollution issues. Together with its large population, China faces significant future challenges. However, China now has the largest economy in the World and will likely continue to grow and prosper from this development. In addition, China has made major strides at developing an environmental framework that can hopefully deal with improving the quality of its air, water, and land from pollution in the future. The combination of a strong and growing economy with a robust set of environmental regulations and political will means that it is likely that China will continue to improve environmentally at fast pace.

14.2.2   Japan

Japan is an island nation comprised of four main islands and thousands of smaller islands just to the east of the western coast of Asia in the Pacific Ocean. The population of Japan is estimated at just over 127 million (United Nations, 2021a). The capital of Japan is Tokyo, which has a population of 36 million residents and is considered the largest urban center in the World (United Nations 2021a). According the United Nations population projects, Japan's population is not expected to grow over the next ten years. In fact, Japan's population is expected to decrease by more than 2 million in the next 10 years (United Nations 2021a).

Japan is a highly developed country with the 4th largest economy in the World behind China, United States, and the European Union. Japan has an estimated GDP of over $5 trillion dollars annually (United Nations 2021a). Japan is ranked the 61st largest country with a land area of 377,973 square kilometers and has 29,751 kilometers of coastline (United Nations 2021a). The population density of Japan is approximately 348 individuals per square kilometer, which is much higher that the worldwide average of 51 and ten times that of the United States, which is 33 individuals per square kilometer (United Nations 2021a). Japan experiences natural disasters in the form of volcanic eruptions, landslides, tsunamis, and typhoons that can lead to significant environmental concerns and lose of life. For example, an earthquake in March 2011 damaged more than 270,000 buildings and was also was responsible for the generation of an estimated 24 million metric tons of waste debris. In addition, the earthquake and subsequent tsunami damaged the Fukushima Nuclear Power Plant that released significant quantities of radiation that had large aerial impacts and will have lasting effects for decades (Japan Ministry of the Environment 2021a).

Japan's economy is primarily production of motor vehicles, electronics, industrial tools, steel and other metals. Japan also has some food production that is comprised of rice, sugar beets, fruits and vegetables, fish and beef (United States Department of State 2021b). Japan has undergone rapid industrialization since the end of World War II in 1945. The United States and Japan have a very close relationship on matters including (United States Department of State 2021b):

• Development assistance	• Health
• Environmental	• Resource protection
• Women's empowerment	• Science and technology
• Infectious disease	• Medicine
• Space exploration	• Education
• International diplomatic initiatives	

14.2.2.1 Environmental Regulatory Overview of Japan

Environmental regulations in Japan are some of the strictest in the World and are the responsibility of the Japan Ministry of the Environment. Environmental regulations in Japan had their beginning in 1967 and were significantly revised in 1994 as Japan was experiencing significant industrialization (Japan Ministry of the Environment 2021b). Japan has significant environmental issues related to air, water, and land. Disposal of solid waste is of particular concern since Japan has a relatively little land compared to other countries and has a large population with a high density and generates large amounts of solid waste (Japan Ministry of the Environment 2021a).

Japan itself points to an incident that occurred in 1896 at the Ashio Copper Mine as the beginning of the environmental movement in Japan. A massive flood from torrential rains occurred in the region of the mine in September 1896 and caused not only the mine to flood but also the Watarase, Tone, and Edo Rivers to overflow their banks. Damage from the flood nearly destroyed 136 towns in the path of the flood and contaminated the entire affected area with heavy metals including many agricultural areas that effectively poisoned the soil and thousands of residents in the path of flood or who ate food from farms within the affected area (Kichira and Sugai 1986 and Japan Ministry of the Environment 2021b). This incident is considered a natural disaster triggered by human actions that impacted an entire region of Japan because of a general lack of appreciation for the environment that included deforestation, poor mining practices, little of no drainage control, and lack of knowledge of potential environmental impact from mining wastes. This incident highlighted the need to place more emphasis on safety and environmental integrity and started Japan's first mass-citizens movement in Japanese history (Japan Ministry of the Environment 2021b).

14.2.2.2   Air

Since the 1980s, Japan has improved its air quality significantly, especially the reduction of Sulphur dioxide which have been reduced by 78%. Air exhaust, especially from diesel engines, has received particular attention in Japan where approximately 80% of the nitrogen oxides and particulate matter originate from diesel engines (Japan Ministry of the Environment 2021b). In addition, in 1986, Japan produced nearly 120,000 metric tons of chlorofluorohydrocarbons (CFC's) or ozone depleting substances and now produces virtually none (Japan Ministry of the Environment 2021b).

An additional concern of air pollution in Japan is air pollution originating in China that affects the air quality in Japan. Sources of pollution in Japan believed to originate in China include photochemical air pollutants that react with sunlight and produce smog and higher ozone levels, nitrogen oxide, acid snow and acid rain, and hydrocarbons (Japan Ministry of the Environment 2021c). As a result of the significant air quality issues experienced in the 1960s and 1970s during a rapid industrial development phase, Japan enacted some of the most strict air pollution control laws in the World (Japan Ministry of the Environment 2021d). Ambient air quality standards for Japan are at http://www.env.go.jp/en/air/aql/aq.html (Japan Ministry of the Environment 2021d).

14.2.2.3   Water

The Ashio Mine incident is often mentioned when discussing water pollution in Japan (Japan Ministry of the Environment 2021d). Japan has made significant progress over the past few decades at improving water quality and the reduction of water pollution, especially heavy metals, as a result of regulations on industrial wastewater. However, other pollutants in water such as volatile organic compounds and other organics have not experienced as much improvement, especially in urban rivers and streams and inland lakes and reservoirs (Japan Ministry of the Environment 2018e).

Pollution of fresh water is not the only type of water pollution that Japan faces. Pollution in the ocean waters surrounding Japan has also been a significant issue, especially since Japan consumes high volumes of fish and other foods that have their origin in the marine environment. Mercury in the form of methylmercury discharged into Minamata Bay for over 30 years from nearby industries contaminated both marine life and residents who relied on the marine environment of Minamate Bay as a source of food (Japan Fact Sheet 2021). Other examples of anthropogenic deterioration of water resources in Japan include (JJapan Ministry of the Environment 2018e):

• Loss of wetlands
• Construction of dams
• Lining riverbanks with cement
• Straightening riving channels

Water discharge regulations in Japan are similar to that of the United States and many developed countries of the World in that Japan requires every facility that discharges industrial wastewater to obtain a permit (Japan Ministry of the Environment 2021f). However, the water quality standards are strict and inspections and enforcement of regulations are usually frequent and strict as well. Japan water quality standards are at http://www.env.go.jp/en/pollution/issues (Japan Ministry of the Environment 2021f).

14.2.2.4   Solid and Hazardous Waste

Solid and hazardous waste in Japan, along with air and water, is perhaps the most important environmental issue in Japan. Being a highly urban and industrialized country with a high relative population and density, which is exacerbated by the fact that Japan is a relatively small country, has placed enormous pressure on Japan to control its solid and hazardous waste (Japan Ministry of the Environment 2021g). Currently, Japan estimates that it produces approximately 50 million metric tons of solid and hazardous waste per year. As a measure to limit the amount of wastes that are disposed of in landfills, Japan incinerates up to 80% of its solid waste and has an aggressive recycling effort for paper, plastic, and metals. As part of recycling efforts, Japan also has aggressive programs in place to reuse material and reduce the amount of waste generated (Japan Ministry of the Environment 2021g).

Of the 50 million metric tons of solid and hazardous waste generated in Japan, approximately 42 million is from industry and the remainder is municipal waste (Japan Ministry of the Environment 2021g). Japan requires specialized air and water pollution control equipment on all incinerators used to burn solid waste. Japan measures dioxin levels in ambient air and water because of its significant use of incinerators used to burn solid wastes and the levels of dioxin have been significantly reduced by well over 95% since 1998 (Japan Ministry of the Environment 2021h). Japan has developed contaminant limits on the leachate from residual waste following the incineration process that is composed of what is termed cinder dust, ash, sludge, and slag and are located at http://www.env.go.jp/en/soil/leachate.html (Japan Ministry of the Environment 2021i).

14.2.2.5 Remediation

During the post-war period, Japan faced the reality of dealing with millions and millions of metric tons of waste and urban waste land. At the time, these wastes were dumped into the ocean, in rivers and piled up in the open causing plagues of flies and mosquitoes and the spread of infectious diseases. The Japanese government made several attempts at passing laws such as the Public Cleansing Act of 1954, and the Act on Emergency Measures of 1963 concerning the development and living environment, and the Basic Act for Environmental Pollution Control enacted in 1967 (Japan Ministry of the Environment 2021h). None of these laws adequately addressed the issue of soil pollution.

It wasn't until the enactment of the Japan Soil Contamination Countermeasures Act of 2002, which had the objective to formulate measures to evaluate the degree to which land in Japan was impacted by contaminants and to provide methods and criteria to prevent further deterioration and restore impacted locations to protect human health did Japan make significant progress at restoring and improving its land environment (Japan Ministry of the Environment 2021h). Cleanup levels for soil in Japan are listed at http://www.env.go.jp/en/soil/leachate.html. (Japan Ministry of the Environment 2018j).

The general conditions under the Soil Contamination Countermeasures Act of 2002 are similar to that of the United States in that Japan requires that those who are responsible for environmental restoration are those who have polluted the environment under a Japanese provision in the regulations termed polluters pay principle.

Japan is heavily dependent on surface water as its primary source of drinking water. In fact, 89% of Japan's drinking water is from surface water sources and 11% is from groundwater. Since Japan is densely populated and a relatively small country by land area and is an island nation, providing a reliable source of drinking water and preventing fresh water sources from contamination is a very high priority (Japan Ministry of the Environment 2021a). Therefore, Japan has an array of over 4,000 monitoring wells located throughout the country to monitoring groundwater and thousands of surface water monitoring points to monitor surface water quality. The purpose of the well network is to act as an early warning system so that potential threats to water quality are address at the earliest stage possible (Japan Ministry of the Environment 2021k).

In addition, Japan has strict soil cleanup values and monitoring points near potential sources of contamination such as landfills and heavily industrialized areas (Japan Ministry of the Environment 2021k). Lastly, Japan has undertaken several water conservation and recycling efforts (Japan Ministry of the Environment 2021k). Groundwater quality values are at http://www.env.go.jp/en/spcl/html. (Japan Ministry of the Environment 2021k). In Japan, soil quality values are the same as groundwater quality values. This is done to protect the groundwater sources (Japan Ministry of the Environment 2021k).

14.2.2.6 Summary of Environmental Regulations of Japan

Japan is one of the most developed nations on Earth and as a result, has experienced significant environmental degradation of its air, water, and land. The contributing factors to environmental degradation in Japan can be traced to wars, lack of knowledge, deforestation, erosion and drainage control, rapid industrializations, and lack of effective pollution control measures. In addition, Japan's geography and geology have greatly contributed to its environmental degradation and loss of life from volcanic eruptions, landslides, tsunamis, earthquakes, and typhoons. However, Japan now has one of the most comprehensive and effective environmental regulatory structure and has improved its air, water, and soil quality. To maintain its quality of life and to sustain its population, Japan must continue to develop methods to continue to protect its air, water, and land. Japan has a regulatory structure similar to that of the United States and air, water, and soil quality criteria are strict and consistent from one media to the next. Japan's inspection and enforcement activities are significant and robust compared to other countries.

14.2.3 INDIA

India is the second most populated country on Earth with an estimated population of over 1.2 billion and is second in population only to China. India is the 7th largest country on Earth with a total area of 3,287,263 square kilometers. The largest river in India is the Ganges which is 2,525 kilometers in length. India has varied geography, temperature and precipitation ranges. India experiences monsoons in the summer months that often flood many urban areas. India's lowest elevation is -2.2 meters and its highest elevation is 8,586 meters (Kanchenjunga), the third highest peak on Earth in the Himalayan mountains along its northern border. India shares borders with Bangladesh, China, Pakistan, Nepal, Myanmar, Bhutan, and Afghanistan. Southern portions of India are located along the coastline of the Indian Ocean that is more than 7,500 kilometers in length (Marsh and Kaufman 2015).

India faces significant pollution issues of its air, water and land. This becomes difficult to manage because of its large human population, limited financial resources, lack of infrastructure and current low technological capabilities. However, between 1984 and 2010 India has made significant progress in addressing environmental issues and improving environmental quality (World Bank 2011). This is reflected in India's increase in life expectancy which was 41 years in 1940 compared to 68 in 2016 (World Bank 2016).

India is one of the top agricultural countries of the World and because of this, India uses large quantities of water for irrigation. Approximately 20% of its population do not have access to usable water and 21% of disease in India can be traced back to unsafe water (World Health Organization 2021b). The Ganges River is considered especially polluted with untreated sewerage discharge to the river at several locations. There is a wide disparity in the standard of living in India, perhaps more than most any other country of the world. The average per person GDP in India is $3,600 (World Bank 2016).

14.2.3.1 Environmental Regulatory Overview of India

One of the most significant environmental incidents in India occurred on December 3, 1984, when more than 36,000 kilograms of methyl isocyanate leaked from a Union Carbide pesticide manufacturing plant in Bhopal, India. At least 3,800 people were immediately killed and injured thousands more (Broughton 2005). The disaster highlighted the need for environmental and safety standards, preventative and engineering strategies, and disaster preparedness.

This incident, perhaps more than any other environmental incident in India, changed the course of environmental regulations in India and in many other parts of the World, including the United States. The disaster also demonstrated that a seemingly local incident of the release of contaminants into the environment was intimately tied to global market economic markets (Broughton 2005). In addition, this incident and many others that have occurred historically throughout the world through history, many of which we have mentioned and discussed in some detail, highlight the need for consistent environmental regulation and enforcement globally.

14.2.3.2   Air

The major sources of air pollution in India are from burning of fuelwood and garbage in cook stoves, burning of biomass and dried livestock waste, burning garbage, automobile exhaust and ineffective air pollution control equipment at industrial facilities. According to the United States Department of State (2021a), Delhi the capital city of India, with its estimated population of 19 million residents, earned the distinction of being the most polluted city on Earth, as air quality for Fine Particulate Matter (PM2.5) reached a level of 1,010 ug/m^3. This is especially disturbing when comparing the detected level with the WHO standard for PM2.5 which is 25 ug/m^3.

In 1981, India enacted the Air Prevention and Control of Pollution Act that had an objective of improving air quality through prevention, control, and abatement. The Air Prevention and Control of Pollution Act has five main purposes that include (India Central Pollution Control Board 2009):

- To advise the central government on any matter concerning the improvement of the quality of the air and the prevention, control, and abatement of air pollution
- To plan and cause to be executed a nation-wide program for the prevention, control and abatement of air pollution
- To provide technical assistance and guidance to the State Pollution Control Board
- To carry out and sponsor investigation and research related to prevention, control and abatement of air pollution
- To collect, compile and publish technical and statistical data related to air pollution
- To develop ambient air quality standards

Ambient air quality standards for India are located at http://www.cpcb.nic.in/upload/latest_final_air_standard.pdf. (India Central Pollution Control Board 2009).

The Air Pollution Control Act of 1981, required that each manufacturing facility evaluate whether there were any air emissions from its operations and if so, whether a permit was necessary. In general, manufacturing facilities with any air emissions likely require a permit. (India Ministry of Environment, Forestry, and Climate Change 2021a).

14.2.3.3   Water

The enactment of the India Water Pollution Control Act of 1974, as amended, prohibited any person or entity from knowingly causing or permitting any poisonous, noxious or pollutant matter to enter any stream, well, or sewer (India Ministry of Environment, Forestry, and Climate Change, 2021b). In 1998, the Act was amended to include effluent limits on industrial discharge of wastewater and required industrial plants to obtain permits to discharge wastewater (India Ministry of Environment, Forestry, and Climate Change, 2021c). Drinking water quality and effluent standards in India are at http://www.environmentallawsofindia.com/water/pollution/act. (India Ministry of Environment, Forestry, and Climate Change 2021c). While India has established criteria for drinking water and industrial effluent, enforcement and infrastructure within India stand as the major impediments to improving overall water quality (Agarwal 2015).

14.2.3.4   Solid and Hazardous Waste

In India, as in most other countries, a solid waste is considered hazardous if it has any of the following characteristics (India Ministry of Environment, Forestry, and Climate Change 2021d):

• Toxic	• Radioactive
• Ignitable	• Reactive
• Corrosive	• Explosive

Similar to that of the United States, India requires as part of its solid and hazardous waste regulations that a detailed evaluation, termed Environmental Impact Assessment (EIA) of any facility that treats, stores, or is the final disposal location for hazardous wastes. The EIA evaluates all aspects of potential impact that the facility may pose including (India Ministry of Environment, Forestry, and Climate Change 2021d):

- Surface and groundwater
- Wetlands
- Forested areas
- Nearby urban areas
- Noise
- Wildlife
- Marine life
- Invertebrates
- Other potential sensitive habitats

For compliance purposes in India, each facility that generates a solid waste must follow the recommended steps (India Ministry of Environment, Forestry, and Climate Change 2021d):

- Identify the point of generation of the waste
- Collect data on the process that generated the waste
- Characterize the waste pertaining to the properties of ignitability, corrosivity, flammability, reactivity, and toxicity
- Evaluate whether the waste was generated under any of the 18 listed waste categories in Table 5.3.4
- Evaluate whether the waste can be recycled, reused, or is considered a solid waste requiring disposal, or a hazardous waste requiring disposal at a licensed TSDF
- Identify the appropriate site for disposal
- Arrange for proper transportation of the waste from a licensed transporter

14.2.3.5   Remediation

In India, remediation standards for cleanup of contaminated sites are generally administered through the local or provincial governmental authority and following the procedures outlined in conducting an Environmental Impact Assessment (EIA) (India Ministry of Environment, Forestry, and Climate Change 2021e). The process of conducting and EIA in India is nearly identical to that of the United States, EU, and many other nations. The process includes following the fundamental steps (India Ministry of Environment, Forestry, and Climate Change 2021e):

- Screening: Considered the initial stage and addresses whether there is a need to conduct an EIA
- Scoping: Once it's established that and EIA is necessary, the Scoping Stage is conducted and determines all relevant environmental concerns
- Public Consultation: Public consultation provides the opportunity for input from any persons of interested stakeholders to provide input through the EIA process
- Investigation: This stage may involve several steps or iterations before the data set is complete and concerns are well understood
- Appraisal: This stage is generally considered a data gap analysis and can either accept data, reject data, or require that more data be collected
- Risk Evaluation: This stage then takes all the data and then conducts a project-specific health risk assessment. The standards used to conduct a health risk assessment are those guidelines established by USEPA or the European Union

- Remediation goals: After each of the stages listed above are completed, remediation goals with specific targets are established

14.2.3.6 Summary of Environmental Regulations of India

India has a large disparity of distribution of wealth and standard of living. The average standard of living for the residents of India is less one twentieth (1/20) that of the United States. Together with its large population of well over 1 billion, lack of infrastructure, and other social and political struggles, the environmental pressures are, at times, overwhelming. However, despite these huge challenges, India has made progress at improving its environment and life expectancy since the Bhopal disaster in late 1984.

India has established a robust set of environmental regulations that compare adequately to the United States, EU, and other rather developed nations. However, lack of infrastructure and enforcement along with social and economic factors discussed above, highlight the need for continued improvement.

14.2.4 SOUTH KOREA

South Korea is located on the Korean Peninsula in north-east Asia. Since 1950, Korea has been divided in half at the 38th parallel as a result of the Korean War. The Amnok River forms the border of North Korea, China, and Russia. The Yellow Sea is located to the east, the East China Sea and Korea Strait is to the south, and the Sea of Japan is located to the east. Korea is approximately 20% of the size of California or about 100,000 square kilometers (United Nations 2012). The devastation caused by the Korean War in the early 1950s is still apparent today. The two Korea's have evolved from a common cultural and historical base into two very different societies with radically dissimilar political and economic systems (Asia Society 2021). North Korea is heavily influenced by Chinese and Soviet/Russian culture and South Korea is heavily influenced by western culture. Today, the population of North Korea is approximately 25 million while the population of South Korea is just over 51 million (United Nations 2021a).

South Korea is mountainous along its western boundary and is much less mountainous along its eastern and southern parts. In fact, South Korea is considered 70% mountainous. Rivers tend to flow from east to west with a general descending topography toward the west and south. South Korea has a total land area of 97,230 square kilometers with a population density of approximately 526 people per square kilometer (United Nations 2021c). Since the cease fire with North Korea in 1953, South Korea has essentially been rebuilt. The capital of South Korea is Seoul which is located in the northwestern portion of the country. Seoul has a population of greater than 10 million residents and more than 25 million is the metropolitan proper. South Korea which was primarily an agrarian society before the Korean War with 75% of its population living in rural areas and farming small plots of land has been completely transformed to an industrial and urban nation. South Korea now has 83% of its population living in urban areas. Since much of the population in Korea is urban, land use is an intermixing of industry, residential and agriculture, which is very different than the United States. This presents a challenge in applying environmental regulations evenly. However, the cost and time spent on transportation is much lower because many people live near where they work, and food is grown nearby (United Nations 2012).

The standard of living in Korea per person is just over $35,000 which is more than twice that of China but is well below that of the United States. South Korea experienced rapid growth in the 1970s and remains under rapid development. The GDP of South Korea is $1.69 trillion which is approximately 12 times less than the United States. As a result of this development, the natural environment suffered deterioration of the air, water, and land and resulted in the passage of numerous environmental laws (United Nations 2012).

14.2.4.1 Environmental Regulatory Overview of South Korea

The United States Environmental Protection Agency (USEPA), is heavily engaged in collaborating with Korea as a result of the Korea-United States Free Trade Agreement (USEPA 2021b).

Through this mechanism, the USEPA and the Korean Ministry of the Environment and partner agencies in both countries have been cooperating to strengthen (USEPA 2021b):

- Environmental governance
- Air quality
- Water quality
- Reduce waste
- Recycling
- Reusable energy
- Protections against toxic pollutants

14.2.4.2 Air

Air quality in South Korea has been improving but is still a major environmental issue, especially in its capital city of Seoul. As with Japan, South Korea experiences increased air pollution from China in addition to its own sources of pollution (Korea Ministry of the Environment 2021a). The major air pollutants in South Korea should include (USEPA 2021b):

- Fine particulate matter (PM2.5)
- Particulate matter (PM10)
- Carbon monoxide
- Nitrogen dioxide
- Ozone
- Lead
- Smog
- Hydrocarbons
- Benzene
- Other volatile organic compounds (VOCs)

South Korea set air quality standards for each of the pollutants listed above in 1983 and set the standard for benzene in 2010 (Korea Ministry of the Environment 2021a).

The Korean Ministry of the Environment has been measuring trends in air pollutants within the country for 15 years and has measured a decreasing trend for lead and particulate matter (PM10). However, Sulphur dioxide, Nitrogen dioxide, and Ozone have either been steady or have increased in concentration in ambient air (Korean Ministry of the Environment 2021a). Ambient air quality standards are similar to the United States and EU and can be viewed at http://www.eng.me.go.kr/eng/web/index.do?menuld/252. (Korean Ministry of the Environment 2021a).

14.2.4.3 Water

South Korea has an aggressive and comprehensive water quality monitoring program with 1,476 water quality monitoring stations throughout the nation which is something learned from Japan. The monitoring stations are located along rivers (697 stations), lakes and marshes (185 stations), agricultural areas (474 stations), and 120 other monitoring stations at other strategic locations. Water is being monitored for dissolved oxygen, total organic carbon, pH and other parameters including volatile organic compounds (VOCs), biochemical oxygen demand *e.coli*, phosphorus, and nitrogen. Groundwater is also monitored twice yearly for 20 different analytes including those listed above at 2,499 locations (Korea Ministry of the Environment 2021b).

Surface water quality has improved by approximately 30% since measurements began in 1997. However, approximately 15% of the surface water regions do not meet water quality objectives. One particular issue with surface water is eutrophication with 33 of the 49 lakes being monitored showing medium levels of contaminants that are attributed to cause eutrophication (Korea Ministry of the Environment 2021b).

Water quality for groundwater in Korea is better when compared to surface water. A major concern for groundwater is properly abandoning wells that are no longer in service. Approximately 43,000 wells have been located and properly abandoned. South Korea developed its water quality criteria based on the USEPA method that uses the Integrated Risk Information System (IRIS) which is based on human health and ecological risks but does not consider mobility or persistence.

Drinking water is supplied to 98.1% of the Korean population as of 2012 (Korea Ministry of the Environment 2021b). Drinking Water Quality Criteria are similar to the United States and EU and are located at http://www.me.go.kr/eng/web/index.do. (Korea Ministry of the Environment 2021b).

In Korea, it is required to obtain an industrial discharge permit for any facility that discharges wastewater. Facilities that propose discharge of specific hazardous substances are not permitted to be located near source water protection areas and designated lakes, which is also the case in United States in what is called well-head protection areas (Korea Ministry of the Environment 2021c). Similar to that of the United States, Korea calculates and manages allowable pollutant load to surface water from water treatment plants under what is termed Total Water Pollution Load Management System (TPLMS). The target and allowable pollutant load is specific to each river watershed and is determined through the nation-wide water quality monitoring program (Korea Ministry of the Environment (2021d).

Effluent standards are calculated for each permit issued and fees are collected if the facility discharges any of the following compounds (Korea Ministry of the Environment 2021e):

1. Organic substances
2. Suspended solids
3. Cadmium and its compounds
4. Cyanide
5. Organo-phosphoric compounds
6. Lead and its compounds
7. Hexavalent chromium compounds
8. Arsenic and its compounds
9. Mercury and its compounds
10. PCBs
11. Copper and its compounds
12. Chrome and its compounds
13. Phenols
14. Trichloroacetatic ethylene
15. Tetrachloroethylene
16. Manganese and its compounds
17. Total nitrogen
18. Total phosphorous

14.2.4.4   Solid and Hazardous Waste

The Waste Control Act of 2016 governs the disposal of solid and hazardous waste in Korea and amends previous versions of 2007, 2011, and 2015. The purpose of the Waste Control Act of 2016 is to contribute to environmental conservation and the enhancement of the people's standard of living by minimizing the production of wastes and disposing of generated wastes in an environmentally-friendly manner (Korea Ministry of the Environment 2021f). Innate in the stated purpose above is the desire to minimize the amount of wastes generated and to improve the quality of its disposal so as not to significantly impact the environment.

The Waste Control Act of 2016 states the following requirements (Korea Ministry of the Environment 2021f):

- Every business entity shall reduce the generation of wastes to the maximum extent possible by improving the manufacturing process, etc. of products and minimize the discharge of wastes by recycling his/her own wastes.
- Every person shall take prior appropriate measures with respect to the discharge of wastes to prevent any harm to the environment or the health of resident.
- Waste treatment shall be properly managed in a manner that reduces their quantities and degree of hazard or otherwise is consistent with environmental conservation and the protection of the people's health.
- Any person with causes environmental pollution by discharging wastes shall be responsible for restoring the affected environment and bear the expenses incurred in restoring the damage caused by such pollution.
- To the extent possible, wastes originated in the Republic of Korea shall be disposed of within the Republic of Korea and the importation of wastes shall be restrained.
- Wastes shall be recycled rather than incinerated or buried in order to contribute to the improvement of resource productivity.

Korea instituted a tracking system as part of its waste management system in 1999 to track the transportation of hazardous wastes (Korea Ministry of the Environment 2021g).

In Korea, a waste is hazardous if it a listed waste and is any of the following (Korea Ministry of the Environment 2021f):

• Explosive	• Flammable
• Radioactive	• Infectious
• Medical waste	• Reactive
• Corrosive	• Characteristic waste

A waste is characteristic hazardous waste if a leachate test indicates a concentration of the listed compounds exceeding concentrations listed at http://www.law.go.kr/DRF/law/wastecontrolact (Korea Ministry of the Environment 2021f).

Listed hazardous wastes in Korean are similar to other nations and include the following (Korea Ministry of the Environment 2021f):

- Waste pesticides or herbicides
- Electronic wastes
- Waste thermometers that contain mercury
- Waste oils
- Waste solvents
- Medical wastes
- Wastewater treatment sludges
- Paint waste
- Waste from pigments, glue, varnish, and printing ink

In efforts to reduce solid wastes, Korea has a vigorous recycling program that also includes food wastes. As stated by the Korean government, increased living standards have resulted in excessive convenience, the use of disposable items, and over-packaged products that cause a waste of resources and generate unnecessary waste and bring about negative impacts to the environment. To combat this, the Korean Ministry of the Environment began a recycling program in 1994 and continues to improve the program through the following (Korea Ministry of the Environment 2021h):

- Reduction of plastics
- Reduction of packaging waste
- Reuse of paper cups and bags
- Introduction of farm produce green packaging guidelines and safety regulations
- Developing technologies for flexible packaging and paper containers
- Recycling of metals
- Recycling of wood products

14.2.4.5 Remediation

Korea has established a nearly identical law as in the United States for dealing with legacy contamination of land, groundwater and surface water. The Korean Liability, Compensation and Relief System for Damages from Environmental Pollution Act of 2016 (Korea Ministry of the Environment 2021h). Soil and Groundwater Cleanup Values are at http://www.eng/me/go.kr/eng/web/law/Actno.166922. (Korea Ministry of the Environment 2021h). In an effort to be proactive and avoid soil and groundwater contamination, the Korea Ministry of the Environment has identified 22,868 specific facilities that are subject to soil and groundwater monitoring that include (Korea Ministry of the Environment 2021i):

- Gasoline service stations
- Certain industrial plants
- Petroleum storage sites
- Abandoned metal, coal, and asbestos mines
- Hazardous waste storage and disposal facilities
- Chemical plants
- Treatment plants
- Power generating plants
- Other ecologically sensitive areas

Each site must conduct tests to ensure that no soil or groundwater is present every 1 to 5 years (Korea Ministry of the Environment 2021j). This type of proactive approach to environmental issues is more advanced that most any country in the world including the United States. Through this program, Korea has identified, investigated, and remediated several sites and continues to be proactive to restore its environment (Korea Ministry of the Environment 2021j).

14.2.4.6 Summary of Environmental Regulations of Korea

Korea has essentially rebuilt its country in the last 70 years due to war. This has provided an opportunity to re-construct a country with less political and social impediments and more opportunity for ease of modernization.

Korea perhaps has the most complete set of environmental regulations that we have examined thus far, even more comprehensive that the United States and the European Union. This is reflected in the fact that Korea regulates its farmland as strict as other locations such as residential and industrial properties. In addition, Korea has robust recycling laws that involve all citizens not just industry or commercial businesses. In addition, from a remediation point of view, Korea monitors its high risk locations for the presence of contamination as an early warning system and also closely monitors is surface water and groundwater resources as well.

Korea has a standard of living of just over $35,000 per person which is well below the average in the United States of $59,500. The population of Korea slightly exceeds 50 million residents, which is approximately 7 times less than the United States. However, Korea is only around 100,000 square kilometers in size which is approximately 100 times smaller in land area than the United States. This means that Korea has a rather large population for its overall land size. The economic, geographical and demographical statistical facts listed above along with its environmental regulations hold insight as to why Korea has such a complete environmental system with close monitoring. With its low relative land area and high population, Korea is forced to protect its environment and must be efficient in its use and care of natural resources. This puts pressure on its economy but Korea has found an effective balance especially since its standard of living is just twice that of China.

14.2.5 SAUDI ARABIA

Saudi Arabia is located in westernmost Asia and occupies the majority of the Arabian Peninsula. The total land area of Saudi Arabia is 2,149,690 square kilometers which makes it the 12th largest country in the World. The population of Saudi Arabia is approximately 30 million (United States Department of State 2021c). Saudi Arabia is bordered by Jordan and Iraq to the north, Kuwait to the northeast, Qatar, Bahrain and the United Arab Emirates to the east, Oman to the southeast and Yemen to the south. It is separated from Israel and Egypt by the Gulf of Aqaba.

Oil was discovered in Saudi Arabia in 1938 and has since developed into the largest oil producer and exporter of oil in the World controlling the second largest oil reserve and 8th largest gas reserves (United States Department of State 2021c). Oil production has provided Saudi Arabia with economic prosperity and substantial political leverage in the region. The political system in Saudi Arabia is considered a monarchy and has no free elections (United State Department of State 2021c).

Saudi Arabia's geography is dominated by the Arabian Desert, associated semi-desert and several mountain ranges. There are very few lakes and no permanent rivers. Average summer temperatures can exceed 45°C (113°F) and have a temperature of 54°C (129°F). In summer, the temperature rarely drops below freezing. The largest city in Saudi Arabia is Riyadh which has an estimated population of 6.5 million residents. Gross domestic product (GDP) of Saudi Arabia is estimated at 646 billion and per capita income is estimated at nearly $21,000 per citizen (United Nations 2002, 2004, and 2021d). Since Saudi Arabia is essentially a country located in a desert, water is a precious resource. More than 50% of fresh water for the country is from desalinization plants located near the Red Sea. The remainder originates from groundwater and some surface water located in the western mountains of the country (United Nations 2021d). The remainder originates from groundwater and some surface water located in the western mountains of the country (United Nations 2021d).

14.2.5.1 Environmental Regulatory Overview of Saudi Arabia

Environmental issues that Saudi Arabia face are closely tied to its oil industry which accounts for most of its economy. Environmental issues in Saudi Arabia are viewed towards the development of the county's oil and gas reserves. Saudi Arabia first passed the General Environmental Regulations and Rules for Implementation in 2001 and formerly adopted environmental laws for ambient air, drinking water, wastewater discharge, solid and hazardous waste, emissions from mobile sources, noise, and pollution prevention and preparedness in 2012. The Presidency of Meteorology and Environment (PME) is responsible for implementing environmental regulations in Saudi Arabia (Chakibi 2013).

14.2.5.2 Air

Air pollution in Saudi Arabia is generally confined to its petroleum and related industries and motor vehicle exhaust. In addition, frequent sand and dust storms also significantly affect ambient air quality (United Nations 2021d). Ambient air standards for Saudi Arabia are similar to the United States and EU and are listed at https://www.pme.gov.sa. (Saudi Arabia Presidency of Meteorology and Environment (2012).

14.2.5.3 Water

Data on water pollution in Saudi Arabia is limited but is believed to be influenced by the petroleum and related industries. Drinking water in Saudi Arabia is supplied by the government at almost no cost to its citizens. Saudi Arabia essentially has adopted water quality criteria established by the World Health Organization (Al-Omran et al. 2014 and World Health Organization 2015b). Saudi Arabia wastewater discharge limits are similar to the United States and EU and are listed at https://www.pme.gov.sa. (Saudi Arabia Presidency of Meteorology and Environment 2012).

14.2.5.4 Solid and Hazardous Waste

Saudi Arabia defines wastes as substances which have been discarded or neglected and can no longer be put to beneficial use. Saudi Arabia defines Hazardous Waste as a type of waste that has characteristics that render them hazardous to human health and include waste that are (Saudi Arabia Presidency of Meteorology and Environment 2012):

• Toxic	• Flammable	• Explosive	• Radioactive
• Corrosive	• Reactive	• Infectious	• Listed wastes

Saudi Arabia also has a hazardous waste category as listed wastes that include any wastes generated by the following (Saudi Arabia Presidency of Meteorology and Environment 2012):

- Medical wastes
- Biocide or phyto-pharmaceutical processes
- Wood preservatives
- Waste mineral oil
- Wastes from heat treatment and steel tempering containing cyanides
- Waste oil
- PCBs
- Waste paints, inks dyes, pigments, lacquers, and varnish
- Waste resins, latex, plasticizers, glues, and adhesives
- Wastes generated from research and development, and teaching arising from activities known to be harmful or suspected to be harmful to humans
- Waste photographic chemicals
- Wastes from metal treatment processes

Saudi Arabia also considers wastes as hazardous if they contain any of the following constituents (Saudi Arabia Presidency of Meteorology and Environment 2012):

• Metal carbonyls	• Beryllium	• Hexavalent chromium
• Copper	• Zinc	• Arsenic
• Selenium	• Cadmium	• Antimony
• Tellurium	• Mercury	• Thallium
• Lead	• Inorganic fluorine	• Inorganic cyanide
• Acidic solutions	• Basic solutions	• Organic phosphorus
• Organic cyanide	• Asbestos	• Phenols
• Ether compounds	• Halogenated organics	• Organic solvents
• Organic halogens	• Dibenzo-furans	• Dibenzo-p-dioxins

14.2.5.5 Remediation

In Saudi Arabia, new developments or expansions of any facility require that the owner conduct a comprehensive Environmental Impact Assessment (EIA). The process begins with notification of the Presidency of Meteorology and Environment and retaining a qualified consultant approved by the regulatory agency. Each project is graded as Category I, II, or III by the regulatory authority according to potential environmental impact with Category I as least amount of potential impact and Category III as the most potential environmental impact (Saudi Arabia Presidency of Meteorology and Environment 2012). Each category must address the following (Saudi Arabia Presidency of Meteorology and Environment 2012):

- Air quality
- The marine and coastal environment
- Surface and underground water
- Flora and fauna
- Land use
- Urban development
- General scenic view

After the EIA is conducted, a risk evaluation is conducted to evaluate whether there are any unreasonable risks posed by the presence of contamination discovered during the investigation process. Cleanup criterion are set by the regulatory authority based on the EIA and risk evaluation (Saudi Arabia Presidency of Meteorology and Environment 2012).

14.2.5.6 Summary Environmental Regulations of Saudi Arabia

Saudi Arabia's published environmental laws are comprehensive and in line with other developed countries. However, information on enforcement and contaminated site remediation is not available and is not readily shared with the outside world. Therefore, it's difficult to evaluate the current status of environmental compliance and sustainability. However, Saudi Arabia does face serious environmental issues related to air and especially water because Saudi Arabia is a desert country and water is scarce and can't afford to be polluted.

14.2.6 Indonesia

Indonesia is the largest country in southeast Asia and is made up of 17,508 islands of which 6,000 are uninhabited, that are spread between the Indian and Pacific Oceans that link the continent of Asia with Australia. Indonesia has five major islands that include Sumatra, Java, Kalimantan, Sulawesi, and Irian Jaya (United Nations 2021e). The land area of Indonesia is characterized as tropical rainforest. Since most of the soil originates from volcanic activity, the soils in Indonesia and considered very fertile. The total land area of Indonesia is 1,904,569 square kilometers and ranks 14th largest in the world. Indonesia shares borders with Malaysia, Papua New Guinea and East Timor (United Nations 2021e).

The population on Indonesia is estimated at 261 million residents making it the 4th most populous nation in the world. The population density of Indonesia is 147 individuals per square kilometer, which is roughly 5 times more densely populated than the United States. Indonesia is home to over 500 languages and dialects. The capital of Indonesia is Jakarta with an estimated population of 9.6 million residents. The gross domestic product of Indonesia is $687 billion and the per capita standard of living is $3,000, which is approximately 20 times less than the United States. Trading partners with Indonesia include Singapore, Japan, and the United States (United Nations 2021e). The significant environmental issues on land in Indonesia include deforestation, air pollution, acid rain, and surface water pollution. Air pollution is especially significant in urban areas, especially Jakarta. The surrounding marine environment have been heavily impacted by poorly planned coastal developments, overfishing, and ocean acidification which has placed 95% of Indonesia reefs under serious threat (United Nations 2021e and Conservation International 2021).

14.2.6.1 Environmental Regulatory Overview of Indonesia

In Indonesia, the Ministry of Environment is responsible for administering and enforcing environmental laws and is also responsible for national environmental policy and planning (Indonesia Ministry of Environment 2021a). The Indonesia Ministry of Environment is collaborating with USEPA to improve air quality, water quality, disposal of wastes, and cleanup sites of environmental contamination in Indonesia (USEPA 2021a).

14.2.6.2 Air

When evaluating air pollution in Indonesia, one must look no further than Jakarta, its capital city of nearly 10 million residents, as the most polluted city in Indonesia. Jakarta routinely has air pollution by Fine Particulate Matter (PM2.5) of 150 ug/m^3 or greater, which is ranked in the World Health Organization Category Ambient Air Quality Index as unhealthy (World Health Organization 2021c). The primary source of the PM2.5 is vehicle exhaust. According to the Indonesia Ministry of Environment, the typical air-polluting industries in Indonesia are manufacturing, mining, energy, oil and gas, and vehicle exhaust (Indonesia Ministry of Environment 2021a). Indonesia ambient air quality standards are listed at https://www.moe.indonesia.org/ambientairstandards. (Indonesia Ministry of Environment 2021b).

In 2009, Indonesia enacted a vehicle emission standard that requires emission test of each vehicle and also set a phase out of the use of leaded gasoline and set a 500 part per million maximum level for sulfur in diesel fuel (Indonesia Ministry of Environment 2021a). Indonesia also enacted a permit process that sets site specific emission standards at industrial facilities that emit pollutants to the atmosphere. The permits require that appropriate air pollution control equipment be installed for each air emission source and that all air pollution control equipment be properly maintain (Indonesia Ministry of Environment 2021a). Burning of garbage and burning forest land for agriculture preparation in Indonesia has historically been a systemic problem. In 2010, the Ministry of Environment started a program to begin to reduce and control this type of air pollution, which often creates a visible and noticeable haze sometimes regionally (Indonesia Ministry of Environment 2021b).

14.2.6.3 Water

The World Health Organization estimated that 59% of residents have access to sanitation and approximately 30% have access to safe drinking water (World Health Organization 2021d). Indonesia laws that govern water were first enacted in 1974. Responsible agencies for use of water in Indonesia are spread across many different government ministries including but not limited to:

- Ministry of Forestry
- Ministry of Finance
- Ministry of Mining
- Ministry of Agriculture
- Ministry of Environment
- Ministry of Commerce
- Ministry of Energy
- Ministry of Industry

Since the government departments responsible for water are spread across several ministries, no one ministry accepts responsibility because of competing interests (World Health Organization

2021d). In general, drinking water quality in Indonesia does not meet national standards on two separate levels (World Health Organization 2021d):

- Indonesia does not supply enough water to its citizens, demand is greater than what can be supplied
- The water quality is not achieved because of pollution

Drinking water quality in Jakarta is especially impacted with poor water quality and scarce availability since 81% of the city's water supply comes from the Citarum River which is heavily polluted with toilet activities, domestic wastewater, trash disposal and other sources of pollution (Water Environment Partnership in Asia 2021). Indonesia lacks basic infrastructure and financial resources to supply safe water to its citizens in the form of (Choudhary 2017):

- Waste water treatment is in some places non-existent
- Basic sanitation to prevent human excrement from contaminating water supplies
- Supplying water filters to citizens
- Pipelines are inadequate
- Demand for water is estimated at 50% of current need

Groundwater in much of Indonesia is also impacted by pollution and is not considered an alternative source of safe drinking water (Choudhary 2017). Water pollution in Indonesia has been traced to health risk and disease that include (Choudhary 2017):

- Gastro-enteritis
- Typhoid
- Cholera
- Paratyphoid
- Dysentery
- Diarrhea

Boiling water as water treatment is also not recommended because it does not remove chemical pollutants (Choudhary 2017). Water quality standards for Indonesia are listed at https://www.aecon.org.indonesia. (Indonesia Ministry of Environment 2021a).

14.2.6.4   Solid and Hazardous Waste

Management of waste in Indonesia is a mounting challenge because of a large population of over 250 million people, lack of infrastructure and, lack of financial resources. This Ministry of Environment published a new regulatory framework in 2014 in an attempt to streamline the regulations and make existing regulations stricter (Indonesia Ministry of Environment 2021a). The revised solid waste regulations also attempt to improve waste disposal methods, specifically to increase the amount of solid water disposed in licensed landfills instead of disposal in streets, rivers and streams, and open burning of waste. Previous to the new solid waste regulations, solid waste was disposed in the following manner as measured in 2008 (Lokahita 2017):

- 68% in landfills
- 16% small scale incineration
- 7% composting
- 5% open burning
- 4% dumped in rivers and streams

Since the regulations were updated, Indonesia has made improvements but these improvements in many circumstances have been met with a negative impact because of a rapid increase in population which overshadows many of the attempted improvements (Lokahita 2017).

14.2.6.5   Remediation

Remediation standards for contaminated sites in Indonesia generally rely on the Dutch Intervention Values and USEPA criteria. The USEPA is collaborating with the Ministry of Environment of Indonesia to assist and educate Indonesia st that at some point they will be able to build a framework of regulations to address legacy sites of environmental contamination. USEPA is currently collaborating with Indonesia on four sites and is working together with the Ministry of Environment to improve emergency response and clean-up procedures (USEPA 2021c).

14.2.6.6   Summary of Environmental Regulations in Indonesia

As stated by USEPA, Indonesia is a key factor in the global environmental arena, with significant importance in ecological resources but with significant challenges to protect and restore its environment (USEPA 2021c). Lack of infrastructure, large population, lack of technology, lack of financial resources, and lack of effective regulations and enforcement have all contributed to environmental deterioration in Indonesia that has led to:

- Poor water quality
- Poor air quality
- Poor waste disposal practices
- Deterioration of the marine environment and reef system
- Poor health
- Disease
- Extinction of species

14.2.7   MALAYSIA

Located near the equator, Malaysia is a country of two parts in southeastern Asia. The land area of Malaysia is 330,603 square kilometers which ranks 66th in size. Malaysia is bordered by Thailand to the north, and Indonesia to the south. Malaysia is a wet and tropical climate and averages approximately 3 meters of rain annually. The population of Malaysia slightly exceeds 31 million residents, has a GDP of 314.5 billion, and a per capita income of nearly $10,000. The capital of Malaysia is Kuala Lumpur with its population of nearly 2 million residents located on the western peninsula region south of Thailand (United Nations 2021a).

14.2.7.1   Environmental Regulatory Overview of Malaysia

Environmental issues in Malaysia include air pollution, water pollution from raw sewage from humans and livestock, and industrial sources, deforestation, and loss of habitat. Industrial sources of water pollution originate from Malaysia's main industries that are tin mining, natural rubber productions and palm oil production (United Nations 2021f).

14.2.7.2   Air

Sources of air pollution in Malaysia include open burning of waste, vehicle exhaust, and industry (Malaysia Department of Environment 2015). In 2015, the Malaysia Department of Environmental developed new ambient air quality criteria. Ambient air quality standards for Malaysia are at http://www.doe.gov.my/portelv1/info-umum/english-ambient-air-standards. (Malaysia Department of Environment 2021a).

In general, Kuala Lumpur has the poorest air quality in Malaysia and largely due to fine particulate matter (PM2.5) (Malaysia Department of Environment 2021b). However, even though the air quality of Kuala Lumpur is often considered poor, it still is routinely much better than Jakarta Indonesia (United Nations 2021f).

14.2.7.3 Water

Sources of water pollution in Malaysia include sewerage directly discharge to water ways, garbage and other solid waste dumped into rivers and streams, discharge from livestock farms including pig and chicken farms, and industrial discharge. Water quality is especially poor in urban areas (Malaysia Department of Environment 2021c). Water quality standards for drinking water in Malaysia at listed at http://www.doe.gov.my/waterquality. (Malaysia Department of Environment 2021c). Malaysia sewerage and industrial effluent discharge standards are listed at http://www.doe.gov.my/discharge-standards/malayasia.htm (Malaysia Department of Environment 2021d).

On occasion, water polluted with ammonia has shut down water treatment plants cutting fresh water to hundreds of thousands of residents in Malaysia. Water supplies are routinely restored in stages after the pollution passes by the water treatment intake locations. Sources of the ammonia have been traced to animal feed lots specifically pig and chicken farms (United Nations 2021f).

14.2.7.4 Solid and Hazardous Waste

Not unlike Indonesia, Malaysia struggles with appropriate disposal of solid and hazardous waste. Although the problem is not at the magnitude as present in Indonesia because Malaysia's population is only 12% of Indonesia's, but the trends and habits are similar (United Nations 2021e and United Nations 2021f).

14.2.7.5 Remediation

Malaysia has set priorities for improving and restoring their environment but achieving success will require changing cultural norms, time, and foreign investment. Currently, Malaysia is focusing on air and water pollution because those are the two environmental issues that are the highest priority for improving human health (Malaysia Department of Environment 2015 and United Nations 2021f). Efforts to improve air quality include (Malaysia Department of Environment 2021a):

- Upgrading transportation infrastructure improve to traffic flow and reduce vehicle emissions
- Restricting emissions from power generating plants and industry through a comprehensive permitting process
- Eliminating the open burning of garbage and solid waste

Efforts to improve water quality include (Malaysia Department of Environment 2021a):

- Improve zoning and land use planning
- Upgrade treatment plants
- Require industrial, agriculture, and livestock to pretreat water before discharge

14.2.7.6 Summary of Environmental Regulations of Malaysia

Malaysia suffers from much the same environmental issues as Indonesia. Those issues of particular concern include urban air pollution, rapid increases in population, providing clean water, sanitation, soil and groundwater contamination, deforestation, and erosion.

14.3 OCEANIA

Oceania is a region made up of thousands of islands in the Central and South Pacifica Ocean. The two largest land masses that make up Oceania include Australia and the microcontinent of New

Zealand. Oceania also include three island regions that include Melanesia, Micronesia, and Polynesia (National Geographic Society 2021a). From a geographic perspective, Oceania is divided into three region types that include (National Geographic Society 2021a):

- Continental Type – this type includes Australia and New Zealand
- High Islands – Also called volcanic islands include Melanesia, Mount Yasur, and Vanuatu
- Low Islands – Also called coral islands include Micronesia and Polynesia

We will evaluate the environmental regulations of Australia and New Zealand located within Oceania, which represent the two largest land masses.

14.3.1 Australia

Australia is the 6th largest country of the world with a land mass of 7,686,850 square kilometers. Australia is generally considered a desert. However, there is a small area located along its northeastern coast that is rainforest (United Nations 1997). Australia's population is estimated at 24.9 million with most of the population living along the eastern coast and the western city of Perth (United Nations 1997). Australia's largest city is Sydney located in the southeastern portion of the country adjacent to the Pacific Ocean. Australia has no land borders with other countries. The Indian Ocean is located along its western shore and the Pacific Ocean is located along its eastern shore (United Nations 1997). The average per capita income in Australia is estimated at $67,442, which ranks 2nd highest in the world just behind Switzerland (World Bank 2021). The Gross Domestic Product of Australia is estimated at $1.69 trillion and ranks 27th in the world. Economic drivers in Australia include mining and tourism along with lessor amounts of farming and livestock. Mining in Australia consists of coal, natural gas, gold, aluminum, iron and other heavy metals, and diamonds and opal (World Bank 2021). From and environmental perspective, Australia is perhaps best known for the Great Barrier Reef which is comprised of over 3,000 individual reef system and coral cays and hundreds of tropical islands along its northwestern coast (United Nations 2021g). The Great Barrier Reef contains the largest collection of coral reefs in the world, with 400 different types of coral, and 1,500 species of fish. The Great Barrier Reef was designated a United Nations World Heritage Site (UNESCO) in 1981 and covers an area of 348,000 square kilometers (United Nations 2021g).

14.3.1.1 Environmental Regulatory Overview of Australia

Environmental issues in Australia are wide ranging and can be summarized as the following (Australia Department of Environment and Energy 2021a and The National Geographic Society 2021b):

- Invasive species. Specifically, the introduction of the rabbit.
- Endangered species. Including the Tasmanian devil and southern cassowary.
- Extinction. More than 50 species of birds, mammals, and marsupials have gone extinct including the Tasmanian Wolf.
- Ocean dumping.
- Land degradation. Caused by land clearing, over grazing, and forest clearing.
- Soil erosion. From poor land use practices.
- Poor water quality. From poor industry practices and land use planning.
- Salinization. From poor water quality irrigation techniques.
- Urbanization.
- Pollution due to mining, including heavy metals, gems, and coal.
- Climate change. Reef in the Great Barrier Reef System are die-off from warmer ocean waters as a result of climate change

14.3.1.2 Air

Australia does not have the level or severity of air pollution that is currently experienced by most other countries. However, Australia does have reduced air quality in several urban locations especially along its east coast (United Nations 2021a). In addition, since Australia does not share a common land boundary with other countries, and most other countries are at a rather large distance, it does not significantly experience air pollution migration from other counties (Australia Department of Environmental and Energy 2021a). In Australia, carbon monoxide, lead, nitrogen oxide, particulate matter, ozone, and sulfur dioxide are regulated and are termed criteria pollutants as they are in the United States. The two criteria pollutants of greatest concern in Australia are particulate matter and ozone (Australia Department of Environmental and Energy 2021b). Ozone is of a particular concern in urban areas and particulate matter is a concern in urban areas and in rural areas of Australia. Rural Australia has a particularly high level of particulate matter because the interior portions of Australia are desert with little or no vegetation and most of the roads are not paved. Therefore, any wind can create particulate matter and road traffic in many instances creates large amounts of particulate matter.

In June 1998, the Australia National Environment Protection Council (NEPC) established Australia's first ambient air quality standards. The NEPC is a statutory body with law making powers that were established under the National Environmental Protection Act of 1994. The Air NEPM set standards for 6 pollutants that are listed at http://www.env.gov.au/protection/air-quality-standards. (Australia Department of Environment and Energy 2021c).

NEPC is also gathering data on five other pollutant it termed "Air Toxics" and are known as "Air Toxics NEPM" to set limits in the future. The Air Toxics NEPM include (Australia Department of Environment and Energy 2021c):

- Benzene
- Formaldehyde
- Benzo(a)pyrene
- Toluene
- Xylenes

Emissions standards for individual facilities are not set by NEPC but rather are set by individual states and territories within Australia, which is similar to that of the United States, and are set under a permit process (Australia Department of Environment and Energy 2021c).

14.3.1.3 Water

Water pollution, including the marine environment, in Australia has historically originated from four main sources that include (Australia Department of Environmental and Energy 2021d and Coolaustralia.org 2021a):

- Dumping of rubbish, sewerage, and industrial waste in the ocean
- Mining waste discharged into rivers and on the land
- Industry
- Urbanization

The King River in Tassie is considered the most polluted River in Australia. For nearly 70 years it was used as a sewer and dumping ground for mining waste from a nearby copper mine. The river is so acidic that nothing can live in the river water (Coolaustralia.org 2021a).

Australia set drinking water quality standards in 2011 and were updated in 2018 (Australia National Health and Medical Research Council 2021). Australia perhaps has the most comprehensive drinking water quality management system of any country we will evaluate. Australia regulates drinking water in many of the same aspects as the United States but continues to update and monitor the quality of the water supplied to the public in a comprehensive manner. Australia evaluates microbial quality of water through evaluating the risk of disease from the following water pathogens (Australia National Health and Medical Research Council 2021):

- Bacterial pathogens
- Protozoa
- Viruses
- Helminths
- Cyanobacteria

Australia also comprehensively evaluates the physical quality, chemical quality, radiological quality, and treatment chemicals of drinking water on a routine bases to ensure the best water quality possible. Water quality limits are listed at https://nhmrc.gov.au. (Australia National Health and Medical Research Council 2021). Effluent discharge of wastewater from industrial sites to any receiving water, including the marine environment, in Australia is regulated through each State or Territory using a site-specific risk assessment approach that is based on internationally-accepted risk assessment approaches (Australia National Health and Medical Research Council 2021). Discharge from ships is regulated by the Australia Maritime Safety Authority under separate permits and regulations (Australian Maritime Safety Authority 2021).

14.3.1.4   Solid and Hazardous Waste

In 2009, Australia set goals for reducing the amount of wastes generated nation-wide. The goal was to create less waste and improve resources (Australia Department of Environment 2009). The need to revise its national policy and improve regulations was because Australia saw an increase in the amount of waste generated nation-wide by 34.7% over a 4 year period from 2002 to 2006 and the amount of hazardous waste doubled during this same period of time (Australia Department of Environment 2009). Since 2009, Australia is now recycling approximately 60% of its solid waste. The waste generated per person in Australia in 2006 was estimated at 2.783 metric tons per person and was estimated at 2.705 metric tons in 2015 and represents a slight decrease nation-wide in the amount of solid waste produced (Australia Department of Environment 2016). In Australia, a solid waste is defined as material or products that are unwanted or have been discarded, rejected, or abandoned. Wastes typically arise from three dominate waste streams that include (Australia Department of Environment 2009 and 2018f):

- Domestic and municipal
- Industrial
- Construction and demolition

Determining whether a solid waste is hazardous or not in Australia is similar to that of the United States and many other countries. In Australia, a solid waste is hazardous if it is a per-listed waste or has the following characteristics (Australia Department of Environment 2021e and 2021f):

• Explosive	• Flammable
• Ignitable	• Gas
• Oxidizing agents	• Corrosive
• Reactive	• Infectious
• Medical	• Toxic
• Radioactive	• Characteristic

A waste is a characteristic hazardous waste if a leachate test of the waste identifies any of the compounds above those listed at http://www.env.gov.au/protection/waste-class.pdf (Australia Department of Environment 2021e). Soil, Sediment, and Water Remediation Values for Select Compounds and Pathways are listed at http://www.environment.gov.au/site-cleanup.pdf (Australia Department of Environmental and Conservation 2010).

14.3.1.5 Remediation

Cleanup or restoration of agricultural land in Australia is perhaps the most significant issue, not industry or urbanization. Approximately 50% of agricultural land in Australia is considered severely degraded because of rising salinity from irrigation, erosion, and vegetative loss. The result has been a reduced capacity for crops which in turn then requires more land for agricultural and continued land degradation. This cycle of agriculture kills farmland, contaminates drinking water, and destroys natural ecosystems if it continues without drastic restoration efforts and modified farming techniques (Coolaustralia.org 2021b).

14.3.1.6 Summary of Environmental Regulations of Australia

Australia has a comprehensive set of environmental regulations and has an enforcement system and culture that is very engaged in protecting human health and the environment. Australia has some disadvantages when it comes to its dry climate, poor soil, and desert geography in that water is more a precious resource than other parts of the world since it is of limited quantity. Therefore, as we discussed earlier in this section, is why Australia must protect its water from contamination. In addition, off Australia's northeast coast is the largest reef system in the world, The Great Barrier Reef. The Great Barrier Reef is experiencing an unprecedented die-off of coral due to a warming of ocean water believed to be directly linked to climate change.

14.3.2 New Zealand

New Zealand is an island country 1,200 kilometers southeast of the coast of Australia in the South Pacific Ocean. New Zealand is comprised of nearly 600 islands, two of which are large and make up most of land mass and are called the North Island and the South Island.

New Zealand is comprised of 267,710 square kilometers of land making it the 75th largest country in the world. The population of New Zealand is 4.8 million. New Zealand's largest city is Auckland with an estimated population of 1.6 million (United Nations 2021h). The Northern Island of New Zealand is characterized as volcanic in nature and the Southern Island is characterized as mountainous in nature. New Zeeland's economy is considered the 53rd largest in the world when measured by gross domestic product, which $200.7 billion in 2016. Per capita income of New Zealand is $37,860 as measured in 2016. New Zealand's economy is considered one of the most globalized and depends greatly upon international trade. Major industries in New Zealand are aluminum mining, food processing, metal fabrication, wood and paper products. Trading partners include Australia, United States, China, Japan, Germany, and South Korea (United Nations 2021h).

14.3.2.1 Environmental Regulatory Overview of New Zealand

Environmental issues facing New Zealand include deforestation, soil erosion, loss of habitat, and surface and groundwater quality degradation. Much of the development in New Zealand has been for farming which has impacted the environment perhaps more than any type of human activity, including industrial impacts. New Zealand actively is involved in sustainability issues and has openly embraced efforts to deal with climate change beginning as early as 1991 (New Zealand Ministry for the Environment 2021a). In New Zealand, the Resources Management Act of 1991 is the main law that regulates discharges into the environment, including air and water. Much of the responsibility for managing discharges and permits has been handed down to district and regional councils. Therefore, just as in the United States, there are differences from district to district in how these issues are handled but the basic goals, objectives, and specifics are the same (New Zealand Ministry for the Environment 2021a).

14.3.2.2   Air

Air quality in New Zealand is similar to that of Australia in that it is most degraded in urban areas but is better than many other urban areas of the world (United Nations 1997). New Zealand's footprint for greenhouse gases is unique in that the majority of greenhouse gases originate from agriculture and not industry (New Zealand Ministry for the Environment 2018b). In fact, New Zealand's industry only accounts for a small percentage, less than 10% of total emissions. Two main sources of greenhouse gases originate from agriculture and that being actual farming which releases greenhouse gases and livestock, especially the dairy industry which produces large quantities of methane from cattle (New Zealand Ministry for the Environment 2021b). New Zealand's ambient air quality standards were first enacted in 1994 and then were updated in 2002 (New Zealand Ministry of the Environment 2021c). The ambient air quality standards are listed at http://www.gw.govt.nz/form-5a (New Zealand Ministry for the Environment 2021c). As with many other countries, air permits are necessary for any industrial emitting source. Permit applications are comprehensive and typically include stack testing or meeting other performance criteria and extensive record keeping to document compliance. Permits limits are low if emitting sources include any compounds listed at http://www.gw.govt.nz/form-5a (New Zealand Ministry for the Environment 2021c).

14.3.2.3   Water Pollution

Water quality in New Zealand rivers has been monitored since 1989 and by international standards is of good quality. However, there are signs of declining quality from increased concentrations of nitrogen from anthropogenic sources, namely agriculture (New Zealand Ministry of the Environment 2021d). Water Quality standards in New Zealand are listed at http://www.mfe.nz/publications/drinking-water. (New Zealand Ministry of the Environment 2021e). Agriculture is the largest land use in the lowlands of New Zealand, particularly dairy and livestock, which has been linked to increased concentrations of nitrogen in soil, surface water, and groundwater. Availability of water has also declined due to increased dairy farming which uses more water in New Zealand than any other activity (New Zealand Ministry of the Environment 2021f). Wastewater discharges in New Zealand require what is termed a resource consent from the district regulatory authority. The resource consent is a discharge permit that makes the discharge subject to facility-specific discharge conditions that take into account the quality of the wastewater and the quality of the receiving body of water (New Zealand Ministry for the Environment 2021f).

14.3.2.4   Solid and Hazardous Waste

In New Zealand, a waste is considered hazardous if it presents some degree of physical, chemical, or biological hazard to people or the environment (New Zealand Ministry for the Environment 2021g). Hazardous waste can be gases, liquids or solids that are:

• Explosive	• Flammable
• Infectious	• Ignitable
• Oxidizing agents	• Corrosive
• Reactive	• Medical
• Toxic	• Eco-toxic
• Radioactive	• Characteristic

Similar to many other countries, hazardous wastes are track from the generator to the transporter to the final disposal site using a manifest system (New Zealand Ministry for the Environment 2021g). New Zealand has developed a Waste List or what is termed the "L Code", which is a method that reflects typical wastes streams in New Zealand. The L-Code provides guidance on identifying wastes in a consistent manner and serves as the basis for record-keeping systems, especially for hazardous wastes. The L-Code is as follows (New Zealand Ministry for the Environment 2021h):

1. Wastes resulting from exploration, mining quarrying, and physical and chemical treatment of minerals.
2. Waste from agriculture, horticulture, aquaculture, forestry, hunting and fishing, food preparation and processing
3. Waste from wood processing and the production of panels and furniture, pulp, paper, and cardboard.
4. Wastes from leather, fur, and textile industries.
5. Wastes from petroleum refining, natural gas purification, and pyrolytic treatment of coal.
6. Wastes from inorganic chemical processes.
7. Wastes from organic chemical processes.
8. Wastes from the manufacture, formulation, supply and use of coating (paints, varnishes and vitreous enamels), adhesives, sealants, and printing inks.
9. Wastes from the photographic industry.
10. Wastes from thermal processes.
11. Wastes from chemical surface treatment and coating of metals and other material: non-ferrous hydro-metallurgy.
12. Wastes from shaping and physical and mechanical surface treatment of metals and plastics.
13. Oil wastes and wastes of liquid fuels.
14. Waste organic solvents, refrigerants, and propellants.
15. Waste packaging; adsorbents, wiping cloths, filter materials and protective clothing not otherwise specified.
16. Wastes not otherwise specified.
17. Construction and demolition wastes (including excavated soil from contaminated sites)
18. Wastes from human or animal health care and/or related research.
19. Wastes from waste management facilities, off-site waste water treatment plants and preparation of drinking water and water use for industrial applications.
20. Municipal wastes (household waste and similar commercial, industrial and institutional wastes) including separately collected fractions.

New Zealand has an extensive reduce, reuse, recycle program and also generates 80% of its electricity from renewable resources that include wind, hydro, and geothermal. The program for reducing waste follows a circular economy approach that is based on three principles which include (New Zealand Ministry for the Environment 2021i):

- Design out waste and pollution
- Keep products and material in use
- Regenerate natural systems

14.3.2.5 Remediation

A Contaminated Sites Remediation Fund (CSRF) was established as part of the Resource Management Act of 1991 that supports regional councils to achieve goals for contaminated land management. Investigation and remediation is conducted using a prioritizing method to determine which sites are funded under the program. The New Zealand Ministry for the Environment assess each application and selects 10 sites that are of the highest priority for funding. Once a site is remediated to an acceptable risk, they are removed from the list and replaced with a new site in need (New Zealand Ministry for the Environment 2021j). New Zealand has developed detailed and comprehensive guidelines for conducting environmental investigations at sites of potential concern with the objective to achieve a nationally-consistent approach to evaluate whether a risk to human health and environment exists (New Zealand Ministry for the Environment 2021k). New Zealand uses a Risk Screening System (RSS) that guides the investigation so that each site is investigated consistently and to appropriately rank a site by risks posed to protect human health and the environment. New Zealand utilizes the USEPA Integrated Risk Information System (IRIS) and the United States Agency for Toxic Substance and Disease Registry to develop minimum risk levels (New Zealand Ministry for the Environment 2021l).

14.3.2.6 Summary of Environmental Regulations of New Zealand

The data certainly demonstrates that New Zealand has a comprehensive, effective, and forward-thinking environmental regulation program and a population that is engaged in protecting its air, water, and land. This is evident in New Zealand's efforts to reduce the amount of solid waste that is generated, and that 80% of its electricity is generated from renewable resources. Challenges for New Zealand include striking a balance with its agricultural activities and environmental degradation of its surface and groundwater and continued population increase and urbanization.

14.4 SUMMARY AND CONCLUSION

Now we have completed our global journey through environmental regulations of many countries of the world. Many countries have a robust set of environmental regulations and that they are very similar to those of the United States. For the most part, Australia, New Zealand, Japan, and Korea not only have robust environmental regulations but they are also effective. In fact, many may be more effective than their counterparts in the United States.

As with North and South America, eastern Europe, and Africa, any countries in Asia struggle with implementing environmental regulations to the extent that they actually work at achieving the underlying purpose of any environmental regulation which is improving human health and the environment. The reasons may vary from country to country throughout the world, but all have common themes that involve war, poverty, corruption, over population, lack of infrastructure, political will, ineffective leadership, and other social issues.

Questions and Exercises for Discussion

1. India has much environmental improvement to conduct before it can focus on sustainability? Do you agree or disagree with this comment and why?
2. In what areas can the countries of Asia improve its environmental regulations? Name at least two areas and explain why for each country in Asia.
3. In your view, what are the environmental priorities for Japan and why is Japan unique compared to other countries of Asia?
4. Of the countries of Asia, which country is the most environmentally impaired and why?
5. Of the countries of Asia, which has the most effective environmental regulations and why?
6. Of the two countries of Oceania, which country has the most effective environmental regulation and why?
7. Of the countries of Asia, provide your view as to what you would recommend to improve the air, water, land, and living organisms for two countries of you choice in Asia.
8. Now that you have read this chapter, what do you think is the most important environmental issue facing Asia and why?

REFERENCES

Agarwal, V. K. 2015. Environmental Law in India: Challenges for Enforcement. *Bulletin of the India National Institute of Ecology. Delhi. India.* Vol. 15. pp. 227–238.

Al-Omran, A., Al-Barakah, F., Altuquq, A., Aly, A., and Nadeem, M. 2014. Drinking Water Quality Assessment and Water Quality Index of Riyadh, Saudi Arabia. *Water Quality Research Journal of Canada.* Vol. 46. no. 1. pp. 287–296.

Asia Society. 2021. *Korean History and Geography.* https://asiasociety.org/education/korean-history-and-political-geography (accessed April 27, 2021).

Australia Department of Environmental and Conservation. 2010. Assessment Levels for Soil, *Sediment and Water. Department of Environment and Conservation. Canberra. Australia.* 56p. http://www.environment.gov.au/site-cleanup.pdf (accessed April 27, 2021).

Australia Department of Environment. 2009. National Waste Policy. Environmental Protection Heritage Council. Canberra, Australia. 22p. http://www.nepc.gov.au/node/849 (accessed April 27, 2021).

Australia Department of Environment. 2016. Australian National Waste Report. Department of Environment. Canberra, *Australia.* 74p. https://www.environment.gov.au/national-waste-report-2016 (accessed April 27, 2021).

Australia Department of Environment and Energy. 2021a. *Publications and Resources.* http://www.environment.gov.au/about-us/publications (accessed April 27, 2021).

Australia Department of Environment and Energy. 2021b. *Air Pollutants in Australia.* http://www.environment.gov.au/airpollution (accessed April 27, 2021).

Australia Department of Environment and Energy. 2021c. *Ambient Air Quality Standards.* http://www.env.gov.au/protection/air-quality-standards (accessed April 27, 2021).

Australia Department of Environment and Energy. 2021d. Guidelines for Risk Assessment of Wastewater Discharges to Waterways. Environmental Protection Agency of Victoria, *Australia.* 36p. https://www.epa.vic.gov.au (accessed Aril 27, 2021).

Australia Department of Environment and Energy. 2021e. *Solid Waste Classification System.* https://www.env.gov.au/protection/waste-class.pdf (accessed April 27, 2021).

Australia Department of Environment and Energy. 2021f. *The Waste and Recycling Industry in Australia.* Report to Australian Government: Canberra, Australia. 152p.

Australia Maritime Safety Authority. 2021. *Discharge Standards.* https://www.amsa.gov.au (accessed April 27, 2021).

Australia National Health and Medical Research Council. 2021. *Australian Drinking Water Guidelines.* Canberra, Australia. 1172p. https://www.nhmrc.gov.au (accessed April 27, 2021).

Broughton, E. 2005. The Bhopal Disaster and Its Aftermath: A Review. *J. of Environmental Health.* Vol. 4. no. 6. United States Institute of Health: Washington, D. C. https://www.ncbi.nlm.nih.gov/pmc/articles/PMC1142333 (accessed April 27, 2021).

Chakibi, S. 2013. Saudi Arabia Environmental Laws. *Journal of Environmental Health and Safety.* http://ehsjournal.org/sanaa-chibi/saudiarabialaws (accessed April 27, 2021).

China Ministry of Ecology and Environment (CMEE). 2021. *Law of the People's Republic of China on the Prevention and Control of Environmental Pollution by Solid Waste. Beijing, China,* https://www.china.org.cn/english/environment/34424.htm (accessed April 27, 2021).

China Ministry of Environment Protection (CMEP). 2014. Five Standards of Environmental Protection. MEP Notice No. 14 of 2014). Beijing, China. 98p.

China Ministry of Environmental Protection. 2021. *New Law on Soil Pollution Prevention.* https://www.china.org.cn/china/2018-09/01/content61417828 (accessed April 27, 2021).

China Research Academy of Environmental Sciences. 2013. *Groundwater Challenges and Environmental Management in China.* Ministry of Environmental Protection: Beijing, China. 28p.

Choudhary, S. 2017. Drinking Water in Indonesia. Indoindians. Jakarta, Indonesia. https://www.indoindians.com/jakartadrinkingwater (accessed April 27, 2021).

Clean Air Alliance of China (CAAC). 2015. *China Air Quality Management Assessment Report.* CAAC Secretariat: Beijing, China. 28p. https://www.iccs.org.cn (accessed April 27, 2021).

Conservation International. 2021. *Indonesia: A Vast and Beautiful Country – At the Crossroads.* https://conservation.org/where/pages/indonesia (accessed April 27, 2021).

Coolaustralia.org. 2021a. *Water Pollution.* https://www.coolaustralia.org/water-pollution-primary (accessed April 27, 2021).

Coolaustralia.org. 2021b. *Land Pollution.* https://www.coolaustralia.org/land-pollution-primary (accessed April 27, 2021).

India Central Pollution Control Board. 2009. *National Ambient Air Quality Standards*. New Delhi. India. 48p. https://www.cpcb.nic.in/upload/latest_final_air_standard.pdf (accessed April 27, 2021).

India Ministry of Environment, Forestry, and Climate Change. 2021a. *Environmental Laws. Air Pollution Control Act of 1981*. http://www.environmentallawsofindia.com/air/pollution/act (accessed April 27, 2021).

India Ministry of Environment, Forestry, and Climate Change. 2021b. *Environmental Laws*. http://www.environmentallawsofindia.com/water/pollution/act (accessed April 27, 2021).

India Ministry of Environment, Forestry, and Climate Change. 2021c. *Effluent Water Discharge Criteria for Industry*. http://www.envfor.nic.in/decision/water-pollution (accessed April 27, 2021).

India Ministry of Environment, Forestry, and Climate Change. 2021d. *Solid and Hazardous Waste Regulations*. http://www.envfor.nic.in/decision/solid-and-hazardous-waste (accessed April 27, 2021).

India Ministry of Environment, Forestry, and Climate Change. 2021e. *Environmental Impact Assessment*. http://www.envfor.nic.in/decision/environmental-impact-assessment (accessed April 27, 2021).

Indonesia Ministry of Environment. 2021a. https://www.aecon.org/indonesia (accessed April 21, 2021).

Indonesia Ministry of Environment. 2021b. *Ambient Air Quality Standards*. https://www.moe.indonesia.org/ambientairstandards (accessed April 27, 2021).

Japan Fact Sheet. 2021. *Environmental Issues: Progress Made but New Challenges Must Be Faced*. http://www.web-Japan.org/environmental/issues (accessed April 27, 2021).

Japan Ministry of the Environment. 2021a. *The Basic Environmental Law and Basic Environmental Plan*. http://www.env.go.jp/en/laws/policy/basic_lp.html (accessed April 27, 2021).

Japan Ministry of the Environment. 2021b. *Air Pollution Control Act*. Tokyo. Japan. http://www.env.go.jp/en/laws/policy/basic_lp.html (accessed April 27, 2021).

Japan Ministry of the Environment. 2021c. *Environmental Air Quality Standards for Ambient Air in Japan*. http://env.go.jp/en/air/aq/aq.html (accessed April 27, 2021).

Japan Ministry of the Environment. 2021d. *Japanese Environmental Pollution Experience*. http://www.env.go.jp/en/coop/experience.html (accessed April 27, 2021).

Japan Ministry of the Environment. 2018e. *State of Japan's Environment at a Glance: Water Pollution*. http://www.env.go.jp/en/water/wq/pollution (accessed April 27, 2021).

Japan Ministry of the Environment. 2021f. *Environmental Issues and Pollution in Japan*. http://www.env.go.jp/en/pollution/issues (accessed April 27, 2021).

Japan Ministry of the Environment. 2021g. *Soil Contamination Countermeasures Act of 2002*. http://www.env.go.jp/en/soilact (accessed April 27, 2021).

Japan Ministry of the Environment. 2021h. *History and Current State of Waste Management in Japan*. Waste Management and Recycling Department: Tokyo. Japan. 31p.

Japan Ministry of the Environment 2021i. *Waste and Recycling Standards for Verification*. http://www.env.go.jp/en/recycle/manage/sy.html (accessed April 27, 2021).

Japan Ministry of the Environment. 2018j. *Soil Pollution Control Values for Soil Leachate*. http://www.env.go.jp/en/soil/leachate/html (accessed April 27, 2021).

Japan Ministry of the Environment. 2021k. Ambient Water Quality Standard for Groundwater and Criteria for Implementation of Soil Pollution Control Measures. *Water Pollution Control Law*. http://www.env.go.jp/en/spcl/html (accessed April 27, 2021).

Kichira, S. and Sugai, M. 1986. *The Origins of Environmental Destruction: The Ashio Copper Mine Pollution Case*. United Nations University: Shinyousha, Tokyo. 167p. https://www.unu.edu.ashiocoppermine (accessed April 27, 2021).

Korea Ministry of the Environment. 2021a. *Air Quality Standards and Air Pollution Levels*. http://www.eng.me.go.kr/eng/web/index.do?menuld+252 (accessed April 27, 2021).

Korea Ministry of the Environment. 2021b. *State of Water Environmental Issue: Republic of Korea*. http://www.wepa-db.net/policies/state/southkorea/southkorea.htm (accessed April 27, 2021).

Korea Ministry of the Environment. 2021c. *Drinking Water Management: Drinking Water Quality Standards and Water Quality Monitoring*. http://www.me.go.kr/eng/web/index.do (accessed April 27, 2021).

Korea Ministry of the Environment. 2021d. *Expansion of the Total Water Pollution Load Management System (TPLMS)*. http://www.me.go.kr/eng/web/indes.do?menuld=274 (accessed April 27, 2021).

Korea Ministry of the Environment. 2021e. *Enforcement Decree of the Water Quality and Aquatic Ecosystem Conservation Act of 2015*. http://www.law.go.kr/eng/web/law/waterquality (accessed April 27, 2021).

Korea Ministry of the Environment. 2021f. *Wastes Control Act of 2016: General Provisions*. http://www.law.go.kr/DRF/law/wastecontrolact (accessed April 27, 2021).

Korea Ministry of the Environment. 2021g. *Hazardous Waste Management Flowchart*. http://www.eng.me.go.kr./eng/web/index.do?menuld=379 (accessed April 27, 2021).

Korea Ministry of the Environment. 2021h. *Korea Waste Prevention and Recycling Policy.* http://www.eng. me.go.kr/eng/web/index.do?menuId=141 (accessed April 27, 2021).

Korea Ministry of the Environment. 2021i. *Liability, Compensation and Relief System for Damages from Environmental Pollution.* http://www.eng/me/go.kr/eng/web/law/Actno.166922 (accessed April 27, 2021).

Korea Ministry of the Environment. 2021j. *Soil and Groundwater Contamination Prevention and Restoration.* http://www.eng.me.go.kr/eng/web/index.do?menuId=311 (accessed April 27, 2021).

Lokahita, B. 2017. *Indonesia Municipal Solid Waste: Regulations and Common Practice.* Tokyo Institute Of Technology. https://www.academia.edu/23109258/indonesia/solidwasteregs (accessed April 27, 2021).

Malaysia Department of Environment. 2015. *Environmental Issues and Challenges – Malaysian Scenario.* http://www.doe.gov.my/environmentalissues (accessed October 7, 2018).

Malaysia Department of Environment. 2021a. *Malaysia Ambient Air Quality Standards.* http://www.doe.gov. my/portelv1/info-umum/english-ambient-air-standards (accessed April 27, 2021).

Malaysia Department of Environment. 2021b. Malaysian Air Pollution Index. *Official Portal of Department of State.* http://www.doe.gov.my/portelv1/info-umum/englisk-api (accessed April 27, 2021).

Malaysia Department of Environment. 2021c. *Water Pollution and Water Quality Standards.* http://www.doe. gov.my/waterquality (accessed April 27, 2021).

Malaysia Department of Environment. 2021d. *Malaysia Sewerage and Industrial Effluent Discharge Standards.* http://www.doe.gov.my/discharge-standards/malaysia.htm (accessed April 27, 2021).

Marsh, W. B. and Kaufman, M. M. 2015. *Physical Geography.* University of Cambridge Press: Cambridge, United Kingdom. 647p.

National Geographic Society. 2017. *China's Solutions to its Air Pollution.* https://www.news. nationalgeographic.com/2017/05/china0air-pollution-solutions-environment (accessed April 2021).

National Geographic Society. 2021a. *Australia and Oceania: Physical Geography.* https://www. nationalgeographic.org/encyclopedia/oceania-physical-geography (accessed April 27, 2021).

National Geographic Society. 2021b. *Corals Are Dying on the Great Barrier Reef.* https://news. nationalgeographic.com/2016/03/160321-coral-bleaching-great-barrier-reef.htm (accessed April 27, 2021).

New Zealand Ministry for the Environment. 2021a. *Environmental Management in New Zealand.* http:// www.mfe.nz/publications/environmental-reporting (accessed April 27, 2021).

New Zealand Ministry for the Environment. 2021b. *Measuring New Zealand's Progress.* http://archive.stats. govt.nz/sustainability-progress (accessed April 27, 2021).

New Zealand Ministry for the Environment. 2021c. *New Zealand Air Discharge Permit Application.* http:// www.gw.govt.nz/Form-5a (accessed April 27, 2021).

New Zealand Ministry for the Environment. 2021d. *Drinking Water for New Zealand.* http://www. drinkingwater.org.nz (accessed April 27, 2021).

New Zealand Ministry for the Environment. 2021e. *Drinking Water Standards.* http://www.mfe.nz/ publications/drinking-water (accessed April 27, 2021).

New Zealand Ministry for the Environment. 2021f. *Wastewater Discharges.* http://www.mfe.nz/publications/ wastewater-discharges (accessed April 27, 2021).

New Zealand Ministry for the Environment. 2021g. *Hazardous Waste Management in New Zealand.* http:// www.oag.govt.nz/2007/cg-2005-06/part11 (accessed April 27, 2021).

New Zealand Ministry for the Environment. 2021h. *Waste List.* http://www.mfe.govt.nz/waste/guidance-and-resources (accessed April 27, 2021).

New Zealand Ministry for the Environment. 2021i. *Waste and Resources.* http://www.mfe.govt.nz/waste (accessed April 27, 2021).

New Zealand Ministry for the Environment. 2021j. *Contaminated Sites Remediation Fund.* http://www.mfe. govt.nz/more/funding/contaminated-stes (accessed April 27, 2021).

New Zealand Ministry for the Environment. 2021k. Contaminated Land Management Guideline. *Revised 2016.* http://www.mfe.govy.nz/sites/land (accessed April 27, 2021).

New Zealand Ministry for the Environment. 2021l. *Contaminated Land Management Risk Screening System.* http://www.mfe.govt.nz/sites/screening-system (accessed April 27, 2021).

Rich, K. H., Huang, W. Z., Zhu, T., Wang, H., Lu Se, O., Ohman-Strickland, P., Zhu, P., and Zhang, J. I. 2010. *Measurement of Inflammation and Oxidative Stress Following Drastic Changes in Air Pollution During the 2008 Beijing Olympics: A panel Study Approach.* Annals of New York Academy of Sciences. August 2010. 1203:160-7.

Saudi Arabia Presidency of Meteorology and Environment. 2012. *General Environmental Regulations.* https://www.pme.gov.sa (accessed February 10, 2019).

Siddiqi, T. A., and Chong-Xain, Z. 1984. Ambient Air Quality Standards in China. *Journal of Environmental Management*. Springer. Vol. 8. no. 6. pp. 473–479.

United Nations. 1997. Australia Country Profile. Department of Policy Coordination. *New York. New York.* http://www.un.org/esa/earthsummit/astra-cp.htm (accessed April 27, 2021).

United Nations. 2002. *Saudi Arabia Country Profile.* Johannesburg Summit: United Nations. New York. New York. 59p.

United Nations 2004. *Kingdom of Saudi Arabia. Public Administration and Country Profile.* Department of Economic and Social Affairs (UNDESA): New York. New York. 16p.

United Nations. 2012. *Country Profile Series: Republic of Korea.* New York, New York. 92p. http://www.un.org.esa/agenda21/natlinfo/rekorea.pdf (accessed April 27, 2021).

United Nations. 2013. *Water Pollution in China.* Food and Agriculture Organization of the United Nations: Rome, Italy. 173p.

United Nations. 2021a. World Population Prospects. United Nations Department of Economic and Social Affairs. *Population Division.* https://esa.un.org/unpd/wpp/data (accessed April 27, 2021).

United Nations. 2021b. *Country Profile for China.* https://www.china.unfpa.org/em/united-nations-population-china-2 (accessed April 27, 2021).

United Nations. 2021c. Human Development Report. *United Nations Development Programme.* http://www.hdr.undr.org/sites/2017 (accessed April 27, 2021).

United Nations. 2021d. *The Demographic Profile of Saudi Arabia with Population Trends.* https://www.unescwa.org/sites/saudiarbiaprofile (accessed April 27, 2021).

United Nations. 2021e. *Country Profile and Review of Indonesia.* http://www.un.org/esa/earthsummit/indon-cp.htm (accessed April 27, 2021).

United Nations. 2021f. *Country Profile and Review of Malaysia.* http://www.un.org/esa/earthsummit/mal.htm (accessed April 27, 2021).

United Nations. 2021g. *Great Barrier Reef: UNESO World Heritage Site.* (http://www.whc.unesco,org/en/list/154 (accessed April 27, 2021).

United Nations. 2018h. *Country Profile of New Zealand.* http://www.un.org/esa/earthsummit/NZ.htm (accessed April 27, 2021).

United States Census Bureau. 2021. *United States Trade in Goods with China 2020.* https://www.census.gov/foreign-trade/balance (accessed April 2021).

United States Department of Commerce. 2018. *Foreign Trade with China 2017.* https://www.commerce.gov/tags/china (accessed April 27, 2021).

United States Department of State. 2021a. *Office of the United States Trade Representative.* https://ustr.gov/nafta (accessed April 27, 2021).

United States Department of State. 2021b. *United States Relations with Japan.* https://www.state.gov/p/eap/ci/ja/ (accessed April 27, 2021).

United States Department of State. 2021c. *Profile and Statistics of Saudi Arabia.* https://www.state.gov/saudiarabia (accessed April 27, 2021).

United States Environmental Protection Agency (USEPA). 2021a. *EPA Collaboration with China.* https://www.epa.gov.international-cooperation/epa-collaboration-china (accessed April 27, 2021).

United States Environmental Protection Agency (USEPA). 2021b. *EPA Collaboration with the Republic of Korea.* https://www.epa.gov.international-cooporation-korea (accessed April 27, 2021).

United States Environmental Protection Agency (USEPA). 2021c. *EPA Collaboration with Indonesia.* https://www.epa.gov.nternational-cooporation-Indonesia (accessed April 2021).

Water Environment Partnership in Asia. 2021. *State of Water Environmental Issues in Indonesia.* https://www.wepa-db.net/policies/state/indonesiad/indonesia.htm (accessed April 27, 2021).

Wencong, W. 2012. *China Unveils Plan to Control Hazardous Waste.* China Daily. Beijing, China. https://www.chinadaily.com.cn/china/2012-11/02content15867364.htm (accessed April 27, 2021).

World Bank. 2011. *Environmental Assessment, Country Data: India.* https://data.worldbank.org/countrydata.india.html (accessed April 27, 2021).

World Bank. 2016. *India in Profile: Economy and Standard of Living.* https://data.worldbank.org/countrydata.india.html (accessed April 27, 2021).

World Bank. 2021. *Australia in Profile. Gross Domestic Product, Economy and Standard of Living.* http://www.data.worldbank.org/countries/australia (accessed April 27, 2021).

World Health Organization (WHO). 2015a. *Guidelines for Drinking-water Quality.* World Health Organization: Geneva, Switzerland. 516p.

World Health Organization 2015b. *Progress on Drinking Water and Sanitation.* World Health Organization: New York, New York. 90p.

World Health Organization. 2021a. *Air Pollution in China: Sources and Amounts.* https://www.who.org/china-air-pollution (accessed April 27, 2021).

World Health Organization. 2021b. *World Health Organization Profile of India.* http://www.who.int/countries/ind/en (accessed April 27, 2021).

World Health Organization. 2021c. *Nine Out of Ten People Breathe Unhealthy Air.* Geneva. Switzerland. https://www.who.int/news-room/detail/02-02-2018 (accessed April 27, 2021).

World Health Organization. 2021d. *United Nations Global Analysis and Assessment of Sanitation and Drinking Water: Indonesia.* https://www.who.org/drinkingwater/Indonesia (accessed April 27, 2021).

15 Evaluation and Status of Global Environmental Regulations

15.1 INTRODUCTION

In this chapter we summarize what we have covered over the previous four chapters and compare and contrast how each country evaluated differs from each other and see how they compare to the United States. It might surprise some of you that there are some countries that have more effective environmental regulations than the United States, Surprising still might be that some countries either lack key environmental laws or the will to enforce the environmental laws they have due to other social or economic challenges or the political realities some countries face. One overriding difficulty many developing countries face is lack of infrastructure, training, and education to properly implement their own environmental regulations. Another overriding factor is climate and climate change which in some instances has greatly increased the negative effects of pollution on the environment and humans.

Lastly, we will examine Antarctica and the world's oceans. Antarctica is a continent that has no permanent human settlements. As a first thought you might ask yourself why should we examine Antarctica, there should not be any significant environmental issues in Antarctica correct? To most of you the answer will be humbling and perhaps even disturbing. We shall have to wade through the information and data presented in this chapter to begin to understand and appreciate the depth and magnitude of the challenges that are confronting humans from an environmental perspective and that are very evident even in Antarctica. We will also briefly discuss the world's oceans which we will learn has been the great dumping ground for human waste for centuries.

15.2 ANTARCTICA

When evaluating environmental regulations of the world, one must not exclude an entire continent in that discussion. Therefore, we shall and must discuss environmental regulations and pollution in Antarctica. Antarctica is a continent on the bottom of the world with a total land area of 14 million square kilometers with 98% of its surface covered in ice. Antarctica shares no border with any other country (see Figure 15.1).

There are no permanent settlements in Antarctica. In 1959, Antarctica was established as an international area dedicated to science and research. Several nations including Argentina, France, Japan, United Kingdom, and the United States signed the Antarctic Treaty in 1959. The treaty contains 14 articles which establish key elements related to the peaceful international coordination on the continent, some of which include (Antarctic and Southern Ocean Coalition 2021):

- Banning military intervention
- Peaceful scientific research
- International information exchange
- No territorial sovereignty

DOI: 10.1201/9781003175810-15

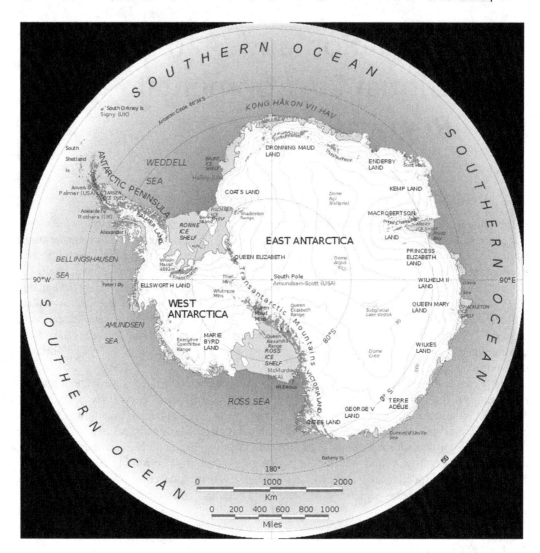

FIGURE 15.1 Map of Antarctica (From National Aeronautics and Space Administration (NASA) 2021. Map of Antarctica. https://nasa.gov/images/contnet/600827main_24.jpg (accessed July 28, 2021).

15.2.1 Pollution In Antarctica

Some pollution issues in Antarctica include (McConnell et al. 2014 and National Aeronautics and Space Administration (NASA) 2014):

- Ozone depletion
- Increased atmospheric CO_2 concentrations
- Solid and hazardous waste generated from research activities and tourism
- Marine pollution from anthropogenic sources such as spills from ships and cargo vessels
- Litter
- Lead
- Invasive species such as dandelions

15.2.2 ENVIRONMENTAL REGULATIONS OF ANTARCTICA

Environmental regulations in Antarctica do exist and are traced back to the Antarctic Treaty of 1959. All activities in Antarctica must be conducted in accordance with the regulations set forth by the Norwegian Polar Institute. For any activity, the following must be conducted (Norwegian Polar Institute 2021):

- Notification of planned activity
- Preparation of an environmental impact assessment
- Preparation of contingency plans
- Obtaining Insurance
- List of experience and equipment
- Preparation of environmental emergency plan
- Waste handling and disposal plan
- Provisions for preventing introduction of alien species
- Preparation of a final report

15.2.3 SUMMARY OF ENVIRONMENTAL REGULATIONS IN ANTARCTICA

It should not come as a surprise after reading this chapter that there are environmental impacts even in Antarctica from anthropogenic sources. This presents an interesting and perhaps conflicting phenomenon to some of you because Antarctica is predominantly devoid of human settlement. It is essentially a wilderness of continental size. So how can Antarctica be polluted and if Antarctica is polluted what does that say about the rest of the Earth. Antarctica is perhaps the best example that pollution does not respect country borders. Evaluating Antarctica has placed a spotlight on the migration of pollution on Earth. One migration pathway that is common for pollution follows Fick's First Law of Molecular diffusion that we covered in Chapter 2 and which is in full display for us to observe and quantify in Antarctica.

15.3 OCEANS

Oceans cover 70% of the surface of the Earth or 360 million square kilometers. In many respects the oceans of the world are the last resting place for most anthropogenic pollution. As an example, an estimated 7 million metric tons of plastic is dumped into the world's oceans every year (see Figure 15.2) (NOAA 2021). This is just a portion of the pollution that enters the oceans every year. The National Oceanic and Atmospheric Administration (NOAA) estimates that billions of pounds of trash and other pollutants enter the oceans each year (NOAA 2021).

15.3.1 OCEAN POLLUTION

The oceans have been the dumping ground for not just our garbage but for some of our most toxic wastes for hundreds of years. The oceans have been treated as the human privy and garbage disposal (see Figure 15.2). Only in the past few decades have the dumping of wastes in the oceans come under some environmental regulations. However, polluting the oceans still continues to this day. The oceans receive pollution from direct discharge or dumping, from deposition from the atmosphere, from rivers and streams that discharge into the oceans, and directly from the land itself.

The oceans were once believed to be so vast and deep that the effects of pollutions would never directly threaten the oceans. This has been proven false. The rate of accumulation of pollution is proportional to human population and increasing energy requirements. The more people, the more pollution ends up in the oceans. The most common types of pollution in the oceans include plastic and other forms of garbage, heavy metals, organic pollutants such as PCBs and oil, radioactive

FIGURE 15.2 Plastic and other garbage on a beach in Hawaii (From National Oceanic and Atmospheric Administration (NOAA) 2021). Ocean Pollution. https://www.noaa.gov/resource-collections/ocean-pollution (accessed July 28, 2021).

substances and waste, biological pollutants, pesticides, herbicides, fertilizers, heavy metals, and untreated raw sewerage (NOAA 2021).

Pollution in the oceans does not all originate from direct discharge but also is delivered through the atmosphere form increased concentrations of carbon dioxide in the atmosphere that is partially adsorbed by the oceans and causes acidification of the surface layers (National Oceanic and Atmospheric Administration 2021). Absorption of carbon dioxide by the oceans has been estimated to be as high as 25%, which represents millions of tons of carbon dioxide absorbed by the oceans every year. The additional carbon dioxide absorbed by the oceans reacts with ocean water and forms an acid called carbonic acid and is lowering the pH of the oceans world-wide. In fact, the oceans are acidifying faster than they have in some 300 million years. The effect of the acidification has been linked to bleaching of coral reefs, including the Great Barrier Reef off the northeastern coast of Australia (Natural Resource Defense Council 2021).

An additional form of pollution in the ocean that is not commonly mentioned is noise. Sound waves travel father and faster in the oceans than they do in the atmosphere. Increased human-induced noise in the ocean is harming numerous marine species because they use sound to navigate, find food, and mate. These crucial activities for marine life survival are having widespread impacts (Natural Resource Defense Council 2021).

15.3.2 ENVIRONMENTAL REGULATIONS OF THE OCEAN

The United States has enacted several laws over the last several decades aimed at protecting the Great Lakes and the Oceans from pollution. As we learned in Chapter 3, the most significant federal laws addressing waterways were first enacted in the United States as far back as 1899 and include (USEPA 2021):

- Rivers and Harbors and Refuse Act of 1899
- Coastal Zone Protection Act of 1972
- Fisheries Conservation and Management Act of 1976
- Global Climate Protection Act of 1987
- Marine Protection Act of 1972

- Marine Mammal Protection Act of 2015
- Ocean Dumping Act of 1988
- Oil Spill Protection Act of 1990

15.3.3 SUMMARY OF ENVIRONMENTAL REGULATIONS OF THE OCEANS

For centuries, the oceans have been subject to the freedom of the seas doctrine which is a principle recognized in the 17th century that limited national rights and jurisdiction over the oceans to a narrow belt of sea surrounding a nation's coastline that typically is 19.3 kilometers (United Nations 2021). Since 1982, the United Nations has been actively involved in establishing protections for the oceans through the adoption of the Law of Sea Convention which established (United Nations 2021):

- Freedom of navigation rights
- Set territorial sea boundaries 19.3 kilometers offshore
- Set exclusive economic zones of up to 321 kilometers offshore
- Set rules for extending continental shelf rights up to 563 kilometers offshore
- Created the International Seabed Authority
- Created other conflict-resolution mechanisms

The United Nations is also involved in protection of the marine environment and biodiversity and reducing pollution from marine shipping especially shipping in the polar regions with the adoption of the Polar Code in 2014. As of 2015, there were nearly 90,000 commercial shipping vessels that carry cargo that were registered worldwide that transported nearly 10 billion metric tons of cargo (United Nations 2021).

15.4 SUMMARY OF ENVIRONMENTAL REGULATIONS AND POLLUTION OF THE WORLD

We have now completed a review of environmental regulations of more than 50 countries and every continent including a brief assessment of Antarctica and the oceans. With respect to pollution, we have learned as perhaps the most significant points are the following:

- No country, continent, or ocean is pollution free
- Pollution caused by humans is everywhere
- Significant environmental degradation caused by humans has touched every country
- Pollution does not respect country borders
- Pollution will continue to migrate in the air, water, and on land
- The oceans have been treated as the human privy and garbage disposal of the world

With respect to environmental regulations, the most significant information we have learned include the following:

- Each country has environmental regulations.
- No two countries are exactly the same.
- Each country has unique attributes whether it be related to its climate, geography, geology, social and economic concerns, or politics which affect the ability of a country to protect human health and the environment that influence many aspects of environmental regulations.
- The platform of environmental regulations of each country are similar to that of the United States or the European Union.

- Although some countries are doing a better job than others, each country struggles with protecting human health and the environment for various reasons two of which seem to be universal and include a growing population and urban expansion.
- Every country has been significantly affected by pollution of the air, water, and land.

The largest polluting activity of land world-wide is agriculture. The largest polluter of air world-wide are motor vehicles that include automobiles, diesel trucks, trains, buses, marine vessels, and aircraft. The largest polluter of water world-wide are biological pollutants largely from human waste and from agricultural activities that include pesticides and herbicides, fertilizers, and erosion (World Health Organization 2013, 2015 and 2016).

The final resting place for much of the pollution of the world are the oceans.

15.4.1 Evaluating and Ranking Each Country's Regulatory Effectiveness

To evaluate the effectiveness of each country's environmental regulations we must examine several factors and assigned integers that appropriately reflect where each country is in the process and to provide an enlightened point of view as to where each country can improve.

This evaluation involves assessing several attributes that we have discussed and highlighted throughout this chapter and include whether each country:

1. Has the appropriate environmental regulations for air, water, land
2. Has appropriately implemented and enforced environmental regulations
3. Has the appropriate infrastructure to support the implementation and enforcement of environmental regulations
4. Has the economic and financial resources to support continued pollution controls
5. Has other social issues that impede or override environmental concerns such as poverty, corruption, violence, unstable political issues, population, and urban constraints.
6. Sustainability initiatives including pollution prevention, recycling, and waste reduction.
7. Geographical, geological, climate factors, and other natural conditions such as natural disasters that impede environmental performance.

To make this evaluation we must first consider how each country has chosen to protect the air, the water, and the land under the first criterion listed earlier. Therefore, the regulations that are most important are associated with protecting the air, water, and land and include:

- Air quality regulations from both stationary and mobile sources
- Water quality regulations for effluent discharges from industrial sources and public-owned treatment works (POTW) and septic tank discharges
- Drinking water quality regulations
- Solid and hazardous waste regulations relating to characterization, generation, transport, and disposal
- Soil, water, and sediment cleanup regulations including investigation and polluters pay principles
- Regulating different land-use designations such as residential, commercial, industrial, and agricultural

The remaining criterion (those listed as 2 through 7 explained earlier) are straightforward qualitative evaluations. Table 4.0 presents the assigned values for each criterion for each country and are ranked from most effective to least affective. Each criterion in Table 7.6 correspond to the list of 1 through 7 described earlier. Each criterion is assigned an integer of 0 through 4 that correspond to the following qualitative assessment category:

- A value of zero (0) indicates that country lack appropriate controls
- A value of one (1) indicates that the country has poor controls
- A value of two (2) indicates that the country has adequate or average controls
- A value of three (3) indicates that the country has good controls

For comparison purposes, the per capita gross domestic product is listed for each country (United States Central Intelligence Agency 2021) and a ratio of GDP to the total score of each country is presented in the last column. As you can see from analyzing Table 4.0, there should be very little surprises when each country is ranked. The countries with the most room for improvement include Turkey, Egypt, Malaysia, Russia, Saudi Arabia, and India. These countries have a host of challenges that include lack of regulations or enforcement of existing regulations, lack of infrastructure, financial hardship, population pressures, and may have to deal with corruption issues. The countries in the middle that include Mexico, Indonesia, Brazil, Argentina, South Africa, Kenya, Tanzania, Peru, China, and Chile are developing nations that are struggling to learn and enforce their environmental regulations, have population pressures, lack infrastructure, and are struggling to deal with cleanup of existing sites of contamination.

The countries that rank the highest include Japan, United States, Canada, Australia, New Zealand, European Union, Switzerland, Norway, and South Korea are all developed nations with financial resources to address environmental issues but still struggle with issues such as population and urbanization, and air and water quality but have all made significant progress compared to other nations. Examining the final column in Table 15.1 provides perhaps the most telling information. The countries that rank the highest have a ratio of GDP to Total Score of less than 2.5. The countries that have the most room for improvement all have a ratio greater than 10, except for Saudi Arabia. Saudi Arabia ranks poorly for environmental controls but high for per capita GDP which indicates that Saudi Arabia has financial resources to improve its environmental controls but chooses to focus their financial resources elsewhere. Egypt and India are the only countries evaluated that have a Total Score to GDP ratio greater than 20. China and Chile are unique when examining their ranking and numerical scores. China and Chile rank in between the low performance and high performance. Knowing that China has been improving its environmental controls indicates that China is transitioning from low to high. Chile appears to be currently more stagnant.

15.4.2 ENVIRONMENTAL CHALLENGES OF EACH COUNTRY

Some of the challenges that each country face that have significantly influenced the score of each country that have been described in this chapter include:

- Turkey is currently facing financial and economic hardship along with a lack of robust environmental standards and enforcement.
- Egypt is experiencing political unrest, a growing population and urbanization, lack of infrastructure, and water shortages.
- Malaysia lacks infrastructure, environmental enforcement, financial resources, and rapid population growth.
- Russia lacks infrastructure, environmental enforcement, financial resources, political framework, and struggles with corruption.
- Saudi Arabia lacks robust regulations and enforcement, some lack of infrastructure but does have financial resources.
- India has large population pressures, lacks infrastructure, and enforcement of regulations.
- Mexico lacks infrastructure, especially with water treatment, and struggles with corruption and lack of complete regulations (i.e., stormwater) and enforcement struggles.
- Indonesia has pressure from a large population, newer regulations, deforestation, and lack of infrastructure

TABLE 15.1

Country Comparison of Environmental Regulation Effectiveness

Country	Criteria							Total Score	GDP Rank	GDP/Score
	1	2	3	4	5	6	7			
Turkey	1	1	0	0	0	0	2	4	69	17.25
Egypt	1	1	1	0	0	1	1	5	105	21
Malaysia	1	1	0	1	1	0	1	5	64	16
Russia	1	1	1	1	1	0	1	6	65	10.8
Saudi Arabia	1	1	1	1	1	0	1	6	20	3.3
India	1	1	0	1	1	1	1	6	132	22
Mexico	1	1	2	1	1	1	1	8	82	10.25
Indonesia	2	1	1	2	1	1	1	9	108	12
Brazil	1	1	1	2	1	1	2	9	95	10.5
Argentina	1	1	1	2	1	1	2	9	95	10.5
South Africa	2	1	2	1	1	2	1	10	103	10.3
Kenya	2	2	1	1	1	1	2	10	154	15.4
Tanzania	2	2	1	1	1	1	2	10	156	15.6
Peru	1	1	1	2	2	1	2	10	104	10.4
China	2	1	2	3	2	1	2	13	94	7.2
Chile	2	2	2	2	3	1	1	13	74	5.7
Japan	3	3	3	3	2	2	1	17	38	2.2
United States	2	3	3	3	2	3	2	18	19	1.05
Canada	2	3	3	3	3	2	2	18	32	1.7
Australia	2	3	3	3	3	2	2	18	27	1.5
New Zealand	3	3	3	3	3	2	2	19	44	2.3
European Union	3	3	3	3	3	3	2	20	43	2.1
Switzerland	3	3	3	3	3	3	2	20	17	0.85
Norway	3	3	3	3	3	3	2	20	12	0.6
South Korea	3	3	3	3	2	3	3	20	42	2.1

- Brazil has population pressure, especially in urban areas, infrastructure, lack of progress in cleaning up contaminated sites, water treatment, deforestation, and corruption
- Argentina lacks infrastructure which places pressure on enforcement, financial resources, water treatment and cleanup of contaminated sites.
- South Africa has a degree of political unrest and financial pressures, climate change, and water issues.
- Kenya has financial and lack of infrastructure resources
- Tanzania has lack of financial and infrastructure resources
- Peru has lack of infrastructure to deal with water contamination issues, especially from mining and lacks financial resources
- China lacks enforcement measures and has population challenges and infrastructure to deal with pollution control measures, especially in the more undeveloped western regions.
- Chile has urban population challenges and water issues in the desert regions of the northern portion of the country
- Japan has geographical and geologic challenges (i.e., island nation that are prone to earthquakes), population pressure, and lack of landfill space.

- United States faces lack of regulations of agricultural areas, significant chemical use of pesticides and herbicides, erosion, deforestation, and pollution from septic systems, and invasive species
- Canada lacks some stormwater controls, and faces habitat destruction from, mining, deforestation, and petroleum development
- Australia faces future water issues, population pressure, and habitat destruction especially the Great Barrier Reef
- New Zealand faces water quality challenges, especially from agriculture and livestock sources
- European Union faces continued challenges from cleaning up historical sites of contamination and population pressure but has done well with regulations and sustainability measures.
- Switzerland has erosion challenges and pollution from agriculture and livestock.
- Norway faces climate change challenges and pollution from agriculture and livestock.
- South Korea faces population pressures and air quality challenges, especially in its capital city of Seoul.

A final point to consider is the objectivity of the evaluation conducted through the selection of countries included in this chapter. As stated at the beginning of this chapter the countries selected include 70% of the human population and 75% of the land area. The countries not selected include countries such as Somalia, Algeria, Libya, Sudan, Democratic Republic of Congo, Angola, Zimbabwe, Botswana, Mozambique, Congo, Syria, Iran, Pakistan, Afghanistan, Venezuela, and many former Soviet-block countries. These countries were not evaluated because of the more severe social, economic, poverty, sickness, and violence they are experiencing which has degraded the natural environment and directly threaten human health. Therefore, when considering environmental degradation and threats to human health and the environment from pollution, the countries examined in this chapter are bias toward better environmental performance.

15.4.3 Summary and Conclusion

Every country we have evaluated have environmental regulations. So, why do so many countries struggle with protecting human health and the environment? The answer for sure is likely very complex, but after we have examined each country and then evaluate each country's environmental effectiveness and compare it to the GDP for each, the answer begins to become clear. Those countries with a low relative GDP struggle with providing its population with basic survival needs than those countries with higher relative GDP. Those countries with a higher relative GDP have more effective environmental regulations because they can afford to better protect human health and the environment.

Antarctica and the Oceans have and will play a significant role concerning measuring future sustainability efforts and survival of our species and many other species on Earth. We have learned thus far that Antarctica and the oceans have become polluted from direct discharge and from migration of contaminates through the air, water, and land. We have also learned that pollution migrates following fundamental laws of physics and that fact is evident by examining the impacts observed in Antarctica and the Oceans. We have also learned that pollution does not respect country borders.

The oceans were the origin of life on Earth and remain the prime source of nutrients and food that sustains life on our planet but have become polluted by our own actions. If not for any other reason than our own survival, we must be diligent at assessing the risks posed by pollutants, enact appropriate protections, and act on those protections to prevent further degradation on a global scale.

To summarize this chapter is to state that we must now accept the reality that humans have adversely impacted the entire Earth and that our efforts to improve our environment since the enactment of environmental regulations in the United States and worldwide have simply not been enough. Earth Scientists are now convinced that out of this fact, we have now moved the needle of geologic time into a new period called the Anthropocene (Smithsonian 2018, Zalasiewicz et al. 2018, and International Union of Geological Sciences (IUGS) 2016). This is significant because the definition of a Geologic Age is a time period that effects the entire Earth and will be recorded in Earth history and cannot be erased.

Now that we have described pollution and its behavior and the examined how numerous countries have enacted environmental regulations and have struggled with their own pollution issues, we will now turn our attention to how to maintain compliance with environmental regulations. The next three chapters outline measures to improve our environment through modifying existing regulations, enacting new regulations, and developing a sustainability program.

Questions and Exercises for Discussion

1. **After reading this chapter, what two items surprised you the most and why?**
2. **After reading this chapter, what are two items that you were the least surprised and why?**
3. **Of the basic themes outlined in Chapter 1, which to you is the most significant after reading this chapter and why?**
4. **After reading this chapter, list your top five priorities for improving environmental regulations of the world and why?**

REFERENCES

Antarctic and Southern Ocean Coalition. 2021. Antarctic Environmental Protection. http://www.asoc.org/advocacy/antarctic-environmental-protection (accessed July 28, 2021).

International Union of Geological Sciences (IUGS). 2016. The Anthropocene Epoch: Adding Humans to the Chart of Geologic Time. *International Geological Congress*. Cape Town. South Africa. https://www.35igc.org/verso/5/scientific-programme (accessed July 21, 2021).

McConnell, J. R., Maselli, O. J., Sigl, M., Vallelongo, P., Neumann, T., Anschutz, H., Bales, R. C., Curran, M. A., Edwards, R., Das, S. B., Kipfstuhl, S., Layman, L., and Thomas, E. R. 2014. Antarctic-Wide Array of High-Resolution Ice Core Records Reveals Pervasive Lead Pollution Began in 1889 and Persist Today. *Scientific Reports*. Vol. 4. no. 5848. 5p.

National Aeronautics and Space Administration (NASA). 2014. Lead Pollution Beats Explorers to the South Pole. https://www.nasa.gov/content/goddard/lead-pollution-in-antarctica (accessed July 28, 2021).

National Aeronautics and Space Administration (NASA). 2021. Map of Antarctica. https://nasa.gov/images/contnet/600827main_24.jpg (accessed July 28, 2021).

National Oceanic and Atmospheric Administration (NOAA). 2021. Ocean Pollution. https://www.noaa.gov/resource-collections/ocean-pollution (accessed July 28, 2021).

Natural Resources Defense Council (NRDC). 2021. Ocean Pollution: The Dirty Facts. https://nrdc.org/stories/ocean-pollution-dirty-facts (accessed July 28, 2021).

Norwegian Polar Institute. 2021. Regulations for Activities in Antarctica. https://www.nplar.no/en/regulations/the-antarctic (accessed July 28, 2021).

Smithsonian. 2018. The Age of Humans: Living in the Anthropocene. *Smithsonian Statement on Climate Change*. https://www.si.edu/newsdesk/releases/smithsonian-statement-climate-change (accessed October 31, 2021).

United Nations. 2021. Oceans and the Law of the Sea. http://www.un.org/en/sections/issues-depth/oceans-and-law-sea (accessed July 28, 2021).

United States Environmental Protection Agency (USEPA). 2021. Laws that Protect Our Oceans. https://www.epa.gov/beach-tech/laws-protect-our-oceans (accessed July 28, 2021).

United States Central Intelligence Agency. 2021. The World Fact Book. https://www.cia.gov/library/publications/the-world-factbook.gos/NA.html (accessed August 30, 2021).

World Health Organization. 2013. Water Quality and Health Strategy 2013–2020. http://www.who.int/water_sanitation_health/dwq/en/ (accessed October 18, 2018).

World Health Organization (WHO). 2015a. *Guidelines for Drinking-Water Quality*. Geneva, Switzerland. 516p.

World Health Organization. 2016. World Health Organization Global Urban Ambient Air Pollution Database. http://www.who.int/phe/health_topics/outdoorair/database.htm (accessed July 28, 2021).

Zalasiewicz, J., Waters, C., Summerhayes, C., and Williams, M. 2018. *The Anthropocene. Geology Today*. Vol. 34. no. 5. pp. 182–187.

16 Environmental Protection and Sustainability at the Global Level

16.1 INTRODUCTION

The United States has made significant progress in restoring the environment and improving sustainability since the USEPA was formed and the Clean Air Act was promulgated in the early 1970s. Since then and over the past 50 years, a cascade of environmental regulations have been promulgated that have addressed impacts to the atmosphere, water, land, and living organism and cultural and historic sites. However, we now know that the United States is not isolated and pollution has no boundaries. We also know through many different scientific studies that our planet's habitability continues to decline. This is evidenced through polluting the air, water, and land, which is evidenced through climate change, species reduction and extinction, ocean pollution, decline in coral reef health, pollution in Antarctica, urban expansion and land use, agricultural practices, forest management, and many others. Therefore, we must act as a global community to more fully understand the impacts humans have made to our planet and act together and diligently through focused efforts to limit the negative effects and restore ecosystems where possible. These actions must be undertaken before it is too late not just for the survival of other species but for the survival of our own species as well.

This chapter is dedicated to understanding, implementing, and improving sustainability at the global level. In addition, we address sustainability methods that can be applied in different settings and locations from urban to agricultural to wilderness areas on land, the oceans and the atmosphere. Many items listed at the global level also apply to actions required at the national and even the individual level but have a different scope and focus. Therefore, there may appear to be duplication but there are differences in scale and application that will be discussed in this chapter and subsequent chapters.

16.2 DEFINING SUSTAINABILITY

USEPA defines **sustainability** as creating and maintaining conditions under which humans can exist in productive harmony to support present and future generations (USEPA 2021a). Note that the definition states "productive harmony." One may ask, in productive harmony with what? The answer is with the environment. Therefore, in order to achieve some level of sustainability we must understand the environment and also understand the aspects of how facility operations impact the environment. Sustainability is then the outcome of analyzing the aspects of facility operations with the natural environment and developing and engineering methods to either minimize or eliminate harmful potential impacts.

Sustainability is based on a simple principle that states that (USEPA 2021b):

Everything we need for our survival and well-being depends, either directly or indirectly, on our natural environment

The United Nations defines sustainability as (United Nations 2021a):

development that meets the needs of the present without compromising the ability of future generations to meet their own needs

DOI: 10.1201/9781003175810-16

Sustainability had its origin in 1969 in the incorporation of the United States National Environmental Policy Act (NEPA) (USEPA 2021c and 2021d):

> *to foster and promote the general welfare, to create and maintain conditions under which humans and nature can exist in productive harmony and fulfill the social, economic, and other requirements of our present and future*

The term "sustainable development" was first used by the United Nations in 1987 (United Nations 2021a).

16.3 SUSTAINABILITY IMPEDIMENTS AND CHALLENGES

Impediments and challenges exist between countries concerning sustainability and include differences in power and responsibility, environmental, economic concerns, population, cultural, and religious differences that must be addressed before sustainability can proceed on a path to become an effective reality on a global level (United Nations 2021a). These are monumental challenges but humans are now facing perhaps the greatest threat to our species. Ironically, that threat has largely been caused by humans. This nothing new, early in the environmental movement in 1971, an Earth Day poster read "We have met the Enemy and He Is Us" (Ohio State University 2021). This is similar to the United States Advisory message that stated People Started Pollution, People Can Stop It." This challenge is also huge opportunity for international cooperation and innovation.

Religious conviction and population challenges are generally not openly discussed and likely represent the most challenging to overcome to an extent is meaningful, especially in western developed countries. This is likely because these two challenges will be interpreted as invading personal freedoms.

16.3.1 POWER AND RESPONSIBILITY

Points of contention are realized between developing and developed countries. It is of concern for developed nations because they are perceived to be imposing their new-found sustainable values upon developing countries. Developed nations have already greatly benefited from exploitation of environmental resources, whereas developing nations have not had the chance or enough time to exploit their natural resources and feel resentment toward developed nations. A significant focus in the sustainable development movement has been on freeing people in developing nations from the bonds of poverty and starvation. Freeing nations from poverty and starvation would mean consuming enormously more energy and resources in order to raise the standard of living of these individuals which number in the billions. One of the most important arguments of sustainability states that the rights provided by governments, such as the right to vote and freedom of speech, are not of much use unless the society provides its citizens with an opportunity of an education, access to food, and employment (United Nations 2021a).

16.3.2 ECONOMIC

This brings us to perhaps one of the two most problematic hurdles in achieving a sustainable world and that is sustainability does not yet appear to be compatible with a capitalistic economic system (United Nations 2021a). An economic system of capitalism relies on an ever-growing economy. How can humans achieve sustainability with a growing population, increasing energy needs, lifting vast human populations (e.g., more than 2 billion people) in undeveloped nations standard of living and achieve a sustainable world where we all live in harmony with our environment. This all must take place in a fragile biosphere called Earth. So far, our economic system continues to grow and gather momentum and as it does it threatens to destabilize the global ecological balance in many significant ways. How can humans modify a capitalistic system that has ecological destructive outcomes and achieve a sustainable planet?

There are specific issues, such as reversing much of the damage to the ozone layer, that have been successful and lends hope and a call to action that not all is lost and that balance can be restored. Afterall, humans are an adaptable species and we are just recently putting our collective intellectual and cooperative skills to work to confront this challenge which we must win.

16.3.3 Religion

The last hurdle relates to religious conviction. In May 2015, Pope Francis of the Roman Catholic Church release an Environmental Encyclical from the Vatican that addressed climate change, global warming, pollution, economics, and population control. The Environmental Encyclical as it is now referred to states that (Pope Francis 2015):

> humans must forcefully reject the notion that we were given dominion over the Earth and absolute dominion over other creatures

This is yet another example of hope and resolve that we must do what must be done and that we all have a part to play as Pope Frances further states that (Pope Francis 2015):

> there is an urgent need for a radical change in the conduct of humanity

16.4 UNITED NATIONS GLOBAL SUSTAINABILITY GOALS

The United Nations has framed sustainability goals termed "Global Goals," which include (United Nations 2021b):

• No Poverty	• Zero Hunger	• Good Health
• Well Being	• Quality Education	• Gender Equality
• Clean Water	• Effective Sanitation	• Affordable Energy
• Clean Energy	• Economic Growth	• Satisfactory Employment
• Industry	• Innovation	• Infrastructure
• Reduced Inequalities	• Sustainable Cities	• Responsible Consumption
• Climate Action	• Life Below Water	• Life on Land
• Peace and Justice	• Partnerships	• Life in the Air

Much of what we have discussed up to this point in this chapter may sound like a bleak outlook from a world view, but let us not ponder and debate the political, social, economic, and religious challenges on a global scale before we examine what can be accomplished at the local level, which as you will see in this chapter, is significant. For instance, sustainability has been a priority in the United States since the USEPA was first formed nearly 50 years ago. This has been achieved through improving air quality, water quality, and land impact through the CAA, CWA, RCRA, and CERCLA and a multitude of other environmental regulations, many of which we have examined. In addition, actions by the United Nations, the World Bank, private nonprofits, thousands of businesses and many other nations have had positive results.

We have covered several subjects within this book that now can be pulled together into a cohesive framework that can be used to implement a comprehensive sustainability program at any level with the end goal of environmental stewardship. The pursuit of sustainability involves many subject areas and levels. In this chapter, we will address actions required at the global level. We will address actions required at the national and individual level in the next two chapters.

Those subject areas that will be addressed in the following sections include:

- Cohesive action from the global community
  - Global Environmental Regulations
  - Obtaining a Sustainable Human Population
  - Modifying the human diet
  - Financial fairness and equity distribution
- Improving environmental regulation at national levels
- Understanding the natural setting of our urban and industrial areas
  - Protecting Sensitive Ecological Areas
  - Smart Urban Development
  - Improving Urban Transportation
  - Improving Building Design
  - Incentivizing multi-family housing
- Knowing the chemicals that are used and how they might cause harm to the environment and humans if they are released
- Knowing what alternative chemicals are available
- How to reduce chemical use
- Preventing Chemical Releases
- Reducing energy use
- Using alternative sources of energy
- Minimizing solid waste, air emissions, and water discharges
- Reducing water use
- Minimizing stormwater runoff
- Repairing environmental damage to the extent possible
- Attaining Environmental Stewardship

16.5   COHESIVE POLICY ACTIONS FROM THE GLOBAL COMMUNITY

At least five policy-oriented actions at the global level are required if sustainability will be effective. These actions include:

- Building a Global Community
- Cohesive environmental policy, regulations, and enforcement
- Controlling the human population
- Modifying the human diet
- Financial fairness and Equity Distribution
- Modifying business models
- Improving our relationship with nature

16.5.1   BUILDING A GLOBAL COMMUNITY

Building a global community is possible and is achieved periodically. An example is the Olympic Games which bring the world together to celebrate sport every 2 years. In addition, banning or limiting the use of CFC's which degraded the ozone layer is another example where the global community cooperated to address an environmental threat. Building a global community on sustainability issues will require significant cooperation on several actions for an extended period of time, and also must include removing nationalistic political, social, and religious boundaries (Harari 2018). Patriotism for a nation, which is common, must graduate to a global level.

16.5.2 COHESIVE ENVIRONMENTAL POLICY, REGULATIONS, AND ENFORCEMENT

Pollution does not respect boundaries. Like all matter, pollution follows the laws of physics and will disperse from a point of origin and given enough time will be detected world-wide. One country can adopt and implement effective sustainability measures but, in the end, will eventually become polluted at similar levels as other countries who do not implement the same measures regardless of financial resources or technology. Therefore, we must develop a global community and enact cohesive environmental policies and enforcement at the international level. Climate change is a good example that action must be taken at a global level since greenhouse gases and a plethora of other pollutants migrate the globe and become ubiquitous. Just in the last few decades there is much less pack ice in the artic and glaciers are shrinking at a significant pace.

16.5.3 CONTROLLING THE HUMAN POPULATION

In Chapter 7 we examined environmental regulations of many countries of the world and discovered that there are at least three factors common to countries that have significant pollution issues and those include (1) lack of financial resources, (2) over population, and, (3) lack of effective environmental regulations.

We have discussed in the previous section that capitalism currently relies on growth for itself to be sustainable. Much of the economic growth that fuels capitalism is achieved through an endless and continued increase in the human population and a growing standard of living, especially in underdeveloped countries. In 1900, the human population was approximately 1 billion and currently is estimated at more than 7 billion (United Nations 2021c and Harari 2015). A continued growing human population is clearly not sustainable and has grabbed the attention of world and religious leaders (Pope Francis 2015).

The question is what is a sustainable population of humans on Earth. Clearly, we have exceeded that balance with our current practices and disregard for our planted. The answer lies in reducing adverse impacts and balancing those achievements with a human population that does not adversely impact future generations and other life on Earth.

Many human technological innovations have been developed through our history that include (1) the agricultural revolution approximately 12,000 years ago, (2) the industrial revolution in the last 300 years, and (3) the information and computer revolution just recently. Each of these technological marvels have enabled humans to establish large cities, have time for recreation, and greatly increase our population (Harari 2015). If humans can solve complex technological unknowns, humans can also solve the human population issue before nature does it for us.

16.5.4 MODIFYING THE HUMAN DIET

Much of the farming in the United States is concentrated in the production of meat (United States Department of Agriculture (USDA) 2011). Meat production in the United States also consumes huge volumes of water and creates significant biological pollution and is also a significant contributor of methane which is a powerful greenhouse gas (USEPA 2021e). Slight modifications in the diet of Americans that include reducing meat and animal fat consumption combined with modifying farming techniques through the addition of urban farms and other measures, can have a significant positive impact on:

- Improving the health of Americans by improving our diet and reducing exposure to contaminants such as pesticides and herbicides
- Reducing greenhouse gas emissions
- Reducing fugitive emissions
- Significantly increasing the amount of natural land

- Decreasing water use
- Decreasing fertilizer, pesticide and herbicide use
- Decreasing exposure to harmful chemicals
- Decreasing biological pollution
- Improving surface water quality

16.5.5 FINANCIAL FAIRNESS AND EQUITY DISTRIBUTION

There is a correlation between effective environmental regulations and standard of living (Rogers 2021). Those countries without financial resources have more significant environmental uses than countries with deeper financial resources. Fortunately, the United Nations and The World Bank are very aware of this situation and have been working diligently to improve the environment of undeveloped nations (United Nations 2021d and World Bank 2021a). In addition, the USEPA is also aware and has been assisting numerous other nations at improving the environmental quality by implementing several environmental initiatives world-wide (USEPA 2021f).

16.5.6 MODIFYING BUSINESS MODELS

A significant potential impediment in achieving sustainability is the current capitalistic economic system. Every country of the world is unique but also shares commonalities which respect to basic economic systems. The current capitalistic economic system must change from growth minded to sustainability minded. Sustainability incentives must be integrated at the international, national, community, individual economic levels.

Thousands of businesses world-wide are adjusting their business models by incorporating sustainability measures and incentives into their culture and using these modifications to promote themselves by meeting the highest verified standards of social and environmental performance, public transparency, and legal accountability to balance profit and purpose. Two widely used methods include ISO 14001 and Certified B Corporations.

The International Organization for Standardization (ISO) 14000 is a family of standards originating in 1992 and has now grown into detailed international standards that specifies requirements for an effective environmental management system (EMS) (American Society for Quality 2021). ISO 14001 is a voluntary standard that examines every aspect of a particular business and evaluates how each aspect potentially impacts the environment. ISO 14001 then examines how to reduce environmental impact by modifying aspects. ISO 14001 examines every step of the manufacturing process from each raw material, energy consumption and type, chemical use, packaging, space, location, transportation, and each type of waste generated (American Society for Quality 2021). ISO 14001 essentially states that everything that is not included in the manufactured product eventually become a waste.

ISO 14001 is widely used throughout the world especially in the automobile manufacturing industry. Automobile manufactures required suppliers to be ISO certified as early as 2002 (Thornton 2001).

ISO 14001 is an environmental management system that has also been used as a tool to evaluate status of environmental compliance but has found more purpose in reducing waste and generating manufacturing efficiencies that translate into business advantages by lowering costs and liabilities and other factors.

Certified B Corporations is an additional method that can steer business toward sustainability. Certified B Corporations believe (B Corporations 2021):

- That they must be the change that people seek in the world
- That people and places matter

- That businesses should aspire to do no harm and benefit all through their products, practices, and profits
- That businesses should act with the understanding that each person depends upon another including future generations

Currently there are hundreds of Certified B Corporations throughout the world with 470 just in the United States (B Corporations 2021).

16.5.7 IMPROVING OUR RELATIONSHIP WITH NATURE

In general, humans have not been kind to the Earth. For the most part, humans have modified the planet to suite their perceived needs and for profit. Some of the numerous negative results include species extinction and global pollution. However, there are actions that can be implemented to improve the health of the planet. Many will be described in the next section and in the two remaining chapters of this book. In basic terms, actions at the global level to improve our relationship with nature should focus on the several items discussed in this section that promote building a global community, which include:

- Cohesive Environmental Policy, Regulations, and Enforcement
- Controlling the Human Population
- Modifying the human diet
- Financial Fairness and Equity Distribution
- Modifying Business Models

In addition, there are methods and examples of protecting those natural areas that include national parks and heritage sites throughout the world that can be improved to prevent degradation in its many forms. In addition, wildlife corridors should be expanded so that the health of existing species can be improved by increasing range and genetic diversity of wildlife (Tallamy 2019). Currently, many national parks operate as genetic islands which threaten the long-term health of species within those islands. By establishing wildlife corridors that could be as small as a vegetated bridge over a highway is an example and provides a means by which wildlife can overcome anthropogenic barriers.

Humans now know that mastering nature can't occur through force, which may have been a belief as recent as 50 year ago, but by understanding. This concept is not new, many naturalists throughout our history have written and documented the destruction of the natural areas world-wide. One response to this destruction, the United States established national parks so that at least some of the most significant natural areas would essentially be persevered for future generations to study and enjoy.

Coordination of several efforts to understand nature and improve the quality of our air, water, land, and living organisms should be conducted at the highest hierarchy global level.

16.6 COORDINATING AND PRIORITIZING SUSTAINABILITY INITIATIVES AT THE GLOBAL LEVEL

From a historical perspective. the global community came together and realized that degradation of the ozone layer in the upper atmosphere and rising carbon dioxide being responsible for climate change were both from anthropogenic sources. This realization stimulated action that has resulted in coordination and cooperation at the global level to address these negative anthropogenic impacts to our planet from the United Nations. Now 195 nations have signed up to find solutions and keep our planet from over heating and restoring the ozone layer, which we covered in previous chapters. Therefore, the United Nations, World Health Organization, World Bank, and every nation must play a role in addressing other potentially catastrophic issues facing our species that threaten our existence that include protecting the air, the water, and the land. If this can be accomplished, then protection of living organisms can be addressed.

The following sections list specific subject areas where either improvement must be instituted or be created to address sustainability at the global level. These subject areas discussed include the following:

- Science-based Regulation Improvements
- Modifying Environmental Enforcement Emphasis and Policy
- Ecological Restoration and Preservation
- Managing Natural Resources
- Improving Water Quality
- Improving Water Quality of the Oceans
- Modifying Agricultural Practices
- Limitations on Herbicides and Pesticides
- Controlling the Use of Fertilizers
- Modify Forest Management Techniques
- Improve Cooking Fuels
- Improving Urban Air Quality
- Banning Harmful Chemicals
- Addressing Emerging Contaminants
- Addressing Invasive Species
- Limiting Species Extinction
- Improving Pollution Prevention Measures
- Modify Urban Land Use Policies
- Addressing Noise Pollution
- Regulating Household Waste
- Addressing Non-point Source Pollution
- Improve Consumer Products and Packaging
- Sustainability Legislation
- Measuring Sustainability Progress

16.6.1 SCIENCE-BASED REGULATION IMPROVEMENTS

We discovered in earlier chapters that contaminants behave differently and are more prone to releases in different geologic, hydrogeologic, geographic, and climatic regions. For instance, regulations specific to geologic risk factors such as earthquakes (e. g., California) or climate change (eastern seaboard and the coast along the Gulf of Mexico) should better reflect those areas where there is the highest risk.

16.6.2 MODIFY ENVIRONMENTAL ENFORCEMENT EMPHASIS AND POLICY

At this point, it is not apparent that a need for modifying USEPA's approach to enforcement is in need of change. Especially since it has been seemingly effective up to this point. However, it's time to move beyond a litigation heavy and punishment-based enforcement model and enter into more of a reform- and repair-based enforcement policy that emphasizes sustainability.

We will learn in the next chapter that some countries are more advanced compared to the United States in how they have chosen to organize and implement their environmental regulations and where they focus internal resources. In addition, the governments of some countries have different attitudes and relationships with industry. The regulatory agencies of some countries have chosen to form working partnerships with industry with the intent to work more closely together to solve complex environmental issues rather than an adversarial approach which much too often results in polarizing points of view and expensive litigation.

The result of working together saves time, effort, and resources and in the long run, has resulted in the fact that some counties are better at protecting human health and the environment than the United States.

16.6.3 ECOLOGICAL RESTORATION AND PRESERVATION

Much has been documented on the destruction of habitat throughout the world from the Amazon rain forest to wildfires in California and wetlands throughout the United States. In addition, most every country in the world can point to an anthropogenic induced environmental that inflicted significant harm to the environment (Rogers 2019b and 2020a).

We will discuss this topic at each hierarchy level because action is required at each level and since action is required at each level. At the global level, we focus sustainability efforts on preserving and expanding National Parks throughout the world. In addition, establishing wildlife corridors that connect National Parks has huge advantages in preserving species and also expanding the genetic gene pool. This is accomplished such that unconnected National Parks do not become genetic islands (Tallamy 2019).

The United Nations already has a program in place that designates areas as World Heritage Sites. Currently there are well over 1,000 World Heritage Sites world-wide (United Nations 2021e). The purpose of World Heritage Sites are appreciate and learn about our legacy from the past, what we live with today, and what we pass on to future generations. Our cultural and natural heritage are both irreplaceable sources of life and inspiration (United Nations 2021a).

16.6.4 MANAGING NATURAL RESOURCES

Many of the countries where there has been an environmental incident that has caused enormous harm to human health and the environment were or are related to natural resource exploitation. Examples include:

- Exxon Valdez oil spill in Alaska
- Deepwater Horizon oil spill in the Gulf of Mexico
- Poor mining techniques throughout the world
- Poor forest management techniques throughout the world
- Hydroelectric power generation throughout the world
- Farming

These examples highlight the need for common regulatory treatment on a global scale for resource exploitation.

16.6.5 IMPROVING WATER QUALITY

According to the United Nations (2015), the most prevalent water quality problem globally is eutrophication, which is a result of high-nutrient loads (mainly phosphorus and nitrogen). The most significant sources of water pollution on Earth is agriculture. Other sources include, domestic sewerage, industrial effluents and atmospheric inputs from fossil fueled burning and wildfires. A significant issue related to water is sanitation and the spread of disease through untreated water with human excrement and from livestock and wildlife called zoonotic pathogens (World Health Organization 2013). In addition, the World Health Organization (2013 and 2015a and 2015b) and the United Nations (2016a, 2016b, and 2016c), estimate that 2.8 billion people do not have access to proper sanitation, while (Baum et al. 2013) estimates that the actual number could be as high as 4.1 billion because most evaluations do not account for all biological disease-causing pathogens.

The WHO lists the top ten most water polluted cities in the World as the following (WHO 2021):

1. Sumgavit, Azerbaijan	2. Linfen, China
3. Tianjing, China	4. Sukinda, India
5. Vapi, India	6. La Oroya, Peru
7. Dzerzhinsk, Russia	8. Norilisk, Russia
9. Chernobyl, Russia	10. Kabwa, Zambia

Contaminants that have affected the most water polluted cities in the World include (WHO 2021):

- Volatile organic compounds (VOCs)
- Oils and fuel
- Radioactive materials
- Garbage
- Zoonotic pathogens
- Biological contaminants including *e coli*
- Pesticides
- Fertilizers
- Plastic
- Other pathogens

In 2011, the WHO published minimum standards for water quality. The WHO Water Quality Standards are intended as a benchmark for those countries that have not established standards of their own and are considered minimum standards (World Health Organization WHO 2015a). However, much needs to be done with respect to improving water quality throughout the planet. Much of the needed action is installation of infrastructure which many poor countries can't afford because the population is increasing at a pace that is faster than infrastructure construction. The causes in many cases are exacerbated by social issues, corruption, war, and poverty which may be wide-spread in some regions (Rogers 2019b).

16.6.6 IMPROVING THE QUALITY OF THE OCEANS

Evaluating the environmental health of the oceans and ocean fisheries is much more difficult because the oceans are large, have many different jurisdictions, and counting the number of fish is difficult even with advanced technologies. However, there is evidence that suggests that ocean fisheries are under stress from overexploitation. This is evidenced by the fact that the numbers of fish caught by species are declining even though the number of fishing vessels have increased and fishing technologies have improved. For example, in 1974, only 10% of ocean fisheries were considered stressed whereas nearly 30% were stressed in 2010 (World Ocean Review 2021). However, the National Oceanic and Atmospheric Administration (NOAA) of the United States released a report in 2017 that concluded that only 15% of fisheries were stressed and represents a decrease over previous evaluations (NOAA 2021).

China catches more than double the amount of fish at nearly 14 million metric tons in 2010 than any other nation. Indonesia and the United States rank second and third, respectively, at just over 4 million tons each (World Ocean Review 2021).

Actions required to improve the health of the oceans will require more study since the oceans are so large and the anthropogenic effects are not fully understood.

16.6.7 MODIFYING AGRICULTURE PRACTICES

Agriculture is now the biggest polluter in the United States and the world (United Nations 2016a). The United Nations estimates that most land pollution is caused by agricultural activities, such as grazing, use of pesticides, fertilizers, irrigation, plowing, and confined animal facilities (United Nations 2016c). Agricultural activities release significant amounts of greenhouse gases from plowing, decaying vegetation, and from livestock. Fugitive emissions and erosion are also significant and release significant amounts or chemicals into the air and the water. In addition, we have also learned that fertilizers are a significant source of pollution to the oceans of the world.

Since environmental regulations have been so successful at reducing the harmful effects from industry in the United States, it is now time to reduce the harmful effects on the environment from agriculture.

A method of farming that reduces environmental impacts is termed **Regenerative Farming.** Regenerative farming is a philosophy of common principles that include re-establishing relationships between humans and land through improving soil health, reducing or eliminating the use of harmful chemicals, growing diverse crops, holistic livestock management, innovative and efficient use of resources, and equitable labor (Natural Resource Defense Council [NRDC] 2021). This type of farming is not new. Indigenous cultures, and even those in the United States in the 19th century, practiced this type of farming before the widespread use of pesticides, herbicides, and extensive use of fertilizers. Current farm and food production in the United States is not designed to prioritize climate, ecosystems water quality, human health, and establishing a healthy and productive relationship with nature (NRDC 2021).

16.6.8   LIMITS ON HERBICIDE AND PESTICIDE USE

Simply put, pesticides and herbicides are manufactured to kill things. USEPA (2021g) defines a **pesticide** as preventing, destroying, repelling or mitigating any pest. An **herbicide** is defined as a substance that is used to kill unwanted plants commonly referred to as weeds. **Pests** are defined as living organisms occurring where they are not wanted or causing damage to crops or humans or other animals (USEPA 2021g). Examples include:

- Insects
- Mice or other animals
- Unwanted plants (weeds)
- Fungi
- Microorganisms, such as bacteria and viruses
- Invasive species that include many plants and animals

Perhaps the most famous of all pesticides is the now banned substance called **DDT** (dichlorodiphenyltrichloroethane). It became famous as an environmental contaminant after the book *Silent Spring* written by Rachel Carson was published in 1962 (Carson 1962). The book catalogued the environmental impacts of the indiscriminant spraying of DDT in the United States and questioned the logic of releasing large amounts of chemicals into the environment without fully understanding their effects on ecology or human health. DDT was widely used in agriculture and by consumers in the 1940s and 1950s. DDT (CAS Registry Number 50-29-3) is now considered a probable carcinogen (Group: A2) by USEPA (2021h) and is no longer used as a pesticide in the United States after it was banned in 1972. The 2001 United Nations Environmental Program meeting held in Stockholm, Sweden (and put into effect in 2004) permitted its use for "vector control"—organisms that produce pathogens, such as mosquitoes) (USEPA 2021h).

Currently, approximately 1 billion pounds of conventional pesticides and herbicides are used to control weeds, insects, and other pests just in the United States (United States Environmental Protection Agency 2017 and USGS 2021).

16.6.9   CONTROLLING THE USE OF FERTILIZERS

Fertilizers are chemical compounds designed to promote plant and fruit growth when applied (USEPA 2021i). The most common fertilizers are Nitrogen (N), phosphorus (P), and potassium (K).

Nitrogen is found primarily in an organic form in soils, but can also occur as nitrate. Because nitrate is very soluble and mobile, it is transported by surface water to rivers, lakes, and streams where it can promote algal growth. In many cases the algal growth is extensive. Nitrate can also contaminate drinking water. Phosphorus occurs in soil in organic and inorganic forms, but being more soluble than nitrate, can also be depleted in soil through surface water runoff. Phosphorus can also promote algal growth in rivers, lakes and streams, because it is a limiting nutrient in fresh water. Potassium (K) in fertilizers is commonly incorporated as potash--an oxide for of potassium that includes the compounds potassium chloride, potassium sulfate, potassium nitrate, and potassium carbonate. The term "potash" comes from the historical practice of extracting potassium carbonate (K_2CO_3) by leaching wood ashes and evaporating the solution in large iron pots. Fertilizers containing potassium generally do not promote algal growth (USEPA 2021i).

The United States Department of Agriculture (USDA) has tracked fertilizer use in the United States since 1960. According to the USDA, approximately 63.5 kilograms (140 pounds) of fertilizer containing N-P-K are applied each year per acre of land farmed and this amount has increased more than 200% since 1960 (United States Department of Agriculture (USDA) 2019). In 2018, USDA estimated that 1.41 million square kilometers (350 million acres) were planted for crops which means that approximate 22.2 trillion kilograms (49 trillion pounds) of fertilizers were applied to farm land in the United States (United States Department of Agriculture (USDA) 2019). In urban areas, fertilizers containing N-P-K are common and routinely applied to residential lawns and golf courses to help maintain green healthy grass and gardens (USEPA 2021i). According to USEPA (2019c), nitrogen, phosphorus, and potassium fertilizers are not currently known to cause cancer. Exposure to high concentrations may cause nausea and vomiting (USEPA 2021j).

The most common fertilizers include Nitrogen (N), phosphorus (P), and potassium (K). A major sink for fertilizers is surface water because they are applied to the soil surface and are considered soluble in water and mobile--especially in the case of nitrates and phosphorus. Once in surface water, nitrate and phosphorus can promote excessive algal growth. Significant algal growth can deplete the dissolved oxygen in surface water and cause suffocation and death to aquatic organisms. The solubility of some fertilizers, combined with the relationship between surface water and groundwater, may lead to groundwater contamination. The natural process of enrichment of surface waters with plant nutrients is termed **eutrophication**. When anthropogenic activities such as fertilization or sewage discharges accelerate this natural process, **cultural eutrophication** occurs (McGucken 2000).

16.6.10 Forest Management

Forest management throughout the planet is considered poor at best. World-wide, forests provide habitats for 80% of amphibians, 75% of bird species, and 68% of mammal species (United Nations 2021f). Approximately one third of the worlds forests have been lost due to multiple anthropogenic affects. Most of the loss of forests has been to land use conversion to agriculture.

From the period from 2011 to 2020, Africa has experienced the most forest loss of 3.9 million hectors with South America next at an estimated loss of 2.6 million hectors (United Nations 2021f). The United Nations has suggested that actions needed to reduce the loss of forests worldwide include modifying human diets, improved governance, and improved and upgraded regulations (United Nations 2021f).

16.6.11 Cooking Fuels

Improving access to affordable and reliable energy services for cooking is essential for poor and developing countries in reducing the harmful health effects from using traditional biomass, including manure and dung, charcoal, coal, and dead vegetation. As of 2011, the United Nations has estimated that 2.64 billion people rely on this cooking method. Besides the obvious increased human health effects, cooking in this manner also emits large quantities of green house gases into the atmosphere (World Bank 2021b).

16.6.12   URBAN AIR QUALITY

The Clean Air Act has made a major impact at improving the air quality throughout the United States, especially with respect to automobile exhaust, ozone, particulate matter, and the other criteria pollutants. According to United States Environmental Protection Agency (USEPA) (2018), compared to emissions in 1980, carbon monoxide emissions have been reduced by 72%, lead by 99%, nitrogen oxides by 61%, volatile organic compounds by 54%, particulate matter (PM10) by 61% and, sulfur dioxide by 89%. However, these improvement achieved in the United States has not had the same effect when applied to many other countries of the world.

Improvements on exhaust from diesel trucks and buses have been slowed and are especially noticeable in urban areas. In addition, other sources of particulate matter are of high concern in urban areas as well and needs improvement. This need is in part the result of an increasing population and that urban areas continue to grow as population centers that will inevitably increase automobile congestion and emissions if not addressed as quickly as possible.

As we shall see in the next chapter, even though the United States is in need of improving urban air, many locations in the rest of the world are far worse when it comes to urban air pollution.

16.6.13   BANNING HARMFUL CHEMICALS

Consistency in banning certain chemicals known to be harmful world-wide is in need of improvement. Knowing that chemical contaminants do not respect country borders begs for a global approach because even if one country bans certain contaminants, the country still may become contaminated from those same contaminants if they migrate from a nearby country that has not banned the same contaminants. The Toxic Substance Control Act (TSCA) is a set of regulations in the United States that provides a mechanism to ban certain harmful contaminants world-wide.

16.6.14   EMERGING CONTAMINANTS

Emerging contaminants are chemicals or substances that have been recently discovered in the environment and, in some instances, have caused harm or have impacted resources to the extent that they can no longer be used for fear of health risks. Emerging contaminants include many compounds including PFAS, PIP, many pharmaceutical compounds, 1,4-dioxane, and others. Many of these emerging compounds are now being detected in the environment and have not been adequately evaluated. This highlights a need for improved prevention measures and the need for improving handling and disposal practices.

16.6.15   ADDRESSING INVASIVE SPECIES

Some consider invasive species a type of emerging contaminant. However, for the purposes of this book we treat them separately because there are so many invasive species and their behavior in the environment to so different compared to chemicals and other substances because their mode of migration is through breeding. Invasive species have been present for the last few hundred years but environmental damage is increasing and in need of further regulatory protections at the global level.

World-wide, there are thousands of invasive species (United Nations 2018). Invasive species are characterized as aggressive when they are transported into areas where they are not naturally occurring usually because they do not have a natural predator or other organisms have not evolved defenses (United Nations 2018 and 2021g and USEPA 2021k).

16.6.16 LIMITING SPECIES EXTINCTION

Species extinction is a global issue that is getting worse. Worldwide, there are thousands of species of animals and plants that are in danger of extinction because of direct human involvement that includes development, habitat loss and destruction, poaching, pollution and other factors. The World Wildlife Fund maintains a general list of those animals that are in danger and also includes an additional category different than that of the United States by the addition of another category called Critically Endangered. Some of the animals currently listed by the World Wildlife Fund as critically endangered or endangered include (World Wildlife Fund 2021):

Amur Leopard	Black Rhino	Bornean Orangutan
Cross River Gorilla	Eastern Lowland Gorilla	Javan Rhino
Malayan Tiger	Sumatran Orangutan	Orangutan
South China Tiger	Sumatran Elephant	Mountain Gorilla
Sumatran Rhino	Sumatran Tiger	Amur Tiger
Asian Elephant	Western Lowland Gorilla	African Elephant
Bluefin Tuna	Blue Whale	Bonobo
Chimpanzee	Borneo Pygmy Elephant	Galapagos Penguin
Hectors Dolphin	Ganges River Dolphin	Indian Elephant
Indochinese Tiger	North Atlantic Wright Whale	Red Panda
Sea Lion	Snow Leopard	Sri Lanka Elephant
Tiger	Black Spider Monkey	African Giraffe

The given list is just to name a few of the more familiar animals that are now considered threatened with extinction. Unless major efforts from the world community are conducted immediately, the loss of these animals and countless others are likely inevitable. The following list includes some North American species that have gone extinct just since 1500 AD and include (International Union on the Conservation of Nature [IUCN] 2021):

Great Auk – 1852	Passenger Pigeon – 1914
California Golden Bear – 1922	Ivory-Billed Woodpecker – 1942
Labrador Duck – 1878	Eastern Cougar – 2011 declared
Cascade Mountain Wolf – 1940	Eastern Elk – 1887
Mexican Grizzly Bear – 1964	Newfoundland Wolf – 1911
Sea Mink – 1860	Caribbean Monk Seal – 1952
South Rocky Mountain Wolf – 1935	Southern California Kit Fox – 1903
Carolina Parakeet – 1918	Dusky Seaside Sparrow – 1987
Heath Hen – 1932	Eskimo Curlew – 1981
Bachman's Warbler – 1988	Tacoma Pocket Gopher – 1970

According to the World Wildlife Fund (2021) and the International Union for Conservation of Nature (2021), hundreds of species of animals are now extinct directly due to humans in several different ways including, hunting, habitat destruction, poisoning, development, and pollution. According to the United Nations, more than 1 million species of animals and plants are threatened with extinction as a direct result of human activities (United Nations 2015).

16.6.17 IMPROVING POLLUTION PREVENTION

The Pollution Prevention Act (PPA) and Oil Spill Prevention Act (OSPA), both of 1990, do not get enough credit in the history of environmental regulations of the United States. The PPA and OSPA were the first real efforts to turn from regulating the environment at the "end of the pipe" to promoting the prevention of pollution from entering the environment. One of the driving factors behind the PPA and OSPA was the realization of how much it cost to clean up the environmental after a release.

It was a sobering lesson that just one incident could cause so much harm, cost billions of dollars, and in some instances could not be cleaned up to acceptable levels because technologies simply did not exist. It was during this time that USEPA and many others discovered that sometimes you can throw money at a contaminated site or incident and make it go away and then there are those where all the money and technology in the world would not be able fix the problem before enormous harm was a reality. Therefore, out of this realization, USEPA began to focus efforts on pollution prevention, in part because it made common sense in that it's proactive at environmental protection and lowers response cost much like fire prevention. As we shall see in the next chapter, it was out of these and similar efforts in the European Union that the term Sustainability took hold.

16.6.18 URBAN LAND USE

One important item that should be added to pollution prevention efforts is urban land use. There is much that we can accomplish at protecting human health and the environment by not locating high risk activities in areas that are especially sensitive to pollution.

Currently, urban planning for industrial activities are near transportation routes and are generally not located near residential areas. This is a good practice. However, there is often a complete disregard for evaluating geologic and hydrogeologic risk factors that can greatly influence the migration of contaminants through air, water, and soil that ultimately lead to human and ecological exposure from these areas. The examples for Woburn Massachusetts, Hinkley Site in California, Toms River, Love Canal, and Times Beach are just a few of hundreds of examples that highlight this risk but has not been adequately addressed with effective legislation for land use planning.

16.6.19 NOISE POLLUTION

An additional type of pollution that is not traditionally considered pollution that must be addressed is noise pollution. Noise pollution is regulated in some countries, such as the European Union, China, and Mexico (Rogers 2019a). One may ask how noise pollution relates to a sustainability issue. Recent research indicates that many animal species react to noise and in fact noise that travels in the oceans is harmful to whales and dolphins that depend on echolocation to survive (National Geographic Society 2021 and United States National Park Service 2021).

16.6.20 HOUSEHOLD WASTE

Some local efforts have been initiated at educating the public on the potential dangers of some households wastes that include, cleaners, solvents, electrical equipment, waste paints, oil and grease, and many other substances. However, much more effort and incentives are needed. The landfills where much of the waste from households are disposed are not as adequately constructed to prevent releases to groundwater from many of the substances that are located within the majority of municipal landfills. In addition, Septic tanks are additional source of pollution to the ground and groundwater and highlight the need for expanding local POTW's so that this untreated waste is not directly discharged into the environment.

16.6.21 NON-POINT SOURCE POLLUTION

As described in the summary of the Clean Water Act, non-point source pollution is in need of reg-ulation, especially from rural and agricultural land to minimize the build-up and discharge of pollutants that include fertilizers, pesticides, herbicides, bacteria, and other pollutants into the waters of the United States and causing massive marine deaths in the Gulf of Mexico and perhaps other coastal locations (e.g., Florida). This will be more difficult to regulate in part because it effects areas that are not within the geographical boundary of the United States. The source of much of this pollution originates from non-point sources located in rural and agricultural land within the interior portions of the country.

Environmental regulations concentrate on urban land which only accounts for 2.5% of developed land in the United States. Other areas that include crop land, range land, and managed forest account for 75% of developed land in the United States and is generally not regulated by the USEPA. USEPA has been successful at improving the environmental quality in urban areas to such a degree that pollution effects from non-point sources, which account for the 75% of other developed land in the United States, is much more noticeable and in need of regulation and protection.

16.6.22 CONSUMER PRODUCTS AND PACKAGING

Plastic has become one of the more important current pollution issues and was the focus of Earth Day 2021 (Earth Day 2021). Plastic has many valuable uses. However, humans have become addicted to single-use of disposable plastic. Around the world, one million plastic drinking bottles are purchased every minute and up to 5 trillion single-use plastic bags are used world-wide every year. In total, half of all plastic produced is designed to be used only once and then thrown away. Plastic waste is now so common in the natural environment that some scientists have even sug-gested that plastic could serve as a geological indicator of the Anthropocene Era (United Nations 2021h). From the 1950s through the 1970s, only a small amount of plastic was produced and was not a significant issue. By the 1990s plastic waste had more than tripled. In the early 2000s plastic output increased more than in the previous 40 years. As of the end of 2018, more than 300 million metric tons of plastic waste is generated per year, which is equivalent to the weight of the entire human population (United Nations 2021h). Although plastic containers and packaging is nearly impossible to regulate by USEPA, there is a clear need to address this issue.

16.6.23 ADDRESSING CLIMATE CHANGE

For the last few decades, the scientific community within the United States and world did not generally have difficulty in accepting the science of climate change, but many governments and politicians did have difficulty. This difficulty did not necessarily focus on disputing the science as much as it focused on dealing with potential economic consequences, fairness between countries, coordinating efforts at the global level, and threats to political systems and careers. This is evi-denced by the fact that the United States did not join the Kyoto Protocol more than 20 years ago. The United States did sign on to the Paris Agreement in 2015 under the President Obama ad-ministration only to be taken back under the Trump administration and then returned under the Biden administration in January 2021. The result has been lost opportunities to lower greenhouse gas (GHG) emissions and will likely be significant because the most recent report on climate change from the Intergovernmental Panel on Climate Change (IPCC) (IPCC 2021), states:

> *"It is unequivocal that human influence has warmed the atmosphere, ocean and land. Widespread and rapid changes in the atmosphere, ocean, cryosphere and biosphere have occurred."*

The given statement has significant implications because it more than suggests that GHG emissions from humans have negatively life on Earth. From a historical perspective, the United States is the world's largest emitter of greenhouse gases (GHG's) (USEPA 2021m). Pressure is mounting for the world to take significant action to reduce GHG emissions so that the consequences can be limited. If not, the result could be cataclysmic.

Many businesses in the United States and the world now measure and track GHG emissions using the GHG Protocol. USEPA has provided guidance in conducting self-assessments and setting GHG reduction targets for businesses (USEPA 2020). GHG reduction targets range between 40 to 50% by the year 2030 and carbon neutral by 2050 (USEPA 2020).

Achieving these goals first requires tracking and measuring GHG emissions for each facility using the GHG Protocol. The second step requires identifying those area where GHG emissions can be reduced. The third step, if necessary, requires providing offsets that remove or reduce carbon emissions, such as planting trees (USEPA 2020).

Reducing GHG emissions have been a priority for many countries including the United States. USEPA has been instrumental at reducing GHG emissions from automobiles and from electrical generating plants. However, much more needs to be conducted and will likely involve (1) further increases in renewable energy sources such as wind, solar, and others, (2) eliminating carbon producing transportation and replacement with electric cars, trucks, trains, and planes, (3) modifying the human diet, (4) reducing GHG emissions from agriculture practices.

16.6.24 SUSTAINABILITY LEGISLATION

Although it's recognized that the United States is in need of a clearer picture of what it means to be sustainable, a world-wide binding agreement or legislation is also needed. The European Union and other countries have established a clearer picture and have set goals for sustainability. There is a need to move beyond reactive measures to proactive sustainability measures to improve our environment in the future. If there are delays in enacting Sustainability measures, it may be too late.

16.6.25 MEASURING SUSTAINABILITY PROGRESS

A measurable sustainability program for any location in the world may seem impossible because of the sheer number of potential variables that greatly increase the complexity and because of the many differences between countries. However, this is just not the case. Sustainability is defined as creating and maintaining conditions under which humans can exist in productive harmony to support present and future generations (United Nations 2021a and USEPA 2021b). The term "productive harmony" applies to the environment. Therefore, sustainability can be reduced to three variables that include (1) understanding the environment, (2) understanding how aspects of human activities impact the environment, and (3) developing and implementing measures to mitigate unacceptable risk (Rogers 2019a and 2019b).

Understanding the geological, hydrological, and ecological environment is the first step in building a sustainability model and is represented by the general term Geological Vulnerability. The second step is evaluating aspects of human activities that affect the environment either through negative or positive outcomes. The third variable is evaluating the effectiveness of risk reduction measures. An equation for sustainability is created by combining these three fundamental concepts through which a sustainability index is the output (see Equation 16.1). The Sustainability Index represents a measure of sustainability for any particular location with a higher value representing increased risk and harm to human health or the environment and therefore, less sustainable (Rogers 2020b):

$$\frac{1}{\text{Geologic Vulnerability} \times \text{Operational Aspects} \times \text{Risk Reduction Measures}} = \text{Sustainability Index} \qquad (16.1)$$

The Sustainability Index can be scaled to evaluate environmental health of a single location or whole ecosystem depending on system inputs and evaluation criterion described in the following sections.

16.6.26 SUMMARY OF IMPROVING OR NEED OF NEW REGULATIONS

The aforementioned sections describe the majority of subject areas where environmental regulations in the United States are either in need of improvement or where new regulations are needed. For reference purposes and to list those areas that have the greatest need for new or improved environmental regulations the following is provided:

- Modify Environmental Enforcement Emphasis and Policy
- Science-based improvements to environmental regulations to account for regional risks and climate change
- Agriculture
- Pesticides and herbicides
- Fertilizers
- Septic tanks
- Invasive species
- Urban Planning and Land use
- Residential and household waste
- Urban air
- Further or new restrictions on harmful chemicals, especially those that are persistent and mobile in the environment
- Emerging contaminants, such as those described in Chapter 2
- Noise
- Plastic
- Sustainability

Many of the items listed fall into the exceptions scenario that we first touched on and explained in the introduction of this chapter. The main theme with the exceptions is that environmental regulations should address all identified risks for a given practice, situation, or exposure scenario. It makes little sense if human health and the environment are protected from 9 out of 10 exposure risks when the remaining risk factor causes harm. This example is realized when considering the lack of regulations of the items listed earlier. Another item worthy of discussion is that there are political factors and areas where there is overlapping jurisdictions that may have negative outcomes that must be addressed when discussing the need for further environmental regulations in the United States. For instance, the United States Department of Agriculture would almost certainly interject an interest in environmental regulation of farm land. In addition, the Department of Interior would also be interested in sustainability regulations as it pertains to national forests and parks and environmental regulations on rangeland.

This will become more evident as we discuss the challenges that are faced by many of the countries of the world in the next chapter.

16.7 SUMMARY AND CONCLUSION

After reading this chapter, you should have a much greater understanding of just how much environmental regulation we have in the United States and why. It should also be apparent that staying in compliance is a complex undertaking requiring much skill and expertise.

There is much we can be proud of when we examine the improvements of our air quality, water quality and clean up of sites of environmental contamination, which now number in the tens of thousands of sites across the United States. Looking back over the sections of different environmental regulations, you may have noticed that much of the legislation was often times a reaction to an incident or environmental tragedy. Laws and regulations originated from many circumstances because of unintentional actions that ended up making quite a mess and causing significant harm to human health, the environment, or both. Some of the incidences that we have touched upon include the Exxon Valdez spill in 1990, Deepwater Horizon in 2010, Love Canal, Bhopal India in 1984, Valley of Drums, and Toms River to the killing and extinction of animal species that also nearly included the American Bison, California Condor, Grizzly Bear, and numerous others. However, this is a reactionary approach not a proactive approach. When we discuss Sustainability later in this book, we will discuss the need to move beyond reactive measures to proactive sustainability measures to improve our environment in the future. If there are delays in enacting Sustainability measures, it may be too late.

The amount and complexity of environmental laws and regulations in the United States is due to the fact that we are an industrialized society and consume enormous amounts of energy and goods and our hunger for those goods and energy continues to rise as we continue to improve our standard of living. The United States produces large volumes of waste of all types and as our population continues to increase, the amount of waste produced will also continue to increase.

These central facts combined with the dynamic and innate complexities of the natural world add an almost infinite number of negative outcomes in a world with over 7 billion people all wanting to improve their standard of living. The fact is, if we do not have effective and comprehensive environmental regulation, we simply threaten our existence and numerous other species as we are slowly devoured in our own garbage.

Our advances in science have created many technologies that have improved our standard of living and in some circumstances have improved our relationship with nature. But our technological advancements have also created many unintended negative side effects that question the benefit of some technologies. An example was presented in Chapter 2 and involves the creation of synthetic compounds that nature has difficulty breaking down. Some of these compounds include polychlorinated biphenyls, chlorinated hydrocarbons, and many pesticides, including DDT which began the environmental movement in the United States in earnest over 50 years ago. This is to say nothing about plastics which we all should know about since we commonly use and discard huge amounts of plastic in our everyday life.

Although efforts are underway to reduce our energy needs and our veracious hunger for consuming nature, one thing seems certain, our thirst for consumer goods and energy needs will continue to grow, and as our population increases our energy needs will increase even faster. This means that environmental regulation will have to grow with us in volume and complexity as we consume more and more of Earth's resources and replace it with our waste.

The next chapter will focus on environmental regulations and pollution worldwide. We will focus on industrialized nations and a few developing nations with large populations including India and China. We will not be examining countries with what are considered extreme hardships that include Somalia, Syria, Afghanistan, Iran and others.

We will evaluate how countries environmental laws compare to the United States. Some findings might be rather surprising in that it will become apparent that a few countries are more advanced at protecting human health and the environment than the United States. Surprising still might be that impression that some countries either lack environmental laws or the will to enforce the environmental laws they have due to other overriding social or economic challenges. As we shall see in the next chapter, it is not enough just to have environmental regulations. Many other factors are required to have an effective environmental regulatory framework that actually accomplishes that ultimate goal of protecting human health and the environment.

Questions and Exercises for Discussion

1. Of the three sustainability impediments and challenges discussed in Section 16.3, which one do you think will be the most difficult to solve and why?
2. Of the three sustainability impediments and challenges discussed in Section 16.3, what actions do you think are necessary to address the most difficult issue you identified in the aforementioned questions?
3. Of the items examined in Section 16.5, which is most important to you and why?
4. Of the more than 20 issues listed in Section 16.6, which five are most important to you and why?
5. Do you have any additions items or issues that you felt were not addressed?

REFERENCES

American Society for Quality (AQS). 2021. ISO 14001: Environmental Management Systems Standard. https://asq.org/quality-resources/iso-14001 (accessed April 19, 2021).

Baum, R., Luh, J., and Bartman, J. 2013. Sanitation: A Global Estimate of Sewerage Connections without Treatment and the Resulting Impact. *Journal of Environmental Science and Technology*. Vol. 47. no. 4. pp. 1994–2000.

B Corporations. 2021. About B Corporations. https://bcorporation.net/about-b-corps (accessed April 19, 2021).

Carson, R. 1962. *Silent Spring*. Houghton Mifflin: Boston, MA

Earth Day. 2021. Earth Day Network Campaign: End Plastic Pollution. https://earthday.or/campaigns/plastic-campaign (accessed April 24, 2021).

Harari, Y. N. 2015. *Sapiens*. Harper Collins Publishers: New York, New York. 443p.

Harari, Y. N. 2018. *21 Lessons for the 21st Century*. Random House Books: New York, New York. 372p.

Intergovernmental Panel on Climate Change (IPCC). 2021. Climate Change 2021: The Physical Science Basis. *IPCC AR6 WGI*. Vol. 6. 3949p.

International Union on the Conservation of Nature (IUCN). 2021. Extinct Species in North America since 1500ad. https://www.iucn.org (accessed April 19, 2021).

McGucken, W. 2000. *Lake Erie Rehabilitated: Controlling Cultural Eutrophication, 1960s–1990s*. University of Akron Press: Akron, OH.

National Geographic Society. 2021. Noise Pollution https://www.nationalgeographicsociety.org/noise-pollution/ (accessed March 31, 2021).

National Oceanic and Atmospheric Administration (NOAA). 2021. Status of Fish Stocks 2017. https://www.fisheries.noaa.gov/status-stocks-2017 (accessed April 19, 2021).

Natural Resource Defense Council (NRDC). 2021. Regenerative Agriculture. https://www.nrdc.org/resources/regenerative-agriculture (accessed November 28, 2021).

Ohio State University. 2021. Earth Day 1971. Walt Kelly, Pogo Collection. *OSU Library*. https://library.osu.edu/site/40stories "We Have Met the Enemy and He Is Us" – Tales from the Vault: 40 Years / 40 Stories (osu.edu) (accessed March 31, 2021).

Pope Francis. 2015. Environmental Encyclical and Care for Our Common Home. *The Vatican*. 184p. https://www.w2.vatican.va/content/francesco/en/encyclicals/documents/papa-francesco-2015024-enciclica-laudato-si.html (accessed April 19, 2021).

Rogers, D. T. 2019a. *Environmental Compliance and Sustainability: Global Challenges and Perspectives*. CRC Press: Boca Raton, FL. 583p.

Rogers, D. T. 2019b. Key Policy Changes Required to Reduce Environmental Degradation of the United States and Beyond. *International Association of Hydrogeologists 46th Congress*. Vol. 1. pp. 529. Malaga, Spain.

Rogers, D. T. 2020a. *Urban Watersheds: Geology, Contamination, Environmental Regulations, and Sustainability*. CRC Press: Boca Raton, FL. 606p.

Rogers, D. T. 2020b. Geological- and Chemical-Based Environmental Risk Factor Sustainability Model. *European Journal of Sustainability Development*. Vol. 9. no. 4. pp. 303–316.

Tallamy, D. W. 2019. *Nature's Best Hope*. Timber Press: Portland, Oregon. 254p.

Thornton, R. V. 2001. ISO 14001 Certification Mandate Reaches the Automobile Industry. *Journal of Environmental Quality Management*. Vol. 10. no. 1. pp. 89–96.

United Nations. 2015. International Decade for Action "Water for Life." United Nations Department of Economic and Social Affairs (UNDESA). *New York. New York.* http://www.un.org/waterforlifedecade/quality.shtml (accessed April 19, 2021).

United Nations. 2016a. *Global Drinking Water Quality Index and Development and Sensitivity Analysis Report.* UNEP Water Programme Office: Burlington, Ontario, Canada. 58p.

United Nations. 2016b. World Air Pollution Status. United Nations News Center. *New York, New York.* https://www.un.org/sustainabledevelopment/2016/09 (accessed April 19, 2021).

United Nations. 2016c. Human Development Report. *United Nations Development Programme.* http://www.hdr.undr.org/sites/2017 (accessed April 19, 2021).

United Nations. 2018. Biodiversity: Invasive Species. *United Nations System-Wide Earthwatch.* https://www.un.org/earthwatch/biodiversity/invasivespecies.html (accessed December 29, 2018).

United Nations. 2021a. What Is Sustainable Development. https://www.un.org/sustainabledevelopment (accessed April 19, 2021).

United Nations. 2021b. Global Sustainability Goals. https://www.undp.org/content/undp/en/home/sustainable-development-goals.html (accessed April 19, 2021).

United Nations. 2021c. World Population Prospects. United Nations Department of Economic and Social Affairs. *Population Division.* https://esa.un.org/unpd/wpp/data (accessed April 19, 2021).

United Nations. 2021d. Funds, Programmes, Specialized Agencies and Others. https://www.un.org/en/sections/about-un/funds-programmes-specialized-agencies-and-others/ (accessed April 19, 2021).

United Nations. 2021e. Description and Statistics About United Nations World Heritage Sites. https://whc.unesco.org/en/about/ (accessed March 31, 2021).

United Nations. 2021f. IAS Network. *Current State of Forests of the World.* https://www.un.org/stateofforests (accessed March 31, 2021).

United Nations. 2021g. Intergovernmental Science-Policy Platform on Biodiversity and Ecosystems Services (IPBES) Summary for Policymakers of the Methodological Assessment Report of the IPBES Scenarios and Models of Biodiversity and Ecosystems Services. https://www.ipbes.net/system/tdf/downloads/pdf/spm_deliverable_3c_scenarios_20161124.pdf?file=1&type=node&id=15245 (accessed April 19, 2021).

United Nations. 2021h. The Facts on Plastic Pollution. https://www.unenvironment.org/beat-plastic-pollution (accessed April 19, 2021).

United States Department of Agriculture (USDA). 2011. Land Use in the United States. https://www.ers.usda.gov/data-products/major-land-use (accessed April 19, 2021).

United States Department of Agriculture (USDA). 2019. *United States Fertilizer Use and Cost.* USDA: Washington, D. C.

United States Environmental Protection Agency. 2017. *Pesticide Industry Usage. USEPA Biological and Economic Analysis Division.* Office of Pesticide Programs: Washington, DC. 24p.

United States Environmental Protection Agency (USEPA). 2018. Air Quality – 2018 National Summary. https://www.epa.gov/air-trends/air-quality-national-summary (accessed April 19, 2021).

United States Environmental Protection Agency (USEPA). 2020. GHG Inventorying and Target Setting. GHG INVENTORYING AND TARGET SETTING SELF-ASSESSMENT: V1.0 (epa.gov) (accessed July 30, 2021).

United States Environmental Protection Agency (USEPA). 2021a. United States Environmental Protection Agency QA Glossary. https://www.epa.gov/emap/archive-emap/web/html (accessed April 19, 2021).

United States Environmental Protection Agency (USEPA). 2021b. What is Sustainability? https://www.epa.gov/sustainability (accessed April 19, 2021).

United States Environmental Protection Agency. 2021c. National Environmental Policy Act. www.epa.gov/NEPA (accessed April 19, 2021).

United States Environmental Protection Agency. 2021d. Summary of Environmental Regulations of the United States. https://epa.gov/summary-environmental-laws (accessed April 19, 2021).

United States Environmental Protection Agency (USEPA). 2021e. Regulations for Greenhouse Gas (GHG) Emissions. https://www.epa.gov/regulations-emissions-vehicles-and-engines/regulations-greenhouse-gas-ghg-emissions (accessed April 19, 2021).

United States Environmental Protection Agency (USEPA). 2021f. International Cooperation. https://www.epa.gov/international-cooperation (accessed April 19, 2021).

United States Environmental Protection Agency (USEPA). 2021g. Pesticides. https://www.epa.gov/pesticides (accessed April 19, 2021).

United States Environmental Protection Agency (USEPA). 2021h. DDT – A brief History and Status. https://www.epa.gov/ingredients-used-pesticide-products/ddt-brief-history-and-status (accessed April 19, 2021).

United States Environmental Protection Agency (USEPA). 2021i. Agriculture Nutrient Management and Fertilizer. https://www.epa.gov/agriculture/agriculture-nutrient-management-and-fertilizer (accessed March 25, 2019).

United States Environmental Protection Agency (USEPA). 2021j. Integrated Risk Information System. https://www.epa.gov/iris (accessed March 24, 2021).

United States Environmental Protection Agency (USEPA). 2021k. Invasive Species. https://www.epa.gov/greatlakes/invasive-species (accessed April 19, 2021).

United States Geological Survey (USGS). 2021. Pesticide and Herbicide Use in the United States. https://www.usgs.gov/centers/oki-water/science/pesticides (accessed March 31, 2021).

United Stated Environmental Protection Agency (USEPA). 2021m. Inventory of U.S. Greenhouse Gas Emissions and Sinks. Inventory of U.S. Greenhouse Gas Emissions and Sinks I Greenhouse Gas (GHG) Emissions I US EPA (accessed February 20, 2021).

United States National Park Service. 2021. Can You Hear Me Now. *Managing Underwater Noise Pollution*. https://npsorg/articles/canyouhearmenow.htm (accessed March 31, 2021).

World Bank. 2021a. The Fight to Improve Lives of the World's Poor. http://www.worldbank.org/en/news/press-release/2013/07/23/improve-lives-world-poor-world-bank-group-delivers-nearly-53-billion-support-developing-countries-fy13 (accessed April 19, 2021).

World Bank. 2021b. Household Cooking Fuel Choice and Adoption of Improved Cookstoves in Developing Countries. *Policy Research Working Paper No. 6903*. https://worldbank.org/curated/en/wps6903 (accessed March 31, 2021).

World Health Organization. 2013. Water Quality and Health Strategy 2013–2020. http://www.who.int/water_sanitation_health/dwq/en/ (accessed April 19, 2021).

World Health Organization (WHO). 2015a. *Guidelines for Drinking-water Quality*. Geneva, Switzerland. 516p.

World Health Organization 2015b. *Progress on Drinking Water and Sanitation*. World Health Organization: New York, New York. 90p.

World Health Organization. 2021a. United Nations Global Analysis and Assessment of Sanitation and Drinking *Water: Indonesia*. https://www.who.org/drinkingwater/Indonesia (accessed April 19, 2021).

World Ocean Review. 2021. The Global Hunt for Fish. https://worldoceanreview.com (accessed April 19, 2021).

World Wildlife Fund (WWF). 2021. Species List of Endangered, Vulnerable, and Threatened Animals. https://www.wwf.org/species/endangered-list (accessed April 19, 2021).

17 Environmental Protection and Sustainability at the National Level

17.1 INTRODUCTION

As stated in the introduction of the previous chapter, sustainability and improving our relationship with nature is at risk because of climate change and that the result may be that much of the near-term sustainability efforts will concentrate on limiting the amount of further deterioration of the environment rather than improving our environment and limiting species loss. Therefore, sustainability efforts to reduce GHG emissions should be accelerated with a sense of urgency.

National level sustainability measures are more specific than those covered at the global level in the previous chapter. Sustainability measures at the national level are separated into three different groups where improvements can be achieved. The first is before development occurs and focuses measures that can be achieved prior to any ecological modifications or disturbance. The second is improvements in environmental regulations. The third focuses on future measures that include improved pollution prevention techniques and proactive measures to improve our environment through sustainability actions.

There is much to be proud of when examining the improvements in the United States compared to 50 years ago that have significantly improved the air quality, water quality, and clean up of sites of environmental contamination, which now number in the tens of thousands of sites across the United States. However, there are several areas where improvements can be realized if some existing regulations were modified or expanded and if some new regulations were developed.

An additional concept that requires attention is our relationship with nature. For many decades, nature was categorized as perhaps the enemy, something that needed to be conquered or tamed. The concept was that progress and development would achieve this desired result. Scientists now realize that this may not have been a good approach and that our relationship with nature must improve. This is what is addressed directly at the national level in the following sections.

17.2 IMPROVING THE NATURAL SETTING OF URBAN AREAS

The natural setting of our urban and industrial areas play a crucial roll in how contaminants behave and how they affect us when they are released into the environment. All of us walk on a history book in each urban area of the world but yet most of us go unaware of its significance and the potential impact it may have on all our lives and it's just centimeters beneath our feet.

The arrangement, thickness, and composition of the soil and sediment layers just centimeters beneath our feet have a profound influence on our lives and all life on Earth. These soil and sediment layers don't just dictate where and how cities are built, where roads are built, and how buildings are constructed. Most significantly, these soil and sediment layers beneath our feet are the meeting place between human civilization and the natural world. This contact is the proving ground and the point at which contaminants released into the environment begin their destructive journey that has now impacted on all living things on Earth. We must understand our natural world in at least the same scientific detail that we need to understand pollution. The reason we must do this is simple, they interact with one other. The near-surface geologic environment has become very important in characterizing and remediating sites of environmental contamination and in evaluating risk to the environment and human health. The reason is simple; the near-surface geologic environment frequently acts as the migration pathway for contaminants to travel from a specific point of release to a point of ecologic or human exposure. In many cases this pathway goes undetected because the contaminants migrate beneath the surface of the ground, and cannot be observed or detected until it is perhaps too late.

17.2.1 GEOLOGIC VULNERABILITY

We do not have control over the natural environment. Therefore, we must understand the natural environment where our urban areas are located and develop methods to minimize or eliminate the potential harmful effects of contaminants upon human health and the environment. A logical first step to this end is through an *understanding* of urban geology, followed by an *evaluation* of the extent that a given urban area's geology influences the migration of contaminants. And, since water plays a critical role in assessing a region's vulnerability to contamination, the analyses performed during the evaluation step require an understanding of water occurring at the Earth's surface and beneath (Rogers 2014). These factors control the severity of the damage and are: (1) the geologic and hydrogeologic environment, (2) the physical chemistry of the contaminants and amounts released, and (3) the mechanism in which the release occurs (Rogers 2018 and 2020a).

Despite the availability of specific methods and procedures, the environmental assessment of many urban areas can become a daunting task. This situation arises because the near-surface geologic deposits in urban areas are poorly understood, difficult to study, complex, have been anthropogenically disturbed, and exhibit high variability over short distances. Therefore, to achieve any level of success in mitigating environmental contamination, it becomes a prerequisite to understand how contaminants migrate in any given urban area which makes it necessary to understand its geology and hydrology (Rogers 2021).

17.2.2 PHYSICAL CHEMISTRY OF POLLUTANTS

If it were not for hazardous substances there would be little need for the majority of environmental regulations. However, that is certainly not the case. There are currently more than 84,000 chemicals in use today and more and more every year. Contaminants are everywhere--in the air, soil, water, inside buildings, and in our homes. Most households contain chemicals that would be considered contaminants if they were released into the environment or disposed of improperly. Therefore, we must understand which chemicals are harmful and we must not use or must limit the use of those chemicals that are most harmful. We have learned that USEPA has not banned any chemicals since 1984, so it's up to industry and the consumer to evaluate the risk and take appropriate action.

In addition, chemicals or substances become pollution when they are released into the environment either inadvertently or improperly - at the wrong place or in the wrong amounts. For example, milk becomes a contaminant when large quantities are released into a stream. Therefore, it's not just what chemicals are used, but preventing releases of most any chemical, especially to sensitive areas (Rogers 2020a and 2020b).

17.2.3 SMART URBAN DEVELOPMENT

The human population is increasing and requires more and more space to live, grow food, and for build roads. The rapid physical growth of urban areas leads to complex processes of landscape transformation which alters the structure and function of ecosystems. Urban invasion into natural areas decreases productivity, decreases infiltration of water, increases impervious surfaces, increases flooding, increases erosion, increases pollutant load, increases energy demand, and decreases species biodiversity. The size of a single family home in the United States has more than doubled in the last 70 years (United States Department of Housing and Urban Development 2021 and United States Census Bureau 2021a).

Many opportunities for smart urban development are available within many cities of the world. Some of the numerous opportunities include the following:

- Less single-family homes and more multi-family units. The average size of a condominium is approximately 130 square meters (1,200 square feet) verses 300 square meters (2,700 square feet) for a single-family home. The average land required for a single-family home in the United States is 815 square meters or 0.25 acres per home. This includes the land required for easements and roads for a typical subdivision (National Association of Home Builders 2021).

  Building large skyscrapers that have hundreds to sometimes over a thousand condominium units have enormous sustainability improvements over the single-family home. Take for example a large skyscraper currently being constructed in Chicago in the United States that will have an estimated 1,200 condominium units on a footprint of approximately 5,000 square meters (2 acres). This represents a 99.5% reduction in occupied land if those 1,200 condominium units were single family homes. There are also significant reductions in the amount of construction material that would be used if each were a single-family home that includes forest materials, insulation, roofing, paint, metal piping, sewers, stormwater, roads and curbs, and sidewalks. There are also significant long-term energy savings including heat and electrical.

- Reducing the size of housing units. The size of a single-family home in the United States has nearly doubled since 1970 and nearly tripled since 1940 (United States Census Bureau 2021b). This is in contrast with the fact that the children per family in the United Sates has decreased by more than 60% from 3.76 children per family in 1940 to 1.58 children per family in 2018 (United States Census Bureau 2021b).

- Living closer to work. The average commute to work in the United States is 20 miles one way and takes nearly 30 minutes (United States Census Bureau 2021c).

- Increase mass transit availability. Mass transit is much more efficient and emits much less pollution than an automobile per person per kilometer travel. Estimates are as high as 95% more efficient when traveling by commuter train compared to an average automobile when considering fuel use alone and not taking into consideration of the cost and inefficiencies of building and maintaining highways (United States Department of Energy 2021 and National Geographic Society 2021).

- Increase other forms of transport that include using scooters, bicycles and walking. Many cities in the United States and throughout the world are improving bike-friendly and foot transport routes to increase these forms of transportation by promoting health and enjoyment benefits.

- Integrating more mixed land use into urban areas. The United States has much to learn from other countries when it comes to this urban development technique. Historically, the United States has not taken an integrated approach to land use. Many European and Asian countries, namely South Korea, have embraced integration of farmland, industry and housing.

- Increasing clean energy production such as solar and wind. Both solar and wind energy have become more efficient and less costly. A key point for solar efficiency is prevent dirt buildup on the solar panels. Wind turbines are available in different engineering configurations that can be applied in small areas and minimize bird deaths (United States Department of Energy 2021).

- Increasing permeable surfaces. To decrease flooding, increasing infiltration of precipitation through several available and simple techniques is widely available and will be discussed later in this chapter.

Currently, examples of sustainable and smart urban development using some or all the items listed are mainly confined to Europe, Asia, and some parts of the United States. Two impediments that have been slow in becoming a reality in the United States concerning smart urban development is the American love affair with the automobile and achieving the American dream in which owning a single-family home has often been utilized as a degree of measure that defines making that dream come true.

17.2.4 RESPECTING NATURE

In the 19th and 20th centuries, many supported the concept of progress, taming and conquering nature, and treating nature as a commodity to be exploited for profit (Tallamy 2019).

We have discussed invasive species and how destructive they can be at reducing native populations of most all form of indigenous life. In addition, climate change has already affected many forms of life such as the Pine Bark Beetle's march north and destroying millions of trees in North America. Other impacts include the reduction in many bee species and significant reduction on the number of Monarch Butterflies, just to name a couple.

At the national level, there is much that can be conducted, a few include (Tallamy 2019):

- Planting native plant species. One example is planting Milkweed which is a plant species required for the survival of the Monarch Butterfly. Native species planting is a trend that is expanding within many state departments of transportation and should be encouraged and funded at higher levels at the national level (United States Fish and Wildlife Service 2021).
- Reducing biological wastelands. Namely lawns which are symbols of wealth and status but are generally biological deserts.
- Establishing biological and wildlife corridors.
- Improving insect habitat. In most ecological systems, insects form the foundation of ecosystem health and without a healthy insect population which provide food through all levels of life, ecosystems may collapse
- Limit or eliminate use of herbicides and pesticides
- Limit or eliminate use of fertilizers
- Restore and expand wetlands
- Practice regenerative farming and gardening techniques. Regenerative farming is a philosophy of common principles that include re-establishing relationships between humans and land through improving soil health, reducing or eliminating the use of harmful chemicals, growing diverse crops, holistic livestock management, innovative and efficient use of resources, and equitable labor (Natural Resource Defense Council [NRDC] 2021). This type of farming is not new. Indigenous cultures, and even those in the United States in the 19th century, practiced this type of farming before the widespread use of pesticides, herbicides, and extensive use of fertilizers. Current farm and food production in the United States is not designed to prioritize climate, ecosystems water quality, human health, and establishing a healthy and productive relationship with nature (NRDC 2021).

17.3 UPGRADING INFRASTRUCTURE

Many cities in the United States are in dire need of infrastructure upgrades that have significant sustainability benefits. Lack of upgrading infrastructure puts millions of people at risk and most of those are located in poor areas. An example is the Flint, Michigan water crisis. A few of the significant infrastructure needs are addressed in the following sections.

17.3.1 Replacing Outdated Water Supply Lines

The Flint, Michigan water crisis began in 2014 when the City of Flint decided to switch its water source from the Detroit River to the Flint River to save money. However, state and local officials and regulators failed to add anti-corrosions chemical to the new water supply which was slightly more acidic that the previous water supply source which was the Detroit River. The result eroded residents water pipes which allowed lead contained in the pipes to leach and contaminate the water that people were using as a source of drinking water (USEPA 2021a).

The Flint, Michigan crisis is a good example that demonstrated a couple important points. Point 1 is the complexity of environmental regulation. Point 2 is that this incident highlighted the need to replace old and compromised water supply lines not just because some are over 100 years old and are deteriorated, but that many contain lead which can put residents at risk, especially children. USEPA has recognized the seriousness of the situation and has made replacing lead-containing water lines a priority (USEPA 2021b).

17.3.2 Upgrading Water Treatment of Potable Water

In part due to the Flint water crisis and the detection of emerging contaminants in drinking water, including some disease-causing pathogens and pharmaceutical drugs, has also highlighted the need for improved treatment of drinking water in the United States. These upgrades are necessary because of a continuing population increase in many areas and the improved analytical procedures in detecting old and new contaminants that present a potential risk to human health (USEPA 2021c).

17.3.3 Upgrading Waste Water Treatment Systems

Improving waste water treatment plants now and in the future will be required due to many reasons that include and increasing population, aging and deteriorated existing waste water treatment plants, new and improved technologies that remove contaminants, increased scientific knowledge on health and environmental risk posed by certain contaminants, and others (USEPA 2021d).

17.3.4 Improving Municipal Landfills

Municipal landfills primarily receive household waste which can contain wastes that would be considered hazardous and would be more strictly regulated if generated from an industrial source. From a sustainability point of view, it would be preferred that households not dispose of certain contaminants in their waste but this sometimes is not the case. Some examples of these wastes include: cleaners, solvents, pesticides, herbicides, batteries, gasoline, and others. Therefore, more strict measures are needed to further prevent hazardous wastes entering municipal landfills to minimize costly cleanup measures and protect human health and the environment. USEPA regulates municipal landfills under RCRA and does require certain construction measures to prevent or minimizes the potential for contaminants to migrate beyond the footprint of the landfill. However, this sometimes is not enough (USEPA 2021e).

17.3.5 Improving Erosion Controls

As climate change continues to place stress on the environment, one sometimes overlooked result of climate change is erosion of sediment during flood events. These sediments have several negative effects that include: contaminants carried in the sediments such as pesticides, pesticides and fertilizers, removal of valuable top soil, and negative impacts to aquatic life (USEPA 2021f).

17.3.6 Upgrading Mass Transit

Mass transit is significantly more efficient than the automobile. However, the United States lacks sufficient infrastructure, namely rail transport, to make mass transit by rail more attractive and convenient to consumers.

17.4 IMPROVING ENVIRONMENTAL REGULATIONS OF THE UNITED STATES

Improving environmental regulation of the United States may seem unnecessary since many believe the environmental regulations of the United States are the most comprehensive, most complex, and most effective. However, there is much that the United States can do to improve their environmental regulations. In addition, and perhaps most importantly, the environmental regulations of the world have been used as a platform for environmental regulations world-wide. In most instances, the rest of the world relies on the United States to lead the way forward with respect to environmental regulations. This is significant because as we know, contaminants do not respect country borders and if the United States forges ahead in improving its environmental regulations, it is only logical that the rest of the world will follow.

17.4.1 Modify Environmental Enforcement Emphasis and Policy

At this point, it is not apparent that a need for modifying USEPA's approach to enforcement is in need of change. Especially since it has been seemingly effective up to this point. However, it's time to move beyond a litigation heavy and punishment-based enforcement model and enter into more of a reform- and repair-based enforcement policy that emphasizes sustainability. In addition, expanding environmental regulations application to other areas that are either not regulated or lightly regulated is long overdue. For instance, according to the United Nations, agricultural practices are the largest polluter on Earth yet environmental regulations of the United States largely exempt agriculture from being regulated. This is no more evident than when examining pesticide and herbicide use and fertilizer use.

We will learn in the next chapter that some countries are more advanced compared to the United States in how they have chosen to organize and implement their environmental regulations and where they focus internal resources. In addition, the governments of some countries have different attitudes and relationships with industry. The regulatory agencies of some countries have chosen to form working partnerships with industry with the intent to work more closely together to solve complex environmental issues rather than an adversarial approach which much too often results in polarizing points of view and expensive litigation.

The result of working together saves time, effort, and resources and in the long run, has resulted in the fact that some counties are better at protecting human health and the environment than the United States.

17.4.2 Science-Based Regulation Improvements

We discovered in earlier chapters that contaminants behave differently and are more prone to releases in different geologic, hydrogeologic, geographic, and climatic regions. For instance, regulations specific to geologic risk factors such as earthquakes (e. g., California) or climate change (eastern seaboard and the coast along the Gulf of Mexico) should better reflect those areas where there is the highest risk.

17.4.3 Improving Urban Air Quality

The Clean Air Act has made a major impact at improving the air quality throughout the United States, especially with respect to automobile exhaust, ozone, particulate matter, and the other criteria pollutants. According to United States Environmental Protection Agency (USEPA) (2018), compared to emissions in 1980, carbon monoxide emissions have been reduced by 72%, lead by 99%, nitrogen oxides by 61%, volatile organic compounds by 54%, particulate matter (PM10) by 61% and, sulfur dioxide by 89%.

Improvements on exhaust from diesel trucks and buses have been slowed and are especially noticeable in urban areas. In addition, other sources of particulate matter are of high concern in urban areas as well and needs improvement. This need is in part the result of an increasing population and that urban areas continue to grow as population centers that will inevitably increase automobile congestion and emissions if not addressed as quickly as possible.

As we shall see in the next chapter, even though the United States is in need of improving urban air, many locations in the rest of the world are far worse when it comes to urban air pollution.

17.4.4 BANNING HARMFUL CHEMICALS

Let's examine the regulations of the United States for just a moment from an overview perspective. The majority of environmental regulations focus on wastes that are generated and whether a facility has a permit. They are most concerned, and for good reason, on air emissions, water discharge, and solid and hazardous waste generation and disposal. The reasoning is that these are the primary pathways for harmful substances to migrate from a point of generation to a point of human exposure where they may cause harm to us or the environment. In general, the regulations do not restrict what substances can be used at a facility. TSCA is the set of regulations that has this authority and to date, TCSA has very few chemicals that have been banned in the United States. You may have wondered and even asked yourself a question. Why didn't we know about the harm certain chemicals have on humans and the environment before they are widely used. This is a very good question and it happens to be a mistake our society has made time and time again. It happened with DDT, PCBs, asbestos, MTBE, and many other compounds that at first, were even advertised as a miracle chemical (i.e., DDT and PCBs), or a miracle substance (i.e., asbestos). But as time marched on and these chemicals or substances were widely used, evidence began to surface that these substances had unintentional and unknown side effects that had a negative effect on human health, the environment, or both. Sometimes there is a latency period, sometimes 20 or more years after exposure, that a symptom or negative effect surfaces.

Currently, there are over 80,000 chemicals in the United States currently being used. USEPA has been successful in restricting only 9 chemicals in its history of over 40 years since promulgated in 1976 and none since 1984. Those substances currently restricted include (Rogers 2019):

- Polychlorinated biphenyls (PCBs)
- Chlorofluorocarbons (CFCs)
- Dioxin
- Asbestos
- Hexavalent chromium
- Four nitrite compounds that are either:
  - Mixed mono and diamides of an organic acid
  - Triethanolanime salt of a substituted organic acid
  - Triethanolanime salt of tricarboxylic acid
  - Tricarboxylic acid

In June 2016, TSCA was amended significantly and gave USEPA additional authority and responsibility to proactively evaluate the risks of all chemical substances. USEPA has identified 10 chemicals as high priority and are currently conducting risk evaluations. They include (Rogers 2019):

- 1,4-Dioxane
- Asbestos
- Cyclic Aliphatic Bromide Cluster
- N-methylpyrrolidone
- Tetrachloroethylene

- 1-Bromopropane
- Carbon Tetrachloride
- Methylene Chloride
- Pigment Violet 29
- Trichloroethylene

An analysis of the given list shows that it includes four substances that belong to the same chemical group called chlorinated volatile organic compounds (CVOCs), and are also known as chlorinated solvents. Those include Carbon Tetrachloride, Methylene Chloride, Tetrachloroethylene, and Trichloroethylene. In addition, 1,4-Dioxane is sometimes associated with the breakdown of Tetrachloroethylene and Trichlorethylene, so perhaps there may be as many as five in the CVOC family of chemicals. Therefore, of the 10 chemicals USEPA has initially placed in its high priority list, half or more belong or are associated with the same group of compounds, CVOCs, or chlorinated solvents.

There is perhaps negative feedback that has occurred by not banning more chemicals in the United States that has resulted in providing a false sense of security concerning chemical use. This sense of false security has developed an attitude of reliance that the USEPA will protect us, after all that is their job right? Many users may say that if a particular chemical was harmful, it would have been banned. This is where we can improve but have not because TSCA was not strong enough when enacted and the burden of scientific proof required to ban a chemical is too high and takes too long to establish. This is evident when we examine some of our example sites in this chapter that include Woburn Massachusetts, Toms River New Jersey, and the Hinckley Site in California.

17.4.5 DEALING WITH EMERGING CONTAMINANTS

Emerging contaminants are chemicals or substances that have been recently discovered in the environment and, in some instances, have caused harm or have impacted resources to the extent that they can no longer be used for fear of health risks. Emerging contaminants include many compounds including PFAS, many pharmaceutical compounds, 1,4-dioxane, and others. Many of these emerging compounds are now being detected in the environment and have not been adequately evaluated. This highlights a need for improved prevention measures and the need for improving handling and disposal practices.

17.4.6 ADDRESSING INVASIVE SPECIES

Some consider invasive species a type of emerging contaminant. However, for the purposes of this book we treat them separately because there are so many invasive species and their behavior in the environment to so different compared to chemicals and other substances because their mode of migration is through breeding. Invasive species have been present for the last few hundred years but environmental damage is increasing and in need of further regulatory protections.

17.4.7 IMPROVING POLLUTION PREVENTION

The Pollution Prevention Act (PPA) and Oil Spill Prevention Act (OSPA), both of 1990, do not get enough credit in the history of environmental regulations of the United States. The PPA and OSPA were the first real efforts to turn from regulating the environment at the "end of the pipe" to promoting the prevention of pollution from entering the environment. One of the driving factors behind the PPA and OSPA was the realization of how much it cost to clean up the environmental after a release.

It was a sobering lesson that just one incident could cause so much harm, cost billions of dollars, and in some instances could not be cleaned up to acceptable levels because technologies simply did not exist. It was during this time that USEPA and many others discovered that sometimes you can throw money at a contaminated site or incident and make it go away and then there are those where all the money and technology in the world would not be able fix the problem before enormous harm was a reality. Therefore, out of this realization, USEPA began to focus efforts on pollution prevention, in part because it made common sense in that it's proactive at environmental protection and lowers response cost much like fire prevention. As we shall see in the next chapter, it was out of these and similar efforts in the European Union that the term Sustainability took hold.

17.4.8 IMPROVING URBAN LAND USE

One important item that should be added to pollution prevention efforts is urban land use. There is much that we can accomplish at protecting human health and the environment by not locating high risk activities in areas that are especially sensitive to pollution.

Currently, urban planning for industrial activities are near transportation routes and are generally not located near residential areas. This is a good practice. However, there is often a complete disregard for evaluating geologic and hydrogeologic risk factors that can greatly influence the migration of contaminants through air, water, and soil that ultimately lead to human and ecological exposure from these areas. The examples for Woburn Massachusetts, Hinkley Site in California, Toms River, Love Canal, and Times Beach are just a few of hundreds of examples that highlight this risk but has not been adequately addressed with effective legislation for land use planning.

17.4.9 REGULATING NOISE POLLUTION

An additional type of pollution we have not addressed in this Chapter is noise pollution. The reason is that noise pollution is largely not regulated in the United States at a federal level. However, as we shall see in the next chapter, noise pollution is a regulated form of pollution in many countries of the world, including the European Union, Mexico, and even China.

17.4.10 NON-POINT SOURCE POLLUTION IMPROVEMENTS

As described in the summary of the Clean Water Act, non-point source pollution is in need of regulation, especially from rural and agricultural land to minimize the build-up and discharge of pollutants that include fertilizers, pesticides, herbicides, bacteria, and other pollutants into the waters of the United States and causing massive marine deaths in the Gulf of Mexico and perhaps other coastal locations (e.g., Florida). This will be more difficult to regulate in part because it effects areas that are not within the geographical boundary of the United States. The source of much of this pollution originates from non-point sources located in rural and agricultural land within the interior portions of the country.

Environmental regulations concentrate on urban land which only accounts for 2.5% of developed land in the United States. Other areas that include crop land, range land, and managed forest account for 75% of developed land in the United States and is generally not regulated by the USEPA. USEPA has been successful at improving the environmental quality in urban areas to such a degree that pollution effects from non-point sources, which account for the 75% of other developed land in the United States, is much more noticeable and in need of regulation and protection.

17.4.11 Consumer Products and Packaging Improvements

Plastic has become one of the more important current pollution issues and was the focus of Earth Day 2021 (Earth Day 2021). Plastic has many valuable uses. However, humans have become addicted to single-use of disposable plastic. Around the world, one million plastic drinking bottles are purchased every minute and up to 5 trillion single-use plastic bags are used world-wide every year. In total, half of all plastic produced is designed to be used only once and then thrown away. Plastic waste is now so common in the natural environment that some scientists have even suggested that plastic could serve as a geological indicator of the Anthropocene Era (United Nations 2021).

From the 1950s through the 1970s, only a small amount of plastic was produced and was not a significant issue. By the 1990s plastic waste had more than tripled. In the early 2000s plastic output increased more than in the previous 40 years. As of the end of 2018, more than 300 million metric tons of plastic waste is generated per year, which is equivalent to the weight of the entire human population (United Nations 2021). Although plastic containers and packaging is nearly impossible to regulate by USEPA, there is a clear need to address this issue.

17.4.12 Improving the Safe Drinking Water Act

Expanding the Safe Drinking Water Act (SDWA) to include more contaminants is necessary due to advances in Toxicity Science and new and emerging contaminants that have been detected in drinking water sources. Improvements in the SDWA should also include additional safeguards to protect drinking water sources such as stream headwaters, and areas of near surface groundwater.

17.4.13 Summary of Improving Environmental Regulations

The sections highlighted earlier describe the majority of subject areas where environmental regulations in the United States are either in need of improvement or where new regulations are needed. For reference purposes and to list those areas that have the greatest need for new or improved environmental regulations the following is provided:

- Modify Environmental Enforcement Emphasis and Policy
- Science-based improvements to environmental regulations to account for regional risks and climate change
- Invasive species
- Urban Planning and Land use
- Urban air
- Improve drinking water
- Further or new restrictions on harmful chemicals, especially those that are persistent and mobile in the environment
- Emerging contaminants
- Noise
- Plastic
- Sustainability

The purpose of environmental regulations is to protect human health and the environment. It makes little sense if human health and the environment are protected from 9 out of 10 exposure risks when the remaining risk factor causes harm. Therefore, environmental regulations should be as comprehensive and complete as possible. Otherwise, they will fail to achieve their intended objective.

This example is realized when considering the lack of regulations of the items listed earlier. Another item worthy of discussion is that there are political factors and areas where there is overlapping jurisdictions that may have negative outcomes that must be addressed when discussing the need for further environmental regulations in the United States. For instance, the United States Department of Agriculture would almost certainly interject an interest in environmental regulation of farm land. In addition, the Department of Interior would also be interested in sustainability regulations as it pertains to national forests and parks and environmental regulations on rangeland.

17.5 ENACTING NEW ENVIRONMENTAL REGULATIONS

Enacting new environmental regulations will be necessary to reduce pollutants released into the environment from sources that either are not regulated or need to be more strictly regulated. Those areas where new regulations are needed include agriculture, public lands, and households.

17.5.1 ADDRESSING AGRICULTURE

Agriculture is now the biggest polluter in the United States and the world (United Nations 2016). The United Nations estimates that most land pollution is caused by agricultural activities, such as grazing, use of pesticides, fertilizers, irrigation, plowing, and confined animal facilities (United Nations 2016). Agricultural activities release significant amounts of greenhouse gases from plowing, decaying vegetation, and from livestock. Fugitive emissions and erosion are also significant and release significant amounts or chemicals into the air and the water. In addition, fertilizers are a significant source of pollution to the oceans of the world.

Since environmental regulations have been so successful at reducing the harmful effects from industry in the United States, it is now time to reduce harm effects on the environment from agriculture.

17.5.2 REGULATIONS OF PUBLIC LAND

Areas of public land addressed here are the National Parks which are managed by the National Park Service, National Forests which are managed by the United States Forest Service, and range land which is largely under the management of the Bureau of Land Management. From an administrative viewpoint, each are regulated separately which sometimes leads to inconsistencies of regulations and environmental protections. Each segment of public land is under stress in some form or another that includes effects from forest harvesting, overgrazing, overuse, erosion, lack of funds, and habitat destruction.

17.5.3 SINGLE FAMILY AND MULTIPLE FAMILY HOUSEHOLDS

Regulating pollution from households has been a point of contention for many decades because of the perception or realization that regulating household wastes may be interpreted as an invasion of privacy. However, this should not be the case if future regulations focus on those wastes that are discarded and therefore, would be ultimately not remain on private property.

17.5.3.1 Household Waste

As stated in Section 17.2.4, municipal landfills primarily receive household waste which can contain wastes that would be considered hazardous and would be more strictly regulated if generated from an industrial source. From a sustainability point of view, it would be preferred that households not dispose of certain materials in their waste but this sometimes is not the case. Some examples of these wastes include: cleaners, solvents, pesticides, herbicides, batteries, gasoline, and others. Therefore, more strict measures are needed to further prevent hazardous wastes entering municipal landfills to minimize costly cleanup measures and protect human health and the environment. USEPA regulates municipal landfills under RCRA and does require certain construction measures to prevent or minimizes the potential for contaminants to migrate beyond the footprint of the landfill. However, this sometimes is not enough (USEPA 2021e).

17.5.3.2 Septic Tanks

USEPA estimates that approximately 20% of homes in the United States use septic systems to treat waste water (USEPA 2021g). When a septic system is not properly maintained, elevated nitrogen and phosphorus can be released and migrate to surface water, groundwater, or both. USEPA estimated that up to 20% of septic systems fail at some point in their operational history and that common causes are age, inappropriate design, and overuse (USEPA 2021g). According to USEPA, the responsibility to properly maintain a septic system is the homeowner (USEPA 2021g).

17.5.3.3 Private Household Water Wells

Private drinking water wells are not regulated by USEPA. In addition, USEPA does recommend criteria or standards for individual drinking water wells. USEPA estimates that more than 13 households rely on private wells for drinking water in the United States (USEPA 2021h). Permits are generally always required for installation of a private well and local regulations vary widely. From a sustainability point of view, overuse of groundwater resources in the United States, predominantly from agriculture, is significant (USGS 2021). The effects from overuse of groundwater resources leads to groundwater depletions which in turn can lead to a reduction of water in streams and lakes, land subsidence, and deterioration of water quality (USGS 2021).

17.5.3.4 Home Heating Oil Tanks

A home heating oil tank is essentially what is considered an above-ground storage tank (AST), or if buried, an underground storage tank (UST). USEPA strictly regulates ASTs and USTs under Subtitle I of RCRA. However, home heating oil tanks are generally exempt under RCRA (United States Environmental Protection Agency (USEPA) 1989). The actual number of home heating oil tanks in the United States is unknown, but is likely significant. Many states have published material to assist homeowners with managing heating oil tanks. Releases from heating oil tanks can cause as significant environmental impact as those regulated under RCRA. In fact, USEPA stated in 1989 that tanks represented the largest risk to groundwater than any other contaminant source (United States Environmental Protection Agency (USEPA) 1989).

17.5.3.5 Construction Materials

Many building materials historically contained hazardous substances including asbestos, lead, polychlorinated biphenyls (PCBs), chlorofluorocarbons (CFCs), and others. These still may be present in older building (USEPA 2021i).

Wood preservative chemicals including pentachlorophenol (PCP), chromate arsenicals, and creosote were largely historically used in industrial settings. Resent wood preservatives for residential use include, alkaline copper quaternary (ACQ), borates, copper azole, copper naphthenate, copper-HDO, and polymeric betaine (USEPA 2021i).

17.5.3.6 Improve Landscaping

Landscaping on public lands, especially along highways, should be improved. These areas represent a large area when considering the thousands of kilometers of interstate highways in the United States. Recent management of these areas is one of many reasons likely responsible for the decline in the Monarch Butterfly decline when milkweed, which the Monarch Butterfly needs for reproduction, was removed in favor of other species, including along roadway easements. Other documented influences for the decline include mortality during migration, and climate change (Michigan State University 2021).

17.5.4　CLIMATE CHANGE REGULATIONS

For the last few decades, the scientific community within the United States and world did not generally have difficulty in accepting the science of climate change, but many governments and politicians did have difficulty. This difficulty did not necessarily focus on disputing the science as much as it focused on dealing with potential economic consequences, fairness between countries, coordinating efforts at the global level, and threats to political systems and careers. This is evidenced by the fact that the United States did not join the Kyoto Protocol more than 20 years ago. The United States did sign on to the Paris Agreement in 2015 under the President Obama administration only to be taken back under the Trump administration and then returned under the Biden administration in January 2021. The result has been lost opportunities to lower greenhouse gas (GHG) emissions and will likely be significant because the most recent report on climate change from the Intergovernmental Panel on Climate Change (IPCC) (IPCC 2021), states:

> *"It is unequivocal that human influence has warmed the atmosphere, ocean and land. Widespread and rapid changes in the atmosphere, ocean, cryosphere and biosphere have occurred."*

The given statement has significant implications because it more than suggests that GHG emissions from humans have negatively life on Earth. From a historical perspective, the United States is the world's largest emitter of greenhouse gases (GHGs) (USEPA 2021j). Pressure is mounting for the world to take significant action to reduce GHG emissions so that the consequences can be limited. If not, the result could be cataclysmic.

Many businesses in the United States and the world now measure and track GHG emissions using the GHG Protocol. USEPA has provided guidance in conducting self-assessments and setting GHG reduction targets for businesses (USEPA 2020). GHG reduction targets range between 40 to 50% by the year 2030 and carbon neutral by 2050 (USEPA 2020).

Achieving these goals first requires tracking and measuring GHG emissions for each facility using the GHG Protocol. The second step requires identifying those area where GHG emissions can be reduced. The third step, if necessary, requires providing offsets that remove or reduce carbon emissions, such as planting trees (USEPA 2020).

Reducing GHG emissions have been a priority for many countries including the United States. USEPA has been instrumental at reducing GHG emissions from automobiles and from electrical generating plants. However, much more needs to be conducted and will likely involve (1) further increases in renewable energy sources such as wind, solar, and others, (2) eliminating carbon producing transportation and replacement with electric cars, trucks, trains, and planes, (3) modifying the human diet, (4) reducing GHG emissions from agriculture practices.

17.6　DEVELOPING SUSTAINABILITY REGULATIONS

Sustainability legislation including GHG regulations and targets, give the United States an opportunity to move its environmental regulations from reactive to proactive. In addition, the United States is in need of a clear picture of what it means to be sustainable. To many, it's a buzz word light on specific goals or even a basic plan for achieving sustainability. The European Union and other countries have established a clear picture and have set goals for sustainability. It's overdue for the United States to formally develop sustainability legislation and set goals at the national level. If there are delays in enacting Sustainability measures, it may be too late.

17.7　SUMMARY AND CONCLUSION

There is much we can be proud of when we examine the improvements of our air quality, water quality and clean up of sites of environmental contamination, which now number in the tens of thousands of sites across the United States. Looking back over the sections of different environmental regulations, you may have noticed that much of the legislation was often times a reaction to an incident or environmental tragedy. Laws and regulations originated from many circumstances because of unintentional actions that ended up making quite a mess and causing significant harm to human health, the environment, or both. Some of the incidences that we have touched upon in earlier chapters include the Exxon Valdez spill in 1990, Deepwater Horizon in 2010, Love Canal, Bhopal India in 1984, Valley of Drums, and Toms River to the killing and extinction of animal species that also nearly included the American Bison, California Condor, Grizzly Bear, and numerous others. However, this is a reactionary approach not a proactive approach. When we discuss Sustainability later in this book, we will discuss the need to move beyond reactive measures to proactive sustainability measures to improve our environment in the future. If there are delays in enacting Sustainability measures, it may be too late.

The amount and complexity of environmental laws and regulations in the United States is due to the fact that we are an industrialized society and consume enormous amounts of energy and goods and our hunger for those goods and energy continues to rise as we continue to improve our standard of living. The United States produces large volumes of waste of all types and as our population continues to increase, the amount of waste produced will also continue to increase.

These central facts combined with the dynamic and innate complexities of the natural world add an almost infinite number of negative outcomes in a world with over 7 billion people all wanting to improve their standard of living. The fact is, if we do not have effective and comprehensive environmental regulation, we simply threaten our existence and numerous other species as we are slowly devoured in our own garbage.

Our advances in science have created many technologies that have improved our standard of living and in some circumstances have improved our relationship with nature. But our technological advancements have also created many unintended negative side effects that question the benefit of some technologies. An example involves the creation of synthetic compounds that nature has difficulty breaking down. Some of these compounds include polychlorinated biphenyls, chlorinated hydrocarbons, and many pesticides, including DDT which began the environmental movement in the United States in earnest over 50 years ago. This is to say nothing about plastics which we all should know about since we commonly use and discard huge amounts of plastic in our everyday life.

Although efforts are underway to reduce our energy needs and our veracious hunger for consuming nature, one thing seems certain, our thirst for consumer goods and energy needs will continue to grow, and as our population increases our energy needs will increase even faster. This means that environmental regulation will have to grow with us in volume and complexity as we consume more and more of Earth's resources and replace it with our waste.

We have spent much of this book examining the environmental regulations of the United States and briefly examined other countries of the world. We have concluded that no country is perfect, some countries are better than others when it comes to effective environmental regulation, and all countries can improve and must if humans desire to inhabit Earth sustainably. We have also discovered that there is a direct correlation between effective environmental regulations and a high standard of living and financial resources, and relatively stable population growth.

Questions and Exercises for Discussion

1. **Of the items discussed in Sections 17.2, 17.3, and 17.4, which five would you choose to strongly implement and why?**
2. **Of the items listed in Section 17.3, which three would you rank as least important and why?**
3. **Do you have any additional items or issues that you felt were not addressed?**

REFERENCES

Earth Day. 2021. Earth Day Network Campaign: End Plastic Pollution. https://earthday.or/campaigns/plastic-campaign (accessed April 24, 2021).

Intergovernmental Panel on Climate Change (IPCC). 2021. Climate Change 2021: The Physical Science Basis. *IPCC AR6 WGI*. Vol. 6. 3949p.

Michigan State University. 2021. Why Is the Eastern Monarch Butterfly Disappearing? https://msutoday.msu.edu/Why is the eastern monarch butterfly disappearing? | MSUToday | Michigan State University (accessed August 15, 2021).

National Association of Home Builders. 2021. Typical America Subdivisions. https://www.nahbclassic.org (accessed April 19, 2021).

National Geographic Society. 2021. Energy Efficiency of Modes of Transportation. https://www.nationalgeographic.com/carbon-footprint-transportation-efficiency (accessed April 19, 2021).

Natural Resource Defense Council (NRDC). 2021. Regenerative Agriculture. https://www.nrdc.org/resources/regenerative-agriculture (accessed November 28, 2021).

Rogers, D. T. 2014. Scientists Call for a Renewed Emphasis on Urban Geology. *American Geophysical Union*. Earth and Space News. Eos., Vol. 95. no. 47. pp. 431–432.

Rogers, D. T. 2018. Derivation of a Comprehensive Environmental Risk Model for Urban Groundwater Protection. *International Association of Hydrogeologists Congress*. Vol. 1. p. 879. Daejeon, Korea.

Rogers, D. T. 2019. *Environmental Complaince and Sustainability: Global Challenges and Perspectives*. CRC Press: Boca Raton, FL. 583p.

Rogers, D. T. 2020a. *Urban Watersheds: Geology, Contamination, Environmental Regulations, and Sustainability*. CRC Press: Boca Raton, FL. 606p.

Rogers, D. T. 2020b. Geological- and Chemical-based Environmental Risk Factor Sustainability Model. *International Journal of Sustainable Development*. Vol. 9. no. 4. Rome, Italy. pp. 303–316.

Rogers, D. T. 2021. Empirical Results of a Geological- and Anthropogenic Contaminant-based Environmental Risk Sustainability Model. *Geological Society of America*. Paper No. 8325. Geological Society of America: Boulder, Colorado.

Tallamy, D. W. 2019. *Nature's Best Hope*. Timber Press: Portland, Oregon. 254p.

United Nations. 2016. Human Development Report. *United Nations Development Programme*. http://www.hdr.undr.org/sites/2017 (accessed April 19, 2021).

United Nations. 2021. The Facts on Plastic Pollution. https://www.unenvironment.org/beat-plastic-pollution (accessed April 19, 2021).

United States Census Bureau. 2021a. Census Bureau Average Population Per Household. https://www.census.gov/population/socdemo/hh-fam/tabHH-6.pdf (accessed April 19, 2021).

United States Census Bureau. 2021b. Median and Average Square Feet of floor Area in New Single-family Homes. https://www.census.gov/const/c25ann/sftotalmedavsqft.pdf (accessed January 6, 2019).

United States Census Bureau. 2021c. Average One-Way Commuting Time and Distance. https://www.census.gov/library/visualization/interactive/travel-time.html (accessed April 19, 2021).

United States Department of Energy. 2021. Energy Efficiency and Renewable Energy. *Alternative Fuels Data Center*. https://www.afdc.energy.gov.data (accessed April 19, 2021).

United States Department of Housing and Urban Development. 2021. Median and Average Square Feet of floor Area in New Single-family Homes. *United States Census Bureau* https://www.census.gov/const/c25ann/sftotalmedavsqft.pdf (accessed April 19, 2021).

United States Environmental Protection Agency (USEPA). 1989. Underground Heating Oil and Motor Fuel Tanks Exempt from Regulations Under Subtitle I of RCRA. USEPA Office of Storage Tanks. United States Environmental Protection Agency: Washington, D.C. 146p.

United States Environmental Protection Agency (USEPA). 2018. Air Quality – 2018 National Summary. https://www.epa.gov/air-trends/air-quality-national-summary (accessed April 19, 2021).

United States Environmental Protection Agency (USEPA). 2020. GHG Inventorying and Target Setting. GHG INVENTORYING AND TARGET SETTING SELF-ASSESSMENT: V1.0 (epa.gov) (accessed July 30, 2021).

United States Environmental Protection Agency (USEPA). 2021a. Flint Drinking Water Crisis. https://www.epa.gov/Flint Drinking Water Response | US EPA (accessed August 15, 2021).

United States Environmental Protection Agency (USEPA). 2021b. Reducing Lead in Drinking Water https://www.epa.gov/Leaders in Reducing Lead in Drinking Water | US EPA (accessed August 15, 2021).

United States Environmental Protection Agency (USEPA). 2021c. Drinking Water Treatment Technologies. https://www.epa.gov/Drinking Water Technologies | US EPA (accessed August15, 2021).

United States Environmental Protection Agency (USEPA). 2021d. Emerging Technologies for Wastewater Treatment. https://www.epa.gov/Emerging Technologies for Wastewater Treatment and In-Plant Wet Weather Management (epa.gov) (accessed August 15, 2021).

United States Environmental Protection Agency (USEPA). 2021e. Municipal Solid Waste Landfills. https://www.epa.gov/Municipal Solid Waste Landfills I US EPA (accessed August 15, 2021).

United States Environmental Protection Agency (USEPA). 2021f. Climate Adaptation and Erosion and Sedimentation. https://www.epa.gov/Climate Adaptation and Erosion & Sedimentation I US EPA (accessed August 15, 2021).

United States Environmental Protection Agency (USEPA). 2021g. The Sources and Solutions: Septic Tanks. https://www.epa.gov/The Sources and Solutions: Wastewater I US EPA (accessed August 15, 2021).

United States Environmental Protection Agency (USEPA). 2021h. Private Drinking Water Wells. https://www.epa.gov/Private Drinking Water Wells I US EPA (accessed August 15, 2021).

United States Environmental Protection Agency (USEPA). 2021i. Overview of Wood Preservative Chemicals. https://www.epa.gov/Overview of Wood Preservative Chemicals I US EPA (accessed August 15, 2021).

United States Environmental Protection Agency (USEPA). 2021j. Inventory of U.S. Greenhouse Gas Emissions and Sinks. *Inventory of U.S. Greenhouse Gas Emissions and Sinks | Greenhouse Gas (GHG) Emissions | US EPA* (accessed February 20, 2021).

United States Fish and Wildlife Service. 2021. Accessing the Status of the Monarch Butterfly. *Assessing the status of the monarch butterfly (fws.gov)* (accessed April 3, 2021).

United States Geological Survey (USGS). 2021. Groundwater Decline and Depletion. https://www.usgs.gov/Groundwater Decline and Depletion (usgs.gov) (accessed August 15, 2021).

18 Environmental Protection and Sustainability at the Individual Level

18.1 INTRODUCTION

Now we turn our attention to sustainability at the local and individual level. This includes actions at a specific site, business, property, park, residential home, subdivision development, condominium, apartment, schools, commercial development, and other type of land use at the parcel level. Subject matter discussed in this chapter include the following:

- Smart urban development
- Evaluating chemical use and risk
- Geologic vulnerability analysis
- Point source pollution prevention using elimination, substitution, prevention, and minimization techniques
- Contaminant Risk Factor (CRF) analysis
- Non-point source pollution prevention techniques
- Financial incentives for pollution prevent efforts
- Building and implementing a sustainability model
- Sustainability measures that can be undertaken at the individual level

18.2 SMART URBAN DEVELOPMENT

The human population is increasing and requires more and more space to live, grow food, and for build roads. The rapid physical growth of urban areas leads to complex processes of landscape transformation which alters the structure and function of ecosystems. Urban invasion into natural areas decreases productivity, decreases infiltration of water, increases impervious surfaces, increases flooding, increases erosion, increases pollutant load, increases energy demand, and decreases species biodiversity. The size of a single family home in the United States has more than doubled in the last 70 years (United States Department of Housing and Urban Development 2021 and United States Census Bureau 2021a and 2021b).

Many opportunities for smart urban development are available within many cities of the world. Some of the numerous opportunities include the following:

- Less single-family homes and more multi-family units. The average size of a condominium is approximately 130 square meters (1,200 square feet) verses 300 square meters (2,700 square feet) for a single-family home. The average land required for a single-family home in the United States is 815 square meters or 0.25 acres per home. This includes the land required for easements and roads for a typical subdivision (National Association of Home Builders 2021).

  Building large skyscrapers that have hundreds to sometimes over a thousand condominium units have enormous sustainability improvements over the single-family home. Take for example a large skyscraper currently being constructed in Chicago in the United

DOI: 10.1201/9781003175810-18

States that will have an estimated 1,200 condominium units on a footprint of approximately 5,000 square meters (2 acres). This represents a 99.5% reduction in occupied land if those 1,200 condominium units were single family homes. There are also significant reductions in the amount of construction material that would be used if each were a single-family home that includes forest materials, insulation, roofing, paint, metal piping, sewers, stormwater, roads and curbs, and sidewalks. There are also significant long-term energy savings including heat and electrical.

- Reducing the size of housing units. The size of a single-family home in the United States has nearly doubled since 1970 and nearly tripled since 1940 (United States Census Bureau 2021). This is in contrast with the fact that the children per family in the United Sates has decreased by more than 60% from 3.76 children per family in 1940 to 1.58 children per family in 2018 (United States Census Bureau 2021b).
- Living closer to work. The average commute to work in the United States is 20 miles one way and takes nearly 30 minutes (United States Census Bureau 2021c).
- Increase mass transit availability. Mass transit is much more efficient and emits much less pollution than an automobile per person per kilometer travel. Estimates are as high as 95% more efficient when traveling by commuter train compared to an average automobile when considering fuel use alone and not taking into consideration of the cost and in-efficiencies of building and maintaining highways (United States Department of Energy 2021 and National Geographic Society 2021).
- Increase other forms of transport that include using scooters, bicycles and walking. Many cities in the United States and throughout the world are improving bike-friendly and foot transport routes to increase these forms of transportation by promoting health and enjoyment benefits.
- Integrating more mixed land use into urban areas. The United States has much to learn from other countries when it comes to this urban development technique. Historically, the United States has not taken an integrated approach to land use. Many European and Asian countries, namely South Korea, have embraced integration of farmland, industry and housing.
- Increasing clean energy production such as solar and wind. Both solar and wind energy have become more efficient and less costly. A key point for solar efficiency is prevent dirt buildup on the solar panels. Wind turbines are available in different engineering configurations that can be applied in small areas and minimize bird deaths (United States Department of Energy 2021).
- Increasing permeable surfaces. To decrease flooding, increasing infiltration of precipitation through several available and simple techniques is widely available.

Currently, examples of sustainable and smart urban development using some or all the items listed are mainly confined to Europe, Asia, and some parts of the United States. Two impediments that have been slow in becoming a reality in the United States concerning smart urban development is the American love affair with the automobile and achieving the American dream in which owning a single-family home has often been utilized as a degree of measure that defines making that dream come true.

18.3 POLLUTION PREVENTION AT THE LOCAL, PARCEL, AND INDIVIDUAL LEVEL

As we previously learned, it is difficult and costly to remediate a contaminant after it has been released into the environment. The contaminant may spread into soil, water, and air, and often causes harm to the environment before cleanup can occur. These reasons underscore why preventing the release of contaminants is a prerequisite for creating a sustainable environment.

Preventing contaminants from entering the environment is called pollution prevention. To be successful, this process includes: reducing or eliminating waste at the source by modifying production processes; promoting the use of non-toxic or less toxic substances; implementing conservation techniques; and reusing materials instead of putting them into the waste stream (USEPA 2021a). Simply put, the USEPA defines **pollution prevention** as any practice that reduces, eliminates, or prevents pollution at its source (USEPA 2021a).

The legal framework for this pollution prevention effort is embodied in the Pollution Prevention Act of 1990, which focused industry on these measures (USEPA 1990):

- Pollution should be prevented or stopped at the source whenever feasible
- Pollution that cannot be prevented should be recycled
- Pollution that cannot be prevented or recycled should be treated in an environmentally safe manner
- Releases of pollution into the environment should be conducted only as a last resort and should be conducted safely

Sources of pollution can be designated as point sources or nonpoint sources (USEPA 2003). **Point source pollution** originates from identifiable sources, such as smokestacks or sewage outfall pipes. **Nonpoint source pollution** emanates from diffuse or unknown sources, and is the leading cause of water pollution in the United States (USEPA 2002). Examples of nonpoint source pollution include:

- Contaminated groundwater from an unknown source
- Contaminated stormwater from runoff originating from parking lots, roads and lawns
- Air deposition of contaminants and particulates
- Erosion
- Runoff from agricultural areas

The road to sustainability must incorporate effective pollution prevention that yields observable results. In the United States, the observation of major pollution events has been a catalyst for significant levels of response (USEPA 2021b). As a corollary to this, observable progress in preventing pollution will likely provide additional incentives to continue those efforts. At the onset and throughout, science must guide the planning process, and the results should be published and open for critical review by scientists, professionals, and the public. Other components of this process include: 1) the maximization of resource efficiency; 2) implementation of existing and developing technical innovations; 3) minimization of use of toxic chemicals; and, 4) education (USEPA 2021b).

The following sections will concentrate on sustainability measures that can be employed at the facility and local level. We will focus our discussion on pollution prevention. As you will see, pollution prevent is the key component on the road to sustainability at the facility and local level. We will discuss a framework for preventing pollution at industrial point sources. Since point and nonpoint sources are characterized by common transport media and transport processes, portions of this framework are then applied to the source reduction efforts for the nonpoint pollution variants of stormwater and erosion.

18.3.1 Implementing Pollution Prevention Techniques – Point Sources

Successful implementation of pollution prevention involves careful planning. Within the broad array of planning venues and forms, (e.g., urban, environmental, strategic, business) there are common threads to the planning process:

- Identification of what you want to do (goals/objectives)
- Collection of data
- Specification of methods for achieving the goals/objectives
- Implementation using the selected methods
- Assessment of the results

The planning process shown here is cyclical. Assessment may lead back to more data collection, or if implementation fails with the methods selected, new methods can be developed and implemented until success is achieved. Sometimes, the outcome changes the entire goal of the project, especially in cases where you bit off more than you could chew. Science should be infused into the planning process wherever appropriate. Accurate measurement is a foundation of good science, so to ensure scientific standards for data collection are met, the procedures should include a statistically sound specification of the sample size.

Another area where science must be incorporated into the execution of a plan is the **experimental design.** Science is fundamentally about identifying and explaining variation, and the experimental design—which is the assignment of subjects to experimental groups—provides the roadmap. Although it may sound obvious, at contaminated sites there two types of locations: contaminated and uncontaminated. Assigning these locations into two groups allows investigators to study the similarities and differences between them. This separation is how we learned that contamination tends to occur more frequently near low points in buildings, and will be described in greater detail later in this chapter. We then use this knowledge to help design the most effective measures for pollution prevention—the "where" of intervention.

Successful source control also requires an understanding of the process producing the pollution. Processes occur over time, so the specification of where to intervene should be accompanied by the proper timing of our pollution prevention efforts—the "when" of intervention. If loading docks are areas in a facility more prone to a contaminant release, then busy times at these locations require special diligence. The initiation of pollution prevention efforts occurs within organizations, and represents a form of change, or innovation. To succeed it important not only to get the science right, the innovation must: 1) be testable and implementable at small-scales; 2) represent an observable improvement over existing conditions; 3) be culturally acceptable (in the corporate/organizational and social senses); 4) be economically feasible; and, 5) be convenient to implement. These five conditions characterize successful innovations (Rogers 1995).

What follows is a planning process based on the experiences of implementing a successful pollution prevention effort at a major manufacturing company (Rogers et al. 2006). This process contains the basic elements of plans, incorporates scientific aspects of the geologic environment and contaminant properties, and recognizes the social context for implementation.

18.3.1.1   Step 1: Establishing Objectives and Gathering Background Data

Establishing objectives or goals for pollution prevention provides a baseline for measuring success. Making decisions with better information can avoid the specification of arbitrary objectives and goals. Information can be obtained with targeted data collection and evaluation, and this will help with the attainment of goals and focus limited resources where they will produce maximum benefit. It is recommended these data are collected:

- Mass and volumes of each type of solid wastes generated, including:
  - Solid wastes such as wood and paper products, plastic, metals, glass etc.
  - Regulated solid wastes
  - Regulated hazardous wastes
- Mass and volumes of liquid wastes generated, including:
  - Stormwater volumes and content
  - Sanitary discharge volumes

- Industrial wastewater volumes and contents
- Regulated liquid wastes in containers
- Regulated liquid hazardous wastes in containers
- Mass and volumes of air emissions generated, including:
  - VOC emissions
  - Particulate emissions
  - Heavy metal emissions
  - All other identifiable air emissions
- Energy
  - Sources, types, consumption, and rates through time
  - Energy loss
- Purchasing habits
  - Bulk containers compared to smaller containers and amounts
  - Types of containers (e.g., gallon containers vs. spray cans)
  - Packaging (plastic vs. cardboard or biodegradable material)
- Production efficiencies
- Product packaging
- Shipping (rail vs. truck)

Where possible, a mass balance should be calculated to ensure the accuracy of the data. Once the data have been collected and evaluated, steps 2, 3, and 4 should be completed before firmly establishing goals and objectives for any pollution prevention program.

18.3.1.2   Step 2: Inventory of Hazardous Substances

The next step in evaluating the need for developing a pollution prevention plan at any location---whether a manufacturing facility or a household--is to inventory and assess hazardous substance use. Chapter 2 covered many common contaminants present in urban areas, including households. As they are often very close to industrial sites, households can assist with any inventory since the average American household stores 3–10 gallons of hazardous materials (Smolinske and Kaufman 2007). To inventory chemicals, CAS Registration Numbers should be used since many products display either trade names or synonyms. If available, Safety Data Sheets (SDS) often provide valuable information.

Inspecting the label on chemical containers is required, especially when the container is a mixture of chemical products. In these cases, the name of the product typically is a trade name and is not very helpful in identifying the specific chemicals contained. Inspection of the label is the only effective method.

Chemicals should be inventoried by chemical group, such as VOCs, PAHs, SVOCs, PCBs, metals, acid, bases, and whether they are present in gas, liquid, or solid form. Transformers containing PCBs should also have appropriate labels as shown in Figure 9.8. After the hazardous substance inventory has been completed, a map should be created showing the following: location of where each hazardous substance enters the property; where the substances are stored before use; where they are used; and, where they are stored after being used and before any residuals are discarded. Since many contaminant release locations occur from low points in buildings, highlighting pits, sumps, trenches, underground storage tanks, and floor drains on the map can also help provide valuable information for pollution prevention efforts (Rogers et al. 2006).

18.3.1.3   Step 3: Assessing CRFs

The hazardous substances having the highest CRFs should now be evaluated. This evaluation helps to prioritize the pollution prevention efforts. Many facilities and households may have more than 100 different hazardous substances. Therefore, prioritizing is critical to achieve maximum benefit. As noted in Chapter 5, chromium VI and DNAPL VOCs have the highest CRFs for groundwater, with PCBs, mercury, chlordane and PAHs having the highest CRFs for soil. Concentrating pollution prevention efforts on those contaminants with the potential to significantly impact groundwater such as chromium VI and DNAPL VOCs will have the most long-term benefit.

18.3.1.4   Step 4: Preliminary Assessment of Geologic Vulnerability

Most urban areas have not been geologically mapped, so the detail necessary to accurately evaluate the vulnerability at any location may not be available (Rogers et al. 2016 and Rogers 2020a). The suggestions offered here can help determine if any given area presents enough potential risk to warrant a more detailed examination, and whether an aggressive pollution prevention initiative should be pursued:

- Source, location, and type of potable water. Contact the local municipality to evaluate the source of potable water for the area in question. In addition, inquire whether there are any groundwater extraction wells of any type within at least a 1 mile radius of the location being evaluated. If any wells exist, request a copy of the installation records for further examination.
- Nearest surface water body. Determine where the nearest surface water body is located with respect to the location being evaluated. Topographic maps (7.5 minute) can help with this procedure.
- Stormwater collection and discharge. The local municipality should have information about stormwater collection, its treatment, and the discharge locations for the area being evaluated.
- General geological conditions. Examining wells records could provide valuable information on soil type, stratigraphy, and depth to groundwater if well records can be obtained in the vicinity of the location being evaluated.

If the source of potable water in the area being evaluated is obtained from groundwater and extraction wells are nearby, environmental risk should be considered high. In this case the geologic vulnerability rating will likely exceed a score of 50, and pollution prevention efforts should become the highest of priorities. The other factors listed earlier will likely require examination and evaluation by a qualified professional before an appropriate geologic vulnerability rating can be determined.

18.3.1.5   Step 5: Preventing Pollution through ESPM Methods

Pollution prevention can be implemented with a stepwise evaluation process that proceeds from the most preventative measure to the least preventative measure. This process is referred to as ESPM, and consists of these steps:

- Elimination: Not using potentially harmful chemicals
- Substitution: Using a less potentially harmful chemical instead of a harmful one.
- Prevention: Using engineering controls and other measures to minimize the potential for a release; employed if eliminating or substituting a potentially harmful chemical is not possible.
- Minimization: Reducing usage of harmful chemicals or a reduction in generated wastes through process changes, recycling, or other methods.

After steps 1 through 4 have been completed, a focused and achievable plan for pollution prevention can be developed and implemented. Facilities where a synergistic effect may be present should receive the highest initial effort characterized by aggressive pollution prevention initiatives for reducing the potential risks (Rogers et al. 2006 and Rogers 2011).

18.3.1.5.1 Elimination

The most aggressive form of pollution prevention is *elimination* of hazardous chemical use. Where possible, elimination of hazardous substance use is the preferred pollution prevention method because it is the easiest to manage and has the greatest benefit to the environment. As noted in Chapters 2 and 3: if there is no hazardous chemical use--there is minimal risk. In most cases, total elimination of hazardous chemical use is usually not possible, so the elimination of chemicals should focus on contaminants with high CRFs. For example, chromium VI and DNAPL VOCs have high CRFs for groundwater and PAHs, chlordane, PCBs, and mercury have high CRFs in soil.

A crucial step in the process of eliminating hazardous substances is to develop a chemical ordering procedure. This procedure protects against unauthorized hazardous substances making their way into operations at the facility without prior knowledge. All proposed chemicals or substances should undergo a review process to evaluate whether they are acceptable for use.

18.3.1.5.2 Substitution

The next most aggressive form of pollution prevention is *substitution*. Substitution involves using alternative chemicals with the goal of greatly reducing risk. For instance, if a facility uses DNAPL VOCs for cleaning, an effective substitute may be citrus-based cleaners. The Solvent Alternatives Guide provides options and guidance for evaluating available chemical substitutes for common solvents (USEPA 2021c).

Other alternatives exist for chemical substitution. For instance, mercury-containing devices such as switches, thermometers, and monometers can be substituted by digital devices. Liquid transformers containing PCBs can be substituted with dry transformers or with transformers not containing any detectable concentration of PCBs. Other examples of substitution include:

- Using paints without VOCs and certain heavy metals
- Using biodegradable oils
- Using paraffin as a lubricant instead of oil

18.3.1.5.3 Prevention

Prevention of contamination entering the environment can be accomplished through engineering controls. Engineering control methods include: release prevention, release detection, release containment, and release cleanup immediately after a spill.

The following examples (denoted by bullets) highlight some ways activities and operations can be modified to minimize the potential for the release of hazardous substances to the environment:

- Evaluating and eliminating potential points of release. Sumps, pits, trenches, floor drains, and chemical storage and usage areas are common points for hazardous substance release. Conducting an inventory of these locations and locations of chemicals present at a facility will assistance in identifying areas where releases may occur.
- Modifying liquid waste storage areas. Storage methods for liquid wastes should include multiple and redundant engineered systems to prevent a release and contain a release if one does occur. These engineered systems typically include:
  - Storing liquids wastes inside and under a roof
  - Coating the floor with epoxy
  - Providing secondary containment if a release were to occur. In this example. drip pans are located beneath each outer container.
  - Locating the liquid waste storage area at a location without floors drains, sumps, trenches, and pits.
  - Using redundant storage containment. Sealed drums containing liquid wastes are located inside the outer containers pictured in this example.

- Properly labeling the contents and potential hazards.
- Sealing sumps, pits, trenches, floor drains, and liquid storage areas with an impervious surface. In most instances, leaks will occur from structures composed of concrete storing or conveying liquid wastes. Leaks can occur from seams, cracks, or in some cases from the direct migration of liquid through the concrete surface itself. These surfaces should be sealed to prevent a release to the subsurface.
- Providing secondary containment. Engineering redundant systems can effectively prevent an uncontrolled release to the environment.
- Spill containment and cleanup. Despite the existence of engineering controls, accidental releases do occur. Therefore, proper response is necessary to prevent the uncontrolled release of a hazardous substance and to protect human health and the environment. Spill stations outfitted with an assortment of tools, containers, personal protective gear, and instructions can ensure small spills of liquids not presenting an immediate threat to human health are addressed quickly and safely. These stations should be located near hazardous substances.
- Education and training. Education and training are critical for the prevention, response, and cleanup of spills. These efforts can also prevent the response to a spill or accidental release of a hazardous substance when evacuation and immediate notification are the necessary courses of action, and qualified emergency personnel are required on site.

18.3.1.5.4   Minimization

Waste minimization involves using less hazardous substances through conservation efforts or process changes. The result is a reduction in the amount of wastes requiring disposal. In some instances, minimization also includes recycling. For instance, reducing the discharge or generation of liquids through process changes, recycling of water, or the evaporation/recycling of liquid or solid waste can greatly lower the volumes and mass of waste streams.

Recycling of metals, plastic, glass, wood products and many other materials can greatly minimize the amount of solid waste generated and disposed of in a landfill. Often, these activities result in significant cost savings. The future in this arena is promising, as the discovery of new options for the beneficial reuse of waste materials parallels the appearance of new waste sources. Minimization is also achieved through awareness, tracking, and accountability.

Energy reduction strategies evaluate the energy types consumed, and their usage rates by location over time. In addition, energy loss and potential recovery/recycling may also come into play.

Purchasing habits can greatly assist minimization efforts. When supplies are purchased in bulk containers, the number of containers requiring disposal is reduced. It is also possible to realize reductions in energy usage and air emissions through production efficiencies. For instance, switching to more efficient gas turbines in certain production processes may save energy, while conducting energy-intensive activities at night when energy demand is low can lower air emissions. At off-peak times power companies are more likely to substitute a non-fossil fuel such as hydroelectric for coal.

Product packaging and transportation methods are also important areas where many improvements can be made from an environmental perspective. For instance, using packaging made from biodegradable material instead of plastic can have a significant positive impact on the environment. In transportation, the use of rail or barges to transport products consumes much less fossil fuel and reduces air emissions.

18.3.1.6   Step 6: Assessing Results

Practicing pollution prevention has qualitative and quantitative benefits that can result in significant financial gains and cost avoidance. The evidence of quantitative pollution prevention benefits can be found by tracking reductions, especially if data have been collected before a pollution prevention program was initiated. Quantitative evaluation is reflected by the reductions of (Rogers 2019a):

- Waste volumes
- Raw materials
- Energy usage

On-site qualitative benefits are realized by (Rogers 1992):

- Release avoidance
- Additional protection of human health and the environment
- Lowering regulatory reporting requirements
- Lowering environmental liability

Across an urbanized watershed, qualitative improvements will be seen in ecosystem health, receiving water quality (surface water and groundwater), soil conditions, and air pollution levels.

Continuous improvement, evaluation, and inspection should be a developed as part of program assessment, because the lack of an effective pollution prevention program can result in significant liability and cost.

One final note: pollution prevention represents an innovation within an organizational context. With this in mind, the key "in-house" actions for achieving environmental project objectives, realizing cost savings, and reducing environmental risk are: (a) establish easily identifiable objectives with the involvement of senior management; (b) perform an effective accounting of cost savings and tracking of other related benefits; and (c) communicate the results across all levels of the organization. In addition, it is beneficial to have compliance personnel trained in environmental science, with these personnel positioned at a high level within the organization (Rogers et al. 2006).

18.3.1.7   Financial Incentives

The major financial incentives for implementing pollution prevention at most industrial facility's, commercial businesses, or even residential homes are the direct and indirect cost savings realized from a successful program. Direct costs savings accrue from reducing the amounts of solid waste generated and water consumed. Indirect cost savings occur when the long-term liabilities associated with the disposed of waste at a licensed facility are lowered because the amount of waste decreases. Other long-term liability reductions such as savings on litigation costs and reduced contaminant releases can also significantly reduce the costs associated with cleanup. Finally, waste reduction often reduces the amount of regulatory reporting requirements, and this outcome can significantly cut costs.

Implementation of an effective pollution prevention plan at any manufacturing facility requires the commitment, input, and cooperation of every employee. Representatives from purchasing, production, maintenance, human resources, and environmental must work together in close cooperation to identify, implement, and measure every aspect of a pollution prevention program. Tracking progress is necessary to sustain a program and create more involvement with its execution and outcomes. Many pollution prevention programs fail because they do not fully quantify their benefits.

An additional item to consider is how the environment wins by implementing a successful sustainability program. This ultimately has the most benefit for everyone, including the planet because we learn to live with nature.

18.3.2   Implementing Pollution Prevention Techniques – Non-Point Sources

Stormwater and erosion are related, as both involve the transport of materials by fluids.

Stormwater originates from precipitation, and is the polluted overland flow of water in urban areas. As precipitation flows over roads, rooftops, parking lots, construction sites, and lawns, it becomes contaminated with oil and grease, pesticides, fertilizers, litter, and pollutants from vehicles. The EPA estimates over 10 trillion gallons of untreated stormwater make their way into U.S. surface waters each year (USEPA 2021d). Dense urbanization exacerbates the problem, since the amount of pollution present in stormwater runoff is correlated with the amount of impervious cover (Schueler 1994). As mentioned earlier, pollutants from farmland are responsible for much of the pollutant load of pesticides, herbicides, and fertilizers that enter the Gulf of Mexico and many other river systems that discharge to the oceans.

The Clean Water Act (CWA) of 1972 gave the USEPA the authority to regulate point source discharges through the National Pollution Discharge Elimination System (NPDES) program. In 1987, a survey of the nation's waters indicated point source control alone was not sufficient to achieve the "fishable and swimmable" goal of the CWA because nonpoint sources, especially agricultural and urban runoff were contributing substantial amounts of pollution (Humenik et al 1987). Also in 1987, the USEPA initiated the NPDES Stormwater program, requiring municipal separate storm sewer systems located in incorporated areas with populations of 100,000 or more to obtain NPDES permits for stormwater discharges (USEPA 2019d). In 1999, Phase II of this program was extended to smaller municipalities and required permit holders to implement post-construction stormwater management programs using Best Management Practices (BMPs). Examples of BMPs for stormwater management include education, road salt management, street cleaning, and erosion control measures, such as silt fences and covering exposed soil (USEPA 2021e).

18.3.3 Applying the Source Control Framework to Non-Point Sources

Since the overland flow of water (stormwater) forms the transport component of erosion, stormwater becomes the focal point of pollution prevention methods. The goal here is not to develop a list stormwater BMPs; the EPA has a "menu of BMPs" designed to help communities with their implementation of the Phase II stormwater rules (USEPA 2021e). Instead, our focus is this: given the limitations of the Phase II Stormwater controls (they only apply to new development or re-development), how can we achieve effective pollution prevention methods for stormwater in older urban areas? To help answer this question, we now apply the pollution prevention framework used for point source control to nonpoint sources as follows:

- Establishing Objectives and Gathering Background Data
- Inventory of Hazardous Substances
- Assessing CRFs
- Preliminary Assessment of Geologic Vulnerability
- Preventing Pollution through ESPM Methods
- Assessing Results

Efforts to reduce erosion and stormwater runoff can be evaluated on several levels. Visually, improvements across a region will be seen in ecosystem health, receiving water quality (surface water and groundwater), and soil conditions. Hydrologically, there would be a lower frequency of floods, and a less "flashy" stream response; that is, a slowing down of the time required to reach peak runoff typical of urban watersheds. Testing of the stormwater runoff would reveal consistently lower amounts of suspended solids, nitrogen, phosphorus, heavy metals, bacteria, and other contaminants responsible for degrading water quality. Groundwater quality would improve based on systematic well testing, and base flow rates would increase as more water was infiltrated. The increased base flow would also help to keep stream channels at higher levels throughout the year, and make aquatic ecosystems less vulnerable to the effects of higher water temperatures.

The results of any sampling/monitoring efforts undertaken to reduce erosion and stormwater within watersheds should be published. This helps educate the public about these efforts and may increase public involvement. It also makes the organizations conducting the work more accountable.

18.4 BUILDING AND MAINTAINING A SUSTAINABILITY MODEL

A sustainability model for any location in the world may seem impossible because of the sheer number of potential variables that greatly increase the complexity and because of the many differences between countries. However, this is just not the case. Sustainability is defined as creating and maintaining conditions under which humans can exist in productive harmony to support present and future generations (United Nations 2020 and United States Environmental Protection Agency [USEPA] 2020). The term "productive harmony" applies to the environment. Therefore, sustainability can be reduced to three variables that include (1) understanding the environment, (2) understanding how aspects of human activities impact the environment, and (3) developing and implementing measures to mitigate unacceptable risk (Rogers 2019a and 2019b).

Understanding the geological, hydrological, and ecological environment is the first step in building a sustainability model and is represented by the general term Geological Vulnerability. The second step is evaluating aspects of human activities that affect the environment either through negative or positive outcomes. The third variable is evaluating the effectiveness of risk reduction measures. An equation for sustainability is created by combining these three fundamental concepts through which a sustainability index is the output (see Equation 1). The Sustainability Index represents a measure of sustainability for any particular location with a higher value representing increased risk and harm to human health or the environment and therefore, less sustainable (Rogers 2020a and 2020b):

$$\frac{1}{\text{Geologic Vulnerability X Operational Aspects X Risk Reduction Measures}} = \text{Sustainability Index} \quad (18.1)$$

The Sustainability Index can be scaled to evaluate environmental health of a single location or whole ecosystem depending on system inputs and evaluation criterion described in the following sections.

To evaluate the environment and geology at the locations where a contaminate release has occurred requires a detailed understanding of the geology and hydrogeology in great detail so that an accurate assessment of contaminate fate and transport can be conducted. This type of assessment is commonly referred to as geologic vulnerability analysis. A logical first step in conducting geologic vulnerability analysis is to understand the near-surface geologic and hydrogeologic environmental on a local and regional scale. This information is commonly presented as a geologic map and a geologic vulnerability map. A geologic map provides detailed information of the subsurface with a perspective and concentration of effort toward providing information helpful to determine how contaminants may migrate in the subsurface environment so that a vulnerability map can be constructed. Geologic vulnerability analysis and mapping have been developed by Murray and Rogers (1999), Rogers 1996, Rogers et al. 2002, 2006, and 2007). Central to this method is a subjective numerical rating system using weighting coefficients for numerous geologic and hydrogeologic parameters (Rogers 1992, Rogers 1996, Murray and Rogers 1999, and Rogers 2002). These factors provide detailed subsurface information relevant to construct a geologic vulnerability map within any urban area and is presented as a matrix of vulnerability factors that are rated through a linear scoring scale (Rogers et al. 2016, Rogers 2020a, and Rogers 2021).

Aspects of operations are divided into two separate activities. The first is evaluating the potential risk posed by contaminants used or stored at any location and the second include aspects of human activities that may increase or decrease the potential for a release to the environment (Rogers 2020a).

Facility risks can be characterized with two measured variables that include Aspect Risks of Operations and Risk Reduction Actions. Much of this information can be obtained from data contained in an environmental audit. The recommended point values for each attribute is presented within a scaled range. Higher point values indicate higher relative risk. Categories included in the checklist follow the environmental audit guidelines and include (Rogers 2014, Rogers 2018 and Rogers 2020a):

- Air
- Solid and Hazardous Waste and Landfills
- Water
- Spills
- PCBs
- Toxic Substance Control Act or Equivalent
- Community Right to Know or Equivalent
- Storage Tanks
- Asbestos
- Regulatory Inspections
- Site Inspection

Typically, an audit is conducted by an environmental professional and much of the information from an environmental can be used to evaluate the environmental risks posed at any facility or property. The objectives of an environmental audit include (1) verifying compliance with environmental requirements, (2) evaluating the effectiveness of in-place environmental management systems, and (3) assessing risks posed from regulated and unregulated materials, chemicals, and practices. From a fundamental point of view, the typical environmental audit encompasses evaluating subject areas that include evaluating how operations at a facility impact the environment through the land, water, and air.

Once a facility has been initially evaluated and scored, the initial score can be compared to subsequent reviews. The results of the comparison will indicate whether the facility is lowering environmental risk and therefore, reducing the potential environmental impact and becoming more sustainable.

18.5   SUSTAINABILITY AT THE INDIVIDUAL LEVEL

There is almost an endless list of sustainability measures that each of us should practice in our daily life. As stated earlier in this book, individual efforts do make a difference. As with governments and large organizations, practicing ESPM methods is advisable even at the individual level. As stated in previous chapters, household waste and households in general are exempt from many environmental regulations in the United States. This is more significant now because as other sectors are regulated, namely industry, they have improved to the point that households now pollute a much greater share compared to levels emitted a few decades ago.

Following is a partial list of individual behavior modifications that each of us can undertake with little effort that together will make a significant impact on moving toward a sustainable future. These items include (USEPA 2021f, United Nations 2021, Tallamy 2019, and Darke and Tallamy 2014):

- Bike to work, to school
- Walk
- Carpool
- Check daily air quality forecasts
- Consider living closer to work
- Install solar house panels
- Compost yard waste and food waste
- Plant a vegetable garden
- Collect rain water from roof drains to water plants and for other outdoor needs
- Drive a more fuel efficient automobile, hybrid, or electric automobile

- Consume less meat
- Consume more plant food
- Use mass transit
- Avoid-thru lines
- Fill fuel tanks in the evening or at night during summer months
- Get regular engine tune ups
- Request flexible work hours
- Dry clothes by hanging them outside
- Consider an electric lawn mower and snow blower
- Lower water heater temperature
- Keep wood stoves and fireplaces well maintained and limit use, if possible
- Do not smoke indoors
- Do not consume water from plastic bottles
- Recycle plastic, paper, glass, and aluminum (contact for municipality for additional information and recycling opportunities)
- Properly dispose of electronic equipment (contact your local municipality for more information concerning other items such as light bulbs, batteries, mercury switches)
- Properly dispose of other household items including cleaners, solvents, paints, and other items (contact your local municipality for a comprehensive local list)
- Utilize reusable shopping bags
- End the use of plastic bags and plastic bottles
- Choose to purchase items with minimal packaging material (remember if you don't consume or use it, it becomes a waste)
- Upgrade appliances with improve energy efficiency
- Install a heat blanket on the hot water heater and lower its temperature 10 to 20 percent
- Use low-flow shower heads
- Take shorter showers
- Wash clothes in cold water
- Purchase organically-grown and eco-friendly food
- Become more ecologically educated
- Encourage and inspire others
- Turn your computer off rather than using sleep mode or screen saver
- Print documents only when required
- Unplug electrical devices such as toasters, printers, chargers, computers and other items when not in use
- Use recycled paper
- Drink more tap water
- Don't drink from disposable cups
- Replace your vehicles air filter every 3,000 miles
- Keep vehicle tires at recommended air pressure
- Do full loads of laundry
- Run the dishwasher only when full
- Limit dry cleaning clothes and other items
- Turn off the water while brushing your teeth
- Receive and pay bills online
- Call companies that participate in junk mail advertisements
- Avoid purchasing food in individual serving sizes
- Avoid purchasing items that are considered single-use
- Turn down the heat or turn air conditioning when not at home
- Lower the thermostat during winter and especially at night and increase the thermostat during the summer

- Contact your local and state-elected officials and encourage them to support sustainable legislation
- Participate in volunteer outdoor clean up efforts in your community
- Plant trees, especially deciduous near your home for shade in the summer
- Plant native vegetation rather than grass
- Use citris-based cleaners rather than volatile organic compounds
- Consider downsizing your home
- Improve insulation of your home
- Purchase more items from garage sales, thrift stores, re-sale, and consignment shops
- Donate more items
- Sell items at a garage sale. resale, or consignment shop
- Purchase bleach-free paper products
- Re-use wrapping paper and envelopes
- Purchase only items that you need
- Collect and donate packing peanuts to local sipping store
- Purchase rechargeable batteries
- Recycle motor oil, ink cartridges, and tires (contact your local municipality for more information)
- Conserve water (contact your local municipality for more information)
- Plant indigenous species
- Plant a butterfly garden
- Let leaf litter and understory plant and vegetative matter remain during late fall and winter to improve the number and health, and variety of native insect species the following spring
- Establish plant and wildlife corridors on individual homesites and coordinate with neighbors to establish larger corridors
- Plant a variety of native tress and understory plants to increase food and shelter options for insects, birds and other wildlife
- Do not cut back flower stems, many are used by insects (including bees) which help support the foundation of the food chain
- Enhance habitat for birds and insects that feed on unwanted species (i.e., mosquito) by installing a bat and swallow house, and dragonfly habitat
- Limit or eliminate pesticide and herbicide application and use
- Limit fertilizer use
- Practice regenerative farming and garden techniques
- Consider collecting rainwater from roofs to water plants and vegetation during dry periods
- Join a local garden club
- Plant additional flower gardens or other native species
- Replace grass with native trees, shrubs, and flowering plants

Embracing some or all of the aforementioned listed items into the everyday life of individuals will have a positive impact on the environment. In addition, incorporating the sustainability measures given in the said list will result in significant cost savings for each individual.

18.6   SUMMARY AND CONCLUSION

Sustainability is complex and will require significant effort at the global, national, local, parcel, and individual level. This chapter has outlined numerous sustainability measures and has presented evidence to support why these measures are important. In summary, in order to support a human population on Earth we must not consume and pollute at current levels. In order to achieve a sustainable living, vibrant, and diverse world instead of a world without humans, we need to diligently work cooperatively toward living in productive harmony with nature instead of trying to change nature.

Questions and Exercises for Discussion

1. Of the items listed in Section 18.5, which have you already implemented?
2. Do you have suggestions for items to be added to those listed in Section 18.5?
3. Now that you have read Chapters 16, 17, and 18 and if you were employed by a municipality, list five ways you would implement to evaluate the need to improve or establish a sustainability program and what areas would be your first focus – air, water, land, or living organisms and why?

REFERENCES

Darke, R. and Tallamy, D. 2014. *The Living Landscape.* Timber Press, Portland, Oregon. 392p.

Humenik, F., Smolen, M., and Dressing, S. 1987. *Pollution from Nonpoint Sources Environmental Science and Technology.* Vol. 23. pp. 737–742.

Murray, K. S. and Rogers, D. T. 1999. Groundwater Vulnerability, Brownfield Redevelopment and Land Use Planning. *Journal of Environmental Planning and Management.* Vol. 42. no. 6. pp. 801–810.

National Association of Home Builders. 2021. Typical America Subdivisions. https://www.nahbclassic.org (accessed April 19, 2021).

National Geographic Society. 2021. Energy Efficiency of Modes of Transportation. https://www.nationalgeographic.com/carbon-footprint-transportation-efficiency (accessed April 19, 2021).

Rogers, E. M. 1995. *Diffusion of Innovations,* 4th ed. The Free Press: NY.

Rogers, D. T. 1992. The Importance of Site Observation and Followup Environmental Site Assessments – A Case Study. Proceedings of the National Ground Water Association Phase I ESA Conference. Orlando, FL. pp. 218–227.

Rogers, D. T. 1996. *Environmental Geology of Metropolitan Detroit.* Clayton Environmental Consultants: Novi, MI.

Rogers, D. T. 2002. The Development and Significance of a Geologic Sensitivity Map of the Rouge River Watershed in Southeastern Michigan, USA. In; Bobrowsky, P. T. editor. *Geoenvironmental Mapping: Methods, Theory, and Practice.* A. A. Balkema Publishers: The Netherlands. pp. 295–319.

Rogers, D. T., Kaufman, M. M. and Murray, K. S. 2006. Improving Environmental Risk Management Through Historical Impact Assessments. *Journal of Air and Waste Management.* Vol. 56. pp. 816–823.

Rogers, D. T., Murray, K. S. and Kaufman, M. M. 2007. Assessment of Groundwater Contaminant Vulnerability in an Urban Watershed in Southeast Michigan, USA. In: Howard, K. W. F. editor. *Urban Groundwater – Meeting the Challenge.* Taylor & Francis: London, England.

Rogers, D. T. 2011. Why Geology and Chemistry Matter in Forming the Foundation in Sustaining Urban Areas. *American Institute of Professional Geologists.* Vol. 49. no. 2. Denver, Colorado.

Rogers, D. T. 2014. Scientists Call for a Renewed Emphasis on Urban Geology. *American Geophysical Union.* Earth and Space News. Eos., Vol 95. no. 47. pp. 431–432.

Rogers, D. T., Kaufman, M. M., and K. S. Murray. 2016. Development of an Environmental Risk Management and Sustainability Model Using Chemistry, Geology, and Human Activity, and Response Inputs. *35th International Geological Congress.* Paper No. 2090. International Union of Geological Sciences: Cape Town. South Africa.

Rogers, D. T. 2018. Derivation of a Comprehensive Environmental Risk Model for Urban Groundwater Protection. *International Association of Hydrogeologists Congress.* Vol. 1 p. 879. Daejeon, Korea.

Rogers, D. T. 2019a. *Environmental Compliance and Sustainability: Global Perspectives and Challenges.* CRC Press: Boca Raton, FL. 583p.

Rogers, D. T. 2019b. Key Policy Changes Required to Reduce Environmental Degradation of the United States and Beyond. *International Association of Hydrogeologists 46thCongress.* Vol. 1. pp. 529. Malaga, Spain.

Rogers, D. T. 2020a. Geological- and Chemical-Based Environmental Risk Factor Sustainability Model. *European Journal of Sustainable Development.* Vol. 9. no. 4. pp. 303–316.

Rogers, D. T. 2020b. *Urban Watersheds; Geology, Contamination, Environmental Regulations, and Sustainability*. CRC Press: Boca Raton, FL. 606p.

Rogers, D. T. 2021. Empirical Results of a Geological- and Anthropogenic Contaminant-Based Environmental Risk Sustainability Model. Geological Society of America. Annual Meeting. Paper No. GSA-8262. Portland, Oregon.

Schueler T. R. 1994. The importance of imperviousness. *Watershed Protection Techniques*. Vol. 1. pp. 1–11

Smolinske, S. and Kaufman, M. M. 2007. *Consumer Perception of Household Hazardous Materials, Clinical Toxicology*. Vol. 45. pp. 1–4.

Tallamy, D. W. 2019. *Nature's Best Hope*. Timber Press: Portland, Oregon. 254p.

United Nations. 2021. What Is Sustainable Development. https://www.un.org/sustainabledevelopment (accessed April 2021).

United States Census Bureau. 2021a. Census Bureau Average Population Per Household. https://www.census.gov/population/socdemo/hh-fam/tabHH-6.pd (accessed April 19, 2021).

United States Census Bureau. 2021b. Median and Average Square Feet of floor Area in New Single-family Homes. https://www.census.gov/const/c25ann/sftotalmedavsqft.pdf (accessed April 19, 2021).

United States Census Bureau. 2021c. Average One-Way Commuting Time and Distance. (https://www.census.gov/library/visualization/interactive/travel-time.html (accessed April 2021).

United States Department of Energy. 2021. Energy Efficiency and Renewable Energy. *Alternative Fuels Data Center*. https://www.afdc.energy.gov.data (accessed April 19, 2021).

United States Department of Housing and Urban Development. 2021. Median and Average Square Feet of floor Area in New Single-family Homes. *United States Census Bureau*. https://www.census.gov/const/c25ann/sftotalmedavsqft.pdf (accessed April 19, 2021).

United States Environmental Protection Agency (USEPA). 1990. Pollution Prevention Act of 1990. *Code of Federal Regulation 42 CFR, Chapter 133*. Washington, D. C.

United States Environmental Protection Agency (USEPA). 2002. National Water Quality Inventory: Report to Congress: 2002 Recycling Report. *USEPA. EPA-841-R-07-001*. Washington, D. C.

United States Environmental Protection Agency (USEPA). 2003. National Management Measures to Control Nonpoint Source Pollution from Agriculture. *USEPA. EPA 841-B-03-004*. Washington, D. C.

United States Environmental Protection Agency (USEPA). 2021a. Pollution Prevention. http://www.epa.gov/p2 (accessed April 19, 2021).

United States Environmental Protection Agency (USEPA). 2021b. United States Environmental Protection Agency QA Glossary. https://www.epa.gov/emap/archive-emap/web/html (accessed April 19, 2021).

United States Environmental Protection Agency (USEPA). 2021c. SAGE 2.1: Solvent Alternatives Guide. https://cfpub.epa.gov/si/si_public_record_report.cfm?Lab=NRMRL&dirEntryId=115684 (accessed April 19, 2021).

United States Environmental Protection Agency (USEPA). 2021d. Nonpoint Source Pollution. https://19january2017snapshot.epa.gov/nps_.html. (A).

United States Environmental Protection Agency (USEPA). 2021e. Stormwater best management practices. http://www.epa.gov/npdes/stormwater/menuofbmps, (accessed April 19, 2021).

United States Environmental Protection Agency (USEPA). 2021f. What is Sustainability? https://www.epa.gov/sustainability (accessed April 20, 2021).

19 Achieving and Maintaining Compliance with Environmental Regulations

19.1 INTRODUCTION

The purpose of this chapter is to provide advice and guidance in achieving compliance with environmental regulations and beyond. The significant lessons or Principles of Environmental Compliance in this chapter include:

1. It all begins with conducting an environmental audit.
2. Being in compliance with environmental regulations does have significant sustainability implications.
3. Work yourself out of environmental regulations.
4. There is little we can control.
5. Limit chemical use when possible.
6. When possible, eliminate chemicals that are very toxic or have a high chemical risk factor.
7. Fully Understand Environmental Permits.
8. Never Accept a Permit Term that the Facility Can't Achieve.
9. Always be accurate and truthful.
10. When in Doubt, It's Usually Always Better to Report a Spill or Other Incident Where Reporting May be Required.
11. The Importance of Waste Characterization and Points of Generation.
12. Any Operational Changes May Require Notice and Permit Changes or a New Permit.
13. Give yourself enough time for collecting additional compliance samples in case a data quality issue arises.
14. Keep well organized and communicate with management and employees regularly.
15. Housekeeping.
16. Signage.
17. Conduct spill drills.
18. Prepare a list of onsite chemicals with corresponding reportable quantities (RQ's) and recommended cleanup procedures.
19. Conduct regular inspections.
20. Proper preparation prevents poor performance.
21. In many instances, proving a negative is required when evaluating a release.
22. When a question arises, consult management, in-house environmental counsel, outside environmental counsel, or the regulatory authority, as appropriate.
23. Lastly, remember that "no matter where you are, it's all the same."

The central focus of this chapter will be conducting compliance audits. Conducting a compliance audit also provides an opportunity to evaluate a facility's environmental health and to identify sustainability opportunities.

DOI: 10.1201/9781003175810-19

19.2   COMPLIANCE OVERVIEW

Environmental compliance is sometimes perceived as just another set of rules and is simple. Just follow the rules. However, the Earth is dynamic. There is change, ever getting more complex as more information and science is discovered, learned, and shared. The weather changes constantly and is perhaps the best example of how to imagine environmental management and risk. It also changes, as does everything. Everything obeys the second law of thermodynamics, namely entropy, in that nature tends to become more complex and disordered as time marches on.

Understanding the natural setting of our urban and industrial areas, knowing the chemicals that are used and how they might cause harm to the environment and humans if they are released, what alternative chemicals are available, how to reduce chemical and energy use, preventing chemical releases to the environment, quantifying the costs involved with cleaning up the environment once a release occurs, and developing environmental stewardship are all subjects that should be taken into account when evaluating the environmental health of a facility and it all begins with first conducting an environmental audit.

19.3   USEPA AUDIT POLICY

In 1986, the United States Environmental Protection Agency (USEPA, published the Environmental Auditing Policy Statement (USEPA 1986). USEPA published the policy statement to encourage the use of conducting "Self Assessments" by the regulated community to help, achieve, and maintain compliance with environmental laws and regulations, and to identify and correct unregulated hazards. In addition, USEPA defined environmental audits as (USEPA 1986):

> *a systematic, documented, periodic, and objective review of facility operations and practices related to meeting environmental requirements*

The policy also identified several objectives for environmental audits that included (USEPA 1986):

1. Verifying compliance with environmental requirements
2. Evaluating the effectiveness of in-place environmental management systems
3. Assessing risks from regulated and unregulated materials and practices

USEPA has published guidelines or protocols for conducting environmental audits in subject areas that include (United States Environmental Protections Agency 2021a):

- Comprehensive Environmental Response, Compensation and Liability Act (CERCLA)
- Clean Water Act (CWA)
- Stormwater under the CWA
- Wastewater under the CWA
- Federal Insecticide, Fungicide, and Rodenticide Act (FIFRA)
- Emergency Planning and Community Right-to-Know (EPCRA)
- Resource Conservation and Recovery Act (RCRA)
- Safe Drinking Water Act (SDWA)
- Toxic Substance and Control Act (TSCA)

The audit protocols in the given list are a checklist for conducting environmental audits on each subject listed and are 1,387 pages in length. Audit protocols are intended to assist the regulated community in developing programs at individual facilities to evaluate their compliance with environmental requirements under federal law. USEPA states that the protocols are intended solely as guidance in this effort. In addition, USEPA states that the regulated community's legal obligations are determined by the terms of applicable environmental facility-specific permits, as well as underlying statutes, and applicable federal, state, and local law (USEPA 2021a and 2021b). Therefore, when examining the audit protocols, the facility must also take into account state, county, municipal, or local regulations that apply in order to fully understand requirements that the facility must comply with before determining whether the facility is in full compliance. This again may require the services of in-house counsel, a qualified and experience environmental attorney, an experience and qualified environmental consultant, and/or all of the aforementioned list. Additionally, contacting the regulatory agency may be justified to clarify certain issues that may arise.

In 1995, USEPA published "Incentives for Self-Policing: Discovery, Disclosure, Correction, and Prevention of Violations," which both reaffirmed and expanded the 1986 policy (USEPA 2021b). USEPA again revised the policy in 2000 and in 2005 (USEPA 2021a). Under USEPA's environmental audit policy gravity-based penalties for violations of USEPA-administered statutes are reduced or completely eliminated if the violations are voluntarily discovered, promptly disclosed to USEPA, and meet a number of other specified conditions. Those other specified conditions include the following (USEPA 2021c):

1. The violation must be systematically discovered, either through (a) an environmental audit, or (b) a compliance management system reflecting the company's actions and methods in preventing, detecting, and correcting violations.
2. The violation must be discovered voluntarily and not by legally mandated sampling or monitoring required by statute, regulation, permit, judicial or administrative order, or consent agreement.
3. The company must fully disclose the specific violation in writing to the appropriate regulatory authority or USEPA within 21 days or within a shorter period of time if necessary (i.e., such as a spill exceeding the corresponding reportable quality in which case the reporting requirement is within 24 hours) after the company discovered that the violation has, or may have occurred.
4. The company must discover and disclose the violation to USEPA prior to (a) the commencement of a federal, state, or local agency inspection or investigation, ot the issuance by the agency of an information request to the company, (b) notice of a citizen suit, (c) the filing of a compliant by a third party, (d) the reporting of the violation to USEPA (or other governmental agency) by a "whistleblower" employee, or (e) imminent discovery of the violation by a regulatory agency.
5. The company must correct the violation within 60 calendar days from the date of discovery, certify as such in writing, and take appropriate action or measures as determined by USEPA to remedy any environmental or human harm due to the violation.
6. The company must agree in writing to take steps to prevent a recurrence of the violation.
7. The specific violation (or closely related violation) cannot have occurred previously within the past three years at the same facility, or within the past five years as part of a pattern of multiple facilities owned or operated by the same entity.
8. The violation cannot be one that (a) resulted in serious harm, or may have presented an imminent and substantial endangerment, to human health or the environment, or (b) violates the specific terms of any judicial or administrative order, or consent agreement.
9. The company must cooperate with USEPA and provide such information as is necessary and requested by USEPA to determine applicability of the Policy.

USEPA has broad statutory authority to request relevant information on the environmental compliance status of regulated entities. However, USEPA believes routine requests for audit reports by the agency could inhibit auditing in the long run, decreasing both the quantity and quality of audits conducted. Therefore, as a matter of policy USEPA will not routinely request environmental audit reports (USEPA 1986).

USEPA's authority to request an audit report, or relevant portions thereof, will be exercised on a case-by-case basis where the Agency determines it is needed to accomplish a statutory mission, or where the Government deems it to be material to a criminal investigation. USEPA expects such requests to be limited, most likely focused on particular information needs rather than the entire report, and usually made where the information needed cannot be obtained from monitoring, reporting, or other data otherwise available to the Agency. Examples would likely include situations where; a company has placed its management practices at issue by raising them as a defense; or state of mind or intent are a relevant element of inquiry, such as during a criminal investigation. This list is illustrative rather than exhaustive, since there doubtless will be other situations, not subject to prediction, in which audit reports rather than information may be required (USEPA 1986).

USEPA acknowledges regulated entities' need to self-evaluate environmental performance with some measure of privacy and encourages such activity. However, audit reports may not shield monitoring, compliance, or other information that would otherwise be reportable and/or accessible to USEPA, even if there is no explicit "requirement" to generate the data. Thus this policy does not alter regulated entities' existing or future obligations to monitor, record, or report information required under environmental statutes, regulations, or permits, or to allow USEPA access to that information. Nor does this policy alter USEPA's authority to request and receive any relevant information, including that contained in audit reports, under various environmental statutes, such as Clean Water Act section 308, and Clean Air Act sections 114 and 208, or in other administrative or judicial proceedings (USEPA 1986).

Regulated entities also should be aware that certain audit findings may by law have to be reported to governmental agencies. However, in addition to any such requirements, USEPA encourages regulated entities to notify appropriate State or Federal officials of findings which suggest significant environmental or public health risks, even when not specifically required to do so (USEPA 1986). Questions and reporting requirements related to this policy and conducting environmental audits should be directed to in-house counsel or to a qualified and experienced environmental attorney.

19.4   THE AUDIT PROCESS

The first step in achieving environmental compliance is to conduct a compliance audit. Conducting a compliance audit will answer some very important questions concerning any operation. There are two basic questions that must be answered right at the beginning to determine if a compliance audit is advisable and include:

- Is the facility subject to any environmental regulations
- Does the facility require an environmentally-related permit for any of its operations

Today, it is rare that a facility doesn't know the answer to those two basic questions, but for sake of argument, let's assume that they do not know and want to do the right thing and evaluate whether they are subject to environmental regulations. Under this scenario, the following questions apply:

- Does the facility store or use any hazardous substances as defined by USEPA in Chapter 6 under RCRA? If the answer is no, then does the facility store of use any hazardous substances as defined by State or local regulations? This question could be important especially in some states that regulate more strictly than the federal regulations, such as California.
- Does the facility operate any equipment that generates any exhaust or has any air emissions?
- Does the facility use or discharge any water other than sanitary?
- Does the facility generate any solid waste?

If the answer is yes to any of the given questions, the facility is likely subject to environmental regulations and may require one or more environmental permits. If questions persist, involvement of company management and in-house counsel, a qualified and experience environmental attorney, an experience and qualified environmental consultant, and/or all of the aforementioned may be necessary. Additionally, contacting the regulatory agency may be justified to clarify specific questions.

19.5 CONDUCTING AN ENVIRONMENTAL AUDIT

Conducting an environmental audit is the first step on a long and winding road toward environmental compliance and ultimately attaining sustainability and environmental stewardship.

Most large companies with several manufacturing locations conduct environmental audits on a regular basis. Some choose to conduct environmental audits more often than others, but most seem to choose conducting an environmental audit every year or every two years. Some companies conduct environmental audits with in-house staff to retain institutional knowledge. Some choose to retain an environmental consulting firm to conduct environmental audits because it's conducted by an independent third party. Some environmental audits are conducted without any prior notice and some are conducted with prior notice. Whichever path is chosen, environmental audits are usually always conduct at the request of counsel.

19.5.1 CATEGORIES OF FINDINGS

When conducting an environmental audit, findings are typically weighted by priority, severity, or category and typically include:

1. Non-compliance
2. Best management practice (BMP)
3. Observation

An issue of non-compliance is the highest priority of any finding and would be considered a regulatory violation. USEPA (2021d) defines an issue of **environmental non-compliance** as not conforming with environmental law, regulations, standards, or other requirements such as an environmental permit. An example of a non-compliance issue would be an improper or unlabeled drum of hazardous waste.

A **best management practice** was first defined by USEPA in 1974 (USEPA 2021d) in the clean water act and has since been adopted by the regulated community. Best management practices are defined for environmental purposes as specific practices that are capable of improving, preventing, and/or minimizing the potential of an occurrence or situation that can lead to environmental non-compliance and/or a release of a hazardous substance to the environment. A best management practice issue, when identified, does not mean that a non-compliant situation exists, but if left unattended, could result in a non-compliance issue in the future. Examples of environmental best management practices include (Connecticut Department of Environmental Protection 2021):

- Storing products and wastes indoors
- Clean catch basins on a regular basis
- Ensuring that lids on dumpsters remain closed

An **observation** is an issue that could be improved but does not reach the threshold of a best management practice. For instance, a better method for keeping track of monitoring results, or improving housekeeping are examples of an Observation.

19.5.2 DOCUMENT LIST

If the environmental audit is announced prior the audit date, a request for applicable documents to be available to be reviewed in advance audit will save time. As applicable to the facility being reviewed, the following list of documents should be available for review during the site visit.

1. Current site map
2. Current organization chart
3. Current process flow diagrams depicting: inputs, process units, waste streams, and other outputs
4. List of processes that have been shut down since the last audit
5. List of processes that are expected to start up with the next 12 to 24 months
6. SPCC Plan
7. Spill reports for the last three years
8. All RCRA manifests since the last audit
9. All special waste notifications and shipping documents since the last audit
10. All analytical tests results and determination for special and hazardous waste
11. All analytical tests for each non-hazardous waste stream
12. Contingency plan
13. Personnel training records since the last audit
14. List of arrangements with local agencies
15. NPDES permit and most recent application
16. Discharge monitoring reports and any associated letters of explanation of permit limit exceedances since the last audit
17. Stormwater plan, intent to comply documents, and all monitoring reports and sample data
18. All notices of violation, noncompliance letters, or consent orders including facility's noncompliance explanations and, if in significant noncompliance, the facility's plan to return to compliance
19. Water flow diagrams and water balances
20. Sketches of all wastewater treatment systems within the facility
21. Names of wastewater analytical laboratory, a list of all analytical methods used for wastewater analysis, and the quality assurance/quality control procedures used for wastewater sample collection and analysis
22. Air permit application modifications
23. All construction and operating permits for air sources
24. Copies of all visible emission observations taken by facility personnel or contractors for the past three years
25. Environmental management plan
26. Any other environmental permit to operate (associated with air, land, or water)
27. Waste shipment summary for the past two years
28. Toxic Release Inventory Reports (Form R) for the past two years

For facilities not in the United States, and as applicable to the specific country, province or state, additional documents to be made available to the reviewers include:

1. Noise studies and noise compliance (e.g., Mexico, China etc.)
2. Environmental Impact Assessment (EIA)
3. Operating Permits
4. European Union Registration, Evaluation, Authorization, and Restriction of Chemicals (REACH) certification status for those locations within the European Union.

19.5.3 Opening Meeting

The following sections describe the environmental review process in greater detail to ensure that a more complete review of the facility compliance status is conducted. In addition, this section also provides guidance for identifying and establishing pollution prevention opportunities to further minimize potential environmental impact in the future.

To improve the efficiency of conducting the environmental review, prior notice to the facility should be made so that all necessary information can be made readily available and appropriate plant personnel are in attendance for the review. Some companies may choose to conduct unannounced audits to emulate an agency inspection, which often are unannounced. Which ever method is used, it's recommended that the facility be reminded that an agency inspection can occur any time on any day and that proper preparation prevents poor performance. Therefore, organization is key and backups systems should be in place in case the facility the facility person in charge of compliance is not present.

A list of documents is provided in Section 6.0 that should be available for review, as applicable to the facility.

An opening meeting should be scheduled in advance of the environmental audit. The purpose of the opening meeting is to:

1. Inform appropriate facility management of the purpose and objectives of the audit.
2. Coordinate logistics and review procedures.
3. Remind facility management with respect to:
   - Internal policy pertaining to environmental audits
   - Report privilege and distribution
   - Procedures to follow during agency inspections and emergency and spill events
   - Periodically review environmental permits
   - Pollution prevention initiatives
   - Preparation for environmental compliance requires an awareness and daily attention to such matters
4. Ensure that priority items are presented to facility management quickly so that appropriate corrective actions can be initiated promptly
5. Remind management ad all personnel that compliance with environmental regulations is not optional
6. Remind management that the process for identified non-compliance items is that they must be addressed immediately and with the utmost priority
7. Remind management that controls must be evaluated and updated as necessary if non-compliance issues are identified to minimize the possibility of a re-occurrence

19.5.4 Environmental Audit Checklist

It is always very helpful to review and become familiarized with the overall facility layout and manufacturing process of the facility before beginning the actual audit begins. This is usually accomplished by reviewing a diagram or figure of the general facility layout and manufacturing process flow and then reviewing a diagram or figure of the facility that identifies (1) locations of all air emission sources, (2) solid and hazardous waste stream sources and points of generation, and (3) water and wastewater flow through the facility, and (4) areas where hazardous substances enter the facility, where they are stored and used within the facility, and where they are stored as wastes before offsite transportation for disposal. The items listed in the checklist in Sections 4.3.2.1 through Section 4.3.2.10 are intended to be used as a guide in conducting the environmental audit. Some of the sections or listed items included in the checklist may not be applicable to a particular facility. Therefore, the checklist should be facility specific and should be used as applicable.

19.5.5 GENERAL AIR BACKGROUND

Regulatory requirements for air vary significantly from facility to facility. Each facility's air emissions are not standardized under similar regulations. Therefore, a country's Federal, State, and Local guidelines that affect each facility should be reviewed independently. In addition, there are specific guidelines under Maximum Achievable Control Technology or MACT that should also be evaluated.

The following items should be considered first before conducting the detailed review of air to become familiarized with the current status of the facility with respect to air status and compliance:

- Evaluate whether the facility is located in an attainment or non-attainment area for specific compounds
- Evaluate whether the facility has any air emissions from industrial sources
- Review whether the facility is a minor, synthetic minor, or major emission source of criteria pollutants and of Hazardous Air Pollutants (HAPS)
- Evaluate whether the facility has emission units that are registered by the State or Local Municipality
- Evaluate whether any changes or modifications to existing equipment have been made or new equipment has been installed that may have increased production and hence may have increased air emissions from any source. This may indicate that a permit or permit modification may have been required.
- Each facility should also have a keen awareness that any process or manufacturing changes or installation of new equipment may have a profound effect on the types, amounts, and chemical and physical characteristics of emissions that are generated which may require a permit or permit modification.
- Review any recent correspondence from or to any regulatory agency
- Review the nature and focus of any regulatory inspections that have been conducted since the last environmental audit.
- Review and evaluate whether the facility has received any NOV's since the last environmental review
- Review and evaluate any corrective actions that have been undertaken to address any NOVs

19.5.5.1 Air Operating Permit
- Review air operating permit, such as Title V (major source), FESOP, SOP
- Review emission source list & flow diagram
- Review emission limits
- Review testing requirements
- Review monitoring requirements
- Review Air Pollution Control (APC) Operation & Maintenance (O & M) plans
- Review Compliance Assurance Monitoring (CAM) plans
- Inspect examples of APC maintenance records
- Evaluate any changes in regulatory status since operating permit issued

19.5.5.2 Construction Permits
- Review permit of any effected emission units
- Discuss any new or modified APC units
- Review permit limits
- Review testing requirements
- Review monitoring requirements
- Review request to modify operating permit

- If NSR/PSD was triggered; review BACT analysis & air modeling
- Review any recent changes to facility which were not covered by a construction permit

19.5.5.3   Air Compliance Documentation
- Discuss recent stack testing with respect to frequency, analytical parameters, analytical methods, results, trends
- Review annual emission inventory
- Review monitoring reports
- Review annual compliance certification and derivation reports
- Obtain copies of any NOV since last review
- Review any agency inspection reports
- Discuss any other significant agency correspondence

19.5.5.4   Air Toxics/Hazardous Air Pollutants (HAPs)
- Review Potential-To-Emit of HAP emissions
- Determine Iron & Steel Foundry MACT Rule Applicability
- Discuss Iron & Steel Foundry Area Source Rule Applicability (Rule under development)
- Review State Air Toxic Program Applicability

19.5.5.5   Compliance with Ozone Depleted Substance (ODS) Requirements
- Review facility's procedures for repair and replacement of ODS containing equipment
- Review list of all ODS containing equipment
- Inspect copies of technician certifications (in-house and/or outside contractor)
- Inspect documentation associated with units containing more than 50 pounds of ODS
- Ensure all ODS containing units are properly evacuated prior to disposal

19.5.5.6   Asbestos Compliance
- Review Asbestos Management Plan, if available
- Review list of asbestos contained and presumed asbestos contained materials (PACM)
- Review internal asbestos inspections
- Review asbestos training records
- Review asbestos abatement activities and documentation since last review, if any
- Review provisions for managing and sampling of PACM

19.5.6   General Water Background

Similar to air regulations, the regulatory requirements for water vary significantly from facility to facility. Each facility's water discharge requirements and permits are not standardized under similar regulations. Therefore, Federal, State, and Local guidelines that affect each facility should be reviewed independently. The following items should be considered first before conducting the detailed review of water to become familiarized with the current status of the facility with respect to water status and compliance:

- Evaluate the volume of water that the facility uses and the source of the water (i.e., onsite water well, city, or other).
- Evaluate the type, volume, and location of the facility's water discharges.
- Evaluate whether the facility currently uses or historically operated a septic system
- Evaluate whether the facility has process waste water discharges.
- Evaluate the overall operation of any wastewater treatment system, especially maintenance concerns and operational concerns.
- Check to make sure that the wastewater system operator is licensed, if necessary.

- Evaluate whether the facility is a Categorical discharger and if so, which one and evaluate any special requirements.
- Evaluate whether the facility has a wastewater discharge permit.
- Review any recent correspondence with the regulatory agency.
- Review the nature and focus of any regulatory inspections have been conducted since the last environmental review.
- Review and evaluate whether the facility received any NOV's since the last environmental review.
- Review and evaluate any corrective actions that have been undertaken to address any NOVs.

19.5.6.1   Water Source(s)
- Evaluate the volume of water that the facility uses on a daily basis.
- Record the sources of water that the facility receives (i.e., onsite water well, city, surface water, or other).
- Evaluate whether the facility has or had an onsite water well.
- Evaluate and review the facility water well permit.
- Evaluate whether the facility has a water usage permit.
- Evaluate whether the facility is located in an area where water rights and consumption is a regulated resource
- Evaluate closure requirements if the facility has an onsite water well that is not currently in use.
- Obtain and review records of any onsite water wells installed since last audit or for historical wells if not previously conducted (i.e., materials of construction, diameter, installation date, depth, strata, pumping rates and capacity, onsite storage capacity, treatment requirements and monitoring).
- Evaluate whether an onsite water source is used as potable water.
- If an onsite water source is used as potable, review all applicable sampling results to evaluate compliance with the Safe Drinking Water Act Maximum Contaminant Levels (MCL's) and other requirements.
- Evaluate how many people the onsite water well serves.
- Review regulatory reporting documents for supplying drinking water, if necessary.

19.5.6.2   Waste Water Streams
- Review list of process and none process wastewater discharges
- Review water/wastewater flow diagrams
- Review all wastewater sources, volume of discharges, and history of discharges
- Review all available charts, graphs, and drawings of wastewater discharges

19.5.6.3   Direct Discharges
- Review list of direct discharges
- Evaluate whether the facility currently uses or historically operated a septic system
- Review septic system permit, if applicable
  Permit Number
  Discharge Limits
  Monitoring requirements
- Review NPDES permit:
  Permit Number
  Discharge Limits
  Monitoring requirements
- Review the most recent (usually 1 year) DMRs (Discharge Monitoring Reports):
  Facility's compliance status

Reports of exceedances and violations
Timeliness
Fees

19.5.6.4   Indirect Discharges (to Sanitary Sewer or POTW)

- Review discharge status of facility
  Industrial User Discharge
  Categorical Discharger, for example:
  - 40CFR 464 Subpart A – Aluminum Casting Subcategory
  - 40CFR 464 Subpart C – Ferrous Casting Subcategory
  - 40CFR 433- Metal finishing
  - Other
- Review Sewer Permit:
  Permit Number:
  Discharge Limits
  Monitoring requirements
- Review the sewer monitoring reports covering the most recent 12 months:
  Identify party responsible for conducting sampling
  Facility's compliance status
  Reports of exceedances and violations
  Timeliness
  Fees

19.5.6.5   Wastewater Monitoring

- Review list of sampling points
- Review sampling procedures
- Review analytical methods
- Review flow meter and sampling equipment calibration records
- Evaluate outside laboratory credentials
- Evaluate whether qualified and authorized personnel are signing regulatory reports
- Review record keeping procedures and organization

19.5.6.6   Stormwater Discharge Associated with Industrial Activity

- Review Stormwater Permit
  Permit Number:
  Type of permit (i.e., General Permit, Individual Permit)
  Discharge limits
- Review Stormwater Pollution Prevention Plan
  Date of the most recent revision:
  Identification and Implementation of Best Management Practices
  Periodic Training
- Review Stormwater monitoring, if required
  Compliance status
  Sampling reports (i.e., DMR)
  NOVs
- Review sampling techniques
- Review laboratory methods
- Review record keeping procedures and organization
- Evaluate whether training records are up to date
- Evaluate whether qualified and authorized personnel are signing regulatory reports
- Review non-stormwater discharge assessment forms

19.5.7 Hazardous Waste

Solid and Hazardous Waste regulations are generally considered the most complex and also vary significantly from facility to facility. To assist in conducting the solid and hazardous waste portion of the environmental review process, this section is divided into several subsections that include (1) hazardous waste, (2) solid waste, (3) universal waste, (4) used oil, (5) company owned or operated landfills, (6) recycling, and (7) pollution prevention plans.

As an additional item to be noted, each facility's solid and hazardous waste requirements and permits may not be standardized under similar regulations. Therefore, Federal, State, and Local guidelines that affect each facility should be reviewed independently.

19.5.7.1 General Background Information

The following items should be considered first before conducting the detail review of solid and hazardous waste to become familiarized with the current status of the facility with respect to status and compliance:

- Evaluate the types of solid waste that the facility generates
- Evaluate whether the facility generates hazardous waste
- Evaluate and record the facility EPA ID number
- Evaluate the current and previous facility generator status (e.g., permit, LQG, SQG, etc.)
- Evaluate whether the facility has a permit to store waste onsite
- Review requirements (40 CFR, 264 and 265 [interim requirements]) and any State and Local requirement if facility is permitted to store wastes onsite
- Evaluate whether the Boiler & Furnace Rule, 40 CFR 266 Subpart H, applies
- Verify that the correct EPA ID number is recorded on all documentation (i.e., manifests)
- Evaluate whether any processes have changed since the last audit that could influence the chemical or physical characteristics of any waste stream. If so, has the associated generated waste been re-characterized. If not, why?
- Review any correspondence from or to any regulatory agency
- Review whether any solid waste inspections have been conducted since the last environmental audit.
- Review and evaluate whether the facility has received any solid waste NOV's since the last environmental audit
- Review and evaluate any corrective actions that have been undertaken to address any NOVs

19.5.7.2 Point of Generation Analysis

- Evaluate and verify that each hazardous waste stream is hazardous by either (1) listing in regulations, (2) laboratory analysis, or (3) knowledge of materials and processes used
- Evaluate whether waste needs to be reclassified or characterized
  1. Evaluate whether any process or manufacturing changes occurred that may impact the classification or chemical characterization of the waste (i.e., installation of new machinery)
  2. Evaluate whether any other changes have occurred that may impact the characterization of the waste (i.e., differences in parent material or re-formulation of bulk product inputs).
  3. Check to ensure that each hazardous waste stream is re-characterized at least every three years or sooner if there is any process changes that could influence the chemistry of the waste material.
- Review waste stream source diagram

19.5.7.3   Inventory of Hazardous Waste

- Ensure inventory of hazardous waste is correct
- Evaluate the appropriateness of generator status
- Check to make sure volumes and weights of each hazardous waste stream are correct and are verified

19.5.7.4   90-Day Storage

- Review inspection documentation
- Verify that no container has accumulated waste for more that 90 days, unless a 30 day extension was granted
- Verify each container and tank is labeled the words HAZARDOUS WASTE and the start accumulation date and applicable waste codes

19.5.7.5   Treatment and Disposal

- Verify wastes is hauled by transporters with valid EPA identification numbers
- Verify wastes is taken to disposal facilities with valid hazardous waste permits
- Review annual waste shipment summaries
- Check to make sure the facility has a tracking procedure to ensure that appropriate actions are undertaken in case a waste manifest is not returned to the facility with the required timeframe.

19.5.7.6   Biennial Reports

- Verify that the biennial report was complete and submitted by March 1
- Verify that copies of the report are kept for three years

19.5.7.7   Manifests

- Inspect for documentation accuracy
- Verify that manifests are used when shipping waste off-site
- Verify that exception reports are filed when a copy of the manifest is not received within the required timeframe of the waste being accepted by the initial transporter (45 days for SQG and 30 days for LQG)
- Check to make sure the facility has a tracking procedure to ensure that appropriate actions are undertaken if a waste manifest is not returned to the facility with the required timeframe.
- Verify that manifests are kept for three years
- Verify that manifests are signed and dated correctly and clearly

19.5.7.8   Storage Areas

- Ensure internal communications system available
- Ensure telephone or two way radio to summon emergency assistance
- Check for portable fire extinguishers
- Inspect spill control equipment
- Inspect decontamination equipment
- Check for fire hydrants
- Some of the earlier-mentioned points may not be necessary depending on the waste
- Determine if equipment is tested and maintained
- Verify sufficient aisle space is maintained
- Verify arrangements with local fire, police and emergency response teams if necessary

19.5.7.9   Personnel Training

- Verify that personnel complete classroom instruction
- Training must include contingency plan implementation, if a LQG

- Training must be completed within six months of employment/assignment
- Verify that annual review training is provided
- Verify that employees do not work unsupervised until training is completed
- Verify specifically that waste storage area managers and hazardous waste handlers have been trained

19.5.7.10   Training Records

- Include job title and description for each employee by name
- Written description of how much training each position will receive
- Documentation of training received by name
- Determine if records on former employees are retained for three years
- Determine if records on current employees are maintained

19.5.7.11   Contingency Plans

- Verify that the contingency plan is designed to minimize hazards from fires, explosions, or releases, if a LQG
- Must describe actions to be taken in an emergency
- Must describe arrangements made local agencies as appropriate
- Must include the name, address and phone number of the emergency coordinator and any alternates
- Must include a list of any emergency equipment, its location and description
- Must include an evacuation plan if required
- Ensure that communications systems are satisfactory and are defined in case of an emergency
- Verify that revisions are maintained and submitted to local emergency services where required
- Verify that adequate isle space is available for unobstructed movement of emergency equipment in case of an emergency
- Verify that the plan is routinely reviewed and updated when regulations change, the plan fails, the facility changes, the emergency coordinators change or the emergency equipment changes
- Review personnel training records to ensure that facility personnel are able to respond effectively during an emergency. This should include but is not limited to (1) inspecting and maintaining equipment, (2) shut off systems, (3) communication and alarms, (4) fire and explosions, (5) contamination, and (6) shut down procedures.

19.5.7.12   Emergency Coordinators

- Verify that at all times there is at least one employee at the facility or on call with responsibility for coordinating emergency response measures
- Verify that he/she is familiar with the facility and the contingency plan
- Verify that he/she has the authority to commit resources to carry out the contingency plan

19.5.7.13   Containers

- Verify that containers are not leaking, bulging, rusting, damaged or dented
- Verify that the storage area is inspected at least weekly
- Verify that empty containers have less the regulated material left in them
- Verify that containers are compatible with the waste
- Verify that containers are kept closed except when adding or removing waste
- Verify that containers are properly labeled and legible

19.5.7.14 Labels

- Verify that every waste container is labeled
- Verify that every waste container is labeled with appropriate accumulation dates, waste identification, and other required information
- Verify whether other labeling requirements apply

19.5.7.15 Satellite Accumulation

- Verify that the satellite accumulation point is at or near the point of generation
- Verify that the containers are in good condition, compatible with the waste and kept closed except when adding or removing waste
- Verify that the containers are marked HAZARDOUS WASTE or other identified marking, as required
- Verify that when waste is accumulated in excess of quantity limitations that the date is marked on the container and within three days the container is moved to the 90 day storage area
- Verify proper containment is provided

19.5.7.16 Restricted Wastes

- Determine by analysis or process knowledge if wastes are restricted from land disposal
- Verify that all notifications and certifications has been made
- Verify that records are kept for three years

19.5.7.17 Record Keeping

- Ensure that records are organized and complete
- Evaluate back-up record keeping systems
- Ensure that al least 3 years of written records are available for inspection

19.5.7.18 Agency Inspections

- When was last agency hazardous waste inspection
- What areas of regulations were the focus
- Were any violations discovered
- Was the violation address re the inspection ended
- Did the agency take any photographs
- What triggered the inspection
- What agency conducted the inspection
- Did the agency issue any Notice of Violation
- How were violations addressed
- Were there any fines or other penalties

19.5.8 Solid Waste

19.5.8.1 Non Hazardous Waste Determination

- Verify that proper waste determinations are performed (at a minimum every three years or more every time the process changes that could result in input changes of waste chemical or physical characteristics)
- Verify that non-hazardous wastes are actually non-hazardous. Ask yourself the question, How do I know its non-hazardous? If you don't know, test it.
- Review solid waste flow diagram
- Review list of solid waste streams

19.5.8.2   Point of Generation Analysis

- Evaluate and verify that each waste stream is not hazardous by either (1) listing in regulations, (2) laboratory analysis, or (3) knowledge of materials and processes used
- Evaluate whether waste needs to be reclassified or characterized
    4. Evaluate whether any process or manufacturing changes occurred that may impact the classification or chemical characterization of the waste (i.e., installation of new machinery, new vendor. ect.)
    5. Evaluate whether any other changes have occurred that may impact the characterization of the waste (i.e., differences in parent material or re-formulation of bulk product inputs).
    6. Check to ensure that each hazardous waste stream is re-characterized at least every three years or sooner if there is any process changes that could influence the chemistry of the waste material.
- Review waste stream source diagram

19.5.8.3   Inventory of Solid Waste

- Ensure inventory of hazardous waste is correct
- Evaluate the appropriateness of generator status
- Check to make sure volumes and weights of each hazardous waste stream are correct and are verified

19.5.8.4   Storage

- Verify that materials are properly stored and labeled
- Verify that no applicable storage time frames are exceeded
- Ensure that storage containers are in good condition
- Ensure that wastes are not mixed together

19.5.8.5   Labels

- Verify that every waste container is labeled
- Verify that every waste container is labeled with appropriate accumulation dates, waste identification, and other required information
- Verify whether other labeling requirements apply

19.5.8.6   Transportation

Verify that materials are hauled according to any applicable state & local requirements

19.5.8.7   Disposal

- Verify that the materials are disposed at an approved site
- Review special waste permits or waste disposal authorizations if applicable
- Review current waste profiles

19.5.8.8   Record Keeping

- Ensure that records are organized and complete
- Evaluate back-up record keeping systems
- Ensure that al least 3 years of written records are available for inspection

19.5.8.9   Agency Inspections

- When was last agency waste inspection
- What areas of regulations were the focus
- Were any violations discovered

- Was the violation address re the inspection ended
- Did the agency take any photographs
- What triggered the inspection
- What agency conducted the inspection
- Did the agency issue any Notice of Violation
- How were the violations addressed
- Were there any fines or other penalties

19.5.9   Used Oil

19.5.9.1   Storage
- Verify that the tanks, containers and fill pipes are labeled USED OIL
- Verify that the tanks or containers are in satisfactory condition
- Ensure that all used oil is placed in appropriate containers (i.e., drip pans beneath machines are not an appropriate storage container

19.5.9.2   Labels
- Verify that every waste container is labeled
- Verify that every waste container is labeled with appropriate accumulation dates, waste identification, and other required information
- Verify whether other labeling requirements apply

19.5.9.3   Transportation
- Verify that the used oil is transported only by transporters with a valid U.S. EPA identification number

19.5.9.4   Recycle
- Verify the final destination of the used oil
- Inspect Certificate of recycling

19.5.10   Universal Waste

19.5.10.1   Identification
- Verify batteries, pesticides, mercury-containing equipment, and lamps are being segregated and properly disposed at a licensed facility.

19.5.10.2   Storage
- Verify that waste is not accumulated for more than the allowed time period
- Ensure that the storage container are appropriate and that the lids are closed

19.5.10.3   Labels
- Verify that all containers are properly marked or labaled

19.5.10.4   Training
- Verify that necessary employees receive required training

19.5.10.5   Transportation
- Verify that the materials are shipped properly and arrive at the final destination

19.5.11   COMPANY OWNED OR OPERATED LANDFILLS

19.5.11.1   Landfill Permit
- Review operational plan
- Review closure/post-closure plan
- Review water quality monitoring plan
- Discuss most recent permit renewal application
- Review list of acceptable/current waste streams

19.5.11.2   Compliance Documentation
- Inspect Landfill Operator Certificates, if required
- Review latest engineering reports
- Review water monitoring reports
- Review closure/post-closure financial assurance
- Obtain copy of any NOV since last review
- Review any agency inspection reports
- Discuss any other significant agency correspondence

19.5.11.3   Other Landfill Related Documents
- Discuss any other significant agency correspondence
- Review closure/post-closure cost estimates
- Review remaining volume/life estimates
- Review recent topographic surveys

19.5.11.4   Beneficial Reuse Efforts
- Discuss facility's efforts towards beneficial reuse of landfill materials
- Discuss any state regulations, policies, guidance regarding foundry waste streams
- Discuss any state regulations, policies, guidance regarding beneficial reuse

19.5.11.5   Recycling and Pollution Prevention
- Evaluate whether the facility has a recycling plan or initiative
- Evaluate what items, products, and materials are or can be recycled
- Evaluate whether items, products, and materials are recycled internally or externally
- Evaluate whether the facility could benefit from a regulatory-sponsored recycling or pollution prevention program (i.e., USEPA's Climate Leaders Program)
- Evaluate the status of the facility's pollution prevention plan and awareness
- Evaluate whether the facility uses chlorinated solvents and enact measures to eliminate the continued use if chlorinated solvents are used or stored at the facility.
- Evaluate whether the facility uses other targeted compounds such as cadmium, mercury, chromium, and lead and evaluate measures to eliminate or reduce use of these compounds.
- Evaluate whether the facility uses large quantities of VOCs other than chlorinated solvents such as xylenes, toluene, MEK, etc. and evaluate initiating a pollution prevention strategy.
- Evaluate the effectiveness of the facility's chemical ordering procedures as a method of pollution prevention and awareness.

19.5.12   SPILLS

19.5.12.1   SPCC Plan
- Review SPCC applicability and current total quantity of regulated liquids stored onsite
- Review SPCC plan
- Review list of potential spill materials

- Review waste and material storage areas with diagram
- Review contingency plans
- Review training records
- Ensure that SPCC plan is signed by a PE
- Ensure that a Certification of Substantial Harm Determination is completed

9.5.12.2 Spill History

- Review any spills since last review
- Review incident reports
- Review NRC or state spill reports
- Inspect spill areas
- Review corrective action
- Discuss historical spills

19.5.13 PCBs

Evaluation and Inventory of Equipment Potentially Containing PCBs

- Record the type (liquid, dry, pole-mounted, pad, PCB, Non-PCB), contents, capacity, age, and location of each transformer at the facility.
- Review list of PCB Equipment including:
  Transformers (>50 ppm, > 500 ppm, Made before July 1, 1979)
  Evaluate ownership records of all transformers on facility property
  Capacitors (Large, High Voltages, Made before July 1, 1979)
  Hydraulic Presses (Made before July 1, 1979)
  Oil-Filled Electrical Switches (Made 1935 – 1979)
  Oil-Filled Electrical Motors (Made before July 1, 1979)
  Oils, Paints, Inks, Sealants (Made 1935 – 1971)
- Compliance Documentation
  Review analysis of oils for the aforementioned units
  Inspect PCB Labeling and Non-PCB Labeling
  Documents regarding storage and disposal of PCB equipment
  Review inspection reports of PCB equipment in use
  Review registration of PCB Transformers
- PCB Spills
  Evaluate whether any recent or historical PCB spills have occurred at the facility
  Obtain copy of any PCB or Non-PCB spill reports

19.5.14 Hazardous Materials (HAZMAT)

- Review DOT requirements for hazardous material shipments
  Registration requirements
  Hazardous Waste Handler Training documentation (every 3 years)
  Management of hazardous materials for shipment
  Recordkeeping
  DOT Security Plan

19.5.15 Emergency Planning and Community Right-to-Know-Act (EPCRA)

- Hazard Communication Program
  Employee training on workplace chemicals, procedures, and recordkeeping

Review MSDS management, updating, chemical listing, tracking, employee access/ availability procedures
- Emergency Planning Notification
Extremely Hazardous Substances threshold planning quantities
Submission of Emergency Response and/or Contingency Plans to LEPC, (verify if plans are up to date and current).
- Tier I, II submission of reports
Review other State requirements for supplemental reporting
Hazardous Chemical Listing submissions (local fire department or fire marshal)
- TRI form R
Review applicability determination (used, manufactured, or processed) and submission of reports

19.5.16 Site Inspection

Following the review of written materials, a site inspection of the entire facility should always be conducted. It is recommended that the site inspection be structured in such a way that it follows the manufacturing process as much as possible. Photographs are recommended to document the current environmental condition of the facility, especially those items that may require a corrective action. It is also recommended that a photograph be taken after any corrective action to document task completion and to demonstrate compliance.

It will be helpful to have a map of the facility while conducting the site inspection that shows the entire property, buildings, and operations that are clearly marked. In addition, it will be helpful to have waste storage areas, air emission points, points of generation of solid and hazardous waste, point of generation of waste water and waste water discharge points, and stormwater outfalls labeled on the map.

During the site inspection, the following items should be observed, evaluated, and noted:

- If there is an item observed during the site inspection or at any time that may present an immediate threat or imminent danger to human health or the environment, follow the organizations emergency notification procedures. Examples may include but would not be limited to the following:
  - Evidence of existing release of hazardous substances or petroleum products
  - Evidence of material threat of a release of hazardous substance or petroleum products
- Other areas to inspect, review, and/or observe:
  - General housekeeping
  - Review history of neighborhood complaints (documented)
  - Vehicle maintenance areas
  - Spill kits location, contents listing and adequacy
  - Emergency Response Phone List – postings, updated, controlled document, etc.
  - Chemical and waste storage areas
  - Labeling, storage, and housekeeping
  - Waste storage areas, piles, drop boxes, covered, leaks, proper contents
  - New product storage areas (oil & chemicals)
  - Universal waste storage practices and labeling
  - Drum management
  - Drum labeling, management, residue issues, secure when not in immediate use, empty drum storage
  - Secondary containment areas
  - Plant and property security – alarm system, guard service, fencing, gates, lighting, access control, etc.

- Roof inspection
- Drainage
- Emission deposits
- Exhaust stack(s) deposits
- Roof vent(s) deposits
- Floor drains identified on plant drawings, purpose, discharge, potential discharge, drain blocker in the event of a spill
- Stormwater conveyance systems preventative maintenance
- Erosion issues
- Catch basins
- Detention ponds
- Outfalls
- Outside processes, if any
- Equipment and product storage potential for contamination
- Utility supply and shutoff valves, identification, security, management, emergency procedures and training
- Compressor discharge management
- Exhaust fan and vent deposits to exterior of the building
- Baghouse inspection, recordkeeping, and preventative maintenance
- Review facility's control of chemicals for chlorinated solvent content
- Secondary container labeling
- Neighboring properties identified and note (1) the potential for hazardous substance use or storage such as ASTs, and drum storage areas, (2) stained soil, (3) topography, (4) drainage patterns, and (5) types of operations
- Property boundaries inspected, non-facility contributions, observations/activity, any issues

19.5.17 Closing Meeting

A closing meeting should be conducted at the end of the environmental audit. The purpose of the closing meeting is to:

1. Inform facility management of the results of the environmental audit
2. Set a timeline for when the draft report will be completed
3. Establish a timeline and assign appropriate personnel for implementing corrective actions as a result of the findings of the environmental audit, if any
4. Remind facility management with respect to:
   - Internal company policy pertaining to environmental audits
   - Report privilege
   - Report distribution,
   - Procedures to follow during agency inspections
   - Procedures to follow during an emergency or spill event
   - Periodically review environmental permits
   - Review company policy for non-routine correspondence with a regulatory agency
   - Review company policy for permit applications
   - Reminding all employees that environmental compliance requires an awareness and daily attention

19.5.18 Report Preparation

A draft report should be prepared soon after the audit has been conducted, certainly no more than a month following the audit. The report should be draft and marked as such when submitted for review. For items that require a corrective action that are discovered during the environmental review, it is recommended that they be separated into either Best Management Practices (BMPs) related issues, Compliance-related issues, or Observations at the end of each section of the report. An example environmental audit report outline is presented as follows:

- Fundamentals
  - Air
  - Water
  - Land
- Conducting the Audit
  - The Opening Meeting
    - Air
    - Water
    - Hazardous Waste
    - Universal waste
    - Non-hazardous waste
    - Spills
    - PCBs
    - Hazardous Materials (HAZMAT)
    - Toxic Substance and Control Act (TSCA)
    - Emergency Planning and Community Right-To-Know-Act (EPCRA)
    - Site Inspection
  - The Closing Meeting
  - Summary and Conclusions
  - Photographs

After the report has been reviewed and finalized, a system should be put in place to ensure that all the recommended items are properly addressed and correct, especially any identified compliance items. As a reminder, any identified compliance items are of the highest priority and must be address immediately. Typically, a spreadsheet is prepared for each item identified in the report as a Compliance, BMP, or Observation that typically includes the following:

- Name of person responsible for supervising or conducting corrective action
- Time period allowed to complete corrective action
- Actions to be taken to complete corrective action
- Completion sign off
- Completion date
- Management sign off

19.6 AGENCY INSPECTIONS

The United States has an established and effective enforcement policy and program. The United States has also consulted with many developing nations in providing assistance in establishing environmental regulations and enforcement programs. Much of the success that the United States has achieved has been, in part, through its environmental enforcement program. The USEPA has also helped themselves by promoting "Self Audits." Self Audits have benefited the regulated community in many ways including education and conducting corrective actions when issues are discovered. Self Audits have benefited USEPA as well by shifting much of the responsibility to the regulated community. However, this is predicated on conducting inspections to ensure compliance.

Inspections by regulatory authorities vary from location to location and state to state. Some inspections may target one particular media such as air, water, or solid and hazardous waste, while others may cover several all at once, but that is rare. Usually, agency inspections will focus on one particular set of regulations. Typically, an agency inspection will consist of the following:

1. Opening meeting. This is to discuss the reason behind the inspection (e.g., routine inspection, or compliant). To discuss the facility layout, production processes, plant history, wastes generated and storage areas, air emission points and stacks, air emission controls, water distribution and water use, water treatment and discharge, and other environmental aspects.
2. Document review. The document review usually follows the opening meeting. Document review usually focuses on demonstrating compliance with applicable environmental permits such as air and water discharge. Document reviews may also include inspection of manifests and waste characterization. Waste characterization may also focus on non-hazardous waste streams to ensure that a non-hazardous waste has been properly characterized. Training records, logs, plans, reports, forms, and correspondence records are also routinely inspected.
3. Site Inspection. From the information obtained from the document review, the agency inspector will have a much more educated perspective on operations and what regulations apply to the facility. The inspector will usually request that a site inspection be conducted to visually inspect the area or areas that the inspection has focused (air, water, solid and hazardous waste, etc.). The regulatory inspector may also ask if photographs can be taken during the site inspection. The regulatory inspector is typically looking to evaluate the consistency between the document review with site observations.
4. Closing meeting. The closing meeting will usually involve whether any violations have been identified and may also involve additional requests for information and documentation that was not readily available at the time of the inspection.

19.7 ENVIRONMENTAL AUDITS AND SUSTAINABILITY

Environmental audits should not be confused with sustainability. An environmental audit is a measure of compliance with environmental laws, regulations, standards, and other requirements. An environmental audit is not a measure of sustainability. However, an environmental audit is a good resource to begin evaluating sustainability. Sustainability is based on a simple principle that states that (USEPA 2021e):

Everything we need for our survival and well-being depends, either directly or indirectly, on our natural environment

USEPA defines **sustainability** as creating and maintaining conditions under which humans can exist in productive harmony to support present and future generations (USEPA 2021e). Note that the definition states "productive harmony." One may ask, in productive harmony with what? The answer is with the environment. Therefore, in order to achieve some level of sustainability we must understand of environment and also understand the aspects of how facility operations impact the environment. Sustainability is then the outcome of analyzing the aspects of facility operations with the natural environment and developing and engineering methods to either minimize or eliminate harmful potential harmful impacts. A good start for understanding facility operations that may negatively impact the environment is with a comprehensive environmental audit. The other key element in the definition of sustainability is understanding the natural environment.

19.8 SUMMARY OF ENVIRONMENTAL COMPLIANCE AND SUSTAINABILITY

The primary purpose of environmental regulation is to protect human health and the environment. We should also have an appreciation of the sheer immensity and often times overwhelming complexity of environmental regulations. An important aspect of environmental regulations and conducting environmental audits is that environmental regulations focus on what is called "end of the pipe" or wastes that are generated. For instance, looking back over this chapter on how to conduct an environmental audit, the regulations focused on the type and amount air emissions, water discharges, and solid and hazardous wastes that were generated. The significant lessons or Principles of Environmental Compliance included:

1. **It all begins with conducting an environmental audit.**

   The first step to compliance is to conduct an environmental audit. An environmental audit is as an assessment of which a facility or organization is observing practices to minimize harm to the environment. USEPA defines an **environmental audit** as a systematic evaluation to determine the conformance to quantitative specifications to environmental laws, regulations, standards, permits, or other legally required documents (USEPA 2021a).

2. **Being in compliance with environmental regulations does have sustainability implications.**

   In this chapter we will begin to explore what it means to look toward environmental stewardship and sustainability as part of environmental compliance with air regulations. Environmental stewardship and sustainability require going beyond environmental compliance and addressing deeper questions about how we may impact the environment. Being in compliance means conducting activities consistent with environmental regulations. It does not ask the question whether you should be doing something else that will enhance your organizational goals. Another words, you have to ask yourself, "Just because you can doesn't always mean you should."

3. **Work yourself out of environmental regulations.**

   This is often difficult and challenging but is well worth the trouble and helps pave the way to environmental stewardship and sustainability.

4. **There is little we can control.**

   Environmental professionals for the most part do not control the location where manufacturing takes place or the type of manufacturing or products that are made. However, environmental professionals typically have input in the chemicals that are used and what management actions and engineering controls to put in place to prevent mismanagement and potentially causing harm to the environment if a release occurs.

5. **Limit chemical use when possible.**

   Remember if there we no hazardous substances there would be little need for environmental regulations or environmental professionals.

6. **When possible, eliminate chemicals that are very toxic or have a high chemical risk factor.**

   Elimination of chemicals provides the highest level of confidence in protecting human health and the environment. Focus should be placed on those chemicals that are highly toxic or have a high chemical risk factor (CRF). Some include the following:

   - Arsenic
   - Lead
   - Mercury
   - Cadmium
   - Volatile organic compounds, especially:
     - Benzene
     - Chlorinated or halogenated VOCs
   - PCBs
   - 1,4-dioxane

- Dioxin
- Cyanide
- PFAS compounds
- Chlordane
- DDT

7. **Fully Understand Environmental Permits.**
   This sounds easy but it is not and many mistakes occur due to numerous reasons. Some include lack of understanding of a permit because it's too long, too complex, or is too technical.

8. **Never Accept a Permit Term that the Facility Can't Achieve.**
   All too often, a permit term may seem easy to comply with but under actual operating conditions the facility can't achieve compliance. Therefore, always collect data to evaluate whether the permit term can be achieved before accepting the permit.

9. **Always be accurate and truthful.**
   There are instances when communicating with a regulatory agency in writing or even verbally, where data can be misinterpreted or a mistake is made, such as a typo. In any instance, it's usually always better to be as accurate as possible and truthful. If a circumstance arises where interpretation of data can be questioned or any other circumstance where data can be challenged, get advice from a knowledgeable, qualified, and reliable source such as an environmental consulting firm, the local environmental authority, or counsel.

10. **When in Doubt, It's Usually Always Better to Report a Spill or Other Incident Where Reporting May be Required.**
    If there is a spill or other incident that requires reporting to the environmental agency within a prescribed time after the incident, it is usually always beneficial to report the incident. The only time when reporting is not advised is when there is absolute certainty that the incident does not require reporting. Under these circumstances detailed documentation should be undertaken to ensure that properly procedures were followed. In addition, know in advance the reporting procedures, company policies on reporting, and seek training so that when an incident does occur, everyone will be prepared.

11. **The Importance of Waste Characterization and Points of Generation.**
    In general, a facility should avoid reliance on generator knowledge to either characterize or partially characterize a waste stream. Reliance should be on laboratory analysis of a representative sample of the waste. In addition, ensuring that an appropriate determination of the point of generation of a waste is equally important to ensure that the waste is properly characterized and representative.

12. **Any Operational Changes May Require Notice and Permit Changes or a New Permit.**
    Each facility should also have a keen awareness that any process or manufacturing changes or installation of new equipment may have a profound effect on the types, amounts, and chemical and physical characteristics of emissions, discharges, and wastes that are generated that may require a permit or permit modification and may render existing waste profiles obsolete and require associated wastes be re-characterized.

13. **Give yourself enough time for collecting additional compliance samples in case a data quality issue arises.**
    Compliance sampling and monitoring should be conducted well in advance of reporting results to the regulatory authority in case a data quality issue is identified during any step of the process.

14. **Keep well organized and communicate with management and employees regularly.**
    Maintaining well organized files for compliance will ensure that there is less confusion and a shorter compliance audit or regulatory inspection. In addition, establishing a yearly

compliance schedule by month and regular communication with management will assist to ensure that ongoing compliance is achieved. Regular communication with employees can lead to identifying areas for improvement and can generate a better environmentally oriented culture.

15. **Housekeeping.**
Maintaining a clean and organized facility will assist at maintaining compliance

16. **Signage.**
Proper label are part of compliance requirements. However, expanding signage in key areas and in waste storage area and compliance monitoring equipment will increase compliance and will educate employees on maintaining compliance.

17. **Conduct spill drills.**
Conducting spill drills will ensure proper response and will provide educational opportunities for everyone involved.

18. **Prepare a list of onsite chemicals with corresponding reportable quantities (RQ's) and recommended cleanup procedures.**
A spill incident is stressful, so calculate RQ's in advance. This will greatly increase a proper response and reporting if required.

19. **Conduct regular inspections.**
It is recommended to conduct regular inspections beyond those that are required under environmental permits. Conducting regular inspections generally has many benefits that include better housekeeping and identifying potential compliance issues.

20. **Proper preparation prevents poor performance.**
Don't wait for a regulatory inspection to improve performance. A Notice of Violation is not something desirable.

21. **In many instances, proving a negative is required when evaluating a release.**
When evaluating the risk posed when a release of hazardous substances has occurred focuses on ensuring that the release will not adversely impact human health or the environment. This places a higher level of inquiry because the perspective is to collect a robust body of scientific data that demonstrates that the release will not result in harming humans or the environment at the highest degree of confidence.

22. **When a question arises, consult management, in-house environmental counsel, outside environmental counsel, an environmental consultant, or the regulatory authority, as appropriate.**
Environmental regulations are complex, dynamic, and sometimes are difficult to interpret. Therefore, engaging others is often the best approach.

23. **Lastly, remember that "no matter where you are, it's all the same."**
We live on one planet with no environmental boundaries and contamination does not respect borders. Another point is that environmental regulations are built on the USEPA or EU platforms and those platforms rely on basic principles that include:
    1. Protect the air
    2. Protect the water, and
    3. Protect the land
    4. Protect living organisms
    5. Protect our cultural and historic places

Environmental compliance is the first and most crucial step in establishing a successful sustainability program. Once compliance is achieved, the focus can easily be transitioned to develop a robust sustainability program. Much of the effort in achieving compliance with environmental laws have sustainability aspects and form the foundation for sustainability initiatives.

Questions and Exercises for Discussion

1. Of the 23 principles of environmental compliance listed in the chapter, which five are the most important in your opinion and why?
2. How often should an environmental audit be conducted at a specific facility?
3. Should an environmental audit be announced prior to conducting an audit?
4. Who should conduct an audit at a specific facility?
5. Who should be the recipient of the audit report and who should the report be addressed?
6. How long should the audit report be retained?
7. How much time do you feel is reasonable to address compliance items that may be identified in an audit report?
8. How much time do you feel is reasonable to address best management practices that may be identified in an audit report?
9. Who should oversee the completion of recommended items identified in an audit report?
10. Should a review of sustainability items be included in an environmental audit? Why or why not?

REFERENCES

Connecticut Department of Environmental Protection. 2021. Environmental Best Management Practices Guide. *Hartford, Connecticut.* 4p. https://www.ct.gov/dep/compliance-assistance (accessed February 15, 2021).

United States Environmental Protections Agency (USEPA). 1986. Environmental Auditing Policy Statement. *Federal Register.* Vol. 51. no. 131. Washington, D. C. Wednesday, July 9, 1986. pp. 25004–25010.

United States Environmental Protection Agency (USEPA). 2021a. Audit Protocols. https://www.epa.gov/compliance/audit-protocols (accessed February 15, 2021).

United States Environmental Protection Agency (USEPA). 2021b. Compliance. https://www.epa.gov/compliance (accessed February 15, 2021).

United States Environmental Protection Agency (USEPA). 2021c. National Menu of Best Management Practices (BMPs) for Stormwater. https://www.epa.gov/npdes/national-menu-best-management-practices-bmps-stormwater (accessed December 1, 2018).

United States Environmental Protection Agency (USEPA). 2021d. United States Environmental Protection Agency QA Glossary. https://www.epa.gov/emap/archive-emap/web/html (accessed February 15, 2021).

United States Environmental Protection Agency (USEPA). 2021e. What is Sustainability? https://www.epa.gov/sustainability (accessed February 15, 2021).

Index